石油管材及装备材料服役行为与结构安全国家重点实验室科研成果汇编

(2021年)

中国石油集团工程材料研究院有限公司
石油管材及装备材料服役行为与结构安全国家重点实验室　编
中国石油集团石油管工程重点实验室
陕西省石油管材及装备材料服役行为与结构安全重点实验室

石油工业出版社

内 容 提 要

本书汇编了中国石油集团工程材料研究院有限公司、石油管材及装备材料服役行为与结构安全国家重点实验室、中国石油集团石油管工程重点实验室和陕西省石油管材及装备材料服役行为与结构安全重点实验室在2021年正式发表在国际国内刊物上的论文，反映了近几年石油管工程的科研成果及进展。内容涉及输送管与管线安全评价、油井管与管柱失效预防、腐蚀防护与非金属材料等方面。

本书可供从事石油管工程的技术人员和石油院校相关专业师生参考。

图书在版编目（CIP）数据

石油管材及装备材料服役行为与结构安全国家重点实验室科研成果汇编.2021年／中国石油集团工程材料研究院有限公司等编.—北京：石油工业出版社，2023.3
ISBN 978-7-5183-5924-0

Ⅰ.①石… Ⅱ.①中… Ⅲ.①石油管道-管道工程-文集 Ⅳ.①TE973-53

中国国家版本馆CIP数据核字（2023）第038915号

出版发行：石油工业出版社
　　　　（北京安定门外安华里2区1号　100011）
　　　　网　址：www.petropub.com
　　　　编辑部：（010）64523687　图书营销中心：（010）64523633
经　　销：全国新华书店
印　　刷：北京中石油彩色印刷有限责任公司

2023年3月第1版　2023年3月第1次印刷
787×1092毫米　开本：1/16　印张：35.25
字数：870千字

定价：200.00元
（如出现印装质量问题，我社图书营销中心负责调换）
版权所有，翻印必究

《石油管材及装备材料服役行为与结构安全国家重点实验室科研成果汇编（2021年）》编辑委员会

顾　问：黄维和　李鹤林　高德利　赵怀斌　张建军

主　任：刘亚旭

副主任：霍春勇　冯耀荣

委　员：（按姓氏笔画排序）

马秋荣　王香增　王铁军　乐　宏　孙　军

刘汝山　闫相祯　汤晓勇　陈　平　张士诚

闵希华　李建军　李贺军　李国顺　周　敏

罗　超　苗长贵　周建良　郑新权　胥志雄

赵新伟　郭文奇　高惠临　崔红升　韩恩厚

魏志平

主　编：马秋荣

副主编：罗金恒　尹成先　韩礼红　戚东涛　池　强
　　　　宫少涛

编辑组：林　凯　林元华　姜　放　房　军　黄桂柏
　　　　陈宏远　冯　春　付安庆　李厚补　马卫锋
　　　　王　鹏

前　言

　　石油管材及装备材料服役行为与结构安全国家重点实验室（以下简称重点实验室）成立于 2015 年 10 月，与中国石油天然气集团有限公司石油管工程重点实验室和陕西省石油管材及装备材料服役行为与结构安全重点实验室两个省部级重点实验室并行运行，是我国在石油管材及装备材料服役安全研究领域的科技创新基地、人才培养基地和学术交流基地。

　　重点实验室依托中国石油集团工程材料研究院有限公司，设置输送管与管线安全评价研究、油井管与管柱失效预防研究、石油管材及装备腐蚀与防护研究、先进材料及应用技术研究 4 个研究方向，围绕我国油气工业发展战略需求，特别是大口径高压输气管道建设和复杂工况油气田勘探开发的技术需求，以石油管材及装备材料服役过程中的断裂、变形、泄漏、腐蚀、磨损、老化等突出失效行为为对象，深入开展石油管材及装备材料服役领域的理论基础和应用基础研究，积累服役性能数据，创新研究试验方法，突破国外技术封锁和壁垒，形成自主创新的技术体系和知识体系，为我国油气重大工程的选材、安全评估与寿命预测提供科学技术支撑。2021 年，重点实验室在研国家重点研发计划项目 3 项、国家级科研课题 33 项、省部级和中国石油天然气集团有限公司科研课题 56 项。获省部级和集团公司级等科技奖励二等奖以上 10 项，其中一等奖 3 项、二等奖 7 项；SCI/EI 收录论文 34 篇；授权专利 56 件，其中发明专利 39 件；制修订国家标准 3 项，发布行业标准 4 项。

　　重点实验室建立了基于应变设计方法及高应变海洋管技术指标体系，形成了非 API 石油专用管服役全过程检测检验和质量管控技术体系，提出了高压高含硫工况非金属管材安全性和可靠性评价技术，解决了西南页岩气井套变问题，为我国石油管材及装备安全运行提供了重要的技术支持。

　　在输送管与管线研究领域，优化土壤约束条件下的断裂控制模型并开发出适用于冻土环境的断裂控制技术，开发了环焊缝应变容量预测模型并完成验证，优化厚壁管材断裂韧性指标，开发出 X80 低温环境用管线钢管产品，初步建立了管道环焊缝强度预测模型及工具；在国内首次建立基于应变设计方法及高应变海洋管技术指标体系；完成了高应变海洋管线管原型技术开发产品试制；成功开发了 L485 高应变管线钢管环焊用焊材，达到高强高韧效果；开发完成了高应变海洋管线管应变评估装置及技术，并通过验证。

在油井管与管柱研究领域，提出面向工况的油井管标准化新体系；西南页岩气井套变控制技术见到实效，并确定进一步推广 10 余口井；完成页岩油气井套管全寿命优化，提出降钢级路线，获得油田用户好评；构建了全井筒密封试验工装，形成套管/水泥环胶结面密封试验能力；储气库等在役井管柱安全评价及寿命预测技术实现了油/套管一体化安全评价，完成 14 口工程井评价；开发了隔热管模拟试验工装与方法，支持了西南油气田预防水合物形成的井下管柱选型；成功开发了特殊螺纹双金属冶金复合油管，顺利通过 B 系 C 系模拟试验；地下煤气化工程用燃烧管开发成功，获得工业样管，提交项目进行实验；形成 3D 打印工艺技术，获得钻机 675 型大钩样品，通过无损检测，材料屈服强度达到 650MPa，比以往提升 15%，质量减轻 25%；余热温差发电技术研制成功，获得两项千瓦级样机。

在腐蚀与防护研究领域，建立了高温高压气井基于全生命周期腐蚀工况的选材图谱；在页岩气管柱及地面管道腐蚀失效方面，明确了微生物是管材腐蚀失效的主控因素之一，初步建立了微生物腐蚀试验方法；攻克了冶金复合管弯管的制造工艺，实现了常温固化无溶剂环氧涂料的中试生产和涂覆工艺；建立了极端环境下 LNG 储罐的概率失效方法及随机载荷作用下传热管换热器的裂纹扩展寿命预测方法；建立了石化装置关键静设备服役性能特征参量甄别与临界值判定方法；研究获得了充氢环境下换热器典型钢材的氢扩散规律及损伤机理，建立了换热器铵盐结晶模型；形成了加氢换热器选材导则及结盐腐蚀工艺控制方案，保障了兰州石化加氢装置连续运行周期延长 30%，同时成果将在长庆石化加氢裂化装置上进行应用。

在非金属与复合材料研究领域，创新开发出聚乙烯管材弹性模量超声波检测技术，形成了适用于在役聚乙烯管道的弹性模量超声波现场测试方法；突破非金属管接头可靠性评价及规范化应用关键技术，建立了玻璃钢管及配件适用性评价技术，确定了连接螺纹的类型和关键参数，建立了柔性复合管接头及密封件适用性评价技术，明确了塔里木油田不同工况下金属接头的选材推荐及密封件材料的适用范围；首次明确高压高含硫工况非金属管选材技术要求，形成了高压高含硫工况非金属管材安全性和可靠性评价技术。

在安全评价与完整性研究领域，在国内首次建立了基于历史数据的环焊缝失效行为智能预测模型，并开发了智能预测应用软件；借助校企联合平台，与西安交通大学合作初步建立了声振融合的动设备故障诊断智能分析模型和特征信号识别方法；研发了聚乙烯管道电熔和热熔接头典型缺陷太赫兹无损检测技术；在国内首次形成了城镇聚乙烯管道完整性管理体系；建立了安全生产标准化管理和钢质管道完整性管理等级相结合的城市燃气管道完整性体系等级评估方法；基于先堵后补理念，针对性建立了集输管道不停输不动火快速堵漏抢维修技术，现场堵漏效果良好。

在失效分析与智能化研究领域，研发突破了125ksi管材高强韧性匹配、螺纹密封/抗扭性能优化两项卡点技术；形成了非API石油专用管服役全过程检测检验和质量管控技术体系；针对海洋石油钻探用管材应用研究工作，完成了全部铝合金钻杆应用研究工作与铝合金套管的结构设计，确立了高强度铝合金钻杆/套管性能评价方法；开发出了720MPa钛合金钻杆，并确立了钛合金钻杆评价方法；国内首次开展的135ksi高强度钛合金钻杆材料性能、实物性能、耐蚀、耐磨及工况适用性系统评价，形成了钛合金钻杆适用性评价方法，为油田超深井勘探开发提供技术保障。

本书是上述成果的总结，收录了重点实验室于2021年在国内外重要刊物和学术会议上发表的52篇论文，并介绍了2021年授权专利及省部级以上获奖成果，反映了重点实验室近期所取得的研究成果。本书可以为从事油气管道工程、油气井工程、石油工程材料、安全工程等方面的工程技术人员、研究人员和管理人员提供参考。

由于编者水平有限，经验不足，加之时间仓促，不足之处在所难免，敬请广大读者批评指正。

编 者

2022年4月

目 录

第一篇 论文篇

一、输送管

Failure Analysis on Girth Weld Cracking of Underground Tee Pipe
.... Cao Jun　Ma Weifeng　Pang Guiliang et al（4）
Service Performance of 3PE Coating for Compressor Export
.... Nie Hailiang　Ma Weifeng　Hu Xu et al（27）
A Novel Test Method for Mechanical Properties of Characteristic Zones
　of Girth Welds .... Nie Hailiang　Ma Weifeng　Xue Kai et al（34）
Root Cause Analysis of Liner Collapse and Crack of Bi-Metal Composite
　Pipe Used for Gas Transmission ...... Zhang Shuxin　Ma Qianzhi　Xu Changfeng et al（46）
Burst Failure Analysis of a HFW Pipe for Nature Gas Pipeline
.... Zhu Lixia　Wu Gang　Luo Jinheng et al（63）
Comparison and Research of Acoustic Emission Testing Standards for
　Atmospheric Storage Tank .... Zhang Shuxin　Wang Weibin　Yang Yufeng et al（71）
Failure Analysis of Tee in Shale Gas Transportation Platform
.... Ji Nan　Feng Jie　Long Yan（83）
Study on Strain Response of X80 Pipeline Steel during Weld Dent Deformation
.... Zhu Lixia　Luo Jinheng　Wu Gang et al（94）
Influence of Boss-Backing Welding to ERW Pipe
.... Luo Jinheng　Zhao Xinwei　Liu Ming et al（107）
Evolution of Grain Boundary α Phase during Cooling from β Phase Field
　in a α+β Titanium Alloy .... Gao Xiongxiong　Zhang Saifei　Wang Lei et al（121）
Impact Assessment of Flammable Gas Dispersion and Fire Hazards from
　LNG Tank Leak .... Li Lifeng　Luo Jinheng　Wu Gang et al（127）

Risk Assessment of Large Crude Oil Tanks Based on Fuzzy Comprehensive
　　Evaluation Method ……………………………… Wu Gang　Li Lifeng　Zhou Huiping et al （149）
北溪管道用感应加热弯管的生产和质量水平 ……………………………… 吉玲康　董　瑾（156）
不同强度匹配X80钢环焊接头力学性能及变形能力
　　……………………………………………………… 何小东　高雄雄　David Han　等（163）
高钢级天然气管道爆炸危害的影响 ……………………… 杨　坤　王　磊　高　琦　等（171）
国内外几种GMAW焊丝强韧性对比试验研究 ………… 李为卫　杨耀彬　何小东　等（179）
热输入量对熔化极气体保护焊缝强韧性的影响 ……… 李为卫　李嘉良　梁明华　等（184）
某凝析油集输管线内腐蚀影响因素分析 ……………… 李　磊　陈庆国　袁军涛　等（189）
某输油管道腐蚀穿孔原因 ……………………………… 吉　楠　廖　臻　朱　辉　等（197）

二、油井管

Numerical Analysis of Casing Deformation under Cluster Well Spatial Fracturing
　　……………………………………… Wang Jianjun　Jia Feipeng　Yang Shangyu et al （206）
Cu_2Se as Textured Adjuvant for Pb-Doped BiCuSeO Materials Leading to
　　High Thermoelectric Performance ………… Jiang Long　Han Lihong　Lu Caihong et al （213）
Microstructure and Improved Thermal Shock Behavior of an In Situ Formed Metal-Enamel
　　Interlocking Coating ……………… Wang Hang　Zhang Chuan　Jiang Chengyang et al （226）
Evaluation of Glass Coatings with Various Silica Content Corrosion in a 0.5mg/L HCl
　　Water Solution ………………………… Wang Hang　Zhang Chuan　Jiang Chengyang et al （241）
Research on Property of Borocarbide in High Boron Multi-Component Alloy with
　　Different Mo Concentration …………… Ren Xiangyi　Han Lihong　Fu Hanguang et al （257）
Effect of Titanium Modification on Microstructure and Impact Toughness of High-Boron
　　Multi-Component Alloy ……………… Ren Xiangyi　Tang Shuli　Fu Hanguang et al （271）
Integrity Analysis of Casing Premium Connection under High Compression Load
　　………………………………………… Wang Peng　Xie Junfeng　Zheng Youcheng et al （286）
Coiled Tubing Plastic Strain and Fatigue Life Model ……………… Wang Xinhu　Tian Tao （292）
Comparison of CrN, AlN and TiN Diffusion Barriers on the Interdiffusion and
　　Oxidation Behaviors of Ni+CrAlYSiN Nanocomposite Coatings
　　………………………………………… Zhu Lijuan　Feng Chun　Zhu Shenglong et al （300）
时效对石油钻杆用Al-7.51Zn-2.37Mg-1.72Cu合金力学及耐热性能的影响
　　………………………………………………… 冯　春　张芳芳　朱丽娟　等（314）
文23储气库注采管柱接头密封性能指标研究 ……… 王建军　孙建华　李方坡　等（323）

油套管用特殊螺纹连接密封完整性探讨 ………………… 王建东　李玉飞　汪传磊　等（329）
NQI 框架下的非 API 油井管质量管控技术 ………………… 朱丽娟　冯　春　韩礼红　等（340）
石墨烯改性环氧涂层在油田注水工况中的适用性研究
　………………………………………………………………… 朱丽娟　冯　春　何　磊　等（348）
三层环氧涂层油管在注水井中的应用研究 ……………… 朱丽娟　冯　春　李长亮　等（353）

三、腐蚀与非金属

Effects of Trace Cl^-, Cu^{2+} and Fe^{3+} Ions on the Corrosion Behaviour of AA6063
　in Ethylene Glycol and Water Solutions …… Fan Lei　Zhang Juantao　Wang Hao et al（360）
Failure Analysis about Girth Weld of Gathering Pipelines
　………………………………………………… Li Fagen　Feng Quan　Fu Anqing et al（387）
Failure Analysis of Cracked and Corroded Tubings in Sudong Block of
　Changqing Oilfield ……………………… Zhao Xuehui　Li Mingxing　Liu Junlin et al（394）
Failure Analysis on Fiber Reinforced Thermoplastic Pipe
　………………………………………… Song Chengli　Liu Xinbao　Bai Zhenquan et al（405）
H_2S Dissociation on Defective or Strained Fe（110）and Subsequent Formation
　of Iron Sulfides: A Density Functional Theory Study
　………………………………………………… Li Fagen　Zhou Zhaohui　He Chaozheng et al（413）
Hydrogen Sulfide Stress Cracking in a Q345R Welded Joint
　………………………………………………………… Han Yan　Luo Jingbin　Fu Anqing et al（424）
Investigations of Polyethylene of Raised Temperature Resistance Service
　Performance Using Autoclave Test under Sour Medium Conditions
　………………………………………… Qi Guoquan　Yan Hongxia　Qi Dongtao et al（430）
Performance Evaluation of Polyamide-12 Pipe Serviced in Acid Oil and
　Gas Environment ……………………… Qi Guoquan　Yan Hongxia　Qi Dongtao et al（442）
Simulation Analysis of Limit Operating Specifications for Onshore Spoolable
　Reinforced Thermoplastic Pipes …… Li Houbu　Zhang Xuemin　Huang Haohan et al（453）
Study on the Hydrogen Embrittlement Susceptibility of AISI 321 Stainless Steel
　……………………………………………… Xu Xiuqing　An Junwei　Wen Chen et al（467）
CH_4 在 PVDF 中的渗透特性及机理 ……………… 李厚补　张学敏　马相阳　等（477）
饱和 CO_2 溶液中 Cl^- 浓度对马氏体不锈钢应力腐蚀敏感性的影响
　………………………………………………………… 赵雪会　刘君林　曾瑞华　等（484）
基于矫顽力的早期蠕变损伤智能诊断方法 …………… 李文升　吕运容　尹成先　等（495）

井下用非金属复合材料连续管研究进展 ……………… 李厚补　张学敏　马相阳　等（502）

四、其他

Analysis of Causes of Burst Failure of a Buffer Tank
　…………………………………… Zhang Shuxin　Luo Jinheng　Wu Gang et al（512）
Study on the Effect of Cement Sheath on the Stress of Gas Storage Well
　………………………… Song Chengli　Liu Xinbao　Li Guangshan et al（525）
Q235B 原油储罐底板腐蚀穿孔原因 ……………… 武　刚　徐　帅　张　楠　等（532）
谐波沉降作用下大型原油储罐变形响应分析 ………… 武　刚　徐　帅　张　哲　等（537）

第二篇　成果篇

一、省部级（含社会力量）科技奖励 …………………………………………（549）
1. 复杂油气井管柱工况模拟试验评价与应用技术 ……………………………（549）
2. 高性能钻杆研发及检测评价技术研究 ………………………………………（549）
3. 油气管道缺陷修复质量检验评价和改进技术研究及应用 …………………（550）

二、授权发明专利目录 ……………………………………………………………（551）

第一篇 论文篇

一、输送管

Failure Analysis on Girth Weld Cracking of Underground Tee Pipe

Cao Jun[1]   Ma Weifeng[1]   Pang Guiliang[2]   Wang Ke[1]
Ren Junjie[1]   Nie Hailiang[1]   Dang Wei[1]   Yao Tian[1]

(1. State Key Laboratory of Performance and Structural Safety for Petroleum Tubular Good Goods and Equipment Materials, Tubular Goods Research Institute of CNPC;
2. Pipe China West Pipeline Company)

Abstract: Three girth weld cracking events were found by X-ray inspection in one natural gas station. To determine the failure cause, one of the three failure events was studied by visual inspection, nondestructive testing, microstructure examination, scanning electronic microscopy (SEM) coupled with energy dispersive spectrometry (EDS), blasting tests, finite element simulations and a series physical and chemical tests. The results revealed that the welding defects were original failure factor, and these welding defects led to crack initiation and propagation through the wall thickness under three aspects of stress concentration effects with internal pressure. The stress concentration effects were originated from the unqualified inner chamfering angle, unequal wall thickness, and inherent shape of tee pipe. The reason causing welding defects of girth weld is that the slag inclusions were not cleaned up timely, and then the slag inclusions led to the existence of welding defects.

Keywords: Girth weld cracking; Welding procedure specification; Welding defects; Stress concentration

1 Introduction

Tee pipe is used to carry and distribute high pressure gas in natural gas station, and it needs a high level of reliability. Since its large diameter and thick wall, the girth welds between tee and straight pipes are unequal wall. The girth welds of unequal wall exist failure risk due to the stress concentration effects at the weld toe.

Cracking[1-5], stress corrosion cracking (SCC), and corrosion are main failure modes in oil and gas pipelines. For the failure mode of cracking, notch effect, external force, hydrogen embrittlement, and weld defects are main failure causes. a crack failure of a high pressure natural gas pipe under split tee was found in girth weld, which concluded that notch effect in the hot tapped hole and the

---

Corresponding author: Cao Jun, caojun1@cnpc.com.cn.

presence of a large periodic stress are the failure causes of the cracking of girth weld[1]. Due to hydrogen-assisted cracking in the highly strained region, a delayed cracking failure event was occurred in a pipe (API 5L X52) of Kuwait[2]. A transverse cracking failure event was occurred in a pipe (API 5L X46) of Brazil[3], and stress raisers, welding defects and corrosion pits, were associated to the cracking nucleation. Since high hardness weakened the hydrogen resistance of the 5Cr steel tee pipe and then led to the hydrogen embrittlement, the cracking of the tee-pipe started from inclusions near the inner surface of the pipe and was accelerated by this hydrogen embrittlement[4]. As the effect of stress relaxation cracking, a girth weld cracking was occurred in 304H stainless teel welds of a chemical reactor serviced at 560℃[5].

A discontinuous change in the geometricalshape was often existed in the pipe system, such as tee pipe and butt welds. Girth butt welds in pipes welded together from pipes with different thickness also formed stress concentration in the root of thin wall side weld. Lotsberg[6] proposed an analytical expression of stress concentration factors (SCF) due to the transition in thickness of girth butt welds, and then the analytical expression was developed by considering comprehensive effect of internal pressure and axial force[7] as well as more geometries of girth weld with unequal thickness[8]. The weldment cracking was easily occurred in the root of weldment due to stress concentration effect and inferior quality for the fillet weld[9]. Meanwhile, few researchers[10-11] studied the effect of stress concentration induced by welding defect on girth weld.

Girth weld is a weak link of oil and gas pipelines, it was found that many failures were related to the girth weld cracking[12-15]. The welding defects, loading conditions, internal pressure, weld strength matching ratio, and yield to tensile strength ratio of base metal are the main influencing factors of girth weld cracking[14,16]. Failure analysis is a significant way to summary failure cause and decrease failure risk. The purpose of this study is to analyze the causes about the girth weld cracking of unequal wall in natural gas station, and then provides an insight of the failure causes of girth weld and the prevention measures.

Recently, three girth weld cracking events between tee and straight pipes in one natural gas station were found during one excavation detection of girth welds, as shown in Fig. 1. To comprehensively analyze the failure cause, one failure event of girth weld cracking between tee (1000mm× 1000mm×1000mm) and straight pipes (1016mm×26.2mm) in one natural gas station was studied by visual inspection, nondestructive testing, microstructure examination, scanning electronic microscopy (SEM) coupled with energy dispersive spectrometry (EDS), blasting tests, finite element (FE) simulations and a series physical and chemical tests. The tee and straight pipes were made of WPHY 70 and API 5L X70 steel, respectively, and the design pressure is 12MPa. Finally, the failure causes of girth weld were analyzed by material properties, weld defects, and stress concentration effects.

2 Experiments and simulation

2.1 Experimental procedure of physical and chemical tests

Since girth weld between tee and straight pipes was failure, the material properties of tee pipe, straight pipe and girth weld needed to be tested. The design criteria, manufacturing standards and

 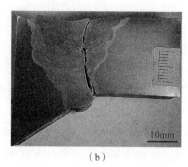 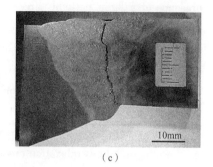

Fig. 1 (a—c) Three girth weld cracking specimens

performance requirements of tee pipe were combined in the standard of CDP-S-OGP-PL-011-2011-2 (General technical conditions for pipe fittings of DN400 and above in oil and gas pipeline engineering) which referenced the standards of MSS SP-75, ASME B16.9, and GB/T 13401. In addition, the design criteria, manufacturing standards, and performance requirements of straight pipe were combined in the standard of CDP-S-OGP-PL-009-2011-2 (General technical specification for steel pipe used in oil and gas pipeline engineering station) which referenced the standards of ISO 3183, GB/T 9711, and API Spec 5L. The welding qualification test standard is Q/SY XQ4—2003 (Specification for weld construction acceptance inspection in west China gas transported to east China pipeline project) which mainly referenced the standard of API STD 1104.

A series of physical and chemical tests were conducted to assess the material properties of tee pipe, straight pipe and girth weld. According to standard of CDP-S-OGP-PL-011-2011-2, the sampling location of tee pipe is shown in Fig. 2(a). Since the failure position [Fig. 2(b)] is far away from S4, the tests of S4 are not considered. From the view of quality problem of tee and straight pipes, the tests of chemical composition, tensile properties, Charpy impact, metallographic structure, hardness were conducted based on CDP-S-OGP-PL-011-2011-2 and CDP-S-OGP-PL-009-2011-2. From the view of welding quality, the tests of tensile properties, nick breaks, side bend, Charpy impact, metallographic structure, and hardness were carried out based on Q/SY XQ4—2003, the sampling position is showed in Fig. 2(c).

The visual inspection was carried out to observe the failure position and macroscopic appearance of failed girth weld. The chemical composition was tested by ARL4460 photoelectric direct reading spectrometer based on ASTM A751-14a. The tensile properties tests were conducted at UTM5305 and SHT 4106 mechanical testing machines based on ASTM A370-18. The Charpy impact energy was tested at PIT302D impact machine based on ASTM A370-18 and GB/T 229. The bending test was conducted by WZW-1000 based on ASTM A370-18. The hardness was measured by KB30BVZ-FA Vickers hardness and RB 2002 Rockwell hardness testing machines based on ASTM E92-17 and GB/T 4340.1. The nick breaks tests were carried out by SHT4106 based on GB/T 2653. In addition, the microstructure was observed by OLS-4100 laser confocal microscope based on ASTM E45-18. The crack appearance and micro area compositions were analyzed by TESCAN VEGA II scanning electron microscope (SEM) coupled with energy dispersive spectrum (EDS).

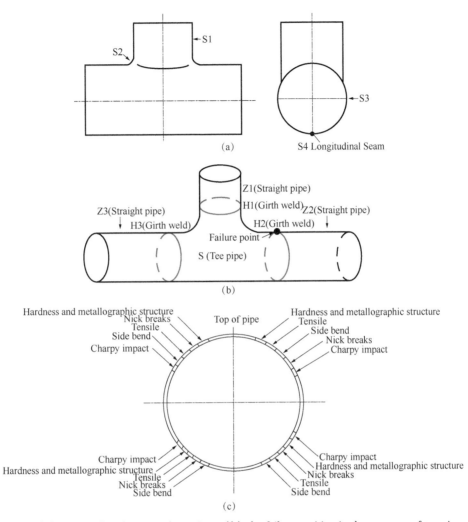

Fig. 2 (a) The sampling location of tee pipe, (b) the failure position in the structure of tee pipe, and (c) the sampling position of girth weld

2.2 Blasting test

To further determine the failure cause of girth weld, the welding process of welding test was same to that of failure girth weld, and the two welding processes (one was performed in failure girth weld, another was performed in the welding test) were conducted by different welders. The blasting test of the welding structure containing tee pipes (1000mm × 1000mm × 600mm), straight pipes (1016mm × 26mm), and three heads were conducted at HYDROSEYS system (Hydraulic test system for steel pipe). The testing and failure tee pipes were same batch and produced from same manufacturers. The welding structure of tee pipe needed to be connected by the hydraulic test pipeline, continuously injected water into the welding structure. After filled the welding structure with water, checked the sealing of the welding structure, and then continuously injected water to boost the pressure according to the water pressure scheme. Finally, the welding structure were continuously to be injected water until it was burst.

2.3 Simulations

Three-dimensional FE models of the welding structure of tee pipe were performed using ABAQUS/

Explicit software. The elasto-plastic hardening material model was used in the FE model, it was expressed as

$$\sigma = \begin{cases} E\varepsilon_e, & \sigma \leq \sigma_s \\ K\varepsilon_p^n, & \sigma \geq \sigma_s \end{cases} \tag{1}$$

Where E is elastic modulus, σ_s is the yield stress, ε_e is the elastic strain, ε_p is plastic strain, K is the strength coefficient, and n is the strain hardening index.

For the sake of simplicity, the constitutive relation of girth weld was considered to be consistent with the constitutive model of straight pipe. The constitutive parameters of straight pipe and tee pipe were obtained by fitting the stress strain curves of tensile tests, the stress strain curve of tee pipe was chosen from the S1-transverse since the yield strength was lowest. The constitutive parameters of tee pipe, straight pipe, and girth weld were listed in Table 1.

Table 1 The constitutive parameters of tee pipe, straight pipe, and girth weld

Material	E(GPa)	K	n
Tee pipe	208.5	280.4	0.295
Straight pipe and girth weld	216.3	839.4	0.0949

The load condition of this simulation only considered the effect of internal pressure from 0 to 50MPa, since one failure event of girth weld cracking was found in the testing pressure of pipe without the stress of piping system and pipe-soil interaction.

In the simulation of stress concentration effect of girth weld from inherent shape of tee pipe, the full-scale welding structure containing tee and straight pipes were adopted in the FE model. The lengths of three straight pipes are 1000mm, respectively. The straight pipes were meshed using 3 dimensional reduced-integration 8-node solid elements (C3D8R) and the tee pipe were meshed using 3 dimensional 10-node modified quadratic tetrahedron elements (C3D10M). The local mesh refinement was designed in the girth weld. The boundary conditions for the full-scale structure part were not constrained.

In the simulations of stress concentration effect of girth weld from unqualified inner chamfering angle θ, a quarter of the pipe with the girth weld was applied in this FE models. The lengths of pipe on both sides of girth weld were 1000mm to avoid end effects. The thicknesses of both sides of girth weld were 26mm and 57mm, respectively. The $X-Y$ and $Y-Z$ surfaces of the geometry were constrained in Z and X directions, respectively. Finer meshes in the girth weld, and coarser meshes farther from girth weld were designed to ensure simulation accuracy and computational efficiency. The minimum mesh size is 0.3mm. There are 9 elements along thin thickness and 28 elements along thick thickness, as shown in Fig. 3.

3 Results

3.1 Visual inspection

The failure girth weld is named as H2, and its location of failure girth weld on tee pipe is

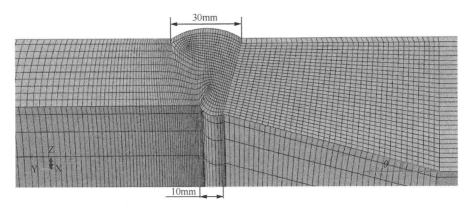

Fig. 3 The mesh design of girth weld

shown in Fig. 2(b). The cracks of H2 girth weld are located in the top of H2 girth weld, as shown in Fig. 4(a) and Fig. 4(b). The leakage point was found through hydrostatic test under 13MPa of pressure, as shown in Fig. 4(c). The cracks were detected in the girth weld through ultrasonic phased array and X-ray tests, and the circumferential lengths of two cracks are 260mm and 63mm, respectively as shown in Fig. 4(d).

Fig. 4 The macroscopic morphology of failure girth weld of tee pipe

3.2 Chemical composition

The chemical compositions of tee and straight pipes were listed in Table 2 and Table 3. As can be seen from Table 2 and Table 3, the chemical compositions of tee and straight pipes meet the standard requirements of CDP-S-OGP-PL-011-2011-2 and CDP-S-OGP-PL-009-2011-2. The chemical composition of girth weld was not required in the standard, so the chemical composition of girth weld was not tested.

Table 2　Chemical composition result of tee pipe　(Wt.%)

Sample	C	Si	Mn	P	S	Cr	Mo	Ni	Nb	V	Ti	Cu	B	Al	N	$CE_{P_{cm}}$
S1	0.071	0.18	1.56	0.0075	0.0022	0.029	0.19	0.21	0.054	0.017	0.010	0.094	0.0001	0.033	—	0.18
S3	0.072	0.18	1.56	0.0078	0.0023	0.029	0.19	0.21	0.055	0.017	0.010	0.094	0.0001	0.034	—	0.18
Requirement	≤0.25	≤0.50	≤1.75	≤0.025	≤0.020	≤0.25	≤0.30	≤1.00	≤0.10	≤0.13	—	≤0.35	≤0.001	—	—	≤0.21
	Cu+Cr+Ni+Mo≤1%															

Note: $CE_{P_{cm}}$ = C+Si/30+Mn/20+Cu/20+Ni/60+Cr/20+Mo/15+V/10+5B.

Table 3　Chemical composition result of straight pipe　(Wt.%)

Sample	C	Si	Mn	P	S	Cr	Mo	Ni	Nb	V	Ti	Cu	B	Al	N	$CE_{P_{cm}}$
Z2	0.022	0.20	1.58	0.012	0.0017	0.043	0.16	0.13	0.047	0.047	0.015	0.034	0.0004	0.036	0.0036	0.13
Requirement	≤0.12	≤0.40	≤1.65	≤0.020	≤0.005	—	—	—	—	—	—	—	—	—	≤0.010	≤0.23

Note: $CE_{P_{cm}}$ = C+Si/30+Mn/20+Cu/20+Ni/60+Cr/20+Mo/15+V/10+5B.

3.3 Tensile property

The tensile properties of tee pipe, straight pipe, and girth welds (H2 and H3) were listed in the Tables 4—6. As can be seen from Tables 5—6, the tensile properties of straight pipe and girth weld were in accordance with the standard requirements of CDP-S-OGP-PL-009-2011-2 and Q/SY XQ4—2003, respectively. The pore defect could be seen in the fracture surface of girth weld (H3), which indicated the quality problem of welding was existed, as shown in Fig.5. The yield and tensile strengths of tee pipe were lower than the standard requirements of CDP-S-OGP-PL-011-2011-2. It can be inferred that the heat treatment process of tee pipe was not controlled strictly. However, the girth weld cracks were located in the side of the thin thickness, as shown in Fig.1. It needed to be studied if the low yield and tensile strengths were related to girth weld cracking.

Table 4　Tensile properties of tee pipe

Sampling location	Sample		Tensile strength R_m (MPa)	Yield strength $R_{t0.5}$ (MPa)	Yield strength tensile strength ratio $R_{t0.5}/R_m$	Elongation $A(\%)$
	Diameter/Width (mm)	Gauge (mm)				
S1-transverse	10	50	539	343	0.64	29.5
S2-transverse	10	50	502	374	0.75	29.5
S3-transverse	10	50	553	363	0.66	27.5
S3-longitudinal	10	50	567	355	0.63	28.5
Requirement			≥570	485~622	≤0.93	transverse≥17 longitudinal≥19

Table 5 Tensile properties of straight pipe

Sample			Tensile strength R_m (MPa)	Yield strength $R_{t0.5}$ (MPa)	Yield strength tensile strength ratio $R_{t0.5}/R_m$	Elongation A (%)
Sampling location	Diameter/Width (mm)	Gauge (mm)				
Z2	12.7	50	686	654	0.95	26
Requirement			570~760	485~630	≤0.93	≥17

Table 6 The tensile properties of H2 and H3 girth welds

Sample		Tensile strength R_m (MPa)	Fracture location
Sampling location	Diameter/Width (mm)		
H2-LS1	25	628	Girth weld
H2-LS2	25	648	Base metal
H2-LS3	25	660	Base metal
H3-LS1	25	682	Girth weld
H3-LS2	25	655	Girth weld
H3-LS3	25	622	Girth weld
H3-LS4	25	664	Girth weld
Requirement		≥570	—

Note: (1) Fracture at base metal and tensile strength meets the requirements;
(2) Fracture at Girth weld and fusion zone, and defects meet the requirements of standard.

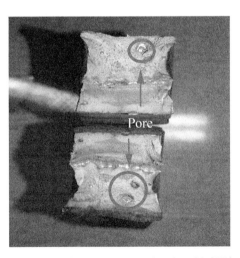

Fig. 5 The pore defects in the fracture surface of girth weld (H3) under tensile test

3.4 Charpy impact test

The absorbed energy of tee pipe, straight pipe and girth welds (H2 and H3) were listed in the Tables 7—10. From the Table 7 and Table 8, the absorbed energy of straight pipe and tee pipe were in accordance with the standard requirements of CDP-S-OGP-PL-009-2011-2 and CDP-S-OGP-PL-011-2011-2. As can be seen from Table 9, one value of absorbed energy of H3 girth weld was lower than the standard requirements of Q/SY XQ4—2003. Since the Charpy impact energy reflects the fracture toughness of material, the results of Charpy impact tests show that the fracture resistance of cracks in the girth weld with no welding defects, tee pipe, and straight pipe almost meet the requirements of standards. However, the dispersion of Charpy impact tests cannot be ruled out, especially in girth weld.

Table 7 Charpy impact test results of straight pipe

Sample		Temperature(℃)	Absorbed energy(J)			
Number	Specification(mm)		Single value			Average value
Z2	10×10×55	−5	482	483	492	486
Requirement		−5	≥40			≥50

Table 8 Charpy impact test results of tee pipe

Sample		Temperature(℃)	Absorbed energy(J)			
Number	Specification(mm)		Single value			Average value
S1	10×10×55	−5	298	305	323	309
S2	10×10×55	−5	301	274	256	277
S3	10×10×55	−5	291	226	257	258
Requirement		−5	≥40			≥50

Table 9 Charpy impact test results of girth weld

Sample		Temperature(℃)	Absorbed energy(J)			
Number	Specification(mm)		Single value			Average value
H2-3-G	10×10×55	−20	272	254	243	256
H2-3-FT	10×10×55	−20	247	213	231	230
H2-3-FZ2	10×10×55	−20	289	345	216	283
H3-0-G	10×10×55	−20	44	208	178	143
H3-0-FT	10×10×55	−20	267	227	225	240
H3-0-FZ3	10×10×55	−20	197	214	194	202
H3-3-G	10×10×55	−20	131	174	205	170
H3-3-FT	10×10×55	−20	269	226	218	238
H3-3-FZ3	10×10×55	−20	142	263	141	182
Requirement		−20	≥56			≥76

Note: "G" represents weld metal zone, "FT" represents fusion zone at the side of tee pipe, "FZ2" represents fusion zone at the side of straight pipe Z2, "FZ3" represents fusion zone at the side of straight pipe Z3.

3.5 Microstructure

The microstructure of tee pipe is composed of polygonal ferrite and some granular bainite, and the grain grade of microstructure meets the standard requirements of tee pipe, as listed in Table 10. Though the standard of tee pipe does not require the special microstructure of material, this microstructure of tee pipe is not normal for WPHY 70 of which microstructure is mainly composed of bainite, as shown in Fig.6(a) and Fig.6(b). These results of microstructure explain why the tensile properties of tee pipe are lower than standard. It pointed that the heat treatment was unqualified.

The microstructure of straight pipe mainly is composed of granular bainite instead of polygonal ferrite, this microstructure is normal for X70, as shown in Fig. 6(c) and Fig. 6 (d).

Table 10 Microstructure of tee pipe
("PF" represents polygonal ferrite. "Bg" represents granular bainite.)

Sample location	Microstructure	Grain grade
S1	PF+Bg	8.5
S2	PF+Bg	8.5
S3	PF+Bg	8.5
Requirement	—	≥6.0

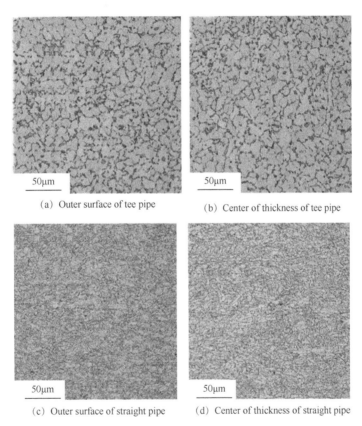

(a) Outer surface of tee pipe   (b) Center of thickness of tee pipe

(c) Outer surface of straight pipe   (d) Center of thickness of straight pipe

Fig. 6 Microstructure of tee pipe for S1 location and Z2 straight pipe

3.6 Hardness test

The hardness of tee pipe, straight pipe and girth welds (H2 and H3) were listed in the Tables 11—12. As can be seen in Tables 11—12, the hardness of tee pipe, straight pipe, and girth weld were in accordance with the standard requirements. It indicates that there is no harmful hardened microstructure in tee pipe, straight pipe, and girth weld.

3.7 Bending test

The bending tests of girth welds show the void found in the sample of H3 girth weld, as shown in Fig. 7. It indicates that the welding quality of girth weld is not stable. The bending tests of H2 girth weld meets the standard requirements, as listed in Table 13.

Table 11 Hardness test results of the tee pipe and straight pipe

Sample	Hardness value (HV10)								
	1	2	3	4	5	6	7	8	9
S1	168	167	167	167	159	169	157	164	164
S2	145	149	147	147	145	147	149	149	152
S3	174	171	173	171	173	167	166	171	170
Z2	234	230	228	221	211	220	223	222	219
Requirement	≤280								

Table 12 Hardness test results of the girth weld
(The locations of 1, 8, 9, 16 are in base material, the locations of 2, 3, 6, 7, 10, 11, 14, 15 are in heat affected zone, and the locations of 4, 5, 12, 13 are in weld metal zone)

Sample	Hardness value (HV10)							
	1	2	3	4	5	6	7	8
H2	231	228	218	216	237	233	235	218
	9	10	11	12	13	14	15	16
	232	228	216	195	193	232	244	207
H3	1	2	3	4	5	6	7	8
	255	210	216	191	185	206	207	190
	9	10	11	12	13	14	15	16
	260	238	237	232	235	232	233	172
Q/SY XQ4—2003	≤265							

Table 13 Bending test results of the girth weld

Number	Sample Specification (mm)	Diameter of bend shaft (mm)	Bend angle (°)	Bend direction	Testing result
H2	300×38×13.5	90	180	Side bend	No crack
	300×38×13.5	90	180	Side bend	No crack
	300×38×13.5	90	180	Side bend	No crack
	300×38×13.5	90	180	Side bend	No crack
H3	300×38×13.5	90	180	Side bend	No crack
	300×38×13.5	90	180	Side bend	No crack
	300×38×13.5	90	180	Side bend	Void (Length is 2.5mm, wide is 1.5mm)
	300×38×13.5	90	180	Side bend	No crack

Note: Based on Q/SY XQ4—2003, the size of crack or other defect in any direction found in the weld and fusion line area on the tensile bending surface of the sample shall not be greater than 1/2 of the nominal wall thickness of the steel pipe and not more than 3mm; the length of the crack generated by the edge of the sample shall not be greater than 6mm in any direction.

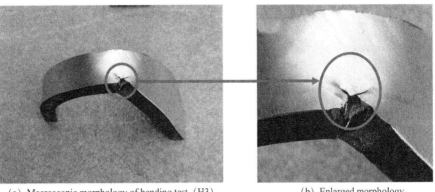

(a) Macroscopic morphology of bending test (H3)　　　(b) Enlarged morphology

Fig. 7　The morphology of bending sample with void for H3 girth weld

3.8　Nick breaks test

Table 14 listed the nick breaks test results of the girth welds (H2 and H3). As can be seen from Table 14, the testing results meet the standard requirements. It indicates that the welding quality of girth weld without failure is acceptable.

Table 14　Nick breaks test results of the girth weld

Sample	Orientation	Specification (mm×mm×mm)	Testing result
H2	Vertical weld	300×25×t	No defects on fracture surface
	Vertical weld	300×25×t	No defects on fracture surface
	Vertical weld	300×25×t	No defects on fracture surface
	Vertical weld	300×25×t	No defects on fracture surface
H3	Vertical weld	300×25×t	No defects on fracture surface
	Vertical weld	300×25×t	No defects on fracture surface
	Vertical weld	300×25×t	No defects on fracture surface
	Vertical weld	300×25×t	No defects on fracture surface
Requirements			(1) The fracture surface of each specimen should be fully welded and fused; (2) The maximum size of any pore should not be greater than 1.6mm, and the cumulative area of all pores should not be greater than 2% of the fracture area; (3) The slag inclusion depth shall be less than 0.8mm, and the length shall not be greater than 1/2 of the nominal wall thickness of the steel pipe and less than 3mm; (4) The distance between adjacent slag inclusions shall be at least 13mm

3.9　Crack analysis

To further analyze the failure cause of girth weld cracking, the four crack specimens were sampled from the region of girth weld cracking, as shown in Fig. 8(a).

The crack surface was opened to observe the actual fracture appearance. The crack surface is semi elliptical and has been corroded, as shown in Fig. 8(b). The cross section of girth weld has almost cracked from inner weld toe of thin-wall of girth weld, as shown in Fig. 8(c).

The black and reddish-brown characters could be observed in the root of girth weld from the fracture surface, as shown in Fig. 8(d). The elemental constituents were analyzed by EDS in Fig. 8 (e). The black character is mainly composed of elements of C, O, Na, Si, S, Mn, Fe, and the reddish-brown character is mainly composed of elements of C, O, Si, S, K, Ca, Mn, Fe. It can be inferred that the O element of black character should come from oxygen in the air with iron at high temperature during welding process. In addition, it can be inferred that the O element of reddish-brown character should come from oxygen in the air after the cracks were formed. The Si, Mn, and S elements should come from slag, which are consistent with the elements of welding wire (C, Si, Mn, P, S, Ni, Cr, Mo, V, Al). Therefore, it is inferred that the black character is slag inclusion in the cracked part of the failure girth weld. The reddish-brown character is rust which was formed after the occurrence of cracks.

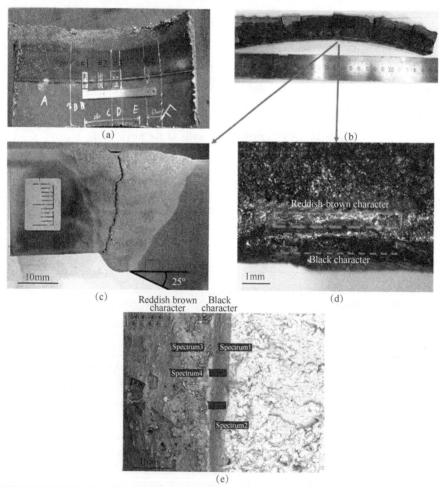

Spectrum	C	O	Na	Si	S	K	Ca	Mn	Fe	Total	Color
Spectrum 1	1.20	45.10	0.57	1.39	0.31			1.91	49.52	100.00	Dark black
Spectrum 2	0.99	48.93		1.59	0.23			2.31	45.96	100.00	Dark black
Spectrum 3	0.49	15.17		0.51	0.33	0.70	0.45	8.28	74.08	100.00	Reddish brown
Spectrum 4	0.22	6.10				0.46		5.69	87.54	100.00	Reddish brown

Fig. 8 (a) The sampling position and number of crack specimens in the H2 failed girth weld, (b) the circumferential surface crack of girth weld, (c) the deepest crack appearance from axial direction, (d) the fracture appearance from circumferential direction, and (e) the EDS result of the fracture surface

Fig. 9 shows the root crack appearance of crack specimen #1. As can be seen from Fig. 9, two thin cracks could be thought as connection of existed defects. That is to say, the two thin cracks formed after the existence of the defects. The thin cracks were formed due to the combined effects of welding defects, stress concentration, and internal pressure. There have some foreign matter inclusions in the defects, it can be inferred that the foreign matter inclusions existed long before the thin cracks[16].

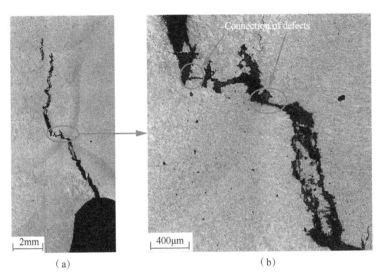

Fig. 9 (a) The root crack appearance of crack specimen #1 and (b) the enlarged crack appearance

Fig. 10 shows the EDS results of elemental composition in the four crack specimens. As can be seen from Fig. 10, the compositions in crack tip of specimen #1, welding defects of specimen #2, #3, and pores of specimen #4 were illustrated, respectively. The main chemical elements are mainly composed of elements of Fe, O, Al, and S, so the element of O might be the high temperature oxide of welding process, and the elements of Al and S might be come from slag inclusion.

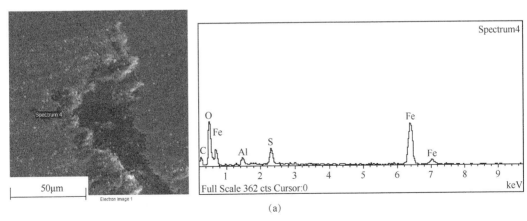

Fig. 10 Energy spectrum analysis of composition in crack for crack specimens
(a) #1, (b) #2, (c) #3, and (d) #4

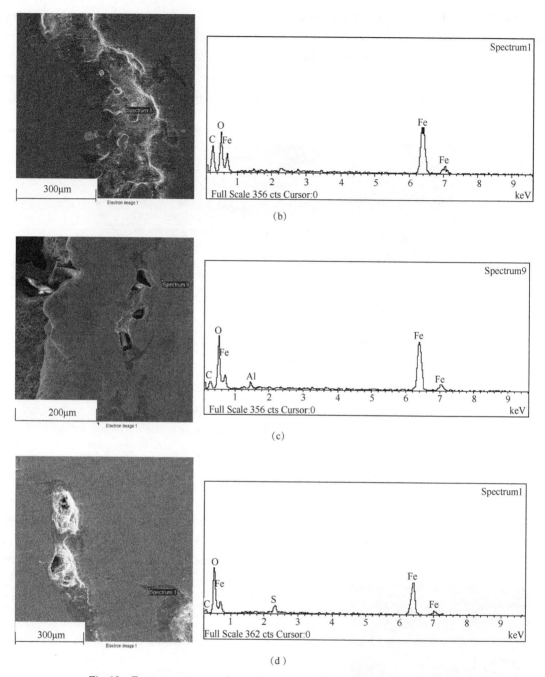

Fig. 10　Energy spectrum analysis of composition in crack for crack specimens
(a) #1, (b) #2, (c) #3, and (d) #4(continued)

Fig. 11 shows the fracture appearance of crack specimen #3 using SEM. As can be seen from the macroscopic fracture appearance of Fig. 11(a), the crack originated in the root of girth weld, as shown in Fig. 11(c). The reason is that the fracture appearance of root region is plane and even, while the fracture appearance of crack growth zone is uneven, and large dimple can be observed, as shown in Fig. 11(b). It reflected that the crack initiated and propagated from the inner wall to the outer wall of girth weld.

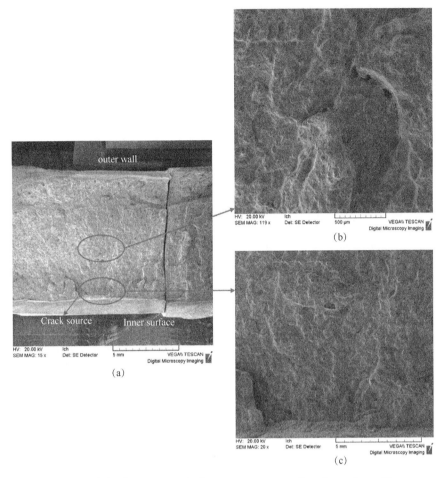

Fig. 11 (a) The macroscopic fracture appearance of crack specimen #3,
(b) the enlarged fracture appearance in center of crack specimen 3#,
and (c) the enlarged fracture appearance in root of crack specimen #3

3.10 Analysis of blasting test

The blasting pressure of testing welding structure is 46MPa, the pressure is far greater than design pressure (12MPa) and testing pressure (18MPa), the pressurizing curve is shown in Fig. 12(a). The blasting positions were located at H1 girth weld and the shoulder of tee pipe, as shown in Fig. 12(b) and Fig. 12(c). The crack striation converges to the junction between H1 girth weld and the shoulder of tee pipe, as shown in Fig. 12(d). It reflected that the welding quality of the girth weld in blasting test was better than that of failure girth weld. Therefore, the failure cause is pointed to the welding quality.

4 Discussion

The failure causes of H2 girth weld were analyzed from the following aspects.

4.1 Material properties

All material properties of straight pipe meet the standard requirements. Therefore, the failure cause of girth weld cracking is irrelevant to the straight pipe.

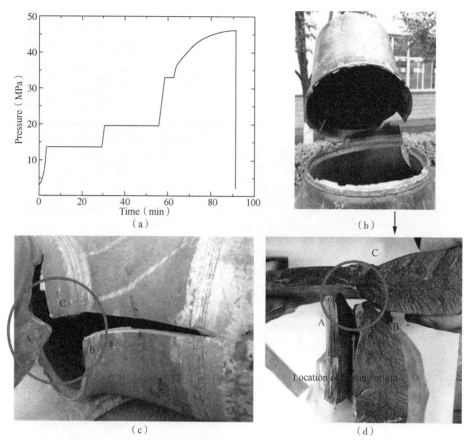

Fig. 12 (a) Hydraulic test curve of welding structure, and the blasting appearance of welding structure at (b) front and (c) side directions, and (d) the fracture appearance of blasting origin region

For the material properties of tee pipe, the chemical composition, hardness, and Charpy impact tests were in accordance with the standard requirements. However, the yield and tensile strengths of tee pipe were lower than the standard requirements, and the microstructure of tee pipe is not normal for WPHY 70, since the microstructure is mainly composed of bainite. The low strength might be caused from unqualified heat treatment process. However, the low strength is not the main failure cause of girth weld, because the blasting pressure of weld structure is far higher than the design pressure, and the tensile properties of testing tee pipe are same to the tee pipe with failure girth weld. It indicates that low strength of tee pipe does not affect the bearing capacity of girth weld and tee pipe largely, since the thickness of tee pipe is 48~65mm. The lowest tensile strength of tee pipe is 502MPa, and the bearing pressure p of this tee pipe could be calculated from

$$p = \frac{2t\sigma_T}{D} \quad (2)$$

where t is thickness, σ_T is tensile strength of tee pipe, D is diameter of pipe. Based on Eq. (2), the bearing pressure p is calculated as 47.4~64MPa.

For the material properties of girth weld, the results of tensile and Nick-Break tests meet the standard requirements. The results of Charpy impact and bending tests of H3 girth weld were not in accordance with standard requirements, and the fracture surface of girth weld (H3) exists pore defects. Combining the welding defects existed in H2 and H3 girth welds, these welding defects reflect the welding quality of failure welding structure of tee pipe is inferior.

4.2 Welding defect

The blasting result reflects the welding procedure specification (WPS) of this tee pipe and straight pipe does not hassystematic problem, and the failure cause of girth weld has been pointed to the welding quality.

H2 and H3 girth welds were welded in same period using same WPS. The 0~1 o'clock region of H2 girth weld has obvious welding defect, as shown in Fig. 10. However, the 0~1 o'clock region of H3 girth weld has no obvious welding defect using X-ray detection, as shown in Fig. 13. Combining the results of the failure H2 girth weld and qualified H3 girth weld, it can be inferred that welding defect might be the failure cause of girth weld. The welding defects are related to the welding technique of welders. In other words, the reason of H2 welding defect is that the welders did not strictly comply with the WPS. The reason of weld defect occurrence is human factor, besides, the three failure girth welds were welded by the same group of welders.

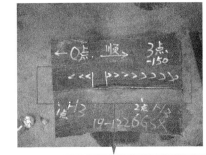

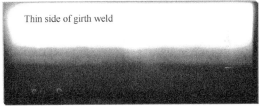

Fig. 13 The X-ray detection result of 0~3 o'clock region of H3 girth weld

The welding parameters of WPS were not carried out by this group of welders, such as welding speed, preheating temperature, welding current and voltage etc. Therefore, the original welding defects might form from above operations. However, another question needs to be answered, the question is how the welding defects make the girth weld crack?

Based on the above EDS analysis of composition in welding defect, the welding defect is mainly slag inclusion. Many researches[17-19] were made to analyze the effect of welding defect on the girth weld cracking. Slag inclusion belongs to weld discontinuity, and this weld discontinuity may be linked with many sources in welding, like improper geometry, wrong welding process and metallurgical discontinuity, and then the crack initiated from weld discontinuity[17]. The harmful slag inclusions were present in the girth weld, resulting in the generation of a certain number of cracks and microcracks in the weld. Owing to the effect of the microcrack tip, the critical stress threshold of the instability was considerably reduced[18]. Multisource cracks easily originated and propagated through the wall thickness under the influence of internal pressure. The presence of welding defects could jeopardize the mechanical integrity of pressurized components[19], so the mechanical integrity of the tee piping system in this study was affected by the welding defects.

In terms of influencing mechanism of welding defects on girth weld cracking, the descriptions are given as follows. Slag inclusion was produced when the previous layer of weld bead was not clean and the slag was not completely discharged. Then, the existence of slag inclusion in the girth weld led that the intergranular bonding force of metal was poor[16]. The defect position easily became the crack initiation point, which led to the crack propagation in the subsequent transportation process. Furthermore, the fracture resistance of the pipes decreased with the increase in initial crack size[20], and the propagation process of crack was accelerated. In addition, the slag inclusion may become a point of weakness, this growing to a semi elliptical surface crack[Fig. 8(b)], resulting in localized thickness reduction and acting as a stress raiser, which was attributed to be possible cause of failure[17].

4.3 Stress concentration effects

Fig. 14(a) shows the stress concentration effects of girth weld with the variation of inner chamfering angle under the inner pressure 18MPa. The stress concentration effect becomes larger and larger with the variation of θ, and the stress concentration region focuses on the inside weld toe of thin wall side.

To understand the stress variation of the stress concentration region, the stress variation of five inner chamfering angles with the increasing inner pressure is observed in Fig. 14(b), and the stress variation with increasing inner chamfering angles at 18MPa of inner pressure is illustrated in Fig. 14(c). The stress concentration increases rapidly with increasing θ from 15° to 25°. The inner chamfering angle θ of failure girth weld is 25°, which is larger than 15° required in the WPS of this girth weld.

The top of girth weld between tee pipe and straight pipe has the stress concentration effect, this phenomenon could be found in this unequal wall thickness girth weld of tee pipe through numerical simulation, as shown in Fig. 15. That is why the cracks of girth weld were easily occurred in the top of girth weld between tee pipe and straight pipe.

SCF is used to analyze the stress concentration of material. Based on the classical shell theory, the estimation formula of the SCF K_t is expressed as[7],

$$K_t = 1 + \frac{2-\nu}{\gamma}\sqrt{\frac{3}{1-\nu^2}}\left(1-\frac{t_1}{t_2}\right) + \frac{3(t_2-t_1)}{t_1}\frac{1}{1+(t_2/t_1)^{2.5}}e^{-\alpha} \quad (3)$$

$$\alpha = \frac{1.82s}{\sqrt{(D-t_1)t_1}}\frac{1}{1+(t_2/t_1)^{2.5}} \quad (4)$$

$$\gamma = \frac{2(t_1^{2.5}+t_2^{2.5})}{t_2^{2.5}-t_2^{0.5}t_1^2}\left[1+\left(\frac{t_1}{t_2}\right)^{1.5}\right]+\left(\frac{t_1}{t_2}\right)^2 - 1 \quad (5)$$

ν is the Poisson's ratio, t_1 is the thickness of small diameter pipe, t_2 is the thickness of large diameter pipe, e is the base of the natural logarithm, D is the pipe outside diameter, s is the weld conical surface length, α is a parameter calculated by s, D, t_1 and t_2, γ is a parameter calculated by t_1 and t_2.

In this study, ν is 0.3, t_1 is 26mm, t_2 is 57mm, D is 1016mm, s is 33.4mm, K_t is calculated as 1.98. It indicates that the stress concentration degree of this grith weld is high, the maximum stress of inside weld toe of thin wall side is twice higher than average stress. The wall thickness ratio and the conical surface length have a great influence on the pipeline, which is the main influencing factor of the stress concentration of the pipeline[10,21].

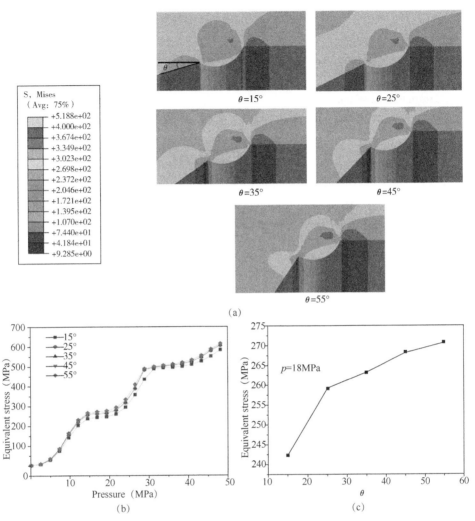

Fig. 14 (a) The stress concentration effects of girth weld with the variation of θ, (b) the stress variation of five inner chamfering angles with increasing inner pressure, and (c) stress variation with the variation of under 18MPa of inner chamfering angle

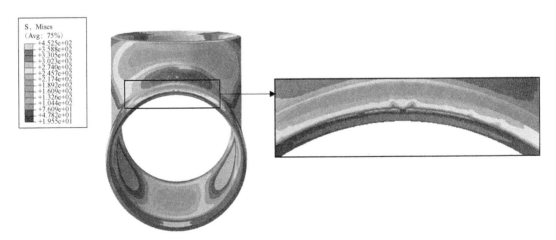

Fig. 15 The stress concentration effects of girth weld of tee pipe on circumferential direction

Three aspects of stress concentration effects were existed at the inner weld toe of thin wall side and the top of girth weld. The first one was induced by the unqualified inner chamfering angle, the second one was originated from the unequal wall thickness, and the last one came from the inherent shape of tee pipe.

4.4 Comprehensive analysis

The welders did not strictly carry out the WPS of the H2 girth weld among tee pipes and straight pipes. That is to say, the welding parameters of WPS were not carried out by this group of welders. Therefore, the slag inclusions were not cleaned up timely. These slag inclusions led to the existence of welding discontinuities. Then, the welding defects were existed in the girth weld. The welding defects easily led to crack since the intergranular bonding force of girth weld is poor. Due to the effect of the crack tip, the critical stress threshold of the instability was reduced. It is hard to found the original welding defects due to the limitation of present nondestructive methods for unequal wall thickness. Therefore, the stress concentration effects originated from the unqualified inner chamfering angle, unequal wall thickness, and inherent shape of tee pipe accelerated the crack initiation and propagation under the existence of welding defects in girth weld and the testing pressure of 18MPa. Finally, these welding defects led to premature failure of girth weld due to the stress concentration effects at the crack tip.

5 Conclusionsand recommendations

The failure analysis of girth weld cracking among tee pipes and straight pipes was performed by visual inspection, nondestructive testing, microstructure examination, scanning electronic microscopy (SEM) coupled with energy dispersive spectrometry (EDS), blasting tests, finite element simulations and a series physical and chemical tests. Based on the analysis of material properties, welding defects and stress concentration effects, the main conclusions are drawn as follows:

(1) The original welding defects were main failure causes of girth weld, and three aspects of stress concentration effects originated from unqualified inner chamfering angle, unequal wall thickness, and the inherent shape of tee pipe accelerated the crack initiation and propagation of girth weld under internal pressure.

(2) The WPS of this girth weld has no systematic problem, and the reason of welding defects of girth weld is that the WPS of this girth weld were not strictly carried out by welders.

(3) The yield and tensile strengths of tee pipe were lower than the standard requirements. However, the failure of girth weld was unrelated to the low strengths of tee pipe.

Based on the above conclusions, several recommendations were then proposed from both design and maintenance point of view.

(1) The design of groove of unequal wall thickness needs to be more reasonable so that reduce the stress concentration effect.

(2) The welding quality of girth weld with unequal wall thickness needs to be strictly supervised and managed.

(3) The nondestructive testing techniques needs to be studied in this girth weld with unequal wall thickness so that these cracks could be found in advance.

Declaration of Competing Interest

The authors declared that there is no conflict of interest.

Funding

The authors are very grateful for the support received from the Natural Science Foundation of Shaanxi Province, China [grant number 2020JQ-934] and Young Scientists Fund of the National Natural Science Foundation of China [grant number 51904332].

References

[1] ASHRAFIZADEH H, KARIMI M, ASHRAFIZADEH F. Failure analysis of a high pressure natural gas pipe under split tee by computer simulations and metallurgical assessment[J]. Engineering Failure Analysis, 2013, 32: 188-201.

[2] SHALABY H, RIAD W, ALHAZZA A, et al. Failure analysis of fuel supply pipeline[J]. Engineering Failure Analysis, 2006, 13(5): 789-796.

[3] AZEVEDO C R. Failure analysis of a crude oil pipeline[J]. Engineering Failure Analysis, 2007, 14(6): 978-994.

[4] YOON K B, BYUN C H, NGUYEN T S, et al. Cracking of 5Cr steel tee-pipe during start-up operation in heavy oil upgrade refinery[J]. Engineering Failure Analysis, 2017, 81: 204-215.

[5] YOON K B, YU J M, NGUYEN T S. Stress relaxation cracking in 304H stainless steel weld of a chemical reactor serviced at 560℃[J]. Engineering Failure Analysis, 2015, 56: 288-299.

[6] LOTSBERG I. Stress concentration factors at circumferential welds in tubulars[J]. Marine Structures, 1998, 11(6): 207-230.

[7] LOTSBERG I. Stress concentration factors at welds in pipelines and tanks subjected to internal pressure and axial force[J]. Marine Structures, 2008, 21(2-3): 138-159.

[8] LOTSBERG I. Stress concentrations due to misalignment at butt welds in plated structures and at girth welds in tubulars[J]. International Journal of Fatigue, 2009, 31(8-9): 1337-1345.

[9] MEGAHED M M, ATTIA M S. Failure analysis of thermowell weldment cracking[J]. Engineering Failure Analysis, 2015, 50: 51-61.

[10] CERIT M, KOKUMER O, GENEL K. Stress concentration effects of undercut defect and reinforcement metal in butt welded joint[J]. Engineering Failure Analysis, 2010, 17(2): 571-578.

[11] OSTSEMIN A A, UTKIN P B. Stress-concentration coefficients of internal welding defects[J]. Russian Engineering Research, 2008, 28: 1165-1168.

[12] MACDONALD K A, CHEAITANI M. Engineering critical assessment in the complex girth welds of clad and lined linepipe materials[C]. International Pipeline Conference, Calgary, 2010, 44212: 823-843.

[13] CARLUCCI A, BONORA N, RUGGIERO A, et al. Crack initiation and growth in bimetallic girth welds[C]. International Conference on Offshore Mechanics and Arctic Engineering, American Society of Mechanical Engineers, San Francisco, 2014, 45462: V06AT04A042.

[14] WU K, LIU X, ZHANG H, et al. Fracture response of 1422-mm diameter pipe with double-V groove weld joints and circumferential crack in fusion line[J]. Engineering Failure Analysis, 2020, 115: 104641.

[15] FU A Q, KUANG X R, HAN Y, et al. Failure analysis of girth weld cracking of mechanically lined pipe used in gasfield gathering system[J]. Engineering Failure Analysis, 2016, 68: 64-75.

[16] DING H, QI D T, QI G Q, et al. Cracking analysis of a newly built gas transmission steel pipe[J].

Engineering Failure Analysis, 2020, 118: 104868.

[17] JHA A K, MANWATKAR S K, NARAYANAN P R, et al. Failure analysis of a high strength low alloy 0.15C-1.25Cr-1Mo-0.25V steel pressure vessel[J]. Case Studies in Engineering Failure Analysis, 2013, 1(4): 265-272.

[18] LIU Q, YU H, ZHU G, et al. Investigation of weld cracking of a BOG booster pipeline in an LNG receiving station[J]. Engineering Failure Analysis, 2021, 122: 105247.

[19] ZANGENEH S, LASHGARI H R, SHARIFI H R. Fitness-for-service assessment and failure analysis of AISI 304 demineralized-water (DM) pipeline weld crack[J]. Engineering Failure Analysis, 2020, 107: 104210.

[20] KRISHNAN S A, NIKHIL R, SASIKALA G, et al. Evaluation of fracture resistance of AISI type 316LN stainless steel base and welded pipes with circumferential through-wall crack[J]. International Journal of Pressure Vessels and Piping, 2019, 178: 104008.

[21] WANG L, TANG Y, MA T, et al. Stress concentration analysis of butt welds with variable wall thickness of spanning pipelines caused by additional loads[J]. International Journal of Pressure Vessels and Piping, 2020, 182: 104075.

本论文原发表于《Materials Science Forum》2021 年第 1035 卷。

Service Performance of 3PE Coating for Compressor Export

Nie Hailiang[1,2] Ma Weifeng[1] Hu Xu[3] Wang Ke[1]
Ren Junjie[1] Cao Jun[1] Dang Wei[1] Huo Chunyong[1]

(1. Institute of Safety Assessment and Integrity, State Key Laboratory for Performance and Structure Safety of Petroleum Tubular Goods and Equipment Materials, Tubular Goods Research Center of CNPC; 2. Northwestern Polytechnical University; 3. National pipe network Group western Pipeline Co., Ltd)

Abstract: The selection of the anticorrosive layer for the compressor outlet pipe is difficult since no standards have included the requirements about high temperature-vibration. In this paper, acceleration tests of temperature-vibration coupling condition are conducted to investigate the applicability and service reliability of three layers of polyethylene anticorrosive layer (3PE). The results showed that 3PE coating has favourable resistance to temperature-vibration damage, which is suitable for the high temperature-vibration condition of compressor outlet pipeline.

Keywords: Station; Compressor outlet; Anticorrosive coating

1 Introduction

In recent years, the failure of theanticorrosive layer at the station compressor outlet has occurred frequently. Serious failure of anticorrosive layer has been found in many excavated buried pipeline of compressor outlet in the west of China. The damaged area can reach tens of centimeters in the annular direction and hundreds of centimeters in the axial direction. The large area of anticorrosive layer failure leads to serious pipeline corrosion[1]. According to the investigation, the compressor outlet is in the state of high-frequency vibration, and the temperature of pipeline can reach up to 70℃. At present, the index requirements about high temperature-vibration of anticorrosive layer have not been included in the relevant standards, so it is urgent to study the applicability and service reliability of anticorrosive layer products in the compressor export, so as to provide a basis for the selection of anticorrosive layer.

There are many researches on the damage mechanism of anticorrosive coating. Worsley et al.[2] believe that porosity is an important factor for the coating's resistance to cathode stripping. The permeability of ions, water and oxygen in the coating depends on porosity, and the higher the

---

Corresponding author: Nie Hailiang, niehailiang@mail.nwpu.edu.cn.

porosity, the stronger the permeability of water and oxygen. Miszczyk and Darowicki[3] believed that there was an obvious correlation between the temperature and the peeling speed of the coating, and the cathode peeling speed of the coating accelerated with the increase of the temperature. Giulia et al.[4] developed a standardized aging test program for durability and stability of coatings. The weathering and aging degradation properties of 4 kinds of standard durable coatings and 7 kinds of super durable coatings were compared, and good results were obtained. Mills et al.[5] proposed the working mechanism of anticorrosive coatings, and introduced in detail a series of relatively simple tests which the author thinks are suitable to measure these properties in a quantitative way, so as to help develop new or better coatings and provide many relevant reference tests for coating testing.

These studies have revealed the damage mechanism of the anticorrosion coating under some environmental conditions, and provided a theoretical basis for the corrosion protection of pipelines, but no relevant reports have been reported on theservice condition of buried pipeline anticorrosion coating under the coupling effect of temperature – vibration. The compressor outlet is under the complicated working conditions of high temperature-vibration, which is easy to cause the failure of the anticorrosive layer of the pipeline. Many qualified anticorrosive layers used in ordinary pipelines will also fail at the compressor outlet.

In this paper, the service reliability of 3-layer polyethylene (3PE) was testedunder temperature-vibration coupling condition. According to the type test stipulated in the standard, the performance of the anticorrosive products is verified to be qualified under the temperature – vibration coupling condition, which provides experimental basis for the selection of anticorrosive layer for compressor outlet pipelines.

2 Specimen Design

The 3-layer polyethylene (3PE) specimens are cut and prepared from the same 3PE pipeline as the second line of west-east gas transmission line. The size and quantity of the specimens is the same with that in the standard. The preparation of anticorrosive layer specimens is shown in Fig. 1.

Fig. 1 Preparation of anticorrosive layer specimens

3 Experimental Design

The research is divided into two stages:

(1) The temperature-vibration coupling experiment stage. The temperature-vibration coupling test system independently developed by our institute was used to accelerate the temperature-vibration coupling test of the anticorrosion layer, so as to simulate the servicing condition of the compressor outlet in station.

(2) The property comparison test. The same type test was carried out for the specimens acted by temperature-vibration and those without coupling temperature-vibration, so as to obtain the property comparison and the law of adhesion degradation.

3.1 Temperature-vibration coupling action test

3.1.1 Experimental platform

In order to make the anticorrosion coating specimens bear certain vibration load at a constant test temperature, the project team developed a set of test equipment that can meet the coupling effect of temperature-vibration, which mainly includes four parts, as shown in Fig. 2:

(1) The temperature part. The conventional electric thermostatic drying heating box of Type 202 produced by Beijing Kewei Yongxing Instrument Co. was used. The temperature range is 0~200℃. A thermostat is used to provide a constant temperature for the specimens.

(2) The vibration part. ET-50-445 electric vibration test system produced by Suzhou dongling vibration instrument co., LTD was used, with the maximum load of 800kg and the maximum acceleration of 1000m/s^2, the vibration frequency ranges from 5 to 3000Hz. This part can provide constant vibration parameters such as vibration frequency and vibration acceleration.

(3) The connection part. The vibration load of the vibrating table is transferred to the specimen placement table in the thermostat through special tools, and the body of the thermostat is isolated from the shaking table, so as to realize the combined action of temperature-vibration and avoid the damage of vibration to the thermostat.

(4) The Real-time control part. This part was used to ensure the constant load on the test plate. A camera is used for real-time monitoring to ensure the continuous and uninterrupted test. The camera is connected to the mobile APP, thus the state of the device can be monitored in real time and problems can be found timely.

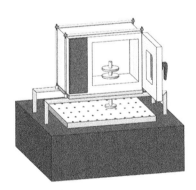

Fig. 2 Experimental platform for temperature-vibration coupling action

3.1.2 Test process

The temperature-vibration coupling equipment was used to simulate the compressor outlet in buried pipeline environment. Vibration parameters are the main frequency (500Hz) and acceleration (80m/s^2) obtained from field tests, and the maximum temperature of the compressor outlet pipe is 70℃.

In order to ensure the consistency between the indoor simulation experiment and the field environment, the influence of soil on the anticorrosive layer is proposed to be added into the simulation. To this end, the researchers extracted about 5kg of soil specimens at Kongquehe compressor station, in preparation for indoor simulation experiments.

The specimens are buried in the sealed crisper containing the soil and water on the site, and fixed on the screw connected with the vibrator through the metal splints, so as to realize the resonance with the vibrator while heating, as shown in Fig. 3.

Fig. 3 Temperature-vibration coupled loading experiment process

After 30 days of continuous and uninterrupted heating and vibration, the specimens were taken out for the next type test and analysis.

3.2 Type tests

After thetemperature-vibration coupling test, the specimens were subjected to the type tests specified in the standards, and the conventional specimen was subjected to the same type tests for comparison. The type test results of the temperature-vibration acted specimens and the conventional specimens were compared to obtain the property degradation rule under the temperature-vibration coupling condition.

In accordance with the standard GB/T 23257—2017 *Polyethylene coating for buried steel pipeline*[6], the thickness test, peeling strength test and bending test of conventional and temperature-vibration coupling acted specimens were respectively carried out.

The thickness measurement is carried out by magnetic thickness meterElcometer 456. The peel strength tests were carried out at 70℃, 60℃ and 50℃ respectively. The test was carried out by tension meter, clamp and electric thermostatic drying heating box. During the experiment, the specimens were preheated for 24h in the electric thermostatic drying heating box to ensure the uniform distribution of the temperature, and then take it out and conduct the peel test immediately. Bending test was carried out by bending machine and freezer at -20℃ and 2.5°.

4 Experimental results and discussion

4.1 Peel strength

The variation of peel strength with temperature is shown in Fig. 4. It can be seen that the peel strength of 3PE anticorrosive coating after temperature-vibration action is about 8N/cm lower than that of conventional 3PE anticorrosive coating at the same testing temperature, but the overall peel strength is still very high (higher than the standard requirement of 70N/cm). The peel strength decreases linearly with the testing temperature.

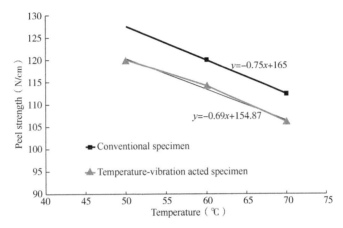

Fig. 4　Variation trend of peel strength with temperature

4.2 Bending test

The specimens after bending loading are shown in Fig. 5. There is no significant difference in the results after bending at −20℃ and 2.5° among conventional specimens and temperature-vibration acted specimens, and both show favourable bending resistance. No cracks or spallation occurred on the surface of the anticorrosive layer, which indicates that 3PE anticorrosive layer has great bending resistance.

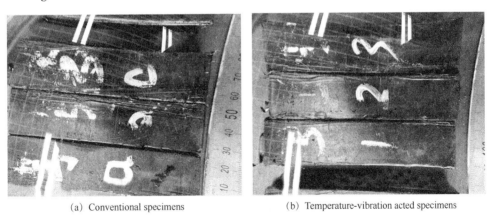

(a) Conventional specimens　　　(b) Temperature-vibration acted specimens

Fig. 5　Curved specimen surfaces

4.3 Thickness variation

The coating thicknesses are shown in Table 1. The thickness of 3PE anticorrosive layer didn't change significantly before and after the temperature-vibration coupling test, indicating that the thickness loss of 3PE anticorrosive layer was not obvious under the temperature-vibration coupling effect.

Table 1 Thickness test results of 3PE

Specimen number	Thickness (mm)					
	Before temperature-vibration action			After temperature-vibration action		
	Point 1	Point 2	Point 3	Point 1	Point 2	Point 3
a	5.37	4.95	4.85	5.19	5.32	5.18
b	5.37	4.86	5.33	5.21	4.72	5.11
c	5.07	5.31	5.36	5.32	5.21	4.99
Average	5.16			5.14		
Standard deviation	0.215			0.176		

Based on the above analysis, after temperature-vibration coupling action, the thickness of 3PE coating basically did not change, resistance to bending performance is favourable, the peel strength under high temperature is a little lower than that of conventional specimens, but a favourable stripping resistance is still maintained, so the 3PE anticorrosion layer is suitable for compressor outlet pipeline.

5 Conclusion

In this study, 3PE anticorrosive coating specimens were tested, the self-developed temperature-vibration coupling action experiment equipment and the soil sample near buried pipeline of compressor outlet were used to complete the indoor acceleration test on compressor outlet buried pipeline.

Through the comparative analysis of various performance indexes of the anticorrosive layers, it is found that the bending resistance of 3PE meets the standard requirements after the temperature-vibration coupling test. With the increase of temperature, the peeling strength of the 3PE anticorrosive layer decreases, but it still keeps a large peeling strength.

The comprehensive comparison shows that the 3PE anti-corrosion layer has favourable anti-temperature-vibration damage performance, which is suitable for compressor outlet buried pipeline anti-corrosion selection.

Acknowledgments

This work was supported by the China Postdoctoral Science Fund (2019M653785) and the National Key R&D Program of China (2016YFC0802101).

References

[1] FENG L, HONG J X, GUO L W, et al. The Discussion of Reasons for the Corrosion of Long-distance Underground Pipelines in Stations and the Corresponding Protection Measures[J]. Total Corrosion Control, 2011, 25(5): 6-9.

[2] WORSLEY D A, WILLIAMS D, LING J S G. Mechanistic changes in cut-edge corrosion induced by variation

of organic coating porosity[J]. Corrosion Science, 2001, 43: 2335-2348.
[3] MISZCZYK A, DAROWICKI K. Effect of environmental temperature variations on protective properties of organic coatings[J]. Progress in Organic Coatings, 2003, 46: 49-54.
[4] GIULIA G, GANZERLA R, BORTOLUZZI M, et al. Accelerated weathering degradation behavior of polyester thermosetting powder coatings[J]. Progress in Organic Coatings, 2016, 101: 90-99.
[5] MILLS D J, SINA S J. The best tests for anti-corrosive paints. And why: A personal viewpoint[J]. Progress in Organic Coatings, 2017, 102: 8-17.
[6] GB/T 23257—2017. Polyethylene coating for buried steel pipeline[S]. Beijing: General Administration of Quality Supervision, Inspection and Quarantine of the People's Republic of China, 2017, 12, 5.

本论文原发表于《Materials Performance》2021年第60卷第6期。

A Novel Test Method for Mechanical Properties of Characteristic Zones of Girth Welds

Nie Hailiang[1]　Ma Weifeng[1]　Xue Kai[2]　Ren Junjie[1]
Dang Wei[1]　Wang Ke[1]　Cao Jun[1]　Yao Tian[1]　Liang Xiaobin[1]

(1. Institute of Safety Assessment and Integrity, State Key Laboratory for Performance and Structure Safety of Petroleum Tubular Goods and Equipment Materials, Tubular Goods Research Center of CNPC; 2. PetroChina Changqing Oilfield Company)

Abstract: The safety evaluation of girth welds is an important issue for steel pipes. Current standards and studies all regard the weld as a uniform material, ignoring its inhomogeneity. In this paper, a specimen processing method for weld characteristic zones is proposed. Based on the natural morphology of the weld structure, specimens are accurately designed and processed for each characteristic zone. In addition, a testing method to determine the tensile properties in the characteristic zones of the weld is also proposed, solving the absence of weld mechanical properties in existing steel pipe safety evaluations. This method was used to test an X80 automatic girth weld, and the results showed that the tensile properties significantly differed among different characteristic zones of the girth weld. The inhomogeneity displayed in this weld demonstrates how a more detailed assessment of weld properties should be considered in practical engineering and safety evaluations.

Keywords: Girth weld; Characteristic zone; Mechanical properties; Inhomogeneity; Test method

1 Introduction

Girth weld failure is an important failuremode of steel pipes. The damage mechanism and safety evaluation of girth welds have become an important research topic[1]. At present, the most common international safety evaluation method for girth welds is by comparing the tensile strength of the weld material with that of the base[2-3]. However, this evaluation method is too conservative for strongly matched welds and unreliable for weakly matched welds, so a better understanding of the tensile properties of pipeline welds is urgently required.

The most reliable evaluation method for girth weld defects is to evaluate the material properties around the defect. However, all relevant standards regard the weld as a uniform material and use a

---

Corresponding author: Nie Hailiang, niehailiang@ mail. nwpu. edu. cn.

tensile specimen containing the whole weld area[4-7]. This method can obtain the macroscopic mechanical properties of a weld, but it cannot distinguish its inhomogeneity. Existing reports indicate that this standard is used to test girth welds to obtain the performance of the overall weld structure. For example, Motohashi and Hagiwara[8] evaluated the effects of strength matching on the fracture performance of girth welded joints by conducting curved wide plate tensile tests based on the assumption of uniform welds. Mathias et al.[9] studied the effects of specimen geometry and loading mode on crack growth resistance curves of a high-strength pipeline girth weld. However, their specimen design was still based on existing standards, and the influence of different characteristic zones of the weld was not studied. Denys et al.[10] developed a strain capacity prediction method adjusted for the heterogeneity of girth welds, allowing for the adoption of existing strain capacity equations for welds connecting homogeneous pipes. However, in the theoretical derivation process, the weld was regarded as a whole structure without considering the difference of mechanical properties in each characteristic zone of the weld.

These assessments of weld properties are insufficient because girth welds have a non-uniform structure. Due to the influence of parameters such as electrode material, welding method, and the welding current of different welding layers, the mechanical properties of the final formed weld show significant differences in different zones. Because each zone has a small range, specimen processing is very difficult. The testing of micro-tensile mechanical properties of girth welds has become a major problem in practical engineering. Tang et al.[11] and Zhu[12] proposed a specimen processing and testing method for evaluating the microzones of girth welds (Fig. 1). This method stratified the weld in the direction of thickness, processing the sheet containing the base metal, the weld, and the heat-affected zone. The thickness of each section was polished to 0.5mm. After the microstructure of the weld was corroded, a tensile specimen of the microzone in each area of the weld was cut by a slow wire. This method cut the weld seam into thin slices along the thickness direction to process micro specimens from each characteristic zone. However, due to the extremely narrow heat-affected zones of the weld, the size of the specimens was too small and specimen surface treatment had a significant impact on the test results. Therefore, a less cumbersome specimen processing method with a strict surface treatment is required.

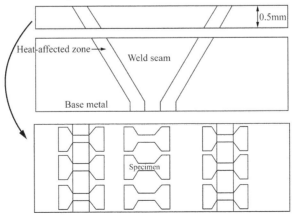

Fig. 1 Diagram of specimen processing method for microzone of girth weld proposed by Tang et al.[11] and Zhu[12]

In this paper, a method for processing tensile specimens in the characteristic zones of girth welds is proposed. Based on the natural morphology of the weld structure, the specimens are accurately designed and processed in each characteristic zone of the weld to maximize the size of each specimen without exceeding the scope of its characteristic zone. This significantly reduces the difficulty of specimen processing and surface treatment in the later stage of analysis. In addition, a testing method to determine the tensile properties in the characteristic zones of the weld is also proposed, solving the problem of unknown weld mechanical properties in weld defect safety evaluations and laying a foundation for improving their accuracy and reliability. Finally, the feasibility of the proposed method is demonstrated by testing an X80 automatic girth weld.

2 Specimen testing method

2.1 Specimen design and processing

In this paper, the processing method for obtaining micro-tensile weld specimens includes the selection of machining sections, microzone division, specimen design, specimen processing, and numbering.

Step 1: Selection of machining sections.

In this step, the weld material was observed and areas with geometric defects such as hutt joints, undercuts, and overlaps were marked. Non-Destructive Testing (NDT) was performed on the welds and defects such as cracks, lack of penetration, and non-fusion defects were marked. The length of each selected specimen processing section along the weld seam was larger than that of the tensile specimen in its characteristic zone.

Step 2: Microzone division.

The weld structure was divided into different zones according to the weld microstructure. A typical weld microstructure is shown in Fig. 2. As can be seen, different areas of the weld show different micro-morphology and grain size, clearly distinguishing different characteristic areas. A common weld structure includes a gap weld, filling weld, root bead, heat-affected zone, and the base metal.

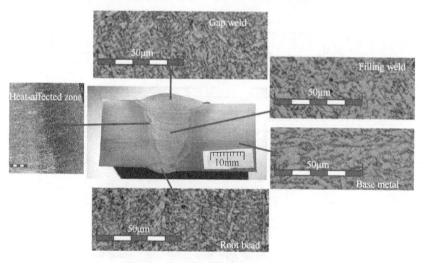

Fig. 2 Typical weld microstructure

The specimen processing method is shown in Fig. 3. The selected machining section was cut perpendicularly to the weld direction into the billet with a length longer than that of the specimen. The two cutting surfaces of the billet were polished to ensure they were smooth and parallel. The gap weld, filling weld, root bead, and heat-affected zones of the weld were revealed by using a weld corrosion solution on an end face of the billet.

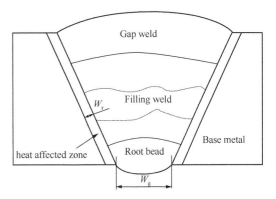

Fig. 3 Schematic diagram of microstructure distribution and dimension measurements of a typical weld microzone

Step 3: Specimen design.

In order to provide a basis for specimen design, the width and length of the heat-affected zone and the height and width of the root bead were measured. The micro-tensile specimens were a flat dog bone shape (Fig. 4). The width of the heat-affected zone W_r was taken as the thickness H of the micro-tensile specimen. In addition, the width of the test section W_b was greater than the thickness H of the specimen and smaller than the width of the root bead W_g. Meanwhile, $\sigma \times W_b \times H$ was less than 80% of the measuring range of the test equipment, where σ is the yield strength of the material in the characteristic zone. The width of the clamping section W was smaller than the width of the root bead. The test section and the clamping section of the specimen were transitioned by a circular arc.

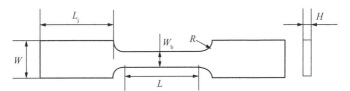

Fig. 4 Schematic diagram of micro-tensile specimen geometry and dimensions

A marker pen was used to mark the arrangement of micro-tensile specimens on the corroded end face of the billet, as shown in Fig. 5. The length direction of the micro-tensile specimens was along the longitudinal direction of the weld seam, and their end faces were located in the corroded surface of the billet. According to the size of the heat-affected zone, gap weld, filling weld, and root bead, the number and location of specimens in each zone were determined.

In the gap weld, filling weld, and root bead, the processing position and quantity of the micro-tensile specimens were determined by drawing a grid using the geometric shape characteristics of each zone.

In the heat-affected zone, the width direction of the specimens was along the contour line. In the gap weld, filling weld, and root bead, the thickness direction of the specimens was along the direction of welding layer accumulation. Using the geometric characteristics of each zone, the tensile specimen processing locations and quantity were determined by the grid method.

Step 4: Specimen processing and numbering.

The specimens were processed by wire cutting. First, the two heat-affected zones of the billet were cut down along the contour by slow wire cutting, then the thin slices of the heat-affected zones

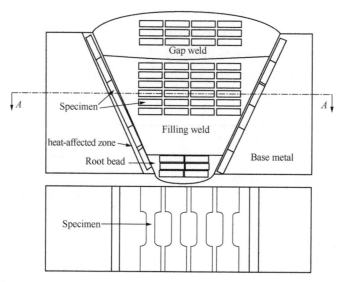

Fig. 5　Schematic diagram of specimen distribution design in typical weld characteristic zones

were cut into micro-tensile specimens by slow wire cutting along the designed position. Each specimen was numbered and marked to identify its position. The remaining billet was then cut by slow wire. When cutting, the grid shape designed in each zone was first cut as a whole, then the slow wire was used to cut each specimen. Each specimen was numbered and marked in the processing sketch.

2.2　Experimental process

A testing method for determining the tensile mechanical properties of the weld characteristic zones is also provided. First, micro-tensile fixtures were designed and processed according to the size of each micro-tensile specimen. Before each test, a small white dot was marked at both ends of each specimen test section to act as the identification point for the optical extension meter. The width and thickness of each specimen testing section were measured with a vernier caliper or micrometer.

During the experiment, each micro-tensile specimen was clamped to the tensile test machine. An optical extensor was used to monitor and record the deformation of the specimens, and the tensile force at both ends of the micro-tensile specimens was detected and recorded by the tensile test machine.

After completion of each test, the recorded data of specimen deformation data and loading history were used to calculate stress-strain curves of the tensile specimens. The tensile mechanical properties of the characteristic zones, including material elastic modulus, yield strength, ultimate strength, and yield strain, were then obtained.

3　Automatic girth weld details

The mechanical properties of an automatic girth weld were studied by using the method proposed in this paper. The weld was a butt girth weld of X80 steel pipe with a pipe diameter of 1422mm and a wall thickness of 21.4mm. Mixed gas protection (80% Ar + 20% CO_2) was used during the welding process. The welding material of the root bead was BOEHLER SG3-P and the welding material of the filling weld and the gap weld was LINCOIN PIPELINER 80Nil.

In order to determine the specimen distribution, the microzone structure of the weld was first revealed by corrosion. The corroded weld structure included the base metal, the heat-affected zone, and the weld zone. The weld zone was further divided into the gap weld, filling weld, and root bead. A schematic diagram of the girth weld microstructure and specimen distribution and zone numbers is shown in Fig. 6. The processed micro-tensile specimen is shown in Fig. 7.

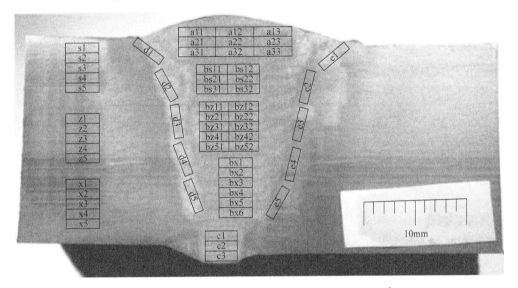

Fig. 6 Photograph showing the characteristic zones and specimen distribution design of the tested automatic welding girth weld

Fig. 7 Photograph of a processed specimen

An Instron 5848 quasi-static test instrument (Fig. 8) was used to carry out the quasi-static tensile experiments at a tensile rate of 0.25mm/min. The tensile load was directly measured by a sensor (range: -2000 to +2000 N) and the tensile strain was measured by an optical extensometer.

4 Results and discussion

4.1 Weld area

The uniaxial tensile test results of the specimens taken from the weld zone are shown in Fig. 9 (a)—(e). Experiments a22, a33, and c1 failed, and no data were obtained for these specimens.

Fig. 8 Instron 5848 quasi-static test instrument

Fig. 9 (a) shows the tensile results of the specimens in the gap weld. Their true stress-strain curves mainly consist of three sections. The first is the elastic section, where stress increases linearly with strain. At about 300~350MPa, the second section (plastic strengthening) is reached. After peak stress is reached, the stress begins to show a downward trend in the third section, and this trend becomes increasingly severe until the specimen fractures. The tensile yield strength of the material in the gap weld is about 300~350MPa, which is the lowest value in the entire weld area.

Figs. 9 (b)—(d) show the tensile results of the specimens in the filling weld. Compared with the gap weld, there is one more plastic platform section in the real stress-strain curve of the material in this area. After the elastic limit is reached, the stress level of the material remains relatively constant, and then it enters the stage of plastic strengthening. The material properties of the filling weld are in between those of the gap weld and the root bead. The yield strength of the material gradually increases from the top to the bottom of the weld area (i. e., 300MPa at the top to 450MPa and then 600MPa at the bottom).

Fig. 9 (e) shows the tensile results of the specimens in the root bead. In this region, the dispersion of the real stress-strain curve is obvious and the yield strength is about 450MPa. The curves show that the c2 specimen was closer to the cover area in a three-stage pattern (elastic, plastic strengthening, and fracture stages), while the c3 specimen was close to the filling weld. That is, a small stress platform section appears between the elastic section and the plastic strengthening section of the c3 curve.

4.2 Heat-affected zone

The uniaxial tensile test results of the heat-affected zone specimens are shown in Figs. 10 (a)—(b). The true stress-strain curves of the five specimens in zone D (specimens d1—d5, to the left of the weld zone) are shown in Fig. 10 (a). The curves show obvious dispersion, and the yield strength is between 400~600MPa. The true stress-strain curves of the five specimens in zone E (specimens e1—e5, to the right of the weld zone) are shown in Fig. 10 (b), also showing

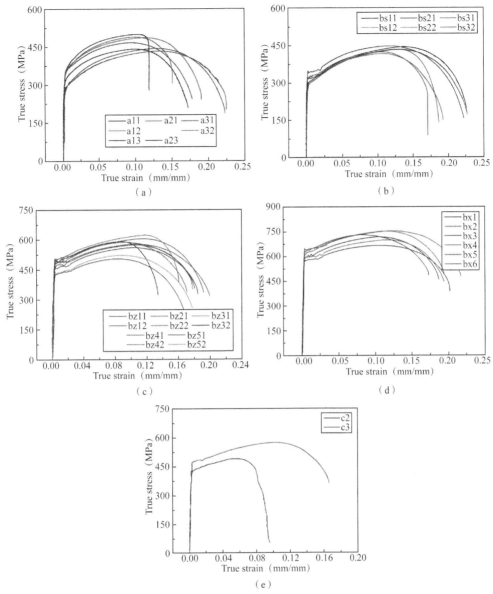

Fig. 9 Uniaxial tensile true stress-strain curves of the materials in the weld zone:
(a) gap weld; (b) upper filling weld; (c) middle filling weld; (d) lower filling weld; (e) root bead

significant dispersion. Among these specimens, the curves of e1, e2, and e3 are in good agreement, with yield strengths close to 300MPa. Specimens e4 and e5 are also in good agreement, with yield strengths as high as 600MPa.

The heat-affected zone is located at the junction of the base metal zone and the weld zone, which may lead to the processing of heat-affected zone specimens near either the base metal zoneor the weld zone and result in the dispersion of the mechanical properties of these specimens. The mechanical properties of the heat-affected zone materials were significantly affected by their location. For example, the e1, e2, and e3 specimens were close to the material of the covered area (a-series specimens) and partially filled area (b-series specimens), and their curves were in good agreement with each other. As discussed, specimens e1, e2, and e3 were very similar and had yield strengths close to 300MPa. The e4 and e5 specimens were closer to the bx-series of specimens in data and position, with yield strengths as high as 600MPa.

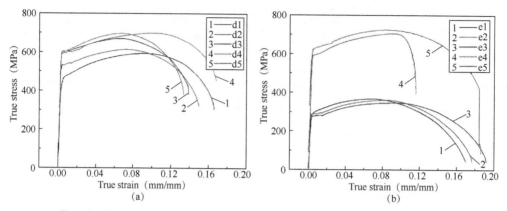

Fig. 10 Uniaxial tensile true stress-strain curves of heat-affected zone materials:
(a) the zone on the left side of the weld (d1—d5); (b) the zone on the right side of the weld (e1—e5)

4.3 Base metal

The uniaxial tensile test results of the specimens taken from the base metal zone are shown in Figs. 11 (a)—(c). The s1, s4, z2, and x3 experiments failed and no data were obtained for these specimens. Overall, the samples in this region have good repeatability and similar yield strengths of about 600MPa. However, their fracture strains are relatively dispersed. This indicates that the mechanical parameters of the material are relatively stable.

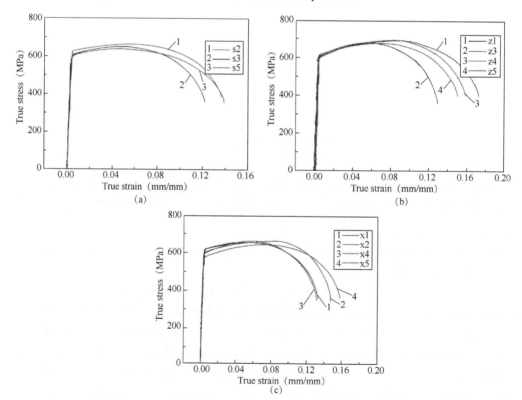

Fig. 11 Uniaxial tensile true stress-strain curves of the base metal: (a) s1—s5; (b) z1—z5; (c) x1—x5

4.4 Difference in mechanical properties in the thickness direction

The stress–strain curves of the weld in the wall thickness direction were compared. Fig. 12 shows the average true stress–strain curves of the five zones in the direction of the welding seam wall thickness (gap weld, upper layer of the filling weld, middle layer of the filling weld, lower layer of the filling weld, root bead). The mechanical properties of the weld zones significantly vary in the direction of thickness. The lowest yield strength is 300MPa, while the highest yield strength is 600MPa. The filling weld demonstrates the best tensile properties, followed by the root bead. The gap weld has the lowest tensile strength.

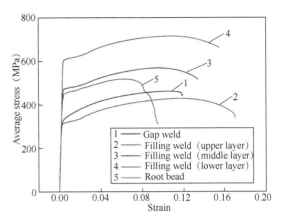

Fig. 12 Stress–strain curves of materials in different characteristic areas in the wall thickness direction of the weld

This difference in stress capacity is the result of the coupling and superposition of the welding heat effect and welding material properties. The cross section of the welding seam is V-shaped, and the root bead is the first weld in the welding process. The root bead has the narrowest width of the welds, so the heat generated by welding is transferred to the base material at a faster speed and then dissipates through heat conduction. Thus, the material properties at the root bead are not significantly affected by its own welding heat. However, with the continuous piling of the filling weld, the heat generated by welding is continuously transmitted to the root bead, which causes repeated heating. Such repeated heat treatment reduces the tensile property of the material at the root bead and the toughness (i.e., the failure strain of root bead is the minimum in Fig. 7). However, this cannot be verified by comparison because the weld material used in the root bead was different than the weld material used in the filling weld and the gap weld.

In thefilling weld, areas closer to the gap weld have a wider weld pass, so the heat generated by welding cannot be easily dissipated through the base metal. On the other hand, areas of the bead closer to the gap weld are not as significantly affected by welding heat, so their material performance is not as significantly affected. Therefore, with a decrease in the distance from the gap weld, the amount of weld bead heating and the heat dissipation rate have opposite competitive effects. The results show that specimens closer to the gap weld have lower yield strengths. Therefore, the influence of the heat dissipation rate on the material performance is greater than the number of repeated heating steps in the filling weld.

The yield strength of the gap weld is only higher than that of the adjacent fillinglayer. Because the material of the gap weld is different from that of the filling weld, it is impossible to determine whether the material itself or the welding heat conduction is the more significant factor in causing this result.

4.5 Difference in mechanical properties of transverse welds

The average true stress–strain curves of the three major zonesof the transverse welds (base metal zone, heat-affected zone, and weld zone) are shown in Fig. 13. As can be seen, the welding

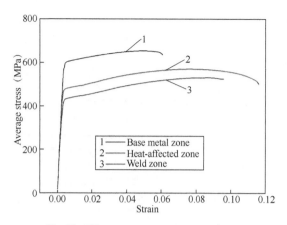

Fig. 13 The average true stress-strain curves of the three major zones of transverse welds

conditions decrease the stress capacity of the material. The yield stress gradually decreases from the base metal zone to the heat-affected zone and then the weld zone. Among these zones, the stress capacity of the base metal zone is significantly higher than that of the heat-affected zone and the weld zone, while the stress capacity of the heat-affected zone is slightly higher than that of the weld zone. The heat-affected zone is the area of the base metal formed by the welding heat, so its performance is lower than that of the base metal after heat treatment.

5 Conclusion

This paper proposes amicro-tensile specimen design and experimental method for testing the mechanical property distribution of the characteristics zones of a girth weld. Based on the differences in metallographic characteristics, a cross-section of the weld was divided into several characteristic zones. Meanwhile, based on the natural morphology of the weld structure, specimens were accurately designed and processed in each characteristic zone to maximize their size without exceeding the scope of each characteristic zone. Therefore, the difficulty of specimen processing and surface treatment in later stages is significantly reduced compared with other testing methods. In addition, a testing method for determining the tensile properties of materials in the weld characteristic zones is proposed to solve the absence of weld mechanical properties in safety evaluations. This method lays a foundation for improving the accuracy and reliability of the safety evaluation of weld defects.

Using the test method proposed in this paper, the characteristiczones of an X80 automatic girth weld were tested to obtain their tensile mechanical properties. The results show that there are significant differences in the material properties of different characteristic zones. The observed inhomogeneity is mainly caused by the combined effect of three factors: the difference in welding materials, the number of welding heat effects, and the rate of welding heat dissipation. It is unreasonable to regard welds as uniform materials and also unreasonable to replace girth weld properties with those of the base metal in traditional evaluation methods. Therefore, the non-uniformity of weld materials should be considered in weld safety evaluations.

Acknowledgements

This work was supported by the Natural Science Foundation of Shannxi Province, China [grant numbers 2021JQ-947, 2020JQ-934]; the China Postdoctoral Science Fund [grant number 2019M653785]; and the Young Scientists Fund of the National Natural Science Foundation of China [grant number 51904332].

References

[1] WANG X, SHUAI J. A calculation method for limit load of the gas pipelines with girth weld surface cracks[J]. Natural Gas Industry B, 2019, 6(5): 481-487.

[2] ANDERSON T L, OSAGE D A. API 579: a comprehensive fitness-for-service guide[J]. International Journal of Pressure Vessels and Piping, 2000, 77(14-15): 953-963.

[3] SY/T 6477—2017. The evaluation method of remaining strength of oil & gas transmission pipeline with the flaw [S]. Chinese Standard, 2017.

[4] API SPEC 5L—2013. Specification for line pipe[S]. American standard, 2013.

[5] ASME A370—2017. Standard test methods and definitions for mechanical testing of steel products[S]. American standard, 2017.

[6] GB/T228.1—2021. Metallic materials-Tensile testing Part 1: Method of test at room temperature[S]. Chinese standard, 2021.

[7] GB/T 2652—2008. Tensile test methods on weld and deposited metal[S]. Chinese standard, 2008.

[8] MOTOHASHI H, HAGIWARA N. Effect of Strength Matching and Strain Hardening Capacity on Fracture Performance of X80 Line Pipe Girth Welded Joint Subjected to Uniaxial Tensile Loading[J]. Journal of Offshore Mechanics & Arctic Engineering, 2007, 129(4): 318-326.

[9] MATHIAS L L S, SARZOSA D F B, RUGGIERI C. Effects of specimen geometry and loading mode on crack growth resistance curves of a high-strength pipeline girth weld[J]. International Journal of Pressure Vessels & Piping, 2013, 111(Complete): 106-119.

[10] HERTELÉ S, VAN MINNEBRUGGEN K, VERSTRAETE M, et al. Influence of pipe steel heterogeneity of the upper bound tensile strain capacity of pipeline girth welds: A validation study[J]. Engineering Fracture Mechanics, 2016, 162(1): 121-135.

[11] TANG Z B, XU F, XU Z J, et al. Research on mechanical properties of micro-zones of welding line[J]. Journal of Mechanical Strength, 2010, 32(1): 58-63.

[12] ZHU M H. Study of test methods of mechanical properties of microspecimen material under high-temperature environment[D]. Xi'an: Dissertation of Northwestern Polytechnical University. 2006.

本论文原发表于《International Journal of Pressure Vessels and Piping》2021年第194卷。

Root Cause Analysis of Liner Collapse and Crack of Bi-Metal Composite Pipe Used for Gas Transmission

Zhang Shuxin[1,2]　Ma Qianzhi[3]　Xu Changfeng[4]　Li Lifeng[1]
Wang Mingfeng[4]　Zhang Zhe[4]　Wang Shuai[1]　Li Lei[1]

(1. Tubular Goods Research Institute, China National Petroleum Corporation &
State Key Laboratory for Performance and Structure Safety of Petroleum Tubular Goods
and Equipment Materials;
2. School of Civil Aviation, Northwestern Polytechnical University;
3. School of Materials Science and Engineering, Xi'an Shiyou University;
4. Operation District of Hutubi Gas Storage, PetroChina Xinjiang Oilfield Company)

Abstract: An underground gas storage company adopted bimetallic composite pipes for gas transmission. The pipelines were inspected with industrial camera, it was found that a large number of pipelines had liner collapsed, and some of collapsed liners had suspected cracks. In order to analyze the reasons for the collapse and cracking of the liner, visual inspection, non-destructive testing, material examination, crack analysis, bend fatigue test and pressure test were conducted. The results showed that the material properties of the bimetal composite pipe meet standard requirements. The external pressure test was imposed on $\phi 168$ composite pipe, the liner collapse under the pressure of 1.76MPa which is lower than the calculated value. The cause of collapse was analyzed from manufacturing process and service condition. There are micro-cracks around the main crack of the liner, the fracture showed fatigue striations and thickness thinning characteristic. The root cause of the liner collapse is that the water enters the interlayer during the manufacturing of bimetallic composite pipe, and the thermal expansion coefficient of base pipe and lining pipe is different. When external anti-corrosion coating was manufactured and temperature changed during operation, collapse tendency of the liner increased. The root cause of the crack of the liner is bending fatigue, and the load originates from pressure fluctuations during operation. In order to avoid such incidents, the online inspection should be performed for the effusion part of the pipeline to check whether the remaining wall thickness of the base pipe meets the requirements for safe operation. The carbon steel pipe with regular pigging was recommended for this application situation.

Keywords: Bimetal composite pipe; Liner collapse; Liner crack; Bend fatigue

---

Corresponding author: Zhang Shuxin, wolfzsx@163.com.

1 Introduction

Bimetal composite pipe is widely used in petrochemical, submarine pipeline industry due to its economic efficiency and good corrosion resistance. The outer pipe known as base pipe adopts welded pipe or seamless pipe made of carbon steel or low alloy steel to bear the pressure. The inner pipe adopts stainless steel, iron-nickel-based alloy, nickel-based alloy or other corrosion-resistant alloy materials to improve the corrosion resistance and erosion resistance of the pipeline. According to the bonding way, the bimetal composite pipe is classified as clad pipe and lined pipe. Clad pipe is that the inner pipe is metallurgically bonded to the base pipe, while lined pipe is that the inner pipe mechanically fit to the base pipe. Due to the simple manufacturing process and low cost of mechanical bimetal composite pipes, it is particularly widely used in marine pipelines, gathering pipelines of China western oil fields. Although the mechanical bimetal composite pipe solves the problem of internal corrosion[1-3], some new problems have arisen. Li F G et al.[4] summarized the failure cases of the mechanical bimetallic lined pipe, the main failure mode of lined pipe is CRA layer collapse, CRA layer corrosion and joint failure. Fu A Q et al.[5] studied the failure case of the girth weld cracking of mechanically lined pipe. It was found that the root cause was due to the poor welding, the martensite structure formed at the weld, the crack initiated from the base metal under the pulling stress, bending stress, and shear stress. Yuan L et al.[6] investigated the liner buckling behavior of the off shore bimetal composite pipe during reeling process, found that the liner would buckle when bent to curvature, and the larger the diameter of the pipe was, the easier the liner would buckle. The internal pressure would prevent the liner from buckling. Vasilikis D et al.[7] build a finite element model to simulate the wrinkling behavior of the lined pipe under bending, and the wrinkled shape of the numerical simulation is consistent with the experimental observation.

At present, the research on bimetal composite pipes mainly focuses on the instability buckling of the liner underdifferent external load[8-10]. There are very few failure cases reported, and no failure cases of liner cracking have been reported.

In this study, the failed bimetallic composite pipe was used as gas transmission pipe by an underground gas storage company. During internal inspection of the pipe with industrial camera, it was found that a large number of pipelines had liner collapsed, and some of collapsed liners had suspected cracks. Two pipes were used to analyze the reasons for the collapse and cracking of the liner.

The macro morphology of bimetal composite pipe is shown in Fig. 1. The specification of 1# pipeline is $\phi 114mm \times (2mm+10mm)$, the length is 1.5m, and the liner is found collapsed without crack. The specification of 2# pipeline is $\phi 168mm \times (2mm+14mm)$, the length is 5.4m, and the liner is found collapsed with suspected cracks. Both pipelines were constructed in May 2013. The base pipe is made of L415QB, the liner is made of 316L stainless steel, and the design pressure is 32MPa. The highest operating pressure in the past three years was 28.6MPa, the operating temperature was 43℃, and the pipeline transportation medium was natural gas, which was composed of CH_4. Both pipelines were used for injecting natural gas into the reservoir on a daily basis and performing gas production operations in winter.

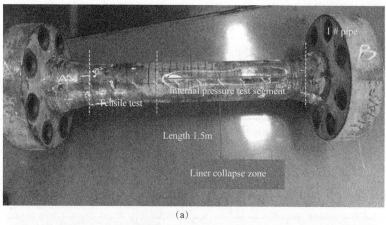

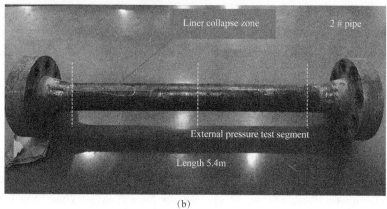

Fig. 1 Macro morphology of bimetal composite pipe (a) 1# pipe (b) 2# pipe

Liner collapse is a common failure mode of bimetallic composite pipes, which will affect the normal operation of pigging and internal inspection. However, liner cracking has not been reported. Once the liner cracks, the medium will enter the interlayer, which may cause galvanic corrosion. Pressure-bearing capacity of the base pipe will reduce, further leakage or even explosion failure happens. Therefore, it is necessary to carry out failure analysis to find the root cause for the collapse and cracking of the liner.

2 Experiment

In order to analyze the cause of the failure, systematic experiments were carried out. The girth welds on both sides of the two pipes were inspected using the radiographic inspection according to standard ASTM E94—2017. The ARL4460 direct reading spectrometer was used to analyze the chemical composition of the base pipe and the liner according to ASTM A751—2014a. According to GB/T 228.1—2010, SHT4106 material testing machine was used to test the tensile performance of the base pipe, and UTM5305 material testing machine was used to test the tensile performance of the liner. According to the GB 232—2010, the WZW-1000 bending tester was used for the bending test of the liner. According to ASTM E3-11 (2017), ASTM E112-13 and ASTM E45—2018a, the MEF4M metallurgical microscope and image analysis system and the OLS4100 laser confocal microscope were used to analyze the microstructure and non-metallic inclusions of the pipes. The transverse and longitudinal samples of the 1# and 2# liner were used for intergranular corrosion test. In order to ensure the reliability of the test, two parallel samples are set for each group of tests. The sample sensitize at 650℃ for 2 hours. After air cooling, the surface was polished, and then the sample was placed in a copper-copper sulfate-16% sulfuric acid etching solution for 16 hours in a boiling state. After the sensitizing process, the sample was bent by WZW-1000 bending tester with a bending shaft diameter of 5mm and a bending angle of 180°. In order to clarify the driving force of the lining collapse and whether the liner collapse can be recovered under the internal pressure, pressure test was performed. The segment of 1# pipe with liner collapsed, while 2# pipe has not liner collapse. The plug was welded at both ends of the 1# pipe sample, and pressure was imposed from one side to reach operating pressure of 29MPa. After the test, the pipe was cut along the axis to observe the collapse condition of the liner. The liner and base pipe of 2# pipe were sealed and welded at each end, and outer wall of the base pipe was drilled to the depth of the liner, and the nozzle was welded to impose the pressure. In order to verify the fracture characteristics, a low-cycle bending test was also carried out, and the liner crack fracture, the bending test fracture, and the tensile test fracture were compared and analyzed.

3 Results

3.1 Visual inspection

Macro morphology of the 1# liner after the pressure test is shown in Fig. 2. It can be seen that after the internal pressure test of the 1# pipe, the collapse did not recover. The collapse was located in the middle of the pipe segment with a length of about 300mm and a width of about 30mm. An axial crack appeared at the largest deformation position in the middle of the collapse. Observed from the inner wall of the liner, the crack was about 25mm long, and from the outer wall, the crack was about 70mm long, and there were micro cracks on both sides of the crack.

Macro morphology of the 2# liner is shown in Fig. 3. The inner wall of the liner of 2# pipe was smooth with no corrosion trace. There was a "heart-shaped" collapse on the liner with length about 620mm. The collapsed bulge was about 25mm in height and 70mm in width. Two longitudinal cracks can be observed at the collapsed part of the liner. Observed from the outer wall of the liner, there were micro cracks on both sides of the crack.

3.2 Non-destructive testing

Radiography testing was carried out by single-wall double shadow method, the exposure time is 1.2min, and each girth weld is equally divided into 6 parts for inspection. The results showed that no weld defects were found in the 4 girth welds of 2 pipelines, and the weld assessment grade was class I.

3.3 Material examination

The chemical composition analysis results of the base pipe and liner of the two pipes are shown in Table 1 and Table 2. The chemical composition of the base tube meets the requirements of the GB/T 9711—2017 standard. The chemical composition of the liner does not meet the requirements of API 5LD—2015 standard for 316L, and the content of Ni element is slightly lower than the lower limit.

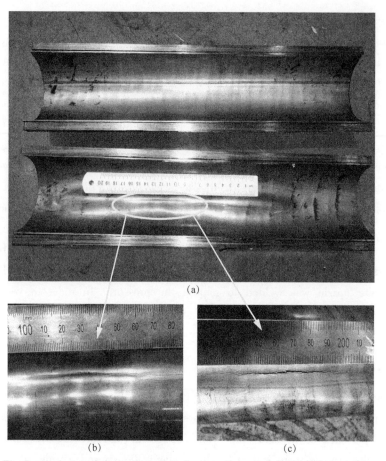

Fig. 2　Macro morphology of 1# pipe after pressure test (a) internal appearance (b) the crack observed from internal wall (c) the crack observed from the external wall

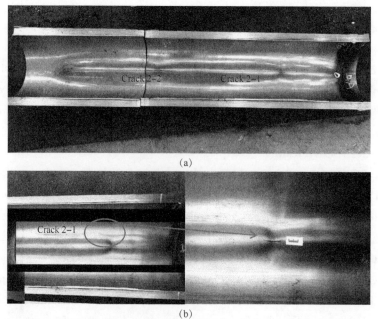

Fig. 3　Macro morphology of 2# pipe (a) internal appearance (b) the crack 2-1 observed from internal wall (c) the crack 2-2 observed from internal wall (d) the collapse appearance

(c)

(d)

Fig. 3 Macro morphology of 2# pipe (a) internal appearance (b) the crack 2-1 observed from internal wall (c) the crack 2-2 observed from internal wall (d) the collapse appearance (continued)

Table 1 The chemical composition of the base pipe (Wt.%)

Sample No.	C	Si	Mn	P	S	Cr	Mo	Ni	Nb	V	Ti	Cu	B	Al	N	CEV*
1# base pipe	0.13	0.35	1.44	0.02	0.0034	0.051	0.0083	0.027	0.043	0.0054	0.029	0.066	0.0003	0.022	0.0058	0.389
2# base pipe	0.12	0.29	1.43	0.019	0.0067	0.048	0.031	0.032	0.039	0.0053	0.022	0.07	0.0003	0.021	0.0057	0.382
GB/T 9711—2017 requirement	≤0.16	≤0.45	≤1.6	≤0.025	≤0.020	≤0.30	≤0.10	≤0.30	≤0.05	≤0.08	≤0.04	≤0.25	—	0.015~0.06	≤0.012	≤0.43

Note: CEV represent carbon equivalent value, $CEV = C + \dfrac{Mn}{6} + \dfrac{Cr+Mo+V}{5} + \dfrac{Cu+Ni}{15}$.

Table 2 The chemical composition of the liner (Wt.%)

Sample No.	C	Si	Mn	P	S	Cr	Mo	Ni	Nb	V	Ti	Cu	B	Al	N
1# liner	0.02	0.59	0.97	0.042	0.003	16.98	2.04	9.94	0.043	0.0054	0.029	0.066	0.0003	0.022	0.017
2# liner	0.014	0.53	1.12	0.026	0.0013	16.46	2.05	9.47	0.039	0.0053	0.022	0.07	0.0003	0.021	0.015
API 5LD—2015 requirement	≤0.03	≤0.75	≤2.00	≤0.045	≤0.030	16.0~18.0	2.0~3.0	10.0~14.0	—	—	—	—	—	—	≤0.10

The test results of the longitudinal tensile properties of the base pipe and liner are shown in Table 3. The tensile properties of the 1# and 2# pipes meet the requirements of GB/T 9711—2017 for L415QB material, and the liner tensile properties meet the requirements of ASTM A312/312A-17 standard for 316L material.

Table 3 Tensile test result

Sample No.	Tensile strength R_m(MPa)	Yield strength $R_{t0.5}$(MPa)	Elongation A(%)
1# base pipe	579	458	39.0
2# base pipe	569	462	38.0
GB/T 9711—2017 requirement	≥525	415~565	≥20
1# liner	710	553	33.0
2# liner	714	571	27.0
ASTM A312/312A-17 requirement	≥485	≥170	—

Table 4 and Fig. 4 show the bending test results of 1# liner. The purpose of the bending test is to investigate whether the liner cracks after bending. The results show that the transverse bending test of the 1# liner meets the requirements of the GB/T 37701—2019 standard, and no cracking occurs.

Table 4 Bend test result

Sample No.	Bending shaft diameter(mm)	Experimental conditions	Result
1# liner	5	Face bend test 180°	No cracks appeared
		Root bend test 180°	No cracks appeared
GB/T 37701—2019 requirement			No cracks appeared

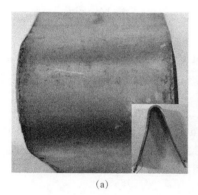

Fig. 4 Bend test (a) Face bend sample after test (b) Root bend sample after test

Fig. 5 shows the results of the transverse and longitudinal intergranular corrosion tests of the 316L liner of 1# and 2# pipe. There were no intergranular corrosion cracks, which met the requirements of GB/T 4334—2020.

Fig. 6 shows the metallographic structure of the base pipe and liner pipe of 1# and 2# pipes. The metallographic structure of the 1# and 2# base pipes is bainite, and the non-metallic inclusions classification of base pipe is A0.5, B0.5, and D0.5. The metallographic structure of the 1# and 2# liner pipes is austenite, the non-metallic inclusions classification of the 1# liner is A0.5, B0.5, and D0.5. No abnormalities occurred.

Fig. 5 Intergranular corrosion test (a) longitudinal sample of 1# liner (b) transverse sample of 1# liner (c) longitudinal sample of 2# liner (d) transverse sample of 2# liner

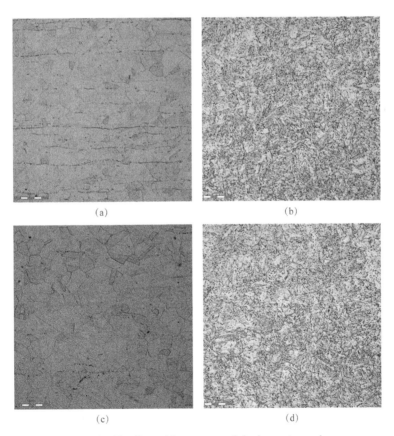

Fig. 6 Metallographic structure of the base pipe and liner pipe (a) 1# liner (b) 1# base pipe (c) 2# liner (d) 2# base pipe

3.4 Crack analysis

The micro structure of the 1#, 2# liner's crack was shown in Fig. 7 and Fig. 8. There are obvious micro structure deformations around cracking positions, and there are several micro-cracks beside the main crack. The micro-cracks originate from the outer wall and are perpendicular to the texture direction. At the same time, there is wall thickness thinning characteristic near the fracture, and the remaining wall thickness of the 1# liner and the 2# liner are 1mm and 0.7mm, respectively.

Fig. 7 Metallographic structure of 1# liner crack (a) 1# liner crack (b) thinning characteristic of the liner

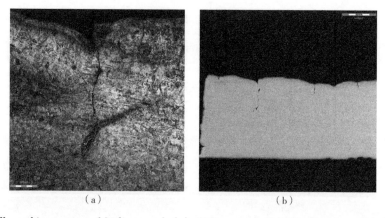

Fig. 8 Metallographic structure of 2# liner crack (a) 2# liner crack (b) thinning characteristic of the liner

Fig. 9 shows the microscopic morphology of the fracture of the 1#liner. The crack originated from the outer wall, showing the characteristics of multi-source cracking from outer wall, the source area is flat, and the propagation zone shows wide fatigue striations which are parallel to the axial direction of the pipe. Under high magnification, "long round" dimples can be seen at the fracture, indicating that the specimen is under shear stress. The width of the dimples is nearly 25μm. The instantaneous breaking zone has a shear lip shape.

3.5 Tensile test fracture

The tensile test fracture of 1# liner was compared with the crack fracture. As can be seen in the Fig. 10, the surface of the tensile fracture is covered with a large number of micro dimples, which is a typical ductile fracture, and it is obviously different from the morphology of the liner crack fracture.

Fig. 9 Scanning electron microscope image of fracture
(a) macroscopic image of the fracture (b) microscopic image of the fracture

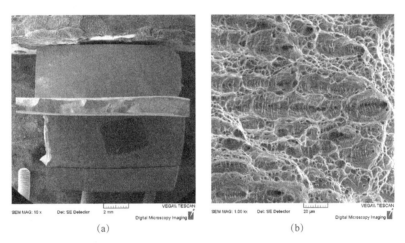

Fig. 10 Scanning electron microscope image of tensile test fracture of
1# liner (a) macroscopic image of the fracture (b) microscopic image of the fracture

3.6 Bend fatigue test and fracture analysis

After 10 times repeated bending tests, the sample broke. The schematic diagram of the bending test is shown in Fig. 11. After the test, the macroscopic appearance of the sample is shown in Fig. 12. It can be seen that there are micro cracks on both sides of the fracture after the bending test, and the fracture is flat, similar to the fracture of the liner crack. Observed from the side view, it can be seen that there is a slight wall thickness reduction in the fracture plastic deformation zone.

Fig. 11 The diagram of the bend fatigue test

The SEM image of the fracture of the fatigue bend test is shown in Fig. 13. Observed by SEM, there are wide fatiguestriations in the fracture propagation area. The width of the fatigue striations is almost the same as the cracked part of the failed liner.

Fig. 12 Macroscopic appearance of the sample after the bend fatigue test

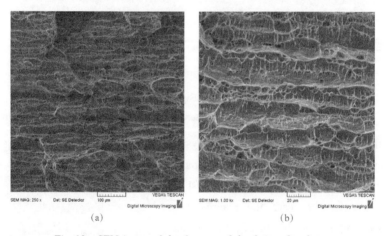

Fig. 13 SEM image of the fracture of the fatigue bend test
(a) macroscopic image of the fracture (b) microscopic image of the fracture

3.7 Pressure test

The internal and external pressure tests were conducted for 1# and 2# pipe segment, respectively. Theliner of 1# pipe segment was collapsed, and the liner of 2# pipe segment was intact. Plugs were welded on both ends of the 1# pipe sample, and the pressure was increased to operating pressure 29MPa from one side. After the test, pressure was released, and the sample was cut along the axis to observe the collapse of the liner. The liner and the base pipe were sealed and welded at each end of the 2# pipe, and then the outer wall of the base pipe was drilled to the depth of the liner, and the nozzle was welded to conduct the pressure test. A video monitor was used to observe the change of the liner during the pressurization process.

In the 1# pipe sample internal pressure test, the pressure was first increased to 12.95MPa, and the pressure was maintained for 18 minutes. The pressure was stable, indicating that the pipe sample was well sealed, and then the pressure was increased to the maximum service pressure of 29.55MPa. After the test, the liner collapsed appearance (Fig. 14) did not change significantly, the liner was not attached to the base pipe.

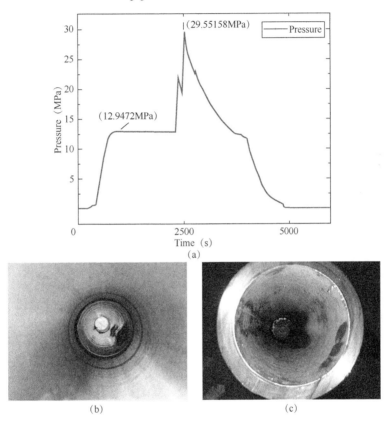

Fig. 14 Internal pressure test of 1# pipe (a) pressure-time curve
(b) the liner collapse condition before test (c) the liner collapse condition after test

In the external pressure test of 2# pipe sample, the pressure was imposed on the outer wall of the liner, and the pressure gradually rose. When it reached 1.76MPa, the liner collapses around the longitudinal weld (Fig. 15), and the collapse shape was same as that of the failed pipe.

4 Discussion

4.1 Root cause analysis of the liner collapse

The liner collapse when the external pressure on the liner is greater than the critical buckling load. According to previous literature, the empirical formula[11] for estimating the critical load of composite pipe liner collapse:

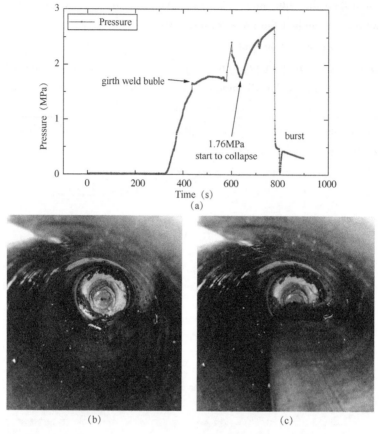

Fig. 15 External pressure test of 2# pipe (a) pressure-time curve (b) the liner start to collapse (c) the liner collapse grew to the maximum

$$p_{\text{critical}} = \frac{14.1\sigma_y}{(D/t)^{1.5}[1+1.2(\delta_0+2g)/t]} \quad (1)$$

D is the outer diameter of the liner; t is the wall thickness of the liner; g is the gap of the base liner; δ_0 is the initial concavity of the liner (0 for seamless pipe), the gap between the liner and base pipe is 0.5mm according to the manufacturing requirement; the liner yield strength (316L) is 330MPa. The following Table 5 shows the critical pressure of the liner collapse:

Table 5 The critical buckling pressure of the liner

Sample	Outer diameter (mm)	Wall thickness (mm)	Criticalbuckling pressure (MPa)
1# φ114mm×(2mm+10mm)	94.3	2	8.98
2# φ168mm×(2mm+14mm)	139.9	2	4.97

When the compressive stress on the outer surface of theliner (i. e. the gap between the base pipe and the liner) or the tensile stress on the inner surface of the liner is greater than the bulging critical load pressure, the collapse and instability will occur. The pressure test results of the 2# φ168mm pipe sample show that the critical buckling pressure is 1.76MPa, which is less than the estimated value of 4.97MPa. The possible reasons are as follows: (1) in the Eq.1, the initial concavity of the liner is taken as 0 for the seamless pipe. In fact, the liner is a longitudinal seam welded pipe. (2) The maximum allowable base/liner gap is 0.5mm after the composite pipe is manufactured, but in the actual presence of a longitudinal weld, the base/liner gap is greater than 0.5mm. (3) During the manufacturing process of the composite pipe, the liner is attached to the base pipe by expanding the diameter. After the pressure is unloaded, there is residual compressive stress on the liner. In summary, when the external pressure on the liner is greater than the sum of the internal pressure and the critical buckling pressure, the collapse can occur. The critical buckling pressure is related with parameters as wall thickness, pipe diameter, lining/base pipe gap, yield strength of the liner. The larger the diameter of the liner, the smaller the critical instability pressure. According to the test results, 100% ϕ168mm pipe (with a servicing time of more than 2 years) collapsed, and 42.9% of the ϕ114mm pipe section collapsed. This phenomenon is consistent with the research results.

The source of the driving force generated by the external pressure of the liner can be analyzed from two aspects: the manufacturing process and the service conditions.

In the manufacturing process of the mechanical bimetal composite pipe, the liner is expanded by hydraulic explosion to fit to the base pipe, as shown in Fig. 16. During the expansion process, the medium water enter the gap due to bad seal. When the product pipe fabricated with three-layer polyethylene (3PE) anticorrosive coating, the pipe need to be heated to 200℃, the water in the gap are heated to form vapor pressure stress. The pressure will act on the outer surface of the liner. Moreover, the thermal expansion coefficient of base pipe and the liner is different, the liner is larger. When the temperature rises, the expansion of the liner pipe is greater than that of the base pipe. The liner is constrained by the base pipe, which will increase the tendency to collapse. Under the dual factors of vapor pressure stress and thermal pressure stress, once the pressure on the outer surface of the thin-walled liner exceeds the bulging critical pressure, it may cause bulging and collapse. Li F. G. et al.[4] injected about 10mL of water into the base pipe/liner gap, and heated it to 250℃ for 5min. Then cooling down, the liner collapse.

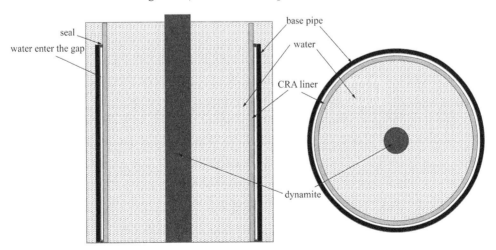

Fig. 16 2D schematic diagram of hydraulic explosion manufacturing process of bimetal composite pipe

During the service process, the mechanical bimetal composite pipe may be subjected to axial compression, bending stress, and cracking of the girth weld, which may cause theliner to collapse.

Lin Y. et al. investigated that under bending load[12], the ellipticity of the inner lining changes, resulting in separation of the base pipe and the liner pipe, and the liner on the compression side produces periodic wrinkles and then collapses. Under compressive load[13], when the strain is small, the liner will have axisymmetric buckling wrinkles, and when the strain is large, the liner will collapse in a "heart shape". While collapse morphology of the liner in this study is different from other researches, the liner did not appear periodic wrinkles. Therefore, the reason for the liner collapse has nothing to do with the axial compression and bending stress.

X-ray inspection showed that the girth welds did not have excessive defects, so the possible situation that corrosion of the girth welds caused the gas to enter the interlayer and cause collapse can be ruled out.

To sum up, the reason for the collapse of the liner may be that water enters the interlayer during the manufacturing process of the bimetal composite pipe, and the thermal expansion coefficient of base pipe and liner is different. When the temperature changed during operation, collapse tendency of the liner increased.

4.2 Root cause of the crack

There are penetrating cracks in the 1# and 2# pipe. Analysis of the causes of crack formation can be conducted from the fracture and environmental factors.

From the results of the bending test, when the pipe sample was bent to 180°, no cracks occurred at the maximum deformation point, and the cracking caused by plastic deformation can be ruled out. From metallographic structure of the crack, there is no stress corrosion cracking characteristics such as "branch cracks", so stress corrosion cracking can be excluded. The fracture surface presents the characteristics of multi-source outer wall cracking, and there are thick fatigue striations in the propagation zone. Under high magnification, there are "long round" dimples in the fracture surface, indicating that the sample is subjected to shear stress, which is consistent to that liner was subjected to bend force. At the same time, from the macroscopic cross-section of the crack, the structure deformation and the thinning of the wall thickness occurred in the cracks. The bending force the liner subjected is similar to the service process of coiled tubing. When the coiled tubing is bent multiple times, the wall thickness will be reduced in the direction of maximum force[14], which is same as the wall thickness reduction at the crack. Studies[15] have shown that when 316L material undergoes low-cycle fatigue, multiple cracks appear on the surface of the sample, while for high-cycle fatigue, plastic flow appears on the surface of the sample. There are multiple cracks at the fracture of the liner, which conforms to the low-cycle fatigue characteristics.

Considering the service situation, the 1# and 2# pipe are used for gas injection-production transmission, and the operating pressure in the past three years was between 15.5 ~ 27.8MPa. When the liner collapses due to instability, pressure fluctuations cause the shape of the collapse change continuously, resulting in low-cycle strain fatigue.

In summary, the cracking of the liner is caused by bending fatigue, and the load originates from pressure fluctuations during operation.

5 Conclusion and recommendation

A gas storage company used an industrial camera to inspect the bimetallic composite pipe, and it was found that a large number of pipeline had liner collapsed, and some of them had suspected cracks. In order to analyze the reasons for the collapse and cracking of the liner, visual inspection, non-destructive testing, material examination, crack analysis, bend fatigue test and pressure test were conducted. The conclusions can be drawn as follows:

(1) The material properties of the bimetal composite pipe meet standard requirements.

(2) Under the operating pressure of the 1# ϕ114mm pipe sample, the collapsed liner did not recover to fit the base pipe, but the liner cracked. The 2# ϕ168mm liner collapsed under an external pressure of 1.76MPa.

(3) The root cause of the liner collapse is that the water enter the interlayer during the manufacturing of bimetallic composite pipe, and the thermal expansion coefficient of base pipe and lining pipe is different. When external anti-corrosion coating was manufactured and temperature changed during operation, collapse tendency of the liner increased.

(4) The root cause of the crack of the liner isbending fatigue, and the load originates from pressure fluctuations during operation.

The main risk of the bi-metal composite pipe is the corrosive liquid into the base/liner interlayer, resulting in galvanic corrosion. Therefore, the online inspection should be performed for the effusion part of the pipeline to check whether the remaining wall thickness of the base pipe meets the requirements for safe operation.

The transmitting gas in gas storage pipeline is clean with little liquid content, the corrosion degree of the medium is relatively low, and therefore, the carbon steel pipe with regular pigging is more suitable for this situation.

Acknowledgements

The authors are grateful to the fund support of National Key R&D Program of China (2017YFC0805804), and all members in Tubular Goods Research Institute who assisted in carrying out this failure analysis study.

References

[1] YANG X L, WANG S H, GONG Y, et al. Effect of biological degradation by termites on the abnormal leakage of buried HDPE pipes[J]. Engineering Failure Analysis, 2021, 124: 105367.

[2] DE FARIAS AZEVEDO C R, BOSCHETTI PEREIRA H, WOLYNEC S, et al. An overview of the recurrent failures of duplex stainless steels[J]. Engineering Failure Analysis, 2019, 97: 161-188.

[3] ZHANG P, SU L, QIN G, et al. Failure probability of corroded pipeline considering the correlation of random variables[J]. Engineering Failure Analysis, 2019, 99: 34-45.

[4] LI F G, LI X J, LI W W, et al. Failure Analysis and Solution to Bimetallic Lined Pipe[C]. Materials Science Forum, 2020, 993: 1265-1269.

[5] FU A Q, KUANG X R, HAN Y, et al. Failure analysis of girth weld cracking of mechanically lined pipe used in gasfield gathering system[J]. Engineering Failure Analysis, 2016, 68: 64-75.

[6] YUAN L, KYRIAKIDES S. Liner buckling during reeling of lined pipe[J]. International Journal of Solids and Structures, 2020, 185: 1-13.

[7] VASILIKIS D, KARAMANOS S A. Wrinkling of Lined Steel Pipes Under Bending[C]. International

Conference on Offshore Mechanics and Arctic Engineering. American Society of Mechanical Engineers, 2013, 55379: V04BT04A027.

[8] HILBERINK A, GRESNIGT A M, SLUYS L J. Mechanical Behaviour of Lined Pipe[J]. Civil Engineering & Geoences, 2011: 401-412.

[9] YUAN L, KYRIAKIDES S. Plastic bifurcation buckling of lined pipe under bending[J]. European Journal of Mechanics-A/Solids, 2014, 47: 288-297.

[10] WANG F C, LI W, HAN L H. Interaction behavior between outer pipe and liner within offshore lined pipeline under axial compression[J]. Ocean Engineering, 2019, 175(MAR.1): 103-112.

[11] VASILIKIS D, KARAMANOS S A. Buckling Design of Confined Steel Cylinders Under External Pressure[J]. Journal of Pressure Vessel Technology, 2011, 133: 011205.

[12] YUAN L, KYRIAKIDES S. Liner wrinkling and collapse of bi-material pipe under bending[J]. International Journal of Solids and Structures, 2014, 51: 599-611.

[13] YUAN L, KYRIAKIDES S. Liner wrinkling and collapse of bi-material pipe under axial compression[J]. International Journal of Solids and Structures, 2015, 60: 48-59.

[14] TIPTON S M. Low-Cycle Fatigue Testing of Coiled Tubing Materials[C]. World Oil Coiled Tubing and Well Intervention Technology Conference, Houston, Texas, USA, 1997: 4-6.

[15] KIM Y, HWANG W. High-Cycle, Low-Cycle, Extremely Low-Cycle Fatigue and Monotonic Fracture Behaviors of Low-Carbon Steel and Its Welded Joint[J]. Materials, 2019, 12: 4111.

本论文原发表于《Engineering Failure Analysis》2022 年第 132 卷。

Burst Failure Analysis of a HFW Pipe for Nature Gas Pipeline

Zhu Lixia[1,2] Wu Gang[1] Luo Jinheng[1] Bai Zhenquan[1]

(1. State Key Laboratory for Performance and Petroleum Tubular Goods and Equipment Materials;
2. School of Materials Science and Engineering, Xi'an University of Technology)

Abstract: The burst failure of a high frequency welded (HFW) pipe used for nature gas pipeline in an oilfield was analyzed systematically by macro analysis, physical and chemical property test, SEM, etc., and the limit internal pressure of the pipeline under operation condition was predicted based on finite element method (FEM). The results showed that the chemical composition and mechanical properties of the pipe meet the requirements of relevant standards. The failure results showed that the dent damage of the straight pipe section at 12 o'clock. In the service of pipeline, the stress in the dent area exceeds the yield strength, which leads to the plastic deformation of the pipeline, resulting in necking and thinning, and the reduction of wall thickness further leads to the decrease of ultimate internal pressure, until the ultimate bearing capacity of the dent area is less than the internal pressure of pipeline operation, resulting in burst. It is suggested to strengthen the supervision of pipeline construction to avoid the pipeline dent damage. Meanwhile, the operation monitoring of the pipeline with dent damage should be strengthened, and timely repair or depressurization operation should be carried out if necessary.

Keywords: Pipeline Steel; Dent; Limit internal pressure; Plastic deformation; Burst

1 Introduction

Pipeline transportation is one of the main ways of oil and gas energy transportation in China. However, with the extension of pipeline service time, the impact of mechanical damage, corrosion and other issues on the pipeline is becoming increasingly prominent, which brings serious threat to the operation safety of the pipeline. Some studies have shown that mechanical damage is the primary cause of pipeline failure[1-2]. It is mainly caused by natural and man-made factors, man-made factors including mechanical damage such as scratches, depressions and boreholes[3] caused by construction, cultivated land or oil and gas theft, which is random and occurs in local locations of pipelines.

In 2019, the pipeline burst at 12 o'clock in an oil field after more than 4 months operation.

Corresponding author: Zhu Lixia, zhulx@cnpc.com.cn.

The pipeline was made of φ219.1mm×5.6mm L415M HFW pipe, and the ordering technical condition is GB/T 9711—2017 *Petroleum and natural gas industries—Steel pipe for pipeline transportation systems*. The design pressure of the pipeline is 5.5MPa, and the actual operating pressure is 1.0~4.8MPa. The failure pipe section is connected with the elbow, which is located on the upward of 42° slope. The position and air flow direction of the pipe section are shown in Fig. 1. In this paper, combined with the field conditions, the cause of the burst failure of the HFW pipe is analyzed in order to provide some reference and guidance for the practical engineering application of the pipeline.

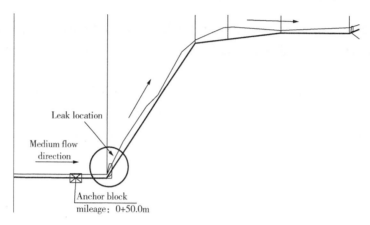

Fig. 1 Location of failure pipeline and air flow direction

2 Analysis method

2.1 Wall Thickness Measurement and Macro Analysis

The macroscopic morphology of the burst failure pipe section is shown in Fig. 2, and the wall thickness around the burst is obviously reduced observed by naked eyes. The initiation of the explosion can not be observed because of the curling and bending of the crack. There is an obvious dent on one side of the burst, with a depth of 22.8mm, and the radius of the depression area is about 40mm.

Fig. 2 Macroscopic morphology of burst samples

After peeling off the anticorrosive coating of the failed pipeline, MX-5 ultrasonic thickness gauge was used to measure the circumferential wall thickness according to GB/T 11344—2021. The measurement results are shown in Fig. 3. The wall thickness of the pipe from 6 o'clock to 8 o'clock meets the requirements of GB/T 9711—2017, and the wall thickness of 11 o'clock to 12 o'clock is lower than the standard. The minimum wall thickness near the burst is 1.10mm, which is located at the edge of the dent. The thickness reduction is 80.4% of the original wall thickness. The wall thickness of the elbow connected with the failure pipe section was also detected. The wall thickness of the elbow met the requirements of SY/T 5257—2012 standard, and there was no thinning.

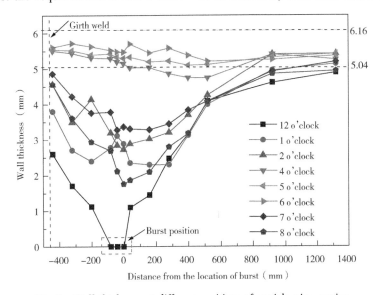

Fig. 3 Wall thickness at different positions of straight pipe section

2.2 Chemical Composition Analysis

Sampling at straight pipe section, the chemical composition was analyzed by ARL 4460 direct reading spectrometer according to GB/T 4336—2016 standard, the results are shown in Table 3. The chemical composition of the pipe body material meets the requirements of GB/T 9711—2017.

Table 1 Results of chemical composition analysis (Wt.%)

Sample	C	Si	Mn	P	S	Cr	Mo	Ni	Nb	V	Ti	Cu	B	$CE_{P_{cm}}$*
Pipe body	0.063	0.26	1.37	0.0078	0.0042	0.021	<0.0009	0.011	0.034	0.0034	0.012	0.023	<0.0001	0.14
Technical requirements	≤0.12	≤0.45	≤1.60	≤0.025	≤0.015	≤0.50	≤0.50	≤0.50	Nb+V+Ti≤0.15			≤0.50	≤0.001	≤0.25

Note: * $CE_{P_{cm}}$ = C+Si/30+(Mn+Cu+Cr)/20+Ni/60+Mo/15+V/10+5B.

2.3 Mechanical Property Analysis

Sampling at straight pipe section, tensile test, Charpy impact test and hardness test were carried out by UTM5305 material testing machine, PIT752D-2 impact testing machine and KB 30BVZ-FA hardness tester, and the results were shown in Table 2. It could be seen from the analysis results that the mechanical properties of the pipe meet the requirements of the technical agreement.

Table 2 Results of mechanical property test of pipe

project	Tensile properties				Impact properties			Hardness
	Specifications (mm)	Yield strength (MPa)	Tensile strength (MPa)	Elongation (%)	Test temperature (℃)	Energy (J)	Shear rate (%)	HV10
Test results	38.1×50	493	563	25.5	20	53	100	184
Technical Requirements	Pipe body	415~565	520~760	≥20	-10	—	—	≤345

2.4 Micro analysis

Samples were taken from normal wall thickness far away from crack for metallographic analysis. The test was carried out according to GB/T 13298—2015, GB/T 10561—2005 and GB/T 4335—2013 standards. The test equipment was MEF4M metallographic microscope and image analysis system. The microstructure of the pipe is acicular ferrite with fine grains.

Fig. 4 Typical Microstructure of the pipe

The microstructure analysis was carried out by taking samples from the fracture (dent area) of the sample. Obvious deformation streamline was found in the structure around the burst opening, and cracks were found on the outer surface of the sample near the fracture surface, as shown in Fig. 8—9. No obvious corrosion trace was observed on the surface of the sample. The appearance of the crack observed by SEM is shown in Fig. 4—6. The edge of the sample is curled to the outer surface, and scratch marks can be observed, so the original fracture morphology can not be observed. Deformation traces can be observed on the outer surface.

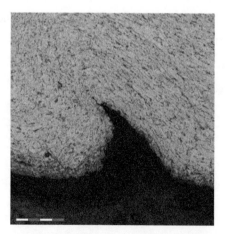

Fig. 5 Microstructure around the crack

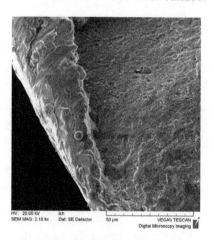

Fig. 6 Crack surface

3 Results and discussion

The physical and chemical analysis results show that the chemical composition and mechanical properties test results of the failed pipe section of ϕ219.1mm×5.6mm L415M HFW pipe meet the requirements of GB/T 9711—2017. The microstructure of the failed pipe section is acicular ferrite with fine grain and no abnormality.

The measurement results of the appearance and geometric dimension of the failed pipe showed that serious wall thickness thinning occurred at the upper end of the pipe body at both sides of the burst from 10 o'clock to 2 o'clock, and the closer to the burst, the greater the wall thickness reduction. In addition, the depth of one side of the dent was 22.8mm. The micro analysis shows that no obvious defects are observed in the burst area, and obvious deformation streamline is found in the burst and dent area, which indicates that the pipeline has obvious plastic deformation. No obvious corrosion trace is observed on the surface of the sample. Combined with the results of macro and micro analysis, it can be preliminarily judged that the failure pipeline burst is due to the plastic deformation caused by insufficient pressure bearing capacity of the pipeline, which leads to the pipe wall thickness thinning and finally burst failure.

The depth of dent damage at 12 o'clock on the outer wall of the failed pipe was 22.8mm, about 10.41% of the outer diameter of the pipe. The evaluation of pipeline with smooth dent is generally based on the maximum depth of the dent. When the depth of the dent reaches the reference value in the standard, it is determined that the dent needs to be repaired or removed. According to API 1156[4] and ASME B31.8[5], the smooth dent depth should not exceed 6% of the outer diameter of the pipeline; Canadian industry standard CSA Z662[6] also stipulates that the smooth dent depth of pipes with external diameter exceeding 105.6 mm shall not be greater than 6% of outer diameter of the pipeline; China's Industry Standard SY/T 6696 stipulates that the dent with a depth greater than 6% of the pipe diameter should be repaired. It can be seen that in the depth based criteria of pipelines with smooth dent, 6% of the outer diameter is taken as the critical value for repairing pipelines with dent. The dent damage depth of the failed pipe section is 22.8mm, reaching 10.41% of the outer diameter of the pipe, which is far beyond the standard critical value and has already met the requirements of repair or replacement.

As the dent is located on the outer wall of the pipeline at 12 o'clock, it is judged that the dent is generated during the process of pipe trench backfilling. The FEM is used to analyze the load of steel pipe in the process of dent deformation. The material properties of ϕ219.1mm×5.6mm L415M welded pipe are as follows: Elastic Modulus was 206GPa and Poisson's Ratio was 0.3. In the simulation, according to the radius of the dent area, the effect of the dent at 12 o'clock position of the pipeline is simplified to a hemispherical shape, as shown in Fig. 7. The true stress−strain model is used for simulation calculation. The process of FEM is divided into three steps: (1) External load: when the internal pressure is 0, the intact pipeline is subjected to external load, which makes the pipeline begin to form dent; (2) Operation pressure loading: after the dent is formed, the normal operation of the dent pipeline is simulated; (3) Based on the failure criterion of net section, the relationship among the depth of dent, initial internal pressure, diameter thickness ratio and

Fig. 7 Geometric model and mesh generation of steel pipe

ultimate bearing pressure is established, and the prediction model of ultimate pressure is established.

The equivalent stress nephogram of the dent area is shown in Fig. 8. It can be seen that obvious stress concentration occurs in the dent area and surrounding area of the pipeline. When the depth of the dent is more than 5mm (2.3% OD), the maximum equivalent stress in the dent reaches 493MPa. After that, with the further increase of the dent depth, the maximum equivalent stress in the dent area exceeds the yield strength of the material, and the steel pipe enters the plastic deformation stage, the wall thickness of the dent area becomes thinner, and the bearing capacity further decreases.

When the internal pressure is 1.1MPa and the dent depth is 22.8mm, the equivalent stress nephogram of the pipeline dent area is shown in Fig. 9. The equivalent stress in the dent area and the surrounding area exceeds 493MPa, which means that the wall thickness thinning has occurred in the larger area. At the same time, it can be seen from the stress nephogram that the maximum equivalent stress on the outer surface of the dent is 646MPa, which has exceeded the tensile strength of the material.

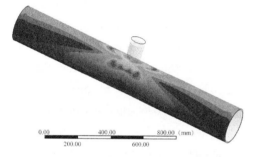

Fig. 8 Equivalent stress nephogram of dent without internal pressure

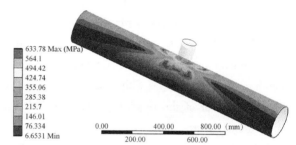

Fig. 9 The equivalent stress nephogram in the dent area (Internal pressure is 1.1MPa, the dent depth is 10.41%)

According to the net section failure criterion of materials, when the equivalent stress at any point in the wall thickness direction of the concerned area is equal to the flow stress, the pipeline reaches the limit state, and the flow stress can be determined according to the actual situation. It is considered that the more reasonable way is to take the flow stress as the tensile strength, that is, when the minimum equivalent stress in the dent reaches the tensile strength of the material, the pipeline will fail. Therefore, the net section failure criterion is used to calculate the ultimate pressure of the pipe with a depth of 22.8mm.

The failure internal pressure calculation formula of intact pipeline can be simplified as follows[7]:

$$p = \sigma_b \frac{2t}{D} \tag{1}$$

Where p is the failure internal pressure of the intact pipeline(MPa); t is the wall thickness of the pipeline(mm); D is the outer diameter of the pipeline(mm); σ_b is the tensile strength of the material(MPa).

Set the limit pressure expression of pipeline with different depth ofdent be the relation function of geometric sensitive parameters such as pipe wall thickness and dent depth.

$$p = a\frac{2\sigma_b t}{D}\left(1+g\frac{p_0}{\sigma_b}\right)^i (1+dh_e)^b \left(e\frac{t}{D}+f\right)^c \qquad (2)$$

Where h_e is the depth of the dent, $h_e = \dfrac{d}{D}$; a, b, c, d, e, f, g, i are the parameters to be regressed.

Through the nonlinear fitting regression, the expression of the limit pressure of the pipeline with dent is obtained as follows:

$$p_{max} = 0.002\sigma_b(1+0.001347p_0)^{-84.0390} \times (1+2.0637h_e)^{-19.5629} \times (0.8952t+12.1169)^{1.3450} \quad (3)$$

Where p_{max} is the ultimate internal pressure of the pipeline; σ_b is the tensile strength of the pipeline; D is the diameter of the pipeline; p_0 is the initial internal pressure of the pipeline.

It is judged that the dent was formed during the construction period, so it is replaced with $p_0 = 0$; t is the wall thickness of the pipeline, and the original wall thickness of the pipeline is 5.6mm; h_e is the depth degree of the dent, which is the ratio of the dent depth d (22.8mm) to the pipe diameter D (219.1mm), which is 10.41%. According to the calculation, when the dent depth is 10.41%, the ultimate internal pressure p_{max} of the pipe with wall thickness of 5.6mm is 1.1411MPa. According to the data, the operating internal pressure of one week before the pipeline failure is shown in Fig. 10. The maximum operating internal pressure is 1.1403MPa (on February 23, 2019), and the average operating pressure is 1.08MPa, which is close to the operation of ultimate bearing internal pressure. Due to the good plasticity and toughness of the pipe section, a lot of plastic deformation occurs in the pipe section with dent damage, and the wall thickness of the dent area and its surrounding area is obviously reduced due to necking. With the decrease of wall thickness, the ultimate bearing internal pressure also decreases, until the ultimate bearing capacity of the concave area of the pipeline with dent is less than the internal pressure of the pipeline, resulting in overload burst.

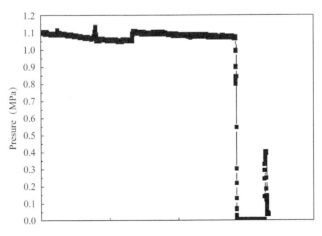

Fig. 10 Operating pressure diagram of one week before failure

4 Conclusion

(1) The chemical composition and mechanical properties of the ϕ219.1mm×5.6mm L415M HFW pipe used in the nature gas pipeline meet the requirements of GB/T 9711—2017 *Petroleum*

and natural gas industries—Steel pipe for pipeline transportation systems. The microstructure of the pipe body is acicular ferrite with fine grains.

(2) The burst failure of ϕ219.1mm×5.6mm L415M HFW pipe is caused by the decrease of pressure bearing capacity caused by dent damage. The failure results from the dent damage at 12 o'clock position of the straight pipe section. The stress in the dent area of the pipeline exceeds the yield strength, which leads to the plastic deformation of the pipeline and the necking and thinning. The reduction of the wall thickness further leads to the decrease of the ultimate bearing pressure, until the ultimate bearing capacity of the concave area of the pipeline with dent is less than the internal pressure of the pipeline in operation, resulting in burst.

(3) It is suggested to strengthen the supervision of pipeline construction to avoid pipeline dent damage. At the same time, the operation monitoring of the pipeline with dent damage should be strengthened, and timely repair or depressurization operation should be carried out when necessary to avoid similar failure accidents.

References

[1] YANG Q, SHUAI J, ZUO S Z. Research actuality of pipelines with dents [J]. Oil & Gas Storage and Transportation, 2009, 28(6): 10-15.

[2] DAWSON S J, RUSSELL A, PATTERSON A. Emerging Techniques for Enhanced Assessment and Analysis of Dents [J]. Journal of Pipeline Engineering, 2008, 7(3): 189-204.

[3] GAO D K. Research on In-line inspection technology for mechanical demage in pipelines [D]. ShenYang: Shenyang University of Technology, 2006.

[4] API PUBL 1156—1999. Effects of smooth and rock dents on liquid petroleum pipelines (phase ii) first edition [S]. USA: API, 1999.

[5] ASME B31.8—2007. Gas Transmission and Distribution Piping Systems [S]. USA: ASME, 2007.

[6] CSA Z662—2007. Oil and gas pipeline systems [S]. Canada: Canadian Standards Association, 2007.

[7] GB/T 9711—2017. Petroleum and natural gas industries—Steel pipe for pipeline transportation systems [S]. Beijing: National Standardization Administration of China, 2017.

本论文原发表于《Materials Science Forum》2021年第1035卷。

Comparison and Research of Acoustic Emission Testing Standards for Atmospheric Storage Tank

Zhang Shuxin[1] Wang Weibin[2] Yang Yufeng[2]
Zhang Qiang[2] Wu Gang[1] Luo Jinheng[1] Dong Xunchang[3]

(1. Tubular Goods Research Institute, China National Petroleum Corporation &
State Key Laboratory for Performance and Structure Safety of
Petroleum Tubular Goods and Equipment Materials;
2. Pipeline R & D Center, PipeChina North Pipeline Company;
3. PetroChina Tarim Oilfield Company)

Abstract: The acoustic emission testing standards ASTM E1930, BS EN 15856, JB/T 10764 and Q/SY GD 0211 for atmospheric storage tank are compared and analysed. Combined with production practice, the revision direction and reference suggestions of acoustic emission testing standards for atmospheric tank were proposed.

1 Introduction

Atmospheric tank is an important storage facility for crude oil and refined oil, which is widely used in petroleum and refining industries. Its volume is generally 2000 ~ 100000m^3, once the leakage occurs, it will cause huge property losses and environmental pollution. At present, the maintenance of storage tank is periodic inspection, which needs to stop production. It is not only time-consuming and laborious, but also has certain safety risks. Therefore, it is of great significance to carry out on-line inspection to evaluate the health status of storage tanks. The on-line inspection technology of storage tank mainly includes internal inspection robot, ultrasonic guided wave, acoustic emission inspection[1-4]. Because the bottom of the tank is usually covered with a layer of oil sludge, the internal inspection robot cannot walk freely, so the technology is still in the research stage, and there is no engineering application case. Ultrasonic guided wave can be used to inspect the inner annular plate from the outer plate, but usually the outer annular plate of storage tank is covered with the outer anti-corrosion coating, so the coating must be removed to carry out the inspection, thus the application is limited. Acoustic emission technology is almost the only practical on-line inspection technology for tank floor.

Acoustic emission testing has developing for a long time. Since 1997, Vallen Company carried out the acoustic emission testing project of storage tank, the testing technology has been widely used

---

Corresponding author: Zhang Shuxin, zhangshuxin003@cnpc.com.cn.

and developed, forming a series of testing standards. However, there are some differences among domestic and foreign standards, and even some clauses are not unified, resulting in the inspection results cannot be compared.

In this paper, the acoustic emission testing standards at home and abroad are compared andanalysed. Combined with the production practice, the revision suggestions of the acoustic emission testing standards for storage tanks are proposed, which is of great significance to further improve the evaluation method of tank floor and guide the on-line inspection and maintenance of storage tanks.

2 The survey of standards at home and abroad for acoustic emission

2.1 foreign acoustic emission test standard

In 1997, ASTM organizations abroad issued the standard of acoustic emission testing for liquid metal tanks of atmospheric/low pressure bearing, ASTM E1930-97[5] *standard test method for examination of liquid filled atmosphere and low pressure metal storage tanks using actual emission*, which was revised in 2002, 2007, 2012 and 2017 respectively. European standard was built based on *acoustic emission detection project for storage tank* project (Contract No. "smt4-ct97-2177") which was carried out in 1997 by Rhine, CESI spa, Vallen, Shell and Dow. In 2010, the acoustic emission testing standard for corrosion defects of metal tanks bearing liquid was issued in BS EN 15856: 2010[6], non-destructive testing—actual emission—General principles of AE testing for the detection of corrosion within metallic bonding fixed with limited.

ASTM E1930 specified personnel qualification, equipment, safety measures, equipment calibration, test steps. EN 15856 specified personnel qualification, equipment, test steps, data analysis and other aspects.

2.2 domestic acoustic emission test standard

Since 1990s, the Institute of special inspection of China, Northeast Petroleum University and CNPC pipeline science and technology center have successively carried out acoustic emission testing research, and formed JB/T 10764—2007[7], non-destructive testing and evaluation methods for acoustic emission of atmospheric pressure metal tanks, Q/SY GD 0211—2011[8], on-line acoustic emission detection and evaluation of bottom plate of vertical cylindrical steel welded tanks.

The contents of JB/T 10764 referred to ASTM E1930-02, and added evaluation methods of tank floor. The acoustic emission parameters, event number and hits number were used to classify the health condition of the tank floor, and the maintenance suggestions of tank are given according to the classification. Q/SY GD 0211—2011 not only refers to ASTM E1930-02, but also absorbs the idea of evaluation and classification in JB/T 10764—2007. In informative Appendix F, the evaluation process of inspection results of acoustic emission signal for the tank floor is specified. The leakage probability level of tank bottom is determined by the sparse degree of event number positioning, and the activity of acoustic source is defined by the number of events per unit per hour and the average energy of the event, the corrosion degree of tank floor is obtained by leakage possibility level and acoustic source activity.

3　Comparison and analysis of standards

3.1　Scope of application

Table 1 lists the scope of each AE standard. It can be seen that all four standards are applicable for the new/in service inspection of liquid metal atmospheric tank. Only JB/T 10764 covers the tanks with medium as gas. In fact, atmospheric tanks are usually not used to store gas. ASTM E1930, BS EN 15856, JB/T 10764 all cover the inspection for tank wall, roof and bottom plate, Q/SY GD 0211 cover tank floor inspection only.

Table 1　Scope of acoustic emission standards

Standard	Scope
ASTM E1930—2017	Suitable for new and in-service storage tanks with liquid medium For flat-bottomed storage tanks, wall and roof can be inspected. Only when the sensor is arranged on the floor, the corrosion can be inspected When the operating pressure is greater than the detection pressure, the inspection is invalid Suitable for carbon steel, stainless steel, aluminium alloy and other metal storage tanks
BS EN 15856: 2010	Suitable for storage tank carrying liquid medium Suitable for corrosion detection of petroleum and petrochemical metal storage tanks, qualitative evaluation and maximum re-service time recommendations can be given For flat-bottomed storage tanks, it can detect the bottom plate and wall below the liquid level, and for floating-roof storage tanks, it can detect the roof of the tank Only active defects can be detected. During the detection process, if the corrosion process stops, then no acoustic emission signal can be detected
JB/T 10764—2007	Suitable for storage tank carrying gas or liquid medium; atmospheric pressure or less than 0.1MPa; low pressure new and in-service metal vertical storage tank; tank wall and tank floor
Q/SY GD 0211—2011	Suitable for the inspection and evaluation of the acoustic emission technology of the corrosion degree of the bottom floor of the vertical metal storage tank, which carry liquids, atmospheric pressure

3.2　Acoustic emission system requirements

Table 2 lists the acoustic emission system requirements of each standard. The standard specified the requirement of acoustic emission inspection system, including sensor, signal line, couplant, preamplifier, power supply signal cable, signal processor. Chinese JB/T 10764—2007 and Q/SY GD 0211—2011 are formulated with reference to ASTM E1930-02, and only have additional agreements on signal acquisition, other acoustic emission system parameter settings are consistent with the ASTM E 1930-02 standard. Since the Q/SY GD 0211—2011 standard is only for the inspection of the tank bottom floor, the sensor resonance frequency setting range is 30~60kHz. BS EN 15856: 2010 involves less requirements for acoustic emission systems, and only specified the sensors and signal acquisition. It requires that the wave propagation mode in the liquid should be used for the inspection of the tank floor, and the resonance frequency of the sensor should be 20~80kHz. In terms of signal acquisition, JB/T 10764—2007 and Q/SY GD 0211—2011 clearly require acquisition amplitude parameters, and BS EN 15856: 2010 does not give clear requirements.

Table 2 Acoustic emission equipment requirements

Standard	Sensor	Signal line	Preamplifier	Signal processor
ASTM E1930—2017	sensor resonance frequency setting for wall inspection 100~200kHz; The resonance frequency setting for tank bottom floor inspection 30~60kHz; The sensitivity should not be greater than 3dB	Typical signal line length is 2m	The noise level should not be greater than 5 microvolts rms; The gain change should not exceed ±1dB	The hit duration should be accurate to ±10μs The defined time of hit should be 400μs
BS EN 15856: 2010	Using waves to propagate in liquids, the resonant frequency is 20~80kHz, and using waves to propagate in metals, the resonant frequency is 100~300kHz	Not required	Not required	Signal acquisition should output at least the following parameters: Acoustic emission hit count, parameters characterizing background noise, peak frequency, duration, rise time, energy, arrival time
JB/T 10764—2007	Basically consistent with the requirements of ASTM E1930—2017, it specifies the signal acquisition output parameters, including count, amplitude, duration, rise time, energy, and arrival time			
Q/SY GD 0211—2011	Consistent with JB/T 10764—2007 requirements			

3.3 Test procedure

In the practice process of acoustic emission test for storage tank, the placement of sensors, instrument calibration, selection of pressurization methods, and detection time are all key parameters. Table 3 lists the acoustic emission test procedure of each standard.

The layout of sensors involves the height, spacing, and number of sensors. In ASTM E1930, the sensor installation spacing needs to be determined based on the attenuation characteristics. The maximum sensor spacing for regional positioning is equal to 1.5 times the attenuation radius, and the maximum sensor spacing for calculation positioning is equal to the attenuation radius. The three standards of BS EN 15856: 2010, JB/T 10764—2007 and Q/SY GD 0211—2011 all suggest that the sensor spacing should not be greater than 13m. The height of the sensor is not clearly specified by ASTM E1930, and the other three standards are different. For crude oil storage tanks containing sludge, only BS EN 15856: 2010, JB/T 10764—2007 recommends that the sensor is higher than sludge. BS EN 15856: 2010 clearly indicates that if a sludge storage tank is contained, there will be a large error in defect positioning.

For the surface treatment of the installation position, only the Q/SY GD 0211—2011 standard requires that it reaches St3 according to the rust removal level, and the other three standards require good coupling.

Testing instrument calibration method, ASTM E1930—2017, JB/T 10764—2007, Q/SY GD 0211—2011 three standards recommend the use of lead break test, the lead break position is required to be at least 10cm away from the sensor, and the peak amplitude requirement of the sensor isdifferent. BS EN 15856: 2010 requires the use of Hsu-Nielsen source to be 50mm from the centre of the sensor for sensitivity testing. The average error of the 4 sensitivity tests should be within

±3dB, which is also the most stringent test.

Pressuring method and inspection time. The pressuring procedure for acoustic emission test is shown in Fig. 1. Generally, the four standards all require testing in a pressurized environment. Only BS EN 15856: 2010 requires that the tank liquid level is only 1m greater than the sensor installation position. Other standards have harsh requirements for pressurized environments. ASTM E1930—2017 and JB/T 10764—2007 require the medium level to be between 75% and 100%, and Q/SY GD 0211—2011 requires the level to be above 85%. Testing time ASTM E1930—2017 does not require standing time. It is sufficient to test at 100% medium level for 30 minutes. JB/T 10764—2007 requires a standing time of at least 2 hours and a testing time of at least 1 hour, but it does not specify a 1h test environment, Q/SY GD 0211—2011 requires standing for at least 12h, testing twice, at least 10h each time.

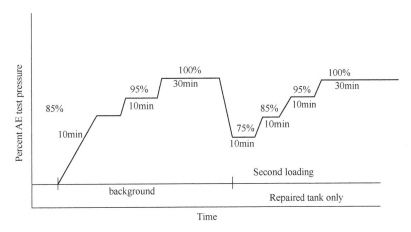

Fig. 1   Pressuring procedure for acoustic emission test

3.4 Result evaluation

ASTM E1930—2017 uses 5 indicators to evaluate the condition of storage tanks. When the indicators reached some extent, other non-destructive testing methods are required to make further inspection. These 5 indicators are the acoustic emission signal, signal duration, number of hits, large amplitude hit, and signal strength. During load holding state, the acoustic emission signal indicates the continuous yield or damage during the creep process, or the continuous change of the defect under the stress level. The signal duration characterizes the overall activity. If the acoustic emission signal consistently exceeds this value, it indicates that the tank is in poor operating condition. The number of hits is mainly used to evaluate the storage tanks in service. When the number of hits is too large, it indicates that the storage tank has more defects. A large amplitude hit indicates that the crack is growing. When it is found that the amplitude is also increasing as the load increases, it indicates that the crack is growing. The signal strength indicates that the defect area responds as the load increases.

BS EN 15856: 2010 requires that the difference in the arrival time of different signals is used to locate acoustic emission events, when the events are adjacent, they can be regarded as the same time and represent the same acoustic emission source. The number of acoustic emission events per unit time (1 hour) in the acoustic emission activity evaluation area is used to rank the storage tank

Table 3 Acoustic emission test procedure

Standard	Sensor height	Sensor spacing	Surface treatment	Sensitivity test	Pressuring method	Test time
ASTM E1930—2017	Not required	For regional positioning, the inspection radius needs to be determined according to the attenuation characteristics. The tank inspection with calculated positioning is adopted, and the maximum allowable sensor spacing is determined according to the attenuation characteristics. For the regional positioning method, the maximum sensor distance is 1.5 times the attenuation radius, and for the calculation positioning method, the maximum sensor distance is equal to the attenuation radius	Clean surface and free of impurities to ensure good coupling	Lead breaking test, the lead is at least 10cm away from the sensor, and the average peak amplitude change does not exceed ±4dB; When event positioning is required, adjacent sensors should detect signals exceeding the threshold, and the positioning accuracy should be within 5% of the sensor spacing	No clear requirements for newly built storage tanks; For in-service storage tanks, the detection level should be between 75% and 100% of the highest operating level	According to the pressurization program
BS EN 15856: 2010	The first row of sensors should be set at a distance of 1m from the bottom plate, and the second layer of guard sensors should be set at 4~6m, directly above the first row of sensors, at least 1m away from the liquid level. When the oil sludge affects the wave propagation, the two rows of sensors must be raised, but this should be negotiated with the tank operator, because this will cause a larger positioning error	Evenly distributed around the tank, avoiding manholes, welds, and the spacing should not exceed 15m. Practice has proved that 13m is more maneuverable. The number of sensors in each row is not less than 6. The distance between the sensor and the weld seam should be more than 200mm	Not required	The Hsu-Nielsen source is set at 50mm away from the center of the sensor for sensitivity testing. The average error of the 4 sensitivity tests should be within ±3dB	The liquid level should be at least 1m above the sensor. For flat-bottomed storage tanks, the storage tank should stand still during testing, with no liquid in or out, and the heater turned off. It is usually sufficient to stand for 24 hours	The most ideal detection environment is: no wind, no rain, no direct sunlight. It can be tested at night. Test for at least 1h. If waveform signals are collected, these signals cover at least 30min. Collecting several times can be used to improve the evaluation results

continued

Standard	Sensor height	Sensor spacing	Surface treatment	Sensitivity test	Pressuring method	Test time
JB/T 10764—2007	The sensors are arranged on the tank wall with the same height as possible, and should be 0.1~0.5m away from the bottom plate, and higher than the sludge depth	When using regional positioning, the inspection radius needs to be determined according to the attenuation characteristics. When the number of events is used for positioning, the maximum allowable sensor distance is determined according to the attenuation characteristics. For the area positioning method, the maximum sensor distance is 1.5 times the attenuation radius, and for the calculation positioning method, the maximum sensor distance is equal to the attenuation radius. The sensor spacing should not be greater than 13m	Clean surface and free of impurities to ensure good coupling	Consistent with ASTM E1930	The detection level should be at 85% to 105% of the maximum operating level. Under special circumstances, the liquid level is more than 1m above the sensor installation position. Before testing, it should be allowed to stand for more than 2h	Test for at least 2h
Q/SY GD 0211—2011	The floating roof storage tank: set on the outer wall of the storage tank at the same height of 0.2~1.0m from the bottom plate; Dome roof storage tank: guard sensor is used, which is set at 1.5~3.0m higher than the tank bottom sensor	Floating roof storage tank: not more than 13m	St3 according to rust removal grade	Lead breaking test, (0.3mm diameter 2H or 0.5mm HB lead core), the pen core is at least 100mm away from the sensor, and the change in the average peak amplitude of all sensors should not exceed ±5dB	The height of the liquid level should be more than 85% of the highest operating liquid level, and it should be left standing at this level for more than 12 hours	Close the inlet and outlet valves to eliminate the source of interference; Test should be carried out at night; Use short-term monitoring and multiple tests, each monitoring time is not less than 10h, and at least two tests

floor's health condition. The rating ranges from "silent acoustic emission source" to "severe acoustic emission source", corresponding to the maximum service period and immediate opening of the tank for maintenance.

JB/T 10764—2007 provides the acoustic emission evaluation method of the storage tank floor in the standard context. Calculation positioning and regional positioning are used to analyse the classification of acoustic emission sources, and further to classify the health condition of the tank bottom plate. Calculation positioning analysis is to obtain the number of events through triangulation analysis of hits. The bottom of the tank is divided into grids. The grids can be square or round, and the side length or diameter should not be greater than 10% of the tank diameter. The number of events per hour E in a grid is used to evaluate the corrosion classification of the bottom plate. The regional positioning analysis is based on the number of hits per hour H collected by each channel toclassify the corrosion level. Furthermore, the maintenance priority order of tank is further analysed according to the analysis result of calculation positioning or regional positioning analysis.

Q/SY GD 0211—2011 provides the acoustic emission evaluation method of the storage tank floor in the informative appendix. The two parameters of leakage probability level and acoustic emission activity are used to describe the results, and then they are placed in a matrix to evaluate the final tank bottom floor rating. Event location is obtained by calculation from the difference of hit arrival time, the degree of event concentration is used to describe the leakage probability level, and the number of events per unit time per area and the average energy per unit time per event is used to describe the acoustic emission activity.

4 Discussion

According to the above comparative analysis, there are big differences between domestic and foreign about acoustic emission inspection and evaluation standards, mainly in the aspects of acoustic emission scope, inspection procedures, and evaluation methods.

4.1 Scope

ASTM E1930, BS EN 15856, JB/T 10764, Q/SY GD 0211 standard test ranges are all covering liquid storage tanks, but not specify which kind of liquid medium. In the actual application process, the liquid medium includes crude oil, gasoline, diesel, and water. The density of different media is different, and the propagation speed of the acoustic signal is different, which will inevitably affect the result. In addition, the bottom of crude oil storage tanks is usually covered with a layer of sludge. Different thicknesses of sludge will dissipate different acoustic emission signals. The result needs to be corrected according to the thickness of the sludge. Sánchez M[9] et al. carried out acoustic emission detection on a gasoline tank with a diameter of 15.68m and a height of 10.52m in 3000m^3, using 8 sensors (35kHz) to be evenly arranged and tested for 1h, and the Vallen AMSY-5 AE system was used for inspection and post-processing. The research found, the cathodic protection system will produce some noise, which will affect the test results. Acoustic emission detected one active leakage defect, and the other one was not detected because the area was covered with sludge and no leakage occurred. Zhang[10] et al. conducted an acoustic emission test on a 20000m³ crude oil storage tank, and found that acoustic emission is an effective online detection technology, which can qualitatively evaluate the corrosion status of the crude oil storage tank bottom plate. Compared with the actual situation, the test evaluation results is conservative, it believes that the acoustic emission rating results of sludge crude oil storage tanks must be revised.

ASTM E1930, BS EN 15856, JB/T 10764, Q/SY GD 0211 standard test ranges are divided according to the location of the storage tank, ASTM E1930, BS EN 15856, JB/T 10764 all cover the tank wall, top and bottom, Only Q/SY GD 0211 is only for the inspection of the bottom of the storage tank. In engineering practice, acoustic emission technology is mainly used for bottom plate corrosion detection, and it is applied in storage tanks of 2000~100000m³. However, for large storage tanks, as the diameter of the bottom plate increases, the attenuation of the acoustic emission signal continues to increase, which will make it difficult for the signal at the centre of the tank to propagate to the tank wall. According to ASTM E1930, the sensor installation distance needs to be determined according to the attenuation characteristics. The maximum sensor distance for regional positioning is equal to 1.5 times the attenuation radius, and the maximum sensor distance for calculation positioning is equal to the attenuation radius. The author measured the attenuation characteristics on gasoline tanks, and found that the attenuation radius was about 5m, which means that the distance between the regional positioning sensors should be 7.5m, and the distance between the calculated positioning sensors should be 5m. In the standard, it is recommended that the maximum sensor spacing should be less than 13m. The radius of the 100000m³ storage tank is about 80m, and the radius of the 20000m³ storage tank is about 40m, which are far greater than the maximum sensor spacing of 13m and the attenuation radius measured by the author of 5m, so the central corrosion signal of the storage tank will not reach the tank wall. However, from the results of the American PAC acoustic emission detection and analysis software, the acoustic emission can detect the corrosion signal of the center of a 100000m³ storage tank, which is inconsistent with the standard and actual situation. For large storage tanks, the acoustic emission detection range can only cover a certain range of the bottom of the storage tank. It is recommended that the acoustic emission detection of large storage tanks can shorten the sensor spacing and increase the number of sensors.

4.2 Inspection procedure

The difference about inspection procedure is the sensitivity test and the pressurization procedure. The sensitivity test is to calibrate the equipment to ensure the repeatability of the test results. The standards all recommend the use of lead break test for calibration, but there is no clear regulation on the size of lead cores, and the requirements for lead-breaking positions are different, so the results of testing according to different standards will be incomparable. It is recommended to clarify the parameters such as the position of the lead, the size of the lead core, and the angle of the lead, as shown in Fig. 2, or use a standard signal generator to make the detection results consistent and comparable.

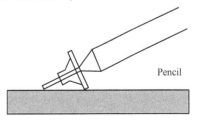

Fig. 2 Lead break test

4.3 Result evaluation

ASTM E1930—2017 adopts five indicators: acoustic emission signal, signal duration, number of impacts, amplitude impact, and signal intensity to evaluate the tank condition; BS EN 15856: 2010 adopts acoustic emission activity/the number of acoustic emission events per unit per hour to rank storage tanks floor condition; JB/T 10764—2007 uses calculation positioning and regional positioning to analyse the classification of acoustic emission sources, and uses the number of events per unit time per unit area and the number of hits per unit time to evaluate the corrosion status of the tank bottom; Q/SY GD 0211—2011 uses two parameters to describe the results of leakage probability level (the intensity of the concentration of acoustic emission events) and the acoustic emission activity (the number of events per unit time, the average energy of the event) to obtain the final tank rating.

Hodaei E[11] et al. conducted acoustic emission detection on a crude oil storage tank with a diameter of 9m and a height of 7.6m with a diameter of 600m^3, 5 sensors were evenly arranged and tested for 1h, and a triangulation method was used to locate and analyse the acoustic emission signal to obtain the tank floor. Distribution of the events was obtained. The study found that the degree of corrosion and thinning of the tank floor is positively correlated with the distribution of the number of events. Yuyama S et al.[12] carried out acoustic emission testing, magnetic flux leakage testing, and ultrasonic testing on nearly 20 storage tanks, and compared the testing results. The results showed that the source positioning results were different from the actual situation, while the concentration of events can effectively reflect the severity of the corrosion of the in-service tank bottom. Therefore, it can be inferred that acoustic emission can qualitatively reflect the corrosion status of the bottom of the tank, but corrosion defect cannot be accurately located.

For the four standards of acoustic emission domestic and abroad, the selected acoustic emission evaluation indicators are not the same, and none of them provide a clear classification method for the classification of storage tanks. For inexperienced inspectors, the results cannot be rated. Given this situation, combined with previous studies, it is recommended to use two parameters, the intensity of acoustic emission events and the number of events per unit time to evaluate the floor condition.

5 Conclusion

The study found that the four domestic and foreign acoustic emission testing standards, ASTM E1930, BS EN 15856, JB/T 10764, and Q/SY GD 0211, have differences inscope, procedures, and evaluation methods. The main conclusions and recommendations are as follows:

(1) ASTM E1930, BS EN 15856, JB/T 10764, Q/SY GD 0211 acoustic emission testing standards all cover liquid storage tanks. While in the actual application process, the liquid medium includes crude oil, gasoline, diesel, water. The density of different media is different, and the propagation speed of the acoustic signal is different, which will inevitably affect the result. It is recommended to give evaluation methods according to the type of media.

(2) For large storage tanks, as the diameter of the bottom plate increases, the attenuation of the acoustic emission signal continues to increase, which will make it difficult for the signal at the

center of the tank to propagate to the tank wall. It is suggested that the acoustic emission detection of large storage tanks can shorten the distance between sensors and increase the number of sensors.

(3) ASTM E1930, BS EN 15856, JB/T 10764, Q/SY GD 0211 all recommend the use of lead-breaking test for sensitivity calibration, but there is no clear regulation on the size of lead-breaking lead cores, and the requirements for lead-breaking positions are different, so It will cause incomparable test results according to different standards. It is recommended to clarify the parameters such as the position of the lead, the size of the lead core, and the angle of the lead, or to use a standard signal generator to make the test results consistent and comparable.

(4) ASTM E1930, BS EN 15856, JB/T 10764, Q/SY GD 0211 acoustic emission standards do not provide a clear classification method for the classification of storage tanks, and for inexperienced inspectors, the results cannot be rated. It is recommended to use two parameters, the intensity of the number of acoustic emission events and the number of events per unit time, to evaluate the results, and to clarify the grading standards.

Acknowledgments

The authors are grateful to the fund support of National Key R&D Program of China (2017YFC0805804).

References

[1] JIANG L L, LI L J, SU B H, et al. Application of acoustic emission technology in the corrosion detection of storage tank bottom plate[J]. Corrosion and Protection, 2021, 42(02): 56-59, 77.

[2] JIANG L L, HAN W L, XU Z P, et al. Research status of acoustic emission on-line detection technology for storage tank floor[J]. Corrosion and Protection, 2016, 37(05): 375-380.

[3] BI H S, LI Z L, CHENG Y P, et al. Discussion on the correlation between corrosion rate of atmospheric storage tank bottom plate and acoustic emission activity[J]. Corrosion and Protection, 2015, 36(06): 573-576, 589.

[4] LI Z L, BI H S, CHENG Y P, et al. Application and prospect of acoustic emission technology in metal corrosion research. Corrosion and Protection, 2014, 35(06): 598-601.

[5] ASTM E1930-97. Standard Test Method for Examination of Liquid Filled Atmospheric and Low Pressure Metal Storage Tanks Using Acoustic Emission[S].

[6] BS EN 15856: 2010. Non-destructive testing-Acoustic emission-General principles of AE testing for the detection of corrosion within metallic surrounding filled with liquid[S].

[7] JB/T 10764—2007. Non-destructive testing. Atmospheric pressure metal storage tank acoustic emission testing and evaluation method[S].

[8] Q/SY GD 0211—2011. On-line acoustic emission detection and evaluation of the bottom plate of a vertical cylindrical steel welded storage tank[S].

[9] SÁNCHEZ M, CARDENAS N, DOMINGUEZ V A. Acoustic emission testing of aboveground petroleum storage tanks: Risk assessment and lessons learned [J]. Process Safety Progress, 2012, 31(2), 159-164.

[10] ZHANG S X, SUN B B, LUO X W, et al. On-line detection and verification of acoustic emission of crude oil storage tank floor. Petroleum Tubular Goods & Instruments, 2020, 6(02): 75-78.

[11] HODAEI E, JAVADI M, BROUMANDNIA A, et al. Evaluation of acoustic emission inspection of oil tank floor via tank bottom plates thickness measurement[J]. Journal of Mechanical Research and Application,

2012, 4(3), 37-44.

[12] YUYAMA S, YAMADA M, SEKINE K, et al. Verification of acoustic emission testing of floor conditions in aboveground tanks by comparison of acoustic emission data and floor scan testing[J]. Materials Evaluation, 2007, 65(9): 929-934.

本论文原收录于 Journal of Physics: Conference Sories(2021年)。

Failure Analysis of Tee in Shale Gas Transportation Platform

Ji Nan[1,2] Feng Jie[3] Long Yan[1,2]

(1. Tubular Goods Research Institute of China National Petroleum Corporation; 2. State Key Laboratory of Performance and Structure Safety of Petroleum Tubular Goods and Equipment Materials; 3. Bohai Equipment Petroleum Special Pipe Co., Ltd.)

Abstract: The work in this paper aiming at making out the causes for internal corrosion and cracking of a tee in shale gas transportation platform. The reasons for internal corrosion and cracking were investigated by macroscopic analysis, physical and chemical property test, metallographic examination, corrosion morphology observation, corrosion product analysis and service condition analysis. The results show that the main reasons for the internal corrosion are based on the synergy action of CO_2 corrosion, SRB corrosion and ersion-corrosion, The crack on the inner wall weld is mainly the result of Sulfate-reducing Bacteria Induced SCC.

Keywords: Tee; Internal corrosion; Sulfate-reducing bacteria; Stress corrosion cracking; Synergistic action

1 Introduction

The progress of the times and the continuous improvement of people's living standards making the proportion of natural gas consumption increasing constantly. The development of the traditional oil and gas field has been unable to meet the increasing demand of energy. In recent years, with the success commercial development of shale gas development in southwest China, making the shale gas become the most reliable type of future energy replacement in China[1-2].

The major component of shale gas is hydrocarbon, methane is in the majority, with small amount of carbon dioxide. Compare with other gas field that contains high amount of carbon dioxide and sulfureted hydrogen, the corrosion of the shale gas is weak, but with the deep development of the shale gas field, the corrosion problems are gradually emerged[3]. According to the statistics data of a shale gas well site platform in southwest[4], from August to November, there occurred seventeen corrosion leakage accidents of the gathering pipeline tee, the majority of the pipelines or pipe fittings are leading to a corrosion perforation failure during a time of two months and one year. The frequent corrosion perforation could not only lead to suspend production of the gas field and

---

Corresponding author: Ji Nan, jinan003@CNPC.com.cn.

result in a huge economic loss, but could also bring a harmful effect to the environment. Corrosion has become the main factor that could influent the stably and efficiently development of the shale gas field, thus aiming at the corrosion accidents in the shale gas field, and making a summarize of the corrosion failure rules, and at the same time, making rational anticorrosion measures, all these are of great significance to the reduction of the pipeline operation risk and service life extension.

In this paper, the internal corrosion and cracking accident of a tee in shale gas transportation platform was studied. through macroscopic analysis, physical and chemical property test, metallographic examination, corrosion morphology observation, corrosion product analysis and service condition analysis. Aiming at making out the cause for the corrosion and cracking of the tee, and putting forward some relevant anticorrosion measures, ensuring the safety operation of the shale gas transportation.

The specification of the tee is $\phi65mm \times 5mm$, the material is L245N, the operation pressure is 4.3MPa, the inner temperature is 30℃. The transportation media is shale gas, containing water, the gas phase composition contain CO_2, and do not contain H_2S and O_2, and the molar content of CO_2 is 1.536%. The ions in the liquid phase are mainly HCO_3^-, SO_4^{2-}, K^+, Na^+, Ca^{2+}, and the pH value is 6.5.

2 Physical and Chemical inspection

2.1 MacroscopicAnalysis

Fig. 1 shows the macroscopic feature of corrosion in the inner wall of the tee. The inner surface of the tee is mainly covered by the black corrosion products and white scale crust. There are a large number of etch pits on the inner wall and weld of the tee, and some are extremely obvious, and the depth of etch pits could as deep as 2.1mm(Fig. 2). The corrosion morphology in the corner of the tee is mainly honeycomb feature and moss shape, which is the typical morphology of the CO_2 corrosion.

2.2 Wall-thickness measurement

The wall – thickness of the tee is measured by using an ultrasonic thickness meter, the measuring position is shown in Fig. 3, and the measuring results are shown in Table 1. The measuring result showed that there is a severely reduction of the wall thickness, and the most reduction is in the 180° position, in which the wall thickness reduction could be as big as 51.6%, considering the tee has only been in service for 165 days, the corrosion rate of the tee could be calculated for 5.73mm/a.

Table 1 Measurement results of tee-junction wall thickness

Position	Position 1	Position 2	Position 3	Position 4	Position 5	Position 6	Position 7	Average value
0°	2.99	3.13	3.42	3.96	4.00	3.05	2.98	
90°	3.00	3.58	3.21	3.99	4.02	3.57	3.42	
180°	3.26	2.88	2.42	3.01	3.46	3.00	2.95	
270°	3.84	3.12	2.95	3.42	3.86	3.46	3.89	

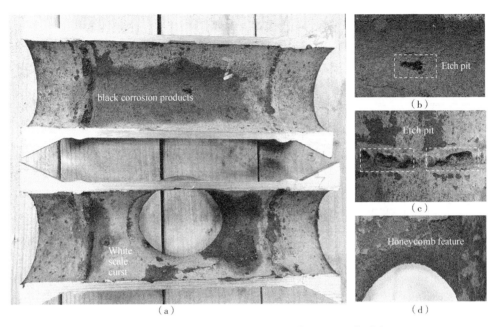

Fig. 1　Macroscopic feature of corrosion in the inner wall of the tee
(a) overall; (b) etch pit on inner wall; (c) etch pit on weld; (d) honeycomb feature in corner

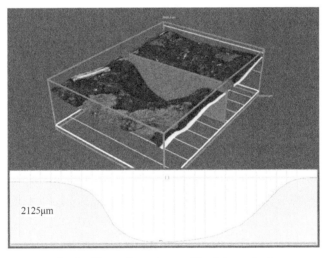

Fig. 2　Three-dimension profile of the etch pit

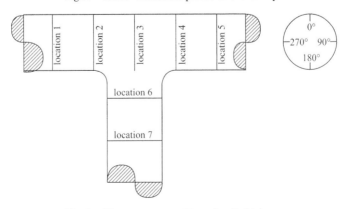

Fig. 3　Measurement position of wall thickness

2.3 ChemicalComposition Analysis

Chemical composition analysis of the tee materialwas carried out according to the test standard ASTM A751-14a, using a direct-reading spectrometer. The analysis results are shown in Table 2. The result showed that the tee material's chemical composition meets the API Spec 5L—2018 standard requirement.

Table 2 Chemical composition analysis result of tee-junction (Wt.%)

Element	C	Si	Mn	P	S	Cr	Mo	Ni	V	Cu
Result	0.17	0.20	0.44	0.014	0.014	0.016	<0.001	0.013	0.0021	0.012
ASTM A234 requirement	≤0.30	≥0.10	0.29~1.06	≤0.050	≤0.058	≤0.040	≤0.015			

2.4 Metallographic Examination

Metallographic examination samples are taken from the etch pits of the inner wall and weld. According to the ASTM E3-11、ASTM E112—2012 and ASTM E45—2013 standards, the metallographic examination was carried out on the MEF4M metallographic microscope. The metallographic examination results of the tee material are shown in Fig. 4. It's microstructure is ferrite and pearlite, the grain size is 9.5 level, the nonmetallic type is thin A0.5、B1.0 and D0.5, there are no excessive defects in the tee material. Fig. 5 shows the metallographic structure of etching pits of tee-junction weld. The microstructure of the weld is acicular ferrite, granular and polygonal ferrite, there could see numbers of cracks starting from the bottom of the etch pits in the weld and extending along the thickness of the tee. The cracks are branching a lot and have an inter-granular propagated feature in some regions. The cracks have a certain characteristic of stress corrosion cracking.

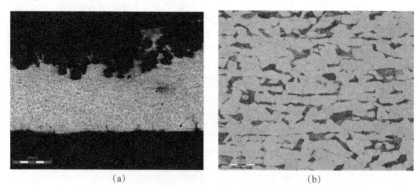

Fig. 4 Metallographic structure of etching pits of tee-junction
(a) pits morphology; (b) micro-structure

2.5 Micro-fractography and corrosion product analysis

2.5.1 Etch pits

The micro-frcatography observation and energy spectrum analysis of the corrosion product was performed on TESCAN VEGAII SEM. The micro-fractography of an etch pit on the inner wall of the tee is shown in Fig. 6. Fig. 6(b) and Fig. 6(c) show that the micro-fractography features of the corrosion products in the flank side of the etch pit (section 1) are all flocculent shaped and the corrosion products in the bottom of the etch pit also exist a micro-feature of globular and rod shaped. The energy spectrum analysis results in Fig. 7 and Table 3 show that the corrosion products are having elementary compositions of C, O, S, Fe, and the elements C, O, S, Fe are having higher content. From the composition of the elements, we can conclude that the corrosion product may be the product of the Fe in the CO_2 corrosion environment—$FeCO_3$.

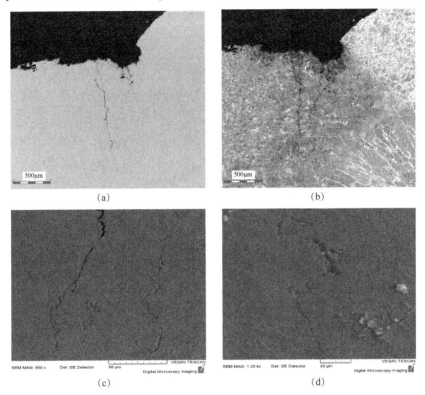

Fig. 5　Metallographic structure of etching pits of tee-junction weld
(a) pits morphology; (b) micro-structure; (c) middle of the crack; (d) tip of the crack

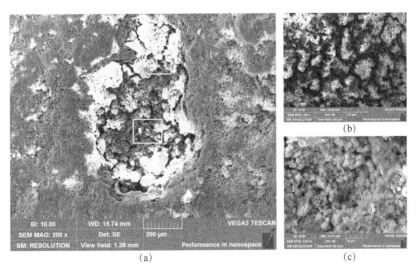

Fig. 6　Micro-morphology of etch pit
(a) overall; (b) corrosion products in section 1; (c) corrosion products in section 2

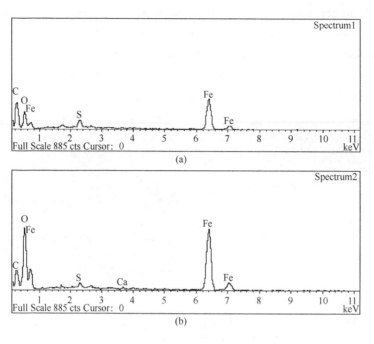

Fig. 7 Energy spectrum analysis of etching pit
(a) area 1 in Fig. 6(b); (b) area 2 in Fig. 6(c)

Table 3 Energy spectrum analysis result of etching pit (Wt. %)

Position	C	O	S	Fe
Area1	23.51	17.10	12.95	36.62
Area2	25.96	28.19	8.68	34.98

2.5.2 Cracks

Fig. 8 and Table 4 show the energy spectrum analysis results of different positions of the crack. The elements in different positions of the crack are all C, O, S, Fe. The existence of the element S in the crack imply that the SCC crack may be induced by sulfuret. Open the crack surface using some mechanical methods, and observed it on the SEM, the micro-fractography of the crack surface was shown in Fig. 9. Fig. 9 shows that there is a intergranular fracture feature in the crack surface. Combine with the energy spectrum analysis results and the micro- fractography of the crack surface, we can conclude that the crack could be a SSC crack.

Table 4 Energy spectrum analysis result of crack in etching pit (Wt. %)

Position	C	O	S	Fe
Etching pit	29.79	21.01	8.18	32.99
Root of crack	36.71	26.02	7.05	30.22
Middle of crack	35.54	34.76	8.98	20.72
Tip of crack	12.80	39.02	7.98	40.2

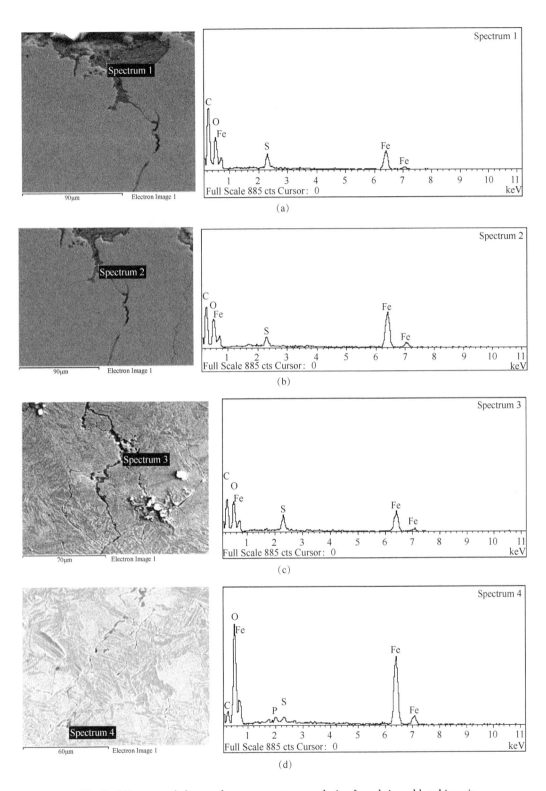

Fig. 8　Micro-morphology and energy spectrum analysis of crack in weld etching pit
(a) etching pit; (b) the root of crack; (c) the middle of crack; (d) the tip of crack

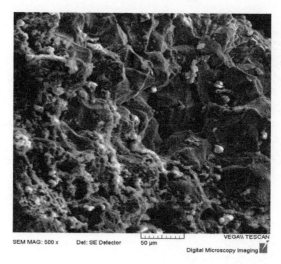

Fig. 9 Micro-morphology of the crack surface in weld etching pit

2.5.3 XRD analysis of corrosion products

The XRD analysis results of corrosion products are shown in Fig. 10. The spectrum in Fig. 10 shows that the corrosion products are consist of $FeCO_3$, FeS, Fe_2O_3, $CaCO_3$. $FeCO_3$ is the corrosion product of iron in CO_2 corrosion environment, FeS is the corrosion product of iron in H_2S corrosion environment, Fe_2O_3 is the corrosion product of iron in O_2 corrosion environment, $CaCO_3$ could be come from formation water in the gas well.

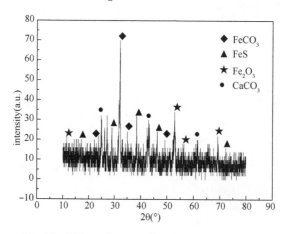

Fig. 10 XRD analysis results of corrosion products

3 Analysis on cause of internal corrosion

The above tests and analysis results show that the corrosion condition of the inner wall of is very severe, and there are numbers of etch pits on it. There are numbers of cracks originating from the bottom of some etch pits of the weld on the inner wall, which show the characteristic of SSC cracking. The XRD analysis results show that the corrosion and cracking causes may be in relationship with CO_2 and H_2S corrosion environment. Considering that the material properties of the tee all meet the standard requirements, the internal corrosion and cracking causes could be as

follows:
3.1 Analysis on cause of internal corrosion

(1) CO_2 corrosion.

From the operation pressure of the tee and the molar content of CO_2 in shale gas, we can calculate that the partial pressure of CO_2 is 0.066MPa. According to the grading of the CO_2 corrosion in some articles[5-6]: $0.021\text{MPa} < p_{CO_2} < 0.21\text{MPa}$ is a moderate corrosion, the corrosion could occurr during the CO_2 corrosion process, the reactions in the anode and cathode are as follows[7]:

cathode:
$$2H_2O + 2e^- \longrightarrow 2HO^- + H_2$$
$$2HCO_3^- + 2e^- \longrightarrow H_2 + 2CO_3^{2-}$$

anode:
$$Fe \longrightarrow Fe^{2+} + 2e^-$$
$$Fe^{2+} + CO_3^{2-} \longrightarrow FeCO_3$$

The corrosion product is $FeCO_3$, which is in accordance with the XRD analysis results of the corrosion product, so the internal corrosion pattern of the tee should be the CO_2 corrosion. The wall thickness measurement results revealed the corrosion rate was 5.73mm/a, the corrosion rate was very high. Obviously, the CO_2 corrosion under a partial pressure of 0.066MPa cannot reach such a high corrosion rate, so there may be other factors that could accelerate the corrosion speed. Based on the analysis of the service condition and the flow state and flow speed of the medium inside the tee, there are two aspects that could accelerate the corrosion rate.

(2) Fluid erosion-corrosion.

The flow speed inside the tee is 9m/s, some research achievements indicate that [8-9]: when the gas flow rate is 1~25m/s, liquid flow rate is 0.1~5m/s, it is very easy to form an impact flow and thus generate high shear force to the internal wall of the tee. The shear force become higher with the increasing of the flow rate and impact angle, and this also lead to an increasing of the corrosion degree. Under the action of the impact flow and shear force, the flow effect inside the tee generate flow, which lead to corrosion, damage and peel off the corrosion product adhered to the internal wall, and expose new metal matrix and directly contact the corrosion medium to accelerate the corrosion.

(3) Sulfate reducing bacteria corrosion.

The XRD analysis results of the corrosion product show that there is FeS in the corrosion product. FeS is the corrosion product of iron in H_2S corrosion environment. From the former analysis, we can know that the shale gas does not contain H_2S gas, so we can conclude that the existence of the H_2S gas is in relationship with the sulfate reducing bacteria in the formation water of the well. The formation water composition analysis results show that there are a number of sulfate reducing bacteria in it, with the content of 7000 individual/mL. It is generally recognized that the suitable breeding environment of sulfate reducing bacteria is pH: 6—9, temperature: 30~35℃ or 55~60℃. The environment inside the tee is just in comply with the environment above. Because there are also SO_4^{2-} ions in the formation water, and the SRB (sulfate reducing bacteria) have the ability of turning SO_4^{2-} as an final electron acceptor, and producing H_2S gas. H_2S gas would finally have a electrochemistry corrosion reaction with the material of the tee, and as a result, generate the corrosion product FeS.

Some research findings reveal that in the CO_2 corrosion environment, the existence of SRB could contribute to the acceleration of the CO_2 corrosion progress[11-13]. In the corrosion environment of CO_2 and SRB, when pH>6, at the beginning, CO_2 corrosion is in the leading place, and the corrosion product is $FeCO_3$, and when the amount of SRB has reached to a certain content, it would turn sulfate into electron acceptor in the organics dissimilation process, and produce sulfide. The cathode reaction is as follows[14]:

$$SO_4^{2-} + 9H^+ + 8e^- \longrightarrow HS^- + 4H_2O$$

The reaction product HS^- could react with the former anode reaction product Fe^{2+}, and finally generate the corrosion product FeS. Besides, research evidences[15-16] show that the existence of SRB could accelerate the process of depolarization in the CO_2 corrosion cathode reaction, and make the pH value in the corrosion environment become lower, and as a result, strengthen the corrosivity of the environment, and thus accelerate the CO_2 corrosion rate.

In summary, the internal corrosion of the tee is the combined action of the CO_2 corrosion, erosion-corrosion and SRB corrosion.

3.2 Cracks in the weld

The cracks in the weld originated from the bottom of the etch pits on the weld, and propagated along the thickness of the tee. The cracks are branching a lot and have an inter-granular propagated feature in some regions. The cracks have a certain characteristics of stress corrosion cracking. The energy spectrum analysis results show that there exists the element of S in the crack surface, from the former discussion, we know that the element S derives from the SRB corrosion process, so the type of the crack could be concluded as the Sulfate – reducing Bacteria Induced SCC[17]. It's formation process could be as follows: when the CO_2 corrosion occurs in the internal of the tee, there may generate numbers of etch pits in the internal wall and weld, which can bring local stress concentration in the bottom place and deposition of the corrosion product. Unlike the etch pits in the internal wall, the weld has generated a shape mutation in the internal wall and brought the throttling effect there, so the corrosion product in the etch pits of the weld were not easy to take away by the gas flow. With the continuous deposition of the corrosion product HS^-, S^{2-} and FeS, the toughness of the material in the etch pits are becoming lower and there is a tendency of cracking. Finally, with the combination of local stress concentration and CO_2-SRB Synergistic corrosion, make the SSC crack originate from that place and continuously propagat.

4 Conclusion

(1) The cause for the internal corrosion of the tee is the combination action of CO_2-SRB Synergistic corrosion.

(2) The cracks in the internal weld of the tee is mainly belong to the Sulfate-reducing Bacteria Induced SCC.

(3) It is recommended to reinforce the frequency of the internal wall of the tee, and at the same time to put fungicide into the pipeline and tee so as to reduce the retention amount of the formation water in the tee, decrease the amount of the sulfate reducing bacteria and form the sustained-release layer.

References

[1] ZOU C N, DONG D Z, WANG Y M, et al. Shale gas in China: Characteristics, challenges and prospects(Ⅰ)[J]. Petroleum Exploration and Development, 2015, 42(6): 689-701.

[2] ZOU C N, DONG D Z, WANG Y M, et al. Shale gas in China: Characteristics, challenges and prospects (Ⅱ)[J]. Petroleum Exploration and Development, 2016, 43(2): 166-177.

[3] LI C, WANG C Q, CHEN X, et al. Corrosion Law and Protection Measures of Shale Gas Well Wellbores[J]. Corrosion&Protection, 2020, 41(1): 35-39.

[4] YUE M, WANG C Y. Analysis on the Corrosion of Shale Gas Well Tubing and Surface gathering pipelines[J]. Drilling&Production Technology, 2018, 41(5): 125-127.

[5] XIAN N, JANG F, RONG M, et al. Corrosion Prediction and Control Measure in CO_2 Gas Field Development[J]. Natural Gas and Oil, 2011, 29(2): 62-66.

[6] DOU G L, CHEN L B, LU C. Analysis of the hazard status of carbon dioxide corrosion in oil and gas field development[J]. Yunnan Chemical Technology, 2018, 45(6): 99-100.

[7] LV X H, ZHAO G X. Corrosion and Protection of tubing and casing material[M]. Beijing: Petroleum Industry Press, 2015: 28-33.

[8] YE F. Analysis on Corrosion Efect of Medium Flow Pattern on Condensate Gas Gathering and Transporation Pipeline[J]. Natural gas and oil, 2009, 27(6): 22-25.

[9] WANG D G, HE R Y, DONG S Y. Advances in Study on Internal Flowage Corrosion of Multiphase Pipelines[J]. Natural gas and oil, 2002, 20(4): 24-28.

[10] XU C M, ZHANG Y H, CHENG G X, et al. Corrosion behavior of 316L stainless steel in the combination action of sulfate-reducing and iron-oxidizing bacteria[J]. Transactions of materials and heat treatment, 2006, 110(4): 104-108.

[11] ZHU L X, LUO J H, LI L F. Cause Analysis for Internal Corrosion Thinning of Corner Elbow for Shale Gas Transportation[J]. Surface Technology, 2020, 49(8): 224-230.

[12] FAN M M. Study of pipeline internal corrosion of sulfate-reducing bacteria in the presence of carbon dioxide[D]. Wuhan: Hua-zhong University of Science and Technology, 2011.

[13] LIU Y X. Influence of sulphate-reducing bacteria on the corrosion bchavior of carbon stccl[D]. Dalian: Dalian University of Technology, 2002.

[14] ZHANG X L, CHEN Z X, LIU H H, et al. Effect of environment factors on the groth of sulfate-reducing bacteria[J]. Journal of Chinese Society for Corrosion and Protection, 2000, 20(4): 224-229.

[15] JIANG J Q, LIU L, XIE J F, et al. Effect of sulfate reducing bacteria from corrosion scale of oil pipeline on corrosion behavior of Q235 steel[J]. Corrosion & protection, 2018, 39(1): 6-16.

[16] QIN S. Research on biomanipulation of oil field effluent containing sulfate reducing bacteria[D]. Wuhan: Huazhong University of Science and Technology, 2011.

[17] XIONG F P, WANG J L, AHMED A F, et al. Research Progress of Sulfate-reducing Bacteria Induced SCC[J]. Corrosion Science and Protection Technology, 2018, 30(3): 213-221.

本论文原发表于《Materials Science Forum》2021 年第 1035 卷。

Study on Strain Response of X80 Pipeline Steel during Weld Dent Deformation

Zhu Lixia[1,2]   Luo Jinheng[1]   Wu Gang[1]   Han Jun[1]
Chen Yongnan[3]   Song Chengli[1]

(1. State Key Laboratory of Performance and Structural Safety for Petroleum Tubular Goods and Equipment Material, CNPC Tubular Goods Research Institute; 2. Xi'an University of Technology; 3. Chang'an University)

Abstract: Mechanical damage is one of the main factors affecting the service safety and service life of pipeline steel. The weld will cause cracking failure due to different strain response with different microstructure during deformation. In order to study the response characteristics and mechanism of the weld during deformation, the strain evolution process of the X80 pipeline steel in the weld zone is studied by finite element analysis(FEA) and experimental prefabrication of pipeline dent. The strain hardening model of the weld zone is established, and the dislocation configuration in different regions is discussed. The results show that the maximum strain gradually transfers from the weld to the heat affected zone (HAZ) when the depth of the dent is greater than 3% outer diameter, and the crack is finally generated at this stress concentration zone. The experiment verifies this phenomenon and shows that the established model can effectively evaluate the strain distribution and variation process of the weld zone with dent. It is believed that the microstructure of the weld is mainly fine ferrite with good strength and toughness; the microstructure of the HAZ is mainly bainitic ferrite with M/A islands distributed, which will hinder the movement of dislocations in ferrite under the external load, resulting in dislocation accumulation and stress concentration. As the depth of the dent increases, the density of dislocation at the grain boundary of the weld and the HAZ increases, and the dislocation slips continuously. The typical dislocation cell substructure appears in the HAZ, which reduces the uniform plastic deformation capacity and deformation capacity. Based on the simulation and experiment, it is proposed that the in-service pipeline should be repaired or replaced when the weld dent depth is larger than 3% outer diameter.

Keywords: X80 pipeline steel; Dent deformation; Microstructure; Strain hardening

1 Introduction

X80 pipeline steel is the mainstreamsteel grade applied in oil & gas pipeline engineering at

Corresponding author: Zhu Lixia, zhulx@ cnpc. com. cn.

present due to its good weldability and low temperature toughness. In the process of construction or service, the pipeline is easy to dent due to external load. When the dent occurs at the weld, the influence of stress concentration will be more significant. Moreover, there are many defects such as pores, inclusions and cracks in the weld, which is more likely to lead to pipeline failure. Therefore, the evaluation and analysis of the dent in weld zone is very important. In order to ensure safe and reliable operation, high grade pipeline steel (such as X80) needs to pay attention to the balance of toughness and plasticity in addition to strength improving. On the other hand, the development of strain based design also puts forward high requirements for the plasticity of pipeline steel. Whether X80 pipeline steel can be widely applied in the future mainly depends on its resistance to large deformation [1-4].

At present, there are many researches on the microstructure and properties of X80 pipeline steel weld[5-7], but there are few studies on the strain characteristics and microstructure of the weld zone. Ci Y and Zhang Z Z [8] studied the microstructure evolution and impact toughness of heat affected zone (HAZ) in X80 high strain pipeline steel by welding thermal simulation technology. Eriksson et al. [9] studied strain hardening caused by tensile stress in steel welding, and concluded that the strain hardening caused by pre-deformation is mainly related to the increment and winding of dislocations, and the greater the pre-deformation, the greater the dislocation density and the greater the strain hardening degree. Han S Y et al. [10] found that continuous yield and low yield ratio are shown on the inner side, while continuous yield and high yield ratio are shown on the outside when studying the spiral forming of X80 pipeline steel plate. For a specific microstructure, the yield strength can be optimized by controlling the forming strain of the pipe, so as to maximize the strain hardening effect. With the development of computer technology, numerical simulation technology has become one of the main methods of buried pipeline research. Although many scholars have carried out numerical simulations on the pipe mechanics under the action of dent, there are few experiments to compare the simulation under different dent depth, especially the influence of the strain variation law on the strain hardening mode after the dent occurs in the weld zone of the pipeline.

In this paper, FEA method was used to study the distribution of strain field in spiral weld zone of X80 pipeline steel, and experiment was carried out to verify the FEA model. The strain distribution of the steel pipe in the dent area of the spiral weld zone is studied. The tensile test and the real-time strain measurement during the prefabricated dent of pipe were carried out to investigate the influence of microstructure on weld dent. The results can provide theoretical support and experimental basis for the evaluation of pipeline steel with weld dent.

2 Material and methods

2.1 Finite element analysis

The research object is X80 SSAW pipe, and the specification and main chemical composition are shown in Table 1. The main mechanical parameters are: Density $\rho = 7.8 \times 10^3 \text{kg/m}^3$, Elastic Modulus $E = 2.1 \times 10^5$ MPa, Poisson's Ratio $\nu = 0.3$, Yield Strength $\sigma_y = 628$MPa. The friction coefficient between the pipe and the surface of the indenter is set as 0.3. When setting the boundary condition, because the bottom of the pipe is in contact with the soil, the circumferential constraint range is 120° along the pipe, and the pipe is divided along the range of 120° at the bottom and completely constrained. 12m long pipe model and a 150mm diameter spherical indenter were created by ABAQUS, then meshed the indenter and the pipe. Considering the contact between the indenter and the pipe and the actual working conditions, the density of the pipe grids gradually increases from both ends to the center. ABAQUS CAE was used to establish the weld zone model of spiral welded pipe, and the spiral scanning mode was used to conduct the weld zone. The outer diameter (OD) of the pipe is 1219mm/ inner diameter is 1182.2mm. There is a spiral seam submerged arc welding on the surface of the pipe model. The model is shown in Fig. 1. In this paper, the effect of different dent depth on the equivalent plastic strain is simulated. The indenter is perpendicular to the pipeline axis, and the dent defect is applied to the pipeline. The variation trend of strain was studied by changing the depth of indenter(1% OD, 2% OD, 3% OD, 4% OD, 6% OD), and the strain data and cloud map generated by different dent depths was recorded.

Table 1 Chemical composition of X80 pipeline steel with spiral welding (Wt. %)

Grade	Diameter	Thickness	C	Si	Mn	Mo	Ni	Cu	Nb+V+Ti
X80	1219	18.4	0.062	0.15	0.182	0.23	0.25	0.25	0.04

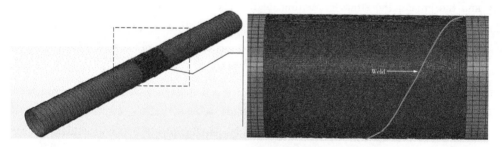

Fig. 1 Model of spiral submerged arc welding pipe

2.2 Dent prefabrication and dynamic strain measurement

In order to verify the FEA results, the weld dent of X80 steel pipe was prefabricated. The test steel pipe is φ1219mm×18.4mm X80 SSAW pipe. The basic parameters of the pipe are shown in Fig. 2(a). The weld dent was pressed on a 1500t composite loading test system, and the defects in the process of dent were detected by ultrasonic testing. The pressing model is shown in Fig. 2(b) and Fig. 2(c). The indenter is hemispherical with a diameter of 150mm. The dent depth of the pipe steel is 1% OD, 2% OD, 3% OD until the depth reaches 3.5% OD, and the cracks as shown in Fig. 2 (d) and Fig. 2(e) were found at the weld toe, so as to explore the influence of different dent depth on the strain. The limit dent depth of the pipe in the weld zone is 42.67mm(3.5% OD). In order to further study the strain evolution characteristics of the weld zone in the dent state, round bar samples are taken from the base metal and weld shown in Fig. 2(a), and the tensile test is carried out. The diameter of tensile specimen $d = 8$mm and gauge distance $L_0 = 30$mm, as shown in Fig. 2 (a). The tensile test was carried out on HT-2402 servo universal testing machine at a tensile rate of 0.5mm/min.

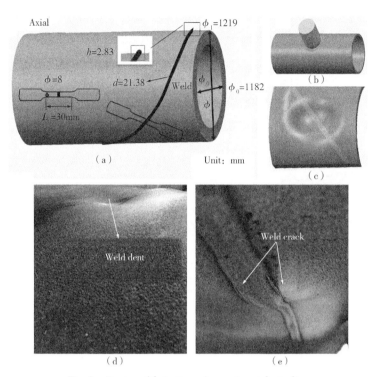

Fig. 2 Dent prefabrication of experimental pipeline
(a) Sampling location; (b, c) Prefabricated model of pipe dent; (d) Weld dent; (e) Crack at inner weld toe

In order to study the strain variation in the process of dent, the strain in the dent area on the inner surface of the steel pipe was measured in real time by using the three-dimensional full field strain measurement and analysis system (XTDIC). In order to study the deformation of the weld and the HAZ during the tension process, XTDIC system was also used to measure and analyze the displacement and strain of the pipeline steel during the tensile test.

2.3 Microstructure and hardness

In order to study the microstructure evolution of the weld zone, the microstructure of weld, HAZ, dent edge and base metal was observed by Leica DMI 8 metallographic microscope. The TEM sample was mechanically polished to 100nm, then the sample was thinned by double-spraying and thinning perforation using a double-spray electrolytic polishing machine (MTP-1A) under liquid nitrogen and alcohol mixture. The internal structure of the samples was observed by transmission electron microscope (JEM-200CX).

The hardness test was carried out using HSV-20 type hardness tester. The load was 200N and the dwell time was 10s. Starting from the center of the weld dent, 5 points were placed near each adjacent position of 10mm, and the average value was taken to measure the Vickers hardness of the sample surface.

3 Results and discussion

3.1 FEA Simulation of strain distribution in weld dent

Fig. 3 shows the FEA results of strain field under different dent depths. It can be seen from

Fig. 3(a) and Fig. 3(b) that when the depth of the dent is less than 3%OD, the maximum strain occurs in the weld, and the strain value gradually decreases when it is far away from the dent center. As the depth of the dent increases, the strain value also increases, and the maximum strain shifts from the weld to HAZ, and the strain at the weld is lower than that in HAZ[Fig. 3(c), Fig. 3(d), Fig. 3(e)]. This is because the grain size in the weld is large, which effectively increases the flow resistance of the grain boundaries and internal dislocations, makes the weld zone has a higher strain hardening ability, which enhances the strength and resistance of the weld joint [11].

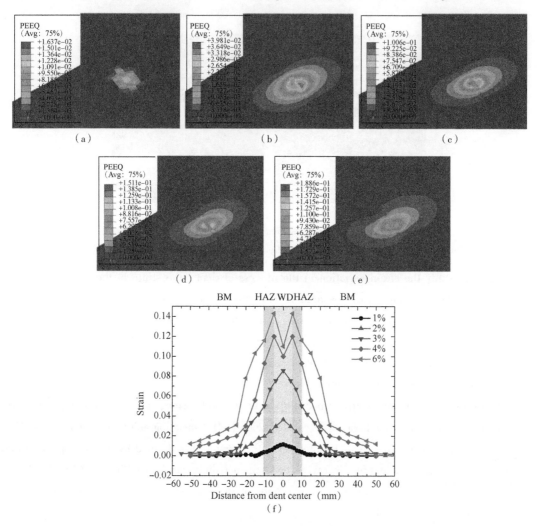

Fig. 3 Strain contour figures and strain curve under different dent depths simulated by FEA
(a)1%OD; (b)2%OD; (c)3%OD; (d)4%OD; (e)6%OD

Fig. 3(f) shows the strain change at different dent depths[the shaded part indicates the weld (WD) and the HAZ, and BM indicates base metal]. As can be seen from the Fig. 3, when the depth of the dent is 1%～3%OD, the maximum strain value of the weld in the center of the dent area increases from 0.01 to 0.087, and the strain value basically increases linearly. When the depth of the dent is 4%～6%OD, the strain value in the center of the dent increases to 0.11, but the maximum strain appears in the HAZ. When the depth of the dent is 6%OD, the maximum strain value is increased to 0.145. This is because the indenter acts on the weld in the initial stage of the dent process. With the depth of the dent gradually increasing to more than 3%OD, the strain transfers to the HAZ where with weak microstructure, and resulting in plastic deformation until the pipeline fails.

3.2 Dynamic strain evolution during weld dent measured

Fig. 4 shows the strain cloud diagrams of the weld zone at different dent depths. With the increase of dent depth, the strain value increases gradually, and the strain affected area expands. When the depth of the dent is less than 3%OD, the maximum strain in the center of the dent increases from 0.029 to 0.138. When the depth of the dent is more than 3%OD, the maximum strain value gradually transferrs from weld zone center to HAZ. As the depth of the dent gradually increases to more than 3.5%OD, the strain concentration in the HAZ at the bottom of the dent increases exponentially to 0.16, resulting in necking failure. The results show that the strain distribution measured by the test is consistent with the strain distribution simulated by FEA, which indicates that the FEA model established in this research can characterize the strain variation trend in actual working conditions.

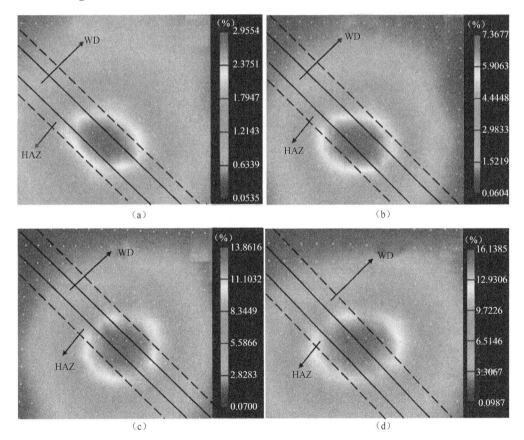

Fig. 4 XTDIC strain contour figures of weld zone with different dent depth
(a)1%OD; (b)2%OD; (c)3%OD; (d)3.5%OD

Fig. 5 shows the comparison of simulated and measured strain values under different dent depths. When the depth of the dent reaches 3.5%OD, cracks are found at the toe of inner weld in dent area, which indicates that the pipeline has failed at this time. The measured strain variation trend is basically consistent with the simulation, and the strain in weld and HAZ is larger, and the strain far away from the center of the dent decreases gradually. The strain affacted area and strain value of the measured pipeline are larger than the simulation results. When the depth of the dent is less than 3%OD, the maximum strain occurs in the center of the dent(i. e. weld seam). With the increase of the dent depth, the increase of the maximum strain also increases. The difference between the simulated and measured strain in the dent center increases from 0. 02 to 0. 05. This is because the FEA can not fully describe the defects and damage of the pipeline under the actual working conditions. Under the simulation conditions, the defect content of the pipeline is less than that of the actual pipeline, and the pipeline has good plasticity. However, combined with Fig. 3, it is found that the measured and simulated results are in good agreement with the numerical trend and variation.

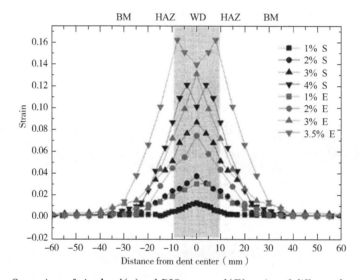

Fig. 5 Comparison of simulated(s) and DIC measured(E) strains of different dent depths

3.3 Strain induced work hardening behavior

The process of dent can be considered as the compressive stress at the contact point between pipe and indenter, and the tensile stress in other places until the specimen yields to necking. Tensile test combined with DIC real-time strain collection was usedto study the dynamic change process of strain in the weld area(the strain distribution contour figures and stress-strain curve are shown in Fig. 6). The uniform deformation stage, weld deformation stage and HAZ necking deformation stage of X80 Pipeline Steel under tensile stress are mainly studied.

In the uniform deformation stage(corresponding to the strain of 1%OD of the dent depth), the strain of the weld and HAZ is basically the same, showing uniform elongation in each zone. With the continuous increase of tensile stress, the compatibility of deformation in the weld and HAZ changes, and the strain begins to gather in the weld zone with higher hardening ability, that is, the weld deformation stage(corresponding to the strain when the dent depth is 2%~3%OD). Because of the good strain hardening ability of the weld, the strain response of the weld is far greater than that of the HAZ. With the continuous increase of stress, the HAZ with weak strain hardening ability can not resist large plastic deformation, so the strain rapidly transfers to HAZ and deforms continuously in HAZ. Therefore, cracks are produced on the surface of the material to release the local stress concentration, which is characterized by necking instability, that is, the necking deformation stage of HAZ(corresponding to the strain when the dent depth is 3%~3.5%OD).

According to Dundu M[12], the strain generated in the uniform plastic deformation stage of the stress-strain curve can form effective strain hardening on the material, as shown in Fig. 2(b). The resistance of weld to plastic deformation is greater than that of HAZ, so the plastic instability of final specimen occurs in HAZ. The above results were verified in the process of dent test. With the increase of dent depth, the strain initially concentrated at the weld, and then gradually transferred to the HAZ to accumulate and increase until the plastic deformation increased to pipeline failure.

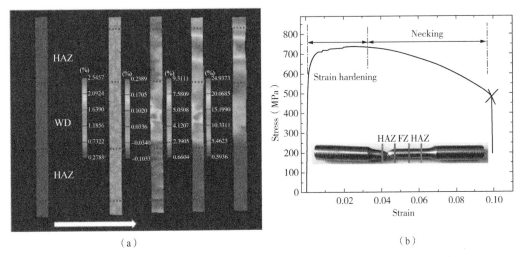

Fig. 6 Stress-strain curve of specimen at spiral seam

Local plastic deformation occurs in the dent process, leading to work hardening, and the work hardening of materials mainly depends on the equivalent plastic strain. Combined with Fig. 3 and Fig. 5, it can be seen that the resistance of pipeline steel to plastic deformation increases with the increase of deformation, which is a typical strain hardening mechanism. As shown in Fig. 7(a), a strong hardening phenomenon occurs in HAZ and WD during the dent process. Compared with the WD, the HAZ with larger equivalent strain has a higher hardness value [Fig. 7(b)]. Meanwhile, the hardness of the base metal is decreased with increasing distance from the center of the depression, reflecting the weaker strain hardening effect.

For most metals and alloys, the strain hardening behavior during uniform plastic deformation can be evaluated by Ludwik or Hollomon model[13], as shown in formula(1), where σ is the true stress, ε is the true plastic strain, K is the strain hardening strength factor, and n is the strain hardening index. Strain hardening index is usually used to characterize the strain hardening ability of plastic deformation in pipeline steels with high deformation. $\ln\sigma - \ln(d\sigma/d\varepsilon_p)$ is usually used to analyze the strain hardening behavior of steel, as shown in formula(2), (3) and (4).

$$\sigma = K \varepsilon_p^n \quad (1)$$

$$\varepsilon_p = \varepsilon_0 + c \sigma^n \quad (2)$$

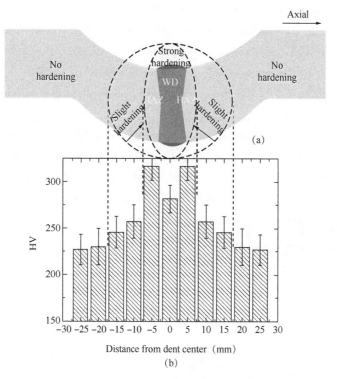

Fig. 7 Strain hardening area and hardness

$$\ln \frac{d\sigma}{d\varepsilon_p} = (1-n)\ln\sigma - \ln(cn) \quad (3)$$

$$\ln\sigma = \ln K + \ln\varepsilon \quad (4)$$

Where ε_p is true plastic strain, ε_0 is initial true strain, n is strain hardening index, C is material constant[14-15]. The strain hardening capacity of steel increases with the increase of n value.

From the above formula, it can beconclude that when plastic deformation occurs, the welded of the pipeline will prevent the continuing development of the plastic deformation, which transferred the deformation to other undeformed parts. With the increase of deformation, the degree of strain hardening in the welded increases, and the deformation area of plastic deformation remaining in the weld decreases, and the strain is rapidly transferred to the HAZ. The harding type of X80 pipeline steel is mixed hardening. With the increase of applied load, the mixed hardening characteristics of X80 pipeline weld will be more obvious and the strain hardening index also increases. In the case of over matching, the strain hardening ability of the base metal is lower than of the weld, which is caused by the chemical composition, crystal structure and microstructure[16].

3.4 Microstructure analysis

In order to further study the strain response during the deformation process of the pipeline steel weld zone, the microstructures of the X80 base metal, weld and heat affected zone of dent area are analyzed, as shown in Fig. 8. The microstructure of the X80 pipeline steel base metal consists of acicular ferrite, polygonal ferrite and granular bainite, as shown in Fig. 8(a). This type of steel with ferrite structure has a certain deformability. The microstructure of the weld is mainly composed of acicular bainite and acicular ferrite. The microstructure is fine and closely arranged. The grains with different sizes occlude and cross distribute with each other to form a fine strip structure, which can effectively resist the slip of dislocations and improve the strength and toughness of weld. On the contrary, the microstructure of HAZ is mainly coarse grain, as shown in Fig. 8(c) (labelled by the arrow), lath bainite, coarse ferrite and distributed with M/A island structure with brittle and ductile structure, which can significantly reduce the ductility and toughness of HAZ of pipeline steel, which is also the starting point of cracks in the process of dent. Han J et al. considered that the M/A island structure, acts as a second phase particle, would block the movement of dislocations in the deformed structure under external loading, causing dislocation accumulation and resulting in stress concentration and cracks [17-18]. Meanwhile, the deformation ability between M/A island structure and ferrite matrix structure is different under external load. The interaction of internal stress and dislocation accumulation may cause the M/A island structure to break or form microcracks. Under the further action of external load, the microcrack will expand and eventually lead to the failure of pipeline steel.

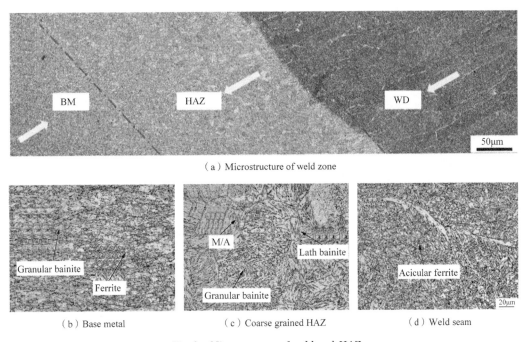

Fig. 8　Microstructure of weld and HAZ

When the pipeline steel is deformed bydent, the dislocation density and aggregation morphology of the weld and HAZ also change correspondingly with the increase of dent depth, as shown in Fig. 9. When the dent deformation is 1% OD, the weld deformation is more significantly and dislocations begin to form, as shown in Fig. 9(a). The dislocations continuously slip and interact to form entanglement near the grain boundary. However, there is almost no plastic deformation in the HAZ, and only a few dislocations slip in the grain. As shown in Fig. 9(b), the existence of a small number of movable dislocations can reduce the initial yield resistance of the material. However, with the increase of dent deformation, multiple dislocations in the weld cross entangle to form a dislocation wall, which effectively strengthens the weld, as shown in Fig. 9(c). The high density dislocation cells formed by the entanglement and aggregation of dislocations in the HAZ become a new obstacle to dislocation slip, which can significantly improve the strength of the material [19-20]. But at the same time, it reduces the ability of uniform plastic deformation and deformation capacity, reduces the stress concentration, relaxes the local stress at the crack tip and limits the crack propagation, damages the defect capacity of the pipe and reduces the bearing capacity.

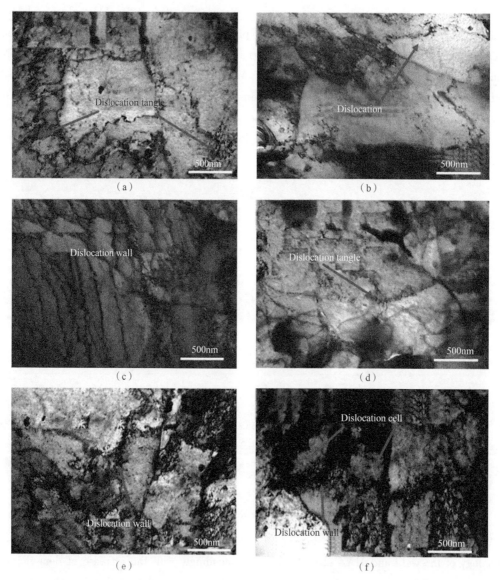

Fig. 9 TEM microstructure under different depths of dent
(a)1%OD-weld; (b)1%OD-HAZ; (c)3%OD-weld; (d)3%OD-HAZ; (e)3.5%OD-weld; (f)3.5%OD-HAZ

When the deformation of the dent increases to 3.5% OD, the dislocation density further increases, and the dislocation density in the weld further increases. The multiple slip caused by dislocation promotes the dislocation crossing on different planes, and the dislocation dense wall appears, as shown in Fig. 9(e). In the process of multiple slip, dislocations cause the slip surface to rotate, which makes the dislocations cross and form dislocation walls. This kind of high density dislocation wall hinders the occurrence of deformation. When the dislocation density increases to a certain extent, the strain hardening degree in the HAZ can not satisfy the requirements of dislocation deformation, and a large number of dislocation cells gather around the M/A island, causing it to break and form microcracks and fail[21-24]. In addition, the change of macro size of weld toe aggravates the stress concentration, which leads to the macroscopic cracks.

4 Conclusions and suggests

(1) In the process of weld dent, the strain distribution of FEA simulation is consistent with measurement results: When the dent depth is less than 3%OD, the maximum strain in the dent area is located at the weld, and the maximum strain transfers to the heat affected zone with the increase of the dent depth. When the dent depth reaches 3.5%OD, the inner weld toe cracks in the dent area. The strain affected area and strain value of the measured are greater than the simulation results due to the fact that the simulated data can not fully consider the actual pipe weld defects.

(2) The microstructure of HAZ of X80 pipeline steel is mainly bainitic ferrite(BF), and the second phase particles M/A aredispersed distributed. With the increase of weld dent depth, the plastic deformation of HAZ increases, the typical dislocation cell like substructure is formed in ferrite grains, and a large number of dislocation cells undertake large plastic deformation. The dislocation movement at the grain boundary of HAZ reduces the uniform plastic deformation capacity and deformation capacity, reduces the stress concentration, relaxes the local stress at the crack tip and limits the crack propagation of pipeline steel, damages the defect capacity of the pipe and reduces the bearing capacity, which finally leading to cracks in the HAZ.

(3) In the management of in-service pipeline with weld dent, when the depth of dent is larger than 3% OD, it is recommended to repair or replace the dent section of pipeline.

References

[1] YOO J Y, AHN S S, SEO D H, et al. New development of highgrade X80 to X120 pipeline steels[J]. Mater Manuf Process, 2011, 26(1): 154-160.

[2] JUN T S, HOFMANN F, BELNOUE J, et al. Triaxialresidual strains in a railway rail measured by neutron diffraction[J]. J. Strain Anal, 2009, 44(7): 563-568.

[3] HWANG B, KIM Y M, LEE S, et al. Correlation of microstructure and fracture properties of API X70 pipeline steels[J]. Metall Mater Trans, 2005, 36(3): 725-739.

[4] HE L L, CHUN Y H, LING K J, et al. Development and Application of High Performance X80 Line Pipe for the 2(nd) West-East Gas Pipeline[C]// International Conference on High Strength Low Alloy Steels, 2011.

[5] LAN L, QIU C, ZHAO D, et al. Analysis of microstructural variation and mechanical behaviors in submerged arc welded joint of high strength low carbon bainitic steel[J]. Materials Science & Engineering A(Structural Materials: Properties, Microstructure and Processing), 2012, 558: 592-601.

[6] HAO S, GAO H, ZHANG X, et al. Study on local embrittlement of welding heat-affected zone in X80 pipeline steels[J]. China Welding, 2011, 20(2): 36-40.

[7] WANG X, SUN X, YONG Q, et al. Effect of VC on toughness of welding heat affected zone in X80 pipeline steel[J]. Heat Treatment of Metals, 2015, 40(4): 7-12.

[8] CI Y, ZHANG Z Z. Simulation study on heat-affected zone of high-strain X80 pipeline steel[J]. Journal of Iron

and Steel Research International, 2017, 24(9): 966-972.

[9] ERIKSSON C L, LARSSON P L, ROWCLIFFE D J. Strain-hardening and residual stress effects in plastic zones around indentations[J]. Materials Science & Engineering: A, 2003, 340(1): 193-203.

[10] HAN S Y, SOHN S S, SHIN S Y, et al. Effects of microstructure and yield ratio on strain hardening and Bauschinger effect in two API X80 linepipe steels[J]. Materials Science & Engineering: A, 2012, 551(31): 192-199.

[11] AFRIN N, CHEN D L, CAO X, et al. Strain hardening behavior of a friction stir welded magnesium alloy [J]. Scripta Materialia, 2007, 57(11): 1004-1007.

[12] DUNDU M. Evolution of stress-strain models of stainless steel in structural engineering applications[J]. Construction & Building Materials, 2018, 165: 413-423.

[13] NI D R, CHEN D L, WANG D, et al. Tensile properties and strain-hardening behaviour of friction stir welded SiCp/AA2009 composite joints[J]. Materials Science & Engineering: A, 2014, 608(7): 1-10.

[14] TOMITA Y. Effect of morphology of second-phase martensite on tensile properties of Fe-0.1C dual phase steels [J]. J. Mater. Sci, 1990, 25(12): 5179-5184.

[15] REED-HILL R E, CRIBB W R, MONTEIRO S N. Concerning the analysis of tensile stress-strain data using log $d\sigma/d\varepsilon_p$ versus log σ diagrams[J]. Metall. Mater. Trans, 1973, 4(11): 2665-2667.

[16] MOTOHASHI H, HAGIWARA N. Effect of Strength Matching and Strain Hardening Capacity on Fracture Performance of X80 Line Pipe Girth Welded Joint Subjected to Uniaxial Tensile Loading[J]. Journal of Offshore Mechanics and Arctic Engineering, 2007, 129(4): 318.

[17] HAN J, LU C, WU B, et al. Innovative analysis of Luders band behaviour in X80 pipeline steel[J]. Materials Science & Engineering A, 2017, 683: 123-128.

[18] SPEICH G R, MILLER R L. In Structure and Properties of DualPhase Steels[M]. New York: AIME, 1979.

[19] ZECEVIC M, KNEZEVIC M. A dislocation density based elasto-plastic self-consistentmodel for the prediction of cyclic deformation: application to AA6022-T4[J]. Int. J. Plast, 2015, 72: 200-217.

[20] ZECEVIC M, KORKOLIS Y P, KUWABARA T, et al. Dual-phase steel sheets undercyclic tension-compression to large strains: experiments and crystal plasticitymodeling[J]. J. Mech. Phys. Solids, 2016, 96: 65-87.

[21] MUGHRABI H. Dislocation wall and cell structures and long-range internal stresses in deformed metal crystals [J]. Acta Mater, 1983, 31: 1367-1379.

[22] PHAM M S, SOLENTHALER C, JANSSENS K G F, et al. Dislocation structureevolution and its effects on cyclic deformation response of AISI 316L stainless steel[J]. Mater. Sci. Eng.: A, 2011, 528: 3261-3269.

[23] YEFIMOV S, VAN DER GIESSEN E. Multiple slip in a strain-gradient plasticity modelmotivated by a statistical-mechanics description of dislocations[J]. Int. J. Solids Struct, 2005, 42: 3375-3394.

[24] WANG H, JING H, ZHAO L, et al. Dislocation structure evolution in304L stainless steel and weld joint during cyclic plastic deformation[J]. Mater. Sci. Eng.: A, 2017, 690: 16-31.

本论文原发表于《Engineering Failure Analysis》2021年第123卷。

Influence of Boss-Backing Welding to ERW Pipe

Luo Jinheng[1] Zhao Xinwei[1] Liu Ming[2] Luo Sheji[3] Hu Meijuan[1]
Wu Gang[1] Li Lifeng[1] Zhu Lixia[1]

(1. State Key Laboratory of Service Behavior and Structure Safety for Petroleum Tubular Goods and Equipment Material, CNPC Tubular Goods Research Institute;
2. School of Materials Science and Engineering, Xi'an Jiaotong University;
3. School of Materials Science and Engineering, Xi'an Shiyou University)

Abstract: Station and valve chamber design often encounter the situation of drilling hole at the main pipeline and welding boss-backing to connect the branch pipe. Boss hole location should generally be at least 100mm away from the longitudinal weld or spiral weld. However, because the electric resistance weld(ERW) is difficult to distinguish in practice, some bosses mounting position coincide with ERW or close to it. In this paper, the influence of boss-backing welding directly on the longitudinal weld to the original residual stresses of ERW pipe was studied. The microstructure of pipe body and longitudinal weld after welding was also analysis. The testing results showed that the overall residual stress of ERW pipe was relatively small. Residual stress at the longitudinal weld region was smaller than that in the pipe body region. After the boss-backing welding, the axial residual stress at the longitudinal weld and the circumferential residual stress at the pipe body region near the intersection increased sharply to 2.5 (444MPa) and 3.8 (433MPa) times, respectively. The invaded width and depth to the ERW pipe after welding were about 15.167mm and 3.376mm. Granular bainite with necklace type M-A constituents could be observed at the invaded zone. It is suggested that small welding heat input should be adopted for boss-backing welding.

Keywords: Station and valve chamber; ERW pipe; Boss-backing; Cross weld; Welding residual stress; Microstructure

1 Introduction

The station and valve chamber are the essential parts in the storage and transportation of oil and gas. Electric resistance welding(ERW) pipe has the advantages of high production efficiency, low cost, high dimensional accuracy and easy automation, and it is widely used in pipeline construction

---

Corresponding author: Luo Jinheng, luojh@cnpc.com.cn.

of oil and gas conveying station/valve chamber[1-3]. In order to meet the installation needs of valves, instruments and flanges, the station and valve chamber design often encounters the situation of making holes on the main pipe and boss-backing welding (branch pipe seat) to connect the branch pipe[4-7], as shown in Fig. 1(a). The opening position should generally be staggered 100 mm from the longitudinal weld or spiral weld of the main manifold[8]. However, in practical operation, because the weld line of ERW pipe is difficult to be identified, some bosses are installed close to or even coincide with the weld line, which may lead to safety risks, as shown in Fig. 1(b).

Fig. 1 Boss connection and ERW welded joint schematic diagram.
(a) welded globe valves and (b) macro morphology of ERW welded joint

The non-compliance points of welding bosses in the stations and valve rooms of pipeline Branch in the western region were once counted by our group, it was found that a total of 42 openings were less than 100mm from the longitudinal weld of the ERW pipe, as shown in Fig. 2 and Table 1.

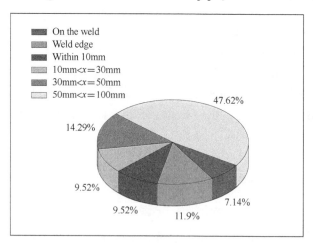

Fig. 2 Distance distribution between non-compliant points and longitudinal weld from Table 1

Some studies have shown that there is a large residual stress in ERW pipe, the residual stress value in some areas reaches the yield limit of the material, and most of the residual stress shows great harm: Such as strength reduction, reducing the fatigue limit, causing stress corrosion and brittle fracture, etc[9-12]. When a cross weld is formed between the bump installation weld and the

ERW straight weld, the extremely complex residual stress distribution near the weld intersection and the change of the structure of the ERW pipe body and the heat affected zone after the multi-pass welding thermal cycle could all affect the overall performance of ERW pipe[13-15], and shorten the service life of the welded pipe and limit its service life.

Table 1 Statistical table of non-compliance points of ERW pipe of convex head opening

Company	No.	Pipe type	Direction of weld (2 o'clock, 8 o'clock)	Name and number of opening instrument/balance valve	Distance (mm)
A	1	φ406.4×12.5	7	PI 1101 (upstream pressure gauge)	90
	2	φ406.4×12.5	10	1106 (throttle globe valve)	90
	3	φ406.4×12.5	1	PI 1102 (Downstream pressure gauge)	10
	4	φ406.4×12.5	10	PI 1101 (upstream pressure gauge)	0
	5	φ406.4×12.5	10	PI 1101 (upstream pressure gauge)	40
	6	φ406.4×12.5	0.94	Pressure gauge PT1101 (JP10061594)	100
	7	φ457×14.2	3	Manual ball valve 1215	60
	8	φ406.4×12.5	0.28	PI1101	30
	9	φ406.4×12.5	—	Downstream lead pipe opening distance	100
B	10	φ457×14.2	0	Bet bypass line, Valve # 1304 balances line 2 connection points	10
	11	φ406×12.5	6	Table PT6401 connection points	3
	12			Quick stop valve 6402# pressure lead pipe connection point	
	13			Quick stop valve 6403# pressure lead line connection point	
	14			PT6406 join points	
	15	—		1103 valve lead pipe at 10 o'clock	40
	16	φ457×14.2	Between 1 and 2	The two connection points of ball-receiving barrel bypass pipeline balances	50
C	17	φ406×12.5	—	PI1101	35
	18		—	PI1101	78
	19		—	PI1101	0
D	20	φ457×14.2	8	1204# Upstream bypass air intake point	110
	21	φ457×14.2	8	1204# Downstream bypass air intake point	25
	22	φ457×14.2	8	1304# Upstream bypass air intake point	90
	23	φ457×14.2	8	1304# Downstream bypass air intake point	70
	24	φ406.4×12.5	8	Cross station 7103# take air point	60
	25	φ406.4×12.5	1	PI1101	60
	26	φ406.4×12.5	1	PI1102	110
	27	φ406.4×12.5	12	PI1102	0
	28	φ457×14.2	1	1204 valve intake point	80

continued

Company	No.	Pipe type	Direction of weld (2 o'clock, 8 o'clock)	Name and number of opening instrument/balance valve	Distance (mm)
E	29	φ406	12	Pressure gauge PT1101	40
	30	φ406	12	Pressure gauge PT1101	40
	31	φ406	11	Pressure gauge PT1101	84
	32	φ406	12	Pressure gauge PT1102	85
	33	φ406	11	Pressure gauge PT1101	100
	34	φ406	12	Pressure gauge PT1102	60
	35	φ406	12	Pressure gauge PT1102	85
	36	φ406	12	Pressure gauge PT1102	10
	37	φ406	12	Pressure gauge PT1102	10
F	38	1102#	0.5	Opening of pressure gauge	19
	39	1102#	0.5	Opening of pressure gauge	15
	40	1102#	11	Opening of pressure gauge	70
	41	1102#	1	Opening of pressure gauge	95
	42	1102#	12	Opening of pressure gauge	4

Therefore, to study the influence of boss-backing welding on the microstructure and properties of ERW pipes is of great significance to ensure the service life of ERW pipes. In this paper, a small hole detection method was used to study the ERW pipe commonly used in the station/valve chamber of a pipeline company, the influence of the boss-backing welding on the ERW pipe straight weld on the original residual stress of the pipe was studied, and the influence of the boss-backing welding on the pipe body and the microstructure of the heat affected zone was also analyzed. The study has important theoretical significance and guiding function for correct design and site construction of boss-backing welding.

2 Experimental

2.1 Materials

The test material is the commonly used ERW pipe in the main pipeline of an oil and gas transmission station/valve room. The steel grade is L415MB and the specification is $\phi406.4\text{mm} \times 12.5\text{mm}$. The welded pipes used in the test were the same steel grade and specification produced by Baosteel, their chemical compositions are shown in Table 2, and the calculated CE_{IIw} and P_{cm} values are also listed in the Table 3. The specimens were transversely cut along ERW pipe, the plate tensile samples and impact samples were machined based on the ASTM A370—2012a and GB T229—2020 standards. The tensile test and impact test were carried out on the MTS C64 universal testing machine and the NC-500 impact testing machine respectively, the tensile and impact properties of the materials are shown in Table 3.

$$CE_{IIw} = C + \frac{Mn}{6} + \frac{(Cr+Mo+V)}{5} + \frac{(Ni+Cu)}{15} \tag{1}$$

$$CE_{P_{cm}} = C + \frac{Si}{30} + \frac{Mn}{20} + \frac{Cu}{20} + \frac{Ni}{60} + \frac{Cr}{20} + \frac{Mo}{15} + \frac{V}{10} + 5B \tag{2}$$

Table 2 Chemical composition of L415MB ERW pipe (Wt. %)

C	Si	Mn	P	S	Cr	Mo	Ni	Nb	V	Ti	Cu	Al	B	CE_{IIw}	C_{eq}	P_{cm}
0.068	0.21	1.25	0.0098	0.0035	0.024	0.002	0.0067	0.041	0.024	0.019	0.017	0.028	0.0002	0.298	0.292	0.152

Table 3 Mechanical properties of L415MB ERW pipe

Specimen	Specification (mm)	Tensile Strength R_m(MPa)	Yield strength 0.5%EUL(MPa)	Elongation A(%)	Impact A_{kv}(J)
Pipe body 90° transverse	38.1×50	570($10^{-3}s^{-1}$)	530	37	207
Welded joint Transverse	38.1	532($10^{-3}s^{-1}$)	—	—	365

The boss used for the test was a welded boss with working pressure of 12 MPa, the material is A350 LF2 and specification is DN400×25mm(Fig. 3).

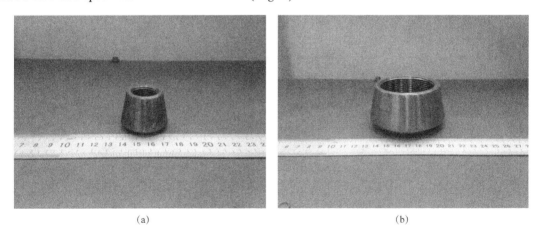

(a) (b)

Fig. 3 Bosses and welded globe valves with bosses
(a)DN400×25mm; (b)DN400×50mm

2.2 Test method

A ϕ25mm diameter hole was opened in straight weld position of ERW pipe through the thermal cutting. According to the welding procedure of the second station/valve chamber of West-to-East gas pipeline of China, the boss was welded by the welding method of GTAW backing and SMAW filling the cover surface. The specific boss-backing welding process is shown in Table 4.

The change of residual stress on ERW pipe before and afterboss-backing welding was measured by CM-1L-32 static resistance strain gauge and BE120-2CA-K strain flower. In the process of testing, the position of the blind hole was reasonably arranged to ensure that the distance between each measuring point was more than 12mm after considering the specific size of the pipe. The effects of thermal cutting and multi-pass welding on the base metal and weld microstructure of ERW pipes were observed by OLS 4100 laser scanning confocal microscope, MEF4M metallographic microscope and image analysis system.

Table 4 The boss-backing welding process

Weld pass	Materials type	Welding materials specifications (mm)	Direct current polarity	Weld Current (A)	Weld voltage (V)	Weld speed (cm/min)	Argon flow rates (L/min)	The length of Tungsten electrode (mm)	Orifice nozzle diameter (mm)	The length of the Argon arc (mm)
Root welding	ER50-6	2.0	DCEN	80~120	10~14	5~10	7~12	6~9	8~10	2~4
		2.5	DCEN	100~160	10~16	5~10	7~12	6~9	8~10	2~4
Fill/cover	E5015	3.2	DCEP	90~140	18~26	5~12				
	E5015	4.0	DCEP	100~150	18~26	8~15				

Note: DCEN indicates that the electrode connected to the welding material is associated with the negative electrode of the power supply; DCEP indicates that the electrode is associated with the positive electrode of the power supply.

The keyhole method is a semi-destructive method for the determination of residual stress. By measuring the strain release caused by the machining of the keyhole, the original residual stress of the keyhole can be converted through the calculation of elastic mechanics. The mechanical effect of the keyhole release method is equivalent to that of the reverse loading. According to the elastic mechanics, the formulas for calculating the principal stress(σ_1, σ_2) and principal stress Angle(γ) of the measured points can be obtained[16]:

$$\sigma_1 = \frac{\varepsilon_1+\varepsilon_3}{4A} + \frac{\varepsilon_1-\varepsilon_3}{4B\cos\gamma} \tag{3}$$

$$\sigma_2 = \frac{\varepsilon_1+\varepsilon_3}{4A} - \frac{\varepsilon_1-\varepsilon_3}{4B\cos\gamma} \tag{4}$$

$$\gamma = \arctan\frac{\varepsilon_1-2\varepsilon_2+\varepsilon_3}{\varepsilon_1-\varepsilon_3} \tag{5}$$

where A and B are referred to as strain release coefficients, and their expressions are as follows[17]:

$$A = -\frac{1+\mu}{2E}\frac{R^2}{r_1 r_2} \tag{6}$$

$$B = -\frac{1}{E}\frac{2R^2}{r_1 r_2}\left[1-\frac{1+\mu}{4}\frac{R^2(r_1^2+r_1 r_2+r_2^2)}{r_1 r_2}\right] \tag{7}$$

where E is modulus of elasticity and μ is Poisson ratio. The calculation formula of welding residual stress is as follows (σ_x is the longitudinal residual stress and σ_y is the circumferential residual stress)[18]:

$$\sigma_x = \frac{\sigma_1+\sigma_2}{2} + \frac{\sigma_1-\sigma_2}{2}\cos\gamma \qquad (8)$$

$$\sigma_y = \frac{\sigma_1+\sigma_2}{2} - \frac{\sigma_1-\sigma_2}{2}\cos\gamma \qquad (9)$$

Then the residual stress values could be calculated by Eqs. 2—9, and the schematic diagram of strain gauge is shown in Fig. 4. During the test process, the measuring point near the weld is X direction along the weld, and Y direction is perpendicular to the weld direction. When calculating the residual stress, positive values indicate tensile stress and negative values indicate compressive stress.

The residual stress test was carried out by blind hole drilling on the drilling machine, as shown in Fig. 5. According to the stress testing standard, the distance between each measuring point should be more than 6 times of the diameter of the small hole, that is, more than 12mm. In the testing process, combined with the specific size of the pipe, the position of the hole was

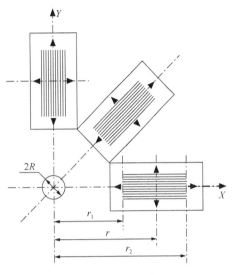

Fig. 4 Schematic diagram of strain gage

reasonably arranged to ensure that the measuring point and the boundary should be kept above 15mm, and the distance between each measuring point should be kept above 12mm.

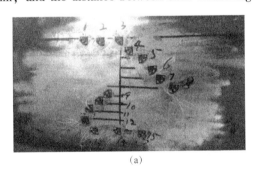
(a)

(b)

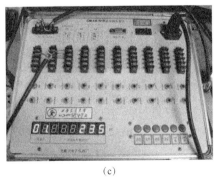
(c)

Fig. 5 Schematic diagram of residual stress testing equipment and testing processes
(a) paste the strain gauge; (b) connect and drill; (c) read

3 Results and analysis

3.1 Residual stress

The original residual stress distribution of ERW pipe is shown in Fig. 6. As can be seen from the Fig. 6, the axial residual stress on the ERW pipe is higher than that of the circumferential residual stress, and the maximum axial and circumferential stresses are 203MPa and 152MPa respectively. There is no apparent stress concentration in the straight weld area. The maximum axial and circumferential residual stresses on the straight weld are 178MPa and 114MPa respectively, which are far lower than the yield strength of the material, and are 34% and 22% of the yield stress of the welded pipe body. Fig. 6(a) shows the comparing magnitude of residual stress on the straight weld of ERW pipe and 100mm away from the straight weld. The axial and circumferential residual stresses in the area 100mm away from the straight weld are slightly larger than those in the straight weld, the maximum axial residual stress difference is 47MPa and the maximum circumferential stress difference is 45MPa. It can be seen from Fig. 6(b), the distribution of residual stress is relatively uniform at various distances from the straight weld, and the rule is still the axial residual stress is higher than that of the circumferential residual stress. The difference between the two residual stresses are small between 45mm and 55mm from the weld, however, the difference increases as the distance continues to increase. The distribution of residual stress on ERW pipe is closely related to its welding and forming process, the sizing and hydraulic test after welding are both important processes that lead to the axial residual stress value being greater than the circumferential residual stress value.

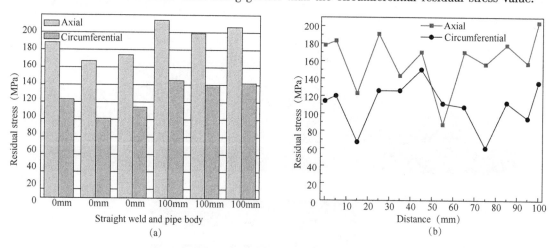

Fig. 6 Initial residual stress distribution of ERW pipe. (a) straight weld and pipe body from 100mm of straight weld and (b) pipe body at various distances from straight weld

The distribution changes of residual stress on the ERW pipe after the boss-backing welding are shown in Fig. 7. It can be seen that the residual stress on the straight weld of ERW pipe near the boss rises steeply after boss-backing welding, and the increase range of axial residual stress is much greater than that of circumferential residual stress. The amplitude of the axial and the annular residual stresses are 444MPa and 201MPa respectively, which are about 84% and 38% of the base metal yield strength, the amplitude of the axial and the annular residual stresses increase by 2.5

and 1.4 times compared with the original peak distribution of ERW pipe. The increase of residual stress value decreases gradually with the increase of distance from the boss. The distribution of residual stress on the ERW pipe at different distances from the welding seam is more uniform, and the overall residual stress value is relatively small. Boss welding changes the original distribution state of residual stress, and makes the residual stress at the edge of the boss rise rapidly. The distance from the boss weld is longer, the influence of welding process on residual stress is smaller, and the increase amplitude of residual stress is smaller. The range of stress concentration on the straight weld is about 92mm, which is consistent with the change rule of ERW straight welded pipe, the residual stress value of the pipe body at the edge of the boss increases rapidly after the boss is welded. However, the increase range of circumferential residual stress in welded pipe body is much greater than that of axial residual stress. The amplitude of the axial and the toroidal residual stresses are 256MPa and 433MPa respectively, which are about 48% and 82% of the yield strength of the base metal. The peak distribution increases by 1.4 and 3.8 times compared with the original state of ERW pipe, and the circumferential stress concentration of the welded pipe body is about 51mm.

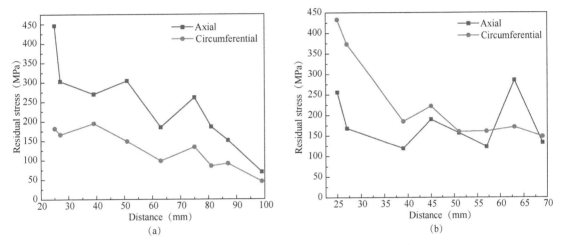

Fig. 7 Residual stress distribution on ERW pipe after boss-backing welding
(a) distance from the central axis; (b) different distance from straight weld pipe body

3.2 Microstructure

3.2.1 Microstructure of ERW

The micostructure of ERW pipe body is a mixture of polygonal ferrite, pearlite and a small amount of granular bainite (PF+P+B grains), with the grain size of 10.6, as shown in Fig. 8 (a). The straight weld microstructure of ERW pipe is a mixture of polygonal ferrite and a small amount of pearlite (PF+P), with the grain size of 10, as shown in Fig. 8(b).

3.2.2 Thermal cutting openings on microstructure

The influence of thermal cutting on the structure of the area near the opening is shown in Fig. 9, the influence range of thermal cutting tissue is funnel-shaped, with the largest surface width of about 3.649mm, and the width of uniform area of 3.523mm. The microstructure near the pore edge is coarse granular bainite structure generated by the transformation of original austenite grains, some grain boundaries can be seen, and the average size of original austenite grains is about

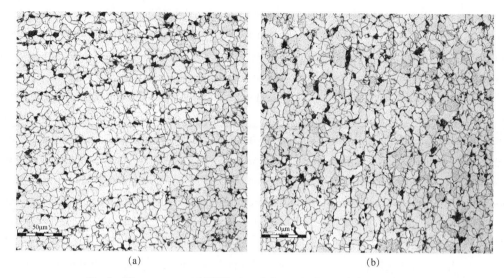

Fig. 8 Microstructure of ERW pipe. (a) pipe body and (b) straight weld

29μm. The large M-A island inside the grain is mostly long strip or round. The microstructure shows that the residence time near the pore edge at high temperature (1100℃) is longer, which results in the significant growth of the original austenite grains. However, the cooling rate is relatively low during the phase transition process, and there is no obvious orientation arrangement of M-A islands in the grain. The appearance of M-A island will result in an increase in the microhardness value of the edge zone. However, M-A island does great harm to impact toughness. It is generally believed that the M-A island is easy to form crack source and crack propagation channels. The larger the M-A island, the more the number of M-A island, the greater the damage to toughness, especially low temperature toughness. The grain size of the fine grain zone in the transition zone shows that it is not much different from that of the base metal, but the change of microstructure shows that the thermal cycle experienced in this zone leads to the relative decrease of the polygonal ferrite content while the increase of the granular bainite content, and the fine and dispersed M-A islands are evenly distributed in the granular bainite grains.

3.2.3 Effect of boss welding on microstructure

The width and thickness of the upper surface of the penetrated ERW pipe after boss-backing welding are 15.167mm and 3.376mm respectively, accounting for about 27% of the total wall thickness. After the boss-backing welding, the original pipe body and straight weld microstructure in the invaded area disappeared, and the microstructure was mainly transformed into the following three types: a small amount of weld microstructure in the surfacing layer, coarse heat-affected zone microstructure and fine grain microstructure, as shown in Fig. 10.

The weld microstructure of the surfacing weld is the coarse columnar grains that nucleate and grow directly on the base metal. The grain morphology of the proeutectoid white ferrite is outlined along the grain boundary of the slender columnar grains, and the fine acicular ferrite structures are found within the grains, as shown in Fig. 10(a). The coarse grain microstructure of the ERW pipe body and straight weld are coarse granular bainite[Fig. 10(b)], a small amount of polygonal ferrite and a mixture of coarse granular bainite[10(c)]. The discontinuous distribution of block M-A at

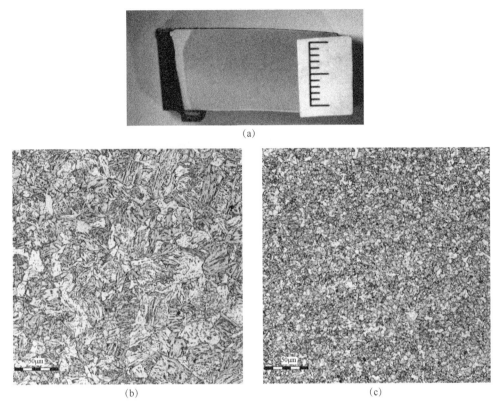

Fig. 9 Influence of thermal cutting openings on microstructure of ERW pipes(a)thermal cutting macroscopic morphology, (b)near-pore coarse-grained microstructure, (c)fine-grained microstructure in transition zone

grain boundary of granular bainite is connected into lines, and a chain structure could be observed. The microstructure of the fine grain zone between the pipe body and the straight weld of the ERW pipe is basically the same, which is a mixture of polygonal ferrite PF, quasi-polygonal ferrite QF and a small amount of pearlite at the grain boundary, as shown in Fig. 10(d).

3.3 Result analysis

After the ERW pipe is welded, the micostructure and properties of the direct weld and heat-affected zone could be improved by online heat treatment, which results in that the residual stress value of the direct weld and its adjacent area is lower than that of the pipe body, and there is no stress concentration in the direct weld area of the ERW pipe compared with the conventional weld. However, after the punch directly opened near the straight weld, the axial residual stress of the straight weld near the cross weld and the circumferential residual stress of the pipe body increase by 2.5 and 3.8 times respectively. That is to say, in the actual operation process, the straight weld in the area near the cross weld of 93mm after the convex welding bears not only the internal pressure, but also the axial and annular tensile stress, which is extremely unfavorable to the safety of actual pipeline operation[19-23].

The original microstructure of ERW pipe is uniform and fine, and the impact toughness of pipe body and weld is high. After the thermal cycle of the convex welding, except for the average

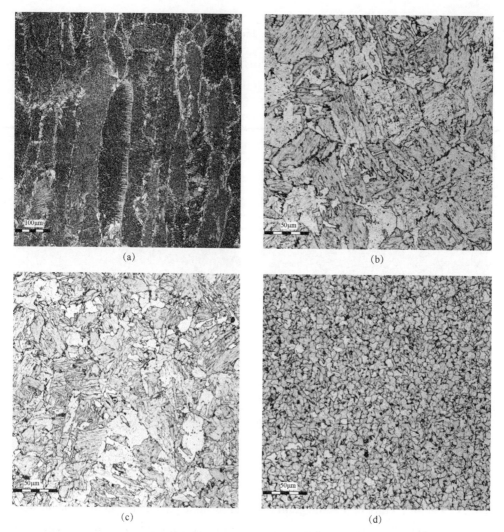

Fig. 10 Changes in the organization of the invaded area after boss-backing welding.
(a) Weld structure of surfacing layer, (b) pipe body, (c) Coarse-grained area at the
tip of straight weld and (d) Fine grain region

microhardness of the fine grain area is 180 HV1, which is the same as the base metal, the hardness values of the three types of microstructure in the 3.376mm thickness area are all higher than those of straight weld zone(203 HV1). The grain size of granular bainite and original austenite in the coarse grain area of ERW pipe and straight weld zone is basically the same, but the morphology, quantity and size of M-A shows different. The chain M-A island at the grain boundary of granular bainite grains in coarse grain zone is easy to form crack source and crack propagation channel, which has great influence on toughness, especially low temperature toughness. Therefore, it is recommended to adopt a smaller welding heat input during the convex welding and to increase the welding cooling rate, which could avoid the formation of chain-like structures in the intrusion area of ERW pipe[24-25].

4 Conclusions

This paper studied the influence of the boss-backing welding on straight weld of commonly used ERW station pipe (L415MB, ϕ406.4mm×12.5mm) residual stress and the microstructure, the main conclusions are as follows:

(1) The overall residual stress on ERW pipe is small, and the maximum axial and circumfluential residual stress is 203MPa and 152MPa, respectively. The residual stress value of the straight weld and its adjacent area is lower than that of the pipe body.

(2) After the boss-backing welding, the axial residual stress of the straight weld near the cross weld and the circumfluential residual stress of the pipe body increase by 2.5 and 3.8 times, which is 444MPa and 433MPa, respectively.

(3) Afterboss-backing welding, the penetration width of the ERW pipe upper surface is 15.167mm and the thickness is 3.376mm. The microhardness of the invasion zone is higher than that of the original pipe body, and the chain-like bainite structure could be observed in the coarse grain zone. Therefore, it is suggested to adopt a smaller welding heat input during welding.

References

[1] ROY A, GUPTA K K, NASKAR S, et al. Compound influence of topological defects and heteroatomic inclusions on the mechanical properties of SWCNTs[J]. Materials Today Communications, 2021, 26: 102021.

[2] WANG R, LUO S, LIU M, et al. Electrochemical corrosion performance of Cr and Al alloy steels using a J55 carbon steel as base alloy[J]. Corrosion Science, 2014, 85: 270-279.

[3] DUAN C H, CAO X K, ZHAO M H, et al. Research on Deformation Prediction Method of Laser Melting Deposited Large-Sized Parts Based on Inherent Strain Method[J]. Key Engineering Materials, 2021, 871: 65-72.

[4] WRAT G, BHOLA M, RANJAN P, et al. Energy saving and Fuzzy-PID position control of electro-hydraulic system by leakage compensation through proportional flow control valve[J]. ISA Transactions, 2020, 101: 269-280.

[5] PLAICHUM S, KAEWVILAI A, PANTONGSUK T, et al. A Novel Ceramic Backing Strip from Metakaolin-Based Geopolymer with Gas Flow Holes for Welding Application[J]. Key Engineering Materials, 2020, 856: 309-316.

[6] PRAKASH S O, KARUPPUSWAMY P, NIRMAL N. Optimal corrosive behaviour on the weldment of AA6063 aluminum alloy by tungsten inert gas(TIG)welding process with backing plates[J]. Metalurgija, 2019, 58(1-2): 91-94.

[7] LI X H. Welding Operation Technique of Welding Electrode Arc Welding Overhead Welding Test Board[J]. Key Engineering Materials, 2020, 861: 122-126.

[8] LV F, HU X, MA C, et al. Failure Analysis on Cracking of Backing Plate of Lifting Lug for Air Preheater[J]. Engineering Failure Analysis, 2020, 109: 104395.

[9] CHEN Z, CHEN X, ZHOU T. Microstructure and Mechanical Properties of J55ERW Steel Pipe Processed by On-Line Spray Water Cooling[J]. Metals, 2017, 7(4): 150.

[10] BI,Z Y, WANG J, JING X T. Effects of Thermo-Mechanical Control Process on the Microstructure and Properties of the Welded Joints of ERW OCTG [C]. Advanced Materials Research, 2011, 189-193: 3564-3569.

[11] WANG L D, TANG D, WU H B, et al. Influence of Sn on mechanical properties and corrosion behaviors of q125-grade tube used for ERW[J]. Journal of South China University of Technology (Natural Science Edition), 2012, 40(1): 94-100.

[12] SONOBE O, HASHIMOTO Y, IGUCHI T, et al. Effect of Mechanical Properties on Formability in Hydroforming of ERW tubes[R]. SAE Technical Paper, 2003.

[13] National Development and Reform Commission. Code for quality acceptance of oil and gas construction engineering station procedure pipeline project[M]. Beijing: Petroleum Industry Press, 2007.

[14] STANKEVICH S, GUMENYUK A, STRASSE A, et al. Measurement of Thermal Cycle at Multi-Pass Layer Build-Up with Different Travel Path Strategies during DLMD Process[J]. Key Engineering Materials, 2019, 822: 396-403.

[15] FENG Z, MA N, TSUTSUMI S, et al. Investigation of the Residual Stress in a Multi-Pass T-Welded Joint Using Low Transformation Temperature Welding Wire[J]. Materials, 2021, 14(2): 325.

[16] TORABI A R, MAJIDI H R, CICERO S, et al. Experimental verification of the Fictitious Material Concept for tensile fracture in short glass fibre reinforced polyamide 6 notched specimens with variable moisture[J]. Engineering Fracture Mechanics, 2019, 212: 95-105.

[17] IURLOVA N A, OSHMARIN D A, SEVODINA N V, et al. Numerical algorithm for searching for layouts of electroelastic bodies with external electric circuits for obtaining the best damping properties[J]. PNRPU Mechanics Bulletin, 2020(3): 108-124.

[18] XU G, GUO Q, HU Q, et al. Numerical and Experimental Analysis of Dissimilar Repair Welding Residual Stress in P91 Steel Considering Solid-State Phase Transformation[J]. Journal of Materials Engineering and Performance, 2019, 28(9): 5734-5748.

[19] HAMLIN R J, DUPONT J N, ROBINO C V. Correction to: Simulation of the Precipitation Kinetics of Maraging Stainless Steels 17-4 and 13-8+Mo During Multi-pass Welding[J]. Metallurgical and Materials Transactions A, 2019, 50(7): 3440-3440.

[20] HAN S W, PARK Y C, KIM H K, et al. Effect of Strain Hardening on Increase in Collapse Pressure during the Manufacture of ERW Pipe[J]. AppliedSciences, 2020, 10(14): 5005.

[21] MURAVYEV V I, BAKHMATOV P V, GRIGOREV V V, et al. Research of the influence of electron beam welding of titanium alloys on hydrogen distribution in the weld[J]. VESTNIK of Samara University Aerospace and Mechanical Engineering, 2019, 18(4): 157-168.

[22] HAN L, PEI K, SUN F, et al. Numerical Simulation on the Effect of Gas Pressure on the Formation of Local Dry Underwater Welds[J]. ISIJ International, 2021, 61(3): 902-910.

[23] SINGH R P, KUMAR S. Effect of current and chemical composition on the hardness of weld in shielded metal arc welding[J]. Materials Today: Proceedings, 2020, 26: 1888-1891.

[24] ZHU Z, WU S, ZHANG C, et al. Length Measurement of Chain-Like Structure of Micron Magnetic Particles Dispersing in Carrier Fluid Effected by Magnetic Field[J]. Journal of Superconductivity and Novel Magnetism, 2021(1-2): 1-12.

[25] HARICHANDRAN G, DIVYA P, YESURAJ J, et al. Sonochemical synthesis of chain-like $ZnWO_4$ nanoarchitectures for high performance supercapacitor electrode application[J]. Materials Characterization, 2020, 167: 110490.

本论文原发表于《Materials Research Express》2021年第8卷第5期。

Evolution of Grain Boundary α Phase during Cooling from β Phase Field in a α+β Titanium Alloy

Gao Xiongxiong[1]   Zhang Saifei[2]   Wang Lei[3]   Yang Kun[1]
Wang Peng[1]   Chen Hongyuan[1]

(1. State key Laboratory for Performance and Structure Safety of Petroleum Tubular Goods and Equipment Materials, CNPC Tubular Goods Research Institute;
2. School of Materials Science and Engineering, Xi'an University of Technology;
3. West Pipeline Company of PipeChina)

Abstract: Evolution of grain boundary α(GB α) in a α+β titanium alloy had been examined by morphology and crystallographic orientation analysis. The results indicate that GB α only retains a Burgers orientation relationship (BOR) with one of the adjacent β grains in most of the prior β/β boundaries. The colony α(single crystallographic variant of parallel α plates) tends to evolve in the adjacent β grain that maintains BOR with GB α, whereas the dendritic α(small protuberances of GB α) tends to grow in the other adjacent β grain that does not maintain BOR with GB α. Additionally, two α colonies developing into two adjacent β grains from GB α keep the same crystallographic orientation as that of GB α when GB α maintain a near BOR with both adjacent β grains in a special prior β/β boundary (60°/<110>). These observations are understood from the perspective of minimum interfacial energy and strain energy.

Keywords: Phase transformation; Grain boundaries; Titanium alloy; Grain boundary α; Crystallographic orientation

1 Introduction

The α+β titanium alloys are commonly used in aircraft, chemical and energy industries due to the high strength to weight ratio, good toughness, and excellent corrosion resistance[1]. Early research suggests that the β heat-treatment for α+β titanium alloys can have adverse effects on strength, ductility[2]. However, it has been recognized that β heat-treatment is more favorable under condition where damage-tolerant properties (fracture toughness, fatigue crack growth rate etc.) dominate[3]. Typical β heat-treatment followed by low cooling produces Widmanstätten α structure consisting of grain boundary α(GB α) and colony α(single crystallographic variant of

---

Corresponding author: Gao Xiongxiong, gaoxiongxiong2012@163.com

parallel α plates)[1].

GB α is an unavoidable microstructural feature in α+β titanium alloys during β→α diffusional phase transformation[4]. Numerous studies have reported that GB α affects morphology and crystallographic orientation of adjacent colony α, which affects the mechanical properties by changing the effective slip length[5-7]. Hence, it is very necessary to understand the evolution of GB α for the control of properties in titanium alloys. In the current study, a dendritic α structure(small protuberance of GB α)is found in a β grain that does not maintain a BOR with GB α. In addition, a special prior β/β boundary is believed to have an important effect on the formation and growth of GB α. Up to now, evolution of GB α is not completely clear during cooling from β phase field in α+β titanium alloys.

2 Materials and experimental procedures

In order to develop a better understanding of the evolution of GB α, a commercial α+β Ti-6Al-4V titanium alloy billet was chosen in this work. With an exact chemical composition(Wt%)of (6.12Al, 4.21V, 0.005C, 0.070Si, 0.0010H, 0.10O, Ti in balance), the β transus of this alloy is 990℃. To study evolution of GB α during cooling from β phase field, heat treatment experiment was performed on a Gleeble-3500 simulator. A Widmanstätten α structure including GB α was obtained by heat treatment at 1020℃ for 10min followed by cooling at rate of 1.5℃/s. The EBSD sample was prepared by vibratory polishing with colloidal silica. EBSD measurements were conducted using a TESCAN MIRA3 XMU scanning electron microscope(SEM).

Themorphology and crystallographic orientation of GB α, adjacent colony α and prior β grains will be examined to study the evolution of GB α during cooling from β phase field.

3 Results and Discussions

3.1 Formation of colony α and dendritic α from GB α at prior β/β boundaries

In Fig. 1(a) and (b), two prior β grains(labeled β1 and β2) can be found and their boundaries are delineated by GB α, as shown by the arrows. Typically, when the β phase is cooled from the β phase field to the α+β phase field, GB α first nucleate and grow along β/β boundaries in a manner similar to the "wetting" of the grain boundary surface[8]. Furthermore, it can be seen from Fig. 1(c) and (d) that the one of the {110} poles and one of the <111> poles of the β1 grain are parallel to(0001)pole and one of <11$\bar{2}$0> poles of the GB α, suggesting that the GB α keeps a BOR with prior β1 grain. This relationship can be explained by the fact that the interfacial energy between the GB α and one of the β grains maintaining a BOR is low for making the nucleation kinetics of GB α favorable[9]. As the temperature decreases further, there is a length of GB α with colony α(labeled α1) and dendritic α growing out into adjacent prior β grains, as shown by the arrows in Fig. 1(a). The pole figure[Fig. 1(d)] shows that there is a superposition of all the poles from GB α, colony α and dendritic α. This implies that they have a similar crystallographic orientation. These findings suggest that the colony α and dendritic α grow from GB α rather than being fresh nucleated GB α, which is generally believed to be related to the instability of α/β boundary[1]. The pole figures [Fig. 1(c) and (d)] analysis shows that the colony α tends to form in the β1 grain that maintains BOR with GB α, whereas the dendritic α does not maintain BOR with β2 grain. It is interesting to note that the length of dendritic α is relatively short. This is because that the interfacial energy between dendritic α and the β2 grain will be relatively high when they do not keep the BOR, which would be energetically unfavorable to the continuous growth of dendritic α[10]. Similar characteristics were also found in other prior β/β boundaries (~ 90%), as schematically shown in Fig. 1(e).

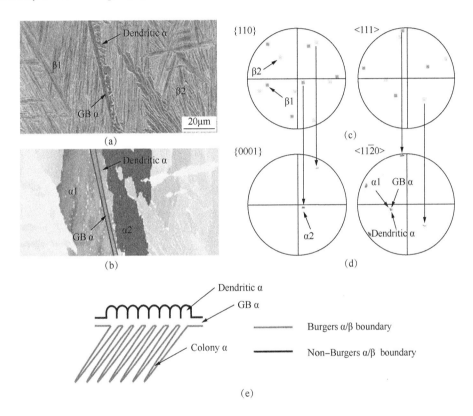

Fig. 1 (a) SEM map of area of interest; (b) IPF maps of highlighted α phases;
(c, d) pole figures of the prior β grains and GB α (including α1 and dendritic α) corresponding area to (b); (e) schematic illustration of formation of colony α and dendritic α

3.2 Formation of two α colonies from GB α at a special prior β/β boundary

As shown in Fig. 2(a) and (b), two α colonies (labeled α1 and α2) grew from the same GB α into two adjacent β grains (labeled β1 and β2), respectively. The hexagonal pole figures show [Fig. 3(b)] that they have a close crystallographic orientation (average misorientation angle about 10°). However, the average misorientation angle between the prior β grains reaches 52.3° (β1: Euler1 = 138.9°, Euler2 = 27.3°, Euler3 = 8.8°; β2: Euler1 = 24.6°, Euler2 = 44.6°, Euler3 = 72.7°). Moreover, it was found from β pole figures in Fig. 3(a) that two β grains share a relatively near {110} pole of β1 and β2 grains [marked by circle in Fig. 3(a), pole A and pole B], which are nearly parallel to the (0001) pole of colony α1 and α2, respectively. The traces of pole A and pole B can be shown in the <111> pole figure of β phase, as shown in Fig. 3(a). In addition, there are only two <111> directions on each {110} plane of β phase. Thus, there are two <111> poles on each trace of pole A and pole B, respectively. The angle between two <111> directions on the same trace is about 70°, which is in good agreement with the theoretical value of 70.5°. Examining the <111> poles within these two traces in Fig. 3(a), it can be found that the <111> directions are rotated by about 60° around common {110} pole. The crystallographic illustration of formation of GB α in a special prior β/β boundary(60°/<110>) is shown in Fig. 3(c). Thus, two <11$\bar{2}$0> poles of GB α are parallel to two <111> poles of β1 and β2 as indicated by arrows in Fig. 3 (a) and Fig. 3(b), respectively. This indicates that the GB α maintains the near BOR with both the adjacent β grains.

The α/β boundarykeeping a BOR is semi-coherent. The inter-planar distance of the (110) plane of β(0.234 nm) is within 1.7% of (0001) plane of α(0.230 nm). The small lattice mismatch between GB α and β minimizes the strain energy at the interface[9]. Thus, in the early stage of phase transformation, the interfacial energy may be minimized when the precipitate can have a semi-coherent interface with both the adjacent β grains[6-7]. Once the GB α is formed, two α colonies may start grow with the near orientation on both sides of the GB α until they meet other plates growing from different directions. As can be found in these pole figures, the two α colonies keep a strict BOR with the adjacent β grains, respectively. However, the (0001) pole for GB α lies between the (0001) poles for the two α colonies, as shown in Fig. 2(d) and Fig. 3(b). In addition, when the line[labeled L in Fig. 2(c)] crossed the α1, GB α and α2, it can be seen from Fig. 3 (c) that a slight change of accumulated misorientations occurs at GB α. The orientation of the GB α can be seen to be a compromise between the orientations of the two α colonies. Hence, GB α is not in a strict BOR with either β grain. The preferred two α colonies grew at GB α with the similar orientation to the GB α, which accommodates the misorientation between the two α colonies by inducing two lower angle boundaries between GB α and each colony, as schematically shown in Fig. 3(d).

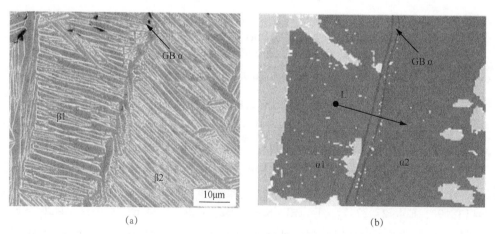

Fig. 2 (a) SEM map of area of interest; (b) IPF maps of highlighted α phases; (c) accumulated misorientation angle along line L in (b); (d) inverse pole figures of α phases corresponding area to (b)

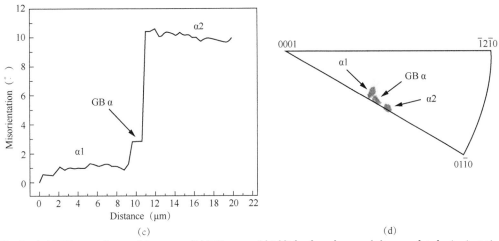

Fig. 2 (a) SEM map of area of interest; (b) IPF maps of highlighted α phases; (c) accumulated misorientation angle along line L in (b); (d) inverse pole figures of α phases corresponding area to (b) (continued)

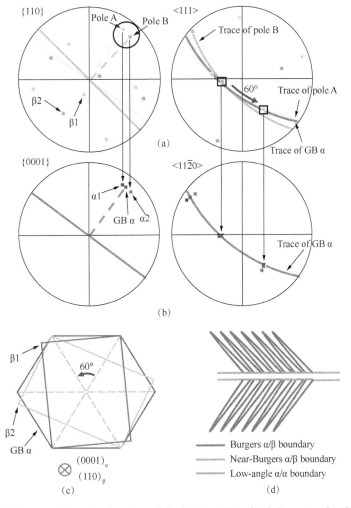

Fig. 3 (a) Pole figures of the prior β grains; (b) pole figure GB α (including α1 and α2) corresponding area to Fig. 2b; (c) crystallographic illustration of formation of GB α in a special prior β/β boundary; (d) schematic illustration evolution of GB α in a special prior β/β boundary

· 125 ·

4 Conclusions

In summary, GB α develops a colony microstructure in the β grain that maintains BOR with GB α. The dendritic α grows in the other adjacent β grain that does not maintain BOR with GB α. In rare case, the GB α maintains the near BOR with both the adjacent β grains in a special prior β/β boundary(60°/<110>) due to low strain energy at the interface. Furthermore, two α colonies that is close to the GB α orientation form preferentially in two adjacent β grains due to low interfacial energy.

Acknowledgements

This work is supported by Natural Science Foundation of Shaanxi province of China(2021JQ-946).

References

[1] LÜTJERING G, WILLIAMS J C. Titanium, 2nd ed. [M]. Springer, Berlin, New York, 2007.
[2] LEYENS C, PETERS M. Titanium and titanium alloys[M]. Wiley-VCH, Weinheim, 2003.
[3] FROES F H. Titanium physical metallurgy processing and applications[M]. Materials Park, 2015.
[4] BANERJEE S, MUKHOPADHYAY P. Phase Transformations: Examples from Titanium and Zirconium Alloys [M]. Elsevier, 2010.
[5] LUTJERING G. Influence of processing on microstructure and mechanical properties of (α+β) tianium alloys [J]. Mater. Sci. Eng. A, 1998, 243(1): 32-45.
[6] SHI R, DIXIT V, VISWANATHAN G B, et al. Experimental assessment of variant selection rules for grain boundary α in tianium alloys[J]. Acta Mater, 2016, 102: 197-211.
[7] LEE E, BANERJEE R, KAR S, et al. Selection of α variants during microstructural evolution in α/β tianium alloys[J]. Philos. Mag, 2007, 87(24): 3615-3627.
[8] PROTASOVA S G, KOGTENKOVA O A, STRAUMAL B B, et al. Inversed solid-phase grain boundary wetting in the Al-Zn system[J]. J. Mater. Sci, 2011, 46(12): 4349-4353.
[9] PORTER D A, EASTERLING K E, SHERIF M Y. Phase Transformations in Metals and Alloys, 3rd ed. [M]. Taylor & Francis Group, Boca Raton, 2009.
[10] ZHEREBTSOV S, SALISHCHEV G, SEMIATIN S L. Loss of coherency of the alphalbeta interface boundary in tianium alloys during deformation[J]. Philosophical magazine letters, 2010, 90(12): 903-914.

本论文原发表于《Materials Letters》2021 年第 301 卷。

Impact Assessment of Flammable Gas Dispersion and Fire Hazards from LNG Tank Leak

Li Lifeng[1,2] Luo Jinheng[1] Wu Gang[1] Li Xinhong[3] Ji Nan[1] Zhu Lixia[1]

(1. State Key Laboratory of Performance and Structural Safety for Petroleum Tubular Goods and Equipment Materials, CNPC Tubular Goods Research Institute;
2. Collage of Pipeline and Civil Engineering, China University of Petroleum(Shandong);
3. School of Resources Engineering, Xi'an University of Architecture and Technology)

Abstract: This study conducts an impact assessment of flammable gas dispersion and fire hazards from LNG tank leak. The release source model is used to estimate LNG release rate. A CFD(Computational Fluid Dynamics) based 3D model is established to simulate dispersion behavior of flammable gas from the phase transformation of LNG. Subsequently, a FDS(Fire Dynamics) based model is built to simulate the pool fire due to LNG tank leak. The impact of gas dispersion and fire on personnel and assets is assessed based on simulation results, which can provide a theoretical basis and method support for major accident assessment of tank leakage in large LNG receiving station. The results show that the dispersion of flammable gas from LNG tank leak has an obvious stage characteristic. The flammable gas reached a steady state around 300s, and the corresponding coverage area is about 16250m^2. The pool fire simulations indicate that the steady flame is formed at 20s. The flames flow along the wind, and the maximum temperature of the fire reaches 670℃, and the maximum thermal radiation reaches 624kW/m^2. According to the fire damage criteria, the pool fire from LNG tank leak may pose a serious threat on the safety of adjacent assets and personnel.

1 Introduction

Liquefied natural gas(LNG) is characteristic by easy to leak and volatile diffusion, flammability andexplosion. Once the LNG leak accident occurs, it will have a catastrophic impact on human life, assets, and the environment. The dispersion law of combustible gas and the consequences of fire and explosion accidents have become the critical of current research to ensure safety operation of LNG storage tank.

Researches on dispersion of flammable gas include field experiments, wind tunnel tests, and

---

Corresponding author: Li Lifeng, 457761448@qq.com.

numerical simulation, which were carried out earlier in foreign countries. Koopman et al. [1] obtained numerous data on the dispersion distance based on Burro series of large-scale LNG leakage and dispersion experiments. Hirst et al. [2] studied the combustion of large gas clouds evolved from LNG and refrigerated liquid propane spills on the sea through Maplin Sands series of experiments. Brown et al. [3] detected the variation of LNG gas concentration and temperature from multiple leakage point through Falcon series of experiments. Based on wind tunnel tests, the impact on LNG release dispersion, e.g., wind speed, vortex induction, overflow rate, leakage amount, etc., and dangerous range can be obtained [4-5].

Numerical simulation has the advantages of low cost, short cycle, simple operation, wide application range, and easy access to comprehensive data. A lot of researches were conducted on LNG leakage and diffusion based on CFD model. Calay et al. [6] used Euler-Lagrangian calculation model to simulate the evaporation and dispersion of LNG gas after injection from a circular hole. Guo et al. [7] used CFD model to study the LNG vapor dispersion law under different atmospheric stability; Luo et al. [8] proposed an integrated multi-phase CFD model to simulate the LNG leakage, evaporation and diffusion process; Saleem et al. [9] proposed a comprehensive dynamic CFD model for large-scale land-based LNG storage tanks; Zhu [10] discussed the mathematical model of the release and dispersion process in LNG leakage accidents. Qi et al. [11] simulated the vapor dispersion of LNG in the atmosphere; Zhuang [12] simulated the wind field and leakage diffusion of LNG full-capacity storage tanks; Zhang Chi et al. [13] studied the influence of factors such as wind field, leak location, leakage volume and other factors on the LNG leakage and diffusion process in the storage tank area; Jiang et al. [14] assessed the impact of cofferdam on LNG leakage; Zhang et al. [15] Analyzed the possible hazards during transportation to LNG carriers after the accident; Zhou et al. [16] studied the characteristics of the three processes of large-scale LNG storage tank leakage, liquid pool evaporation, and gas cloud diffusion at receiving stations; Yang et al. [17] established a three-dimensional numerical model for a large LNG storage tank in a LNG receiving station in southern China.

In addition, a lot of studies were devoted to the risk assessment of accident consequences. Li et al. [18] simulated the thermal response process of vertical full-capacity storage tank under different fire conditions. Sun et al. [19] Xie et al. [20] studied the features of large LNG pool fire. Di et al. [21] studied the development law of the flame in the set LNG fire scene. Further, the hazardous area caused by LNG leakage accident, e.g., fire, explosion, etc., can be simulated [22-24]. Based on hazardous features and main dangerous accidents, the consequence simulation and quantitative assessment on various accident can be obtained [25] and used to assess the risk of LNG storage system [26]. Some researches focus on the effect on fire or explosion accidents, Baalisampang et al. [27] proposed a methodology to model an integrated impact of evolving accident scenarios. Pio et al. [28] evaluated the effect of released fuel and its composition on the thermochemical characteristics of the small-scale LNG pool fire. Lv et al. [29] developed a correlation of the maximum explosion overpressure in the LNG storage tank area based on the momentum conservation equation and the deduced factors in the explosion test. Jujuly et al. [30] studied the effect of environmental conditions on the domino of an LNG pool fire. Wei. [31] studied the impact of wind seeped and leakage location to accidents caused by fire and explosion.

Although a lot ofprogress on assessments of LNG storage tank accidents have been made, most previous studies mainly focused on LNG leakage or fire and explosion accidents. It indicates a relatively independence between them in which there is little correlation. Thus, the cascading disaster-causing process mechanism of LNG storage tank from leakage to fire and explosion cannot be described. Process characteristics of LNG storage tank leakage fire and explosion accident still need further study. In addition, the influence of the evaporation phase variation on the dispersion process is not taken into consideration using simplified the leakage source conditions, which fails to accurately characterize the LNG dispersion process by obtained leakage rate. Therefore, the law of the phase change process caused by the temperature needs to be further studied.

This paper proposes an impact assessment of flammable gas dispersion and fire hazards from LNG tank leak. Based on previous research results, the phase transformation and the mechanism of cascade disaster of large-scale LNG storage tank leakage is fully taken into consideration and systematically studied to assess accident impacts. Firstly, the mechanism of LNG phase change and dispersion cascading pool fire was revealed by identifying the location distribution characteristics of large LNG storage tanks and the phase change process of leakage sources. Secondly, the risk factors of disaster chain are analyzed to obtain the impact of LNG leakage disasters on personnel, equipment, and environmental safety. Finally, a large-scale LNG leakage chain disaster assessment method based on the phase transformation dispersion cascade process of pool fire disasters is formed, which provides the theoretical basis and method support for assessing significant leakage accidents of storage tanks in large-scale LNG receiving stations.

2 Methodology

Fig. 1 presents the flowchart of impact assessment of flammable gas release, dispersion and fire hazards from LNG tank leak. The main steps of the methodology include: (1) Establishing release source model; (2) Mathematical model of gas evaporation and dispersion; (3) Mathematical model of fire.

2.1 Release source model

This paper mainly studies the continuous leakage of the liquid phase space in the lower part of the LNG storage tank and the dispersion behavior of the LNG after the formation of a stable liquid pool on the ground. LNG storage tanks of liquidleakage rate can be calculated by the Eq. (1):

$$Q_L = C_d A \rho_1 \sqrt{2gh + \frac{2(p_t - p_0)}{\rho}} \tag{1}$$

Where Q_L is the leakage mass flow, kg/s; A is the hole area, m^2; C_d is the liquid leakage coefficient; ρ_1 is the liquid density in the storage tank, kg/m^3; h is the height of the leakage location from the fluid level in the tank, m; p_t is the pressure in the storage tank, Pa; p_0 is the pressure of the external environment, Pa.

When the liquid in the storage tank leaks to the ground, it will spread aroundwith the leak source as the circle's center. If there are no obstacles in the dispersion process, the radius of the liquid pool will continue to increase and reach the maximum at a certain moment. The relationship

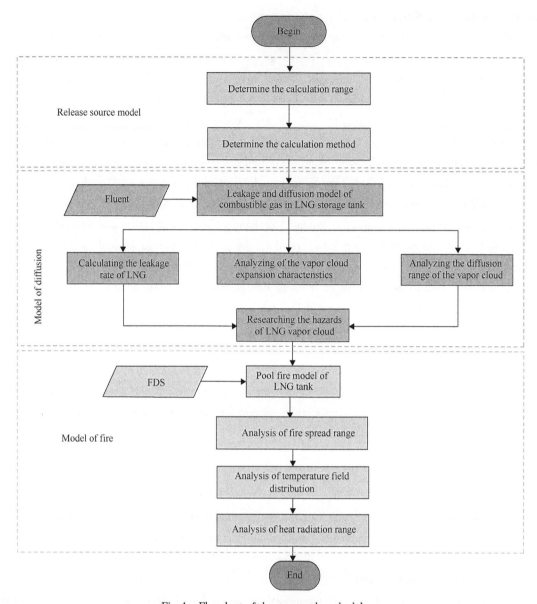

Fig. 1 Flowchart of the proposed methodology

between the radius r of the liquid pool and the time t can be calculated by the Eq. (2):

$$r_{(t)} = \left(\frac{t}{\sqrt[3]{\frac{9\pi\rho_2}{32gQ}}} \right)^{\frac{3}{4}} \tag{2}$$

Where Q is the mass flow rate of the liquid that leaks to the ground to form the liquid pool, kg/s; ρ is the density of the liquid LNG, kg/m³.

The LNG in the liquid pool will continue to diffuse in the atmosphere through evaporation to produce LNG vapor. Due to the low temperature of the leaked LNG, it forms a significant temperature difference with the external environment. When LNG leaks to the ground, the temperature difference with the ground will cause the liquid pool to evaporate. The main form of evaporation is heat conduction evaporation. The evaporation rate caused by heat conduction is calculated by Eq. (3):

$$Q_1 = \frac{kB(T-T_0)}{H\sqrt{\pi\alpha t}} \quad (3)$$

Where, k is the thermal conductivity of the substance, J/(m·s·K); B is the area of the liquid pool, m^2; T is the ambient temperature, K; T_0 is the boiling point of the liquid under normal pressure, K; H is the latent heat of vaporization of the liquid, J/kg; α is the thermal dispersion coefficient, m^2/s; t is the evaporation time, s.

2.2 Mathematical model of gas evaporation and dispersion

Based on fluid mechanics theory, Fluent as a tool for LNG leak diffusion process simulation build LNG storage tanks of flammable gas dispersion model. The study of LNG leakage dispersion behavior is mainly based on two aspects: the calculation of LNG leakage rate and the analysis of Characteristics of vapor cloud expansion of LNG leak. According to the leakage rate, the scope of LNG leakage and the scope of vapor cloud formation after LNG leakage, the hazard of vapor cloud after LNG leakage is studied.

Fluent is used to simulate the flow process of fluids. The flow of fluids is governed by the laws of conservation of physics, and the fluid flow process is calculated through control equations. Since the LNG in this paper gasifies and then forms a mixture with air after the leak, the Mixture theoretical model is selected. The governing equations include mass conservation equation, momentum conservation equation and energy conservation equation. These three conservation equations can be expressed in a unified mathematical expression that is the governing Eq. (4).

$$\frac{\partial}{\partial t}(\rho\varphi) + \mathrm{div}(\rho\bar{u}\varphi) = \mathrm{div}(\Gamma\mathrm{grad}\varphi) + S \quad (4)$$

Where, ρ is the density; φ is the general variable; Γ is the dispersion coefficient; S is the source term.

The Realizable k-model introduces variables related to rotation and curvature in the turbulence intensity coefficient equation. This model can consider the anisotropy in the turbulent flow process and can be effectively used for curved wall flow and curved streamline flow. The expression of the Realizable k-turbulence model is as follows:

K equation:

$$\frac{\partial(\rho k)}{\partial t} + \frac{\partial(\rho k u_i)}{\partial x_i} = \frac{\partial}{\partial x_i}\left[\left(\mu + \frac{\mu_t}{\sigma_k}\right) \cdot \frac{\partial k}{\partial x_i}\right] + G_k + G_b - \rho\varepsilon - \gamma_M \quad (5)$$

ε equation:

$$\frac{\partial(\rho k)}{\partial t} + \frac{\partial(\rho k u_i)}{\partial x_j} = \frac{\partial}{\partial x_j}\left[\left(\mu + \frac{\mu_t}{\sigma_\varepsilon}\right) \cdot \frac{\partial k}{\partial x_j}\right] + \rho C_1 S_c - \rho C_2 \frac{\varepsilon^2}{k + \sqrt{v\varepsilon}} + C_{1\varepsilon}\frac{\varepsilon}{k}C_{3\varepsilon}G_b \quad (6)$$

Where, $C_{1\varepsilon}$, $C_{3\varepsilon}$, C_1, C_2, σ_k, σ_ε are constants. $C_1 = \max\left[0.43, \frac{\eta}{\eta+5}\right]$.

G_k represents the turbulent energy term due to the existence of the velocity gradient:

$$G_k = -\rho\overline{u_i'u_j'}\frac{\partial u_j}{\partial u_i} \quad (7)$$

G_b represents the turbulent energy item due to buoyancy:

$$G_b = \beta g_i \frac{\mu_i}{p \tau_i} \frac{\partial T}{\partial x_i} \quad (8)$$

Where, g_i represents the component of gravity in the direction; β represents the coefficient of thermal expansion, $\beta = -\frac{1}{\rho}\left(\frac{\partial \rho}{\partial T}\right)$; $G(Gas) = -g_i \frac{\mu_\gamma}{p_{\gamma_i}} \frac{\partial \rho}{\partial x_i}$.

Where p_{γ_i} is the Prandtl number, from the K equation, it can be found that the turbulent kinetic energy growth trend mainly appears in the unstable layer. For stable layers, buoyancy tends to suppress turbulent flow.

The dispersion behavior of LNG leakage belongs to the flow of multi-component substances. The mass content of each substance can be predicted by iteratively solving the conservation equation of transport and dispersion of each substance. The material transport and dispersion equation are shown in Eq. (9).

$$\frac{\partial}{\partial_t}(\rho h) = \nabla \cdot (\rho \bar{v} \gamma_i) = -\nabla \cdot \bar{J}_i \quad (9)$$

Where, γ_i is the mass content of substance i, $\bar{J}_i$ is the mass dispersion rate of the substance. In the turbulent flow process, the mass dispersion equation is expressed as Eq. (10):

$$\bar{J}_i = -(\rho D_{i,m} + \frac{\mu_i}{S_{ct}}) \nabla \gamma_i \quad (10)$$

LNG is a liquid state, which is transferred with the heat of the air and gasified into natural gas. The LNG evaporation phase change model was written in FLUENT. The model proposed by Lee W H is the most widely used. The mass transfer equation of the two phases is as follows:

$$\begin{aligned} T_{mix} > T_{sat}, \quad m_{l \to v} = b \alpha_l \rho_l \frac{(T - T_{sat})}{T_{sat}} \\ T_{mix} < T_{sat}, \quad m_{v \to l} = b \alpha_v \rho_v \frac{(T_{sat} - T)}{T_{sat}} \end{aligned} \quad (11)$$

Where, T_{mix} is the unit temperature of the mixing zone, K; T_{sat} is the saturation temperature, K; $m_{l \to v}$ is the phase change rate of the liquid phase into the gas phase, kg/(m³·s); $m_{v \to l}$ is the phase change rate of the gas phase into the liquid phase, kg/(m³·s); b is a factor that controls the intensity of the phase transition. According to Schepper et al.'s simulation of the hydrocarbon feedstock flow and evaporation process, b is taken as 0.1 s^{-1}; α_l, α_v are the surface tensions of the liquid and gas phases, respectively, N/m; ρ_l, ρ_v is the density of liquid phase and gas phase respectively, kg/m³.

2.3 Mathematical model of fire

FDS is a fire of driven fluid flow dynamics software. It uses FDS to model sea surface gas fire accidents, simulates fire and smoke formation by solving low Mach number NS equations, and predicts heat flow and the concentration of toxic substances produced by the fire. The large eddy simulation (LES) method considers the turbulence characteristics in the heat flow process. During the fire development process, the smoke and heat flow obey the conservation of mass, momentum, energy and component transport equations:

$$\frac{\partial \rho_g}{\partial t}+\nabla \cdot \rho_g u=0 \tag{12}$$

$$\frac{\partial}{\partial t}(\rho_g u)+\nabla \cdot \rho_g uu+\nabla p=\rho_g f+\nabla \cdot \tau_{ij} \tag{13}$$

$$\frac{\partial}{\partial t}(\rho_g h_e)+\nabla \cdot \rho_g h_e u=\frac{Dp}{Dt}+\vec{q'''}-\nabla \cdot q+\phi \tag{14}$$

$$\frac{\partial}{\partial t}(\rho_g Y_i)+\nabla \cdot \rho_g Y_i u=\nabla \cdot \rho_g D_i \nabla Y_i+\vec{m_i'''} \tag{15}$$

Where, ρ_g is gas density, kg/m^3; t is time, s; u is the velocity vector, m/s; ∇ is the Laplace operator; p is pressure, Pa; f is the external force vector, N; τ_{ij} is the viscosity vector, Pa · s; h_e is the enthalpy value of leakage gas components, kJ/kg; $\vec{q'''}$ is the heat release rate per unit volume of gas, kW/m^3; q is radiant heat flux, kW/m^2; φ is the dissipation rate, kW/m^3; Y_i is the mass fraction of the i-th component of the leakage gas; D_i is the dispersion coefficient of the i-th component, m^2/s; $\vec{m_i'''}$ is the unit volume generation rate of the i-th component, kg/(m^3 · s).

Thermal radiation is a vital damage index of fire. The control equation of FDS thermal radiation calculation is as follows:

$$\vec{q_r'''}\equiv -\nabla \cdot \vec{q_r''}(x)=k(x)[U(x)-4\pi I_b(x)] \tag{16}$$

$$U(x)=\int_{4\pi}I(x,s')\mathrm{d}s' \tag{17}$$

Where, $k(x)$ is the absorption coefficient; $I_b(x)$ is the source term; $I(x,s)$ is the solution of the non-scattering grey gas radiation transport equation.

3 Model establishment

3.1 LNG tank farm

At present, my country's LNGreceiving stations generally adopt the LNG full-capacity storage tank type and most of the completed receiving stations have a capacity of 160,000m^3. The storage tank type is a full-capacity storage tank; the storage capacity is 16×10^4m^3; the storage temperature is −162 ℃; the maximum liquid level is 34.6m; The minimum liquid level is 2.9m; the normal working pressure range is 0.3~0.7MPa. Take a specific LNG receiving station as the research object for analysis. The overall overview of LNG receiving station is shown in Fig. 2.

3.2 Dispersion model of combustible gas leakage from LNG tank

3.2.1 Determine the calculation domain of leakage dispersion

The LNG storage tank is in a completely open

Fig. 2 Overall overview of LNG receiving station

environment, but due to the necessity and time problems, it is not necessary to simulate the wind field of the entire atmosphere and the gas dispersion in the simulation process. In this paper, the size of the calculation area was set as 640m×300m×100m. The LNG tank area model is established in the calculation domain, including six storage tanks. The upper storage tanks are numbered #1, #2 and #3 from left to right. The lower storage tanks are numbered #4, #5, and #6 from left to right.

The numerical model adopts the finite volume method, and the process of meshing is to discretize the computational domain. The quantity and quality of meshing determine the computational time and precision, and the meshing should meet the requirements of computational accuracy, computational time, computer configuration and other aspects. Since the vicinity of the leak location is a vital calculation domain for phase change and gas dispersion, and the calculation importance of the calculation domain far away from the leakage dispersion location gradually decreases, so this paper uses a gradual unstructured grid to mesh the entire calculation domain. The size of the grid at the site of the leak (initial grid size) is 0.05m, the expansion coefficient of the grid is 1.2, and the maximum grid size is 4.5 m. The meshing of the entire computational domain is shown in Fig. 3. The grid division near the leak is shown in Fig. 4, and the total number of grids reaches 4027679.

Fig. 3　Overall grid model of the LNG receiving station　　Fig. 4　Mesh division near the leakage port

3.2.2　Boundary conditions

3.2.2.1　Setting of wind speed

LNG vapor cloud dispersion is mainly driven by wind. Wind direction determines the dispersion direction of methane vapor, and wind speed affects the dispersion speed of methane vapor. Because the wind is affected by the surface conditions and atmospheric temperature, the change of wind speed gradient caused by atmospheric height should be considered in the actual engineering calculation. Wind speed profile exponential equation is usually used to describe wind speed gradient changes in the environment, as shown in Eq. (18).

$$u_z = u_{10} \left(\frac{Z}{10} \right)^{0.091} \tag{18}$$

3.2.2.2　Boundary conditions and parameter setting for wind field calculation

(1) Boundary condition setting

TheFluent's solution process is that the data is extended from the boundary or boundary surface to the whole calculation region. Reasonable boundary conditions are the key to ensure the accuracy and correctness of the model. The setting of boundary conditions should conform to the actual situation of simulation.

When the wind field in the whole calculation domain is simulated, the boundary conditions are set as shown in Table 1.

Table 1 LNG storage tank leakage and dispersion calculation area boundary types

Boundary name	Boundary type	Boundary name	Boundary type
The left side of computing domain(wind inlet)	VELOCITY_INLET	Tank wall	WALL
Leakage	WALL	The computational domain top	SYMMETRY
Ground	WALL	The front side of computational field	SYMMETRY
The right side of the computational domain(outlet)	OUT_FLOW	The backside of the computational field	SYMMETRY

(2) Pressure-velocity coupling

After the windfield is stable, the transient calculation method is used to solve the velocity distribution in the calculation domain, and SIMPLE can provide a more conservative computational convergence. Therefore, the SIMPLE algorithm is used as the pressure and velocity coupling method of the wind field.

3.2.2.3 Boundary conditions and solution parameters of LNG evaporation and dispersion

(1) Boundary condition setting

The LNG storage tank leakage dispersion boundary condition setting only needs to change the leakage portin the boundary type of the wind field calculation area, as shown in Table 1. the mass flow is 436.6 kg/m^3, and the other boundary types remain the same.

(2) LNG material definition

LNG is a liquid mixture, and the meaning does not exist in the material library. Therefore, the user needs to define it by himself. When determining the material, the relevant physical and chemical property data of the substance is required. The basic parameters of LNG materials are shown in Table 2.

Table 2 LNG physical and chemical parameters

Physical and chemical parameters	Numerical value	Physical and chemical parameters	Numerical value
Density(kg/m^3)	430	Thermal conductivity[W/(m·K)]	0.21
Kinematic viscosity[kg/(m·s)]	0.0001183	Saturated vapor pressure(Pa)	Piecewise function representation
Latent heat of vaporization(J/kg)	509332	Heat release during pyrolysis(J/kg)	0
Vaporization temperature(K)	90.7	The surface tension of the droplet(N/m)	0.0133
Specific heat capacity[J/(kg·K)]	2055	Boiling point(K)	111.66

Note: The saturated vapor pressure in the table is a value that changes with temperature, which is defined by a piecewise linear function. As shown in Table 3.

Table 3 LNG saturated vapor pressure piecewise linear function setting

Split point	1	2	3	4	5	6
Temperature(K)	90.7	92	94	96	98	100
Saturated vapor(Pa)	11719	13853	17679	22314	27877	34495
Split point	7	8	9	10	11	12
Temperature(K)	102	104	106	108	110	111.6
Saturated vapor(Pa)	42302	51441	62063	74324	88389	101325

Define the properties of the LNG material and write the function file by writing the Fluent database file. Place it in the Fluent working directory, the system will automatically load this database, and new materials can be provided.

(3) The setting of solution model

The LNG release and dispersion process involve three-phase air, LNG and methane, and this paper chooses a Mixture model to solve the entire leakage dispersion process. The transient calculation is used to simulate the leakage dispersion, to observe the situation of LNG leakage dispersion at any moment. PISO algorithm is used for pressure and velocity coupling, which is suitable for transient calculation, ensuring the convergence of the model and speeding up the calculation speed.

3.3 Pool fire model of LNG tank

3.3.1 Determine the calculation domain of pool fire

According to the tank farm layout in 3.2.1, a fire simulation model of the LNG receiving station tank farm is established in Pyrosim. The model is shown in Fig. 5. The leakage port is located on the wall of #4 tank which is 6 m above the ground and is closest to #5 tank. Liquid LNG flows out from the leak port to the bottom and generates a liquid pool. Since the temperature of the liquid LNG is very low and the outside temperature forms a large temperature difference, the liquid pool itself will continue to evaporate due to heat conduction. The radius of the ground liquid pool reaches a stable maximum when the release mass rate is equal to the evaporation rate of the liquid pool. The model in this paper selects the largest liquid pool when it reaches a steady-state and this liquid pool is used in the ignition source for fire analysis.

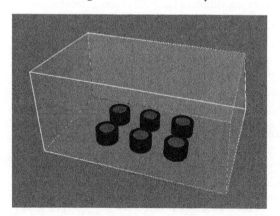

Fig. 5 Tank farm model of the LNG receiving station

3.3.2 The LNG tank fire simulation parameters

(1) Fire atmospheric conditions setting

The atmospheric conditions are set by the simulation settings of gas release and dispersion, and the environmental conditions of LNG tank are set according to the actual conditions. As shown in Table 4.

Table 4 Setting of atmospheric environment parameters of the LNG tank farm

Wind speed(m/s)	Wind direction	Temperature gradient(℃/m)	Atmospheric stability	Ambient temperature(K)
5	S	0.015	better	300

(2) Reaction setting

LNG mainly contains methane, and it is used as a reactant to participate in the fire reaction.

(3) Mesh division and distribution of measuring points

The boundary size of the computational domain is determined to be 640m×400m×400m by preliminary calculation, taking into account the size of the flame and its influence range. The appropriate grid size is determined through many calculations to ensure numerical calculation prediction. The result is independent of the number of grids, considering the computational efficiency of numerical simulation. Finally, the total number of grids in the model is determined to be 2 660 000.

In simulation, we are concerned about the distribution of temperature and thermal radiation near the fire, monitoring points are set at some positions of adjacent tanks to detect the real-time thermal radiation, and the corresponding temperature field distribution is obtained by analyzing the section of calculation domain. Fig. 6 shows the slices of the LNG tank fire simulation calculation and the distribution of monitoring points.

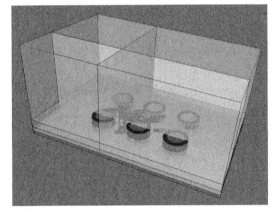

Fig. 6 Grid division of the LNG storage tank fire simulation and distribution of monitoring points

4 Results and discussion

4.1 Calculation results of release

It is assumed that the leak is located at the height of 6m above the ground. According to the relevant data on accidents in the chemical industry from 1949 to 1988 in the "Analysis and Prevention of Chemical Equipment Accidents", the general continuous leak has a small aperture of 100mm and a large aperture of 200mm. In this example, 200mm large apertures are selected as the leakage apertures. The leakage rate, the maximum radius of the liquid tank, the time it takes for the liquid pool to reach the maximum radius, and the evaporation rate caused by heat conduction is calculated as shown in Table 5 when the evaporation time is 100s.

Table 5 Leak calculation results

Leakage aperture (mm)	Leakage rate (kg/s)	Maximum radius time(s)	Maximum liquid pool radius(m)	Evaporation rate(kg/s)
200	436.6	65.3	41.8	675.2

4.2 Results of wind field and dispersion process

4.2.1 Analysis of wind field in tank farm

Fig. 7 shows the cloud map of wind speed distribution in the LNG tank farm. The initial ambient wind speed is 5m/s. From the horizontal wind speed distributed cloud image, we can see the wind speed distribution at different horizontal positions 35m above the ground. From the horizontal wind speed distributed cloud image, it can find that the wind field skips both sides of the storage tank at a relatively high speed, because the wind is in the arc tank of the storage tank. The flow around the wall makes the wind speed on both sides of the tank wall reach about 8m/s. The vertical wind speed distribution cloud chart shows the wind speed at different heights in the vertical direction. Due to the positive blocking effect of the storage tank, the wind disturbance and swirling flow appear between the two storage tanks, resulting in the obvious weakening of the wind field between the two storage tanks. The wind speed was only 1m/s, and it even appeared in some locations. The air retention phenomenon causes the wind speed to be 0m/s. Because wind also circulates at the arc-shaped top of the storage tank, the wind speed in a large area above the top of the storage tank is relatively high, reaching about 8m/s. There is no obstacle after the most downstream storage tank, so the low wind speed area in the downwind direction of the storage tank is extended.

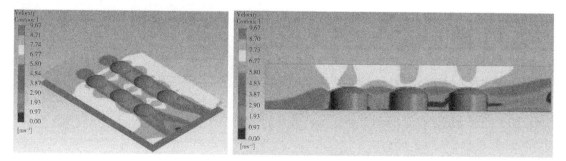

Fig. 7 Horizontal and vertical distribution of wind speed in LNG tank farm

4.2.2 Analysis of LNG evaporation and dispersion process

The hazards of LNG leakage and dispersion are mainly determined by the hazards of its combustion and explosion. The potential hazard of LNG combustion and explosion mainly depends on the dispersion range of LNG vapor cloud leakage within the explosion limit. Once the methane air mixed vapor cloud within the explosion limit meets the ignition source, it is likely to cause fire and explosion accidents, which will harm the surrounding personnel and equipment. The harm scope of vapor cloud reaching explosion limit is analyzed and studied, and scientific suggestions are put forward to reduce the harm of LNG dispersion leakage.

Fig. 8(a—f) is the distribution map of methane vapor cloud with a volume concentration of 5% at different times of LNG leakage. The dispersion behavior of methane LEL (Lower Explosion Limited) vapor cloud at the lower explosion limit can be obtained. During the expansion process of methane vapor cloud, due to the disturbing effect of wind, it is easy to accumulate and form vapor cloud in areas with relatively slow wind speed. Due to the barrier effect of the storage tanks, vapor cloud accumulation occurred in the space between #5 and #6 storage tanks. With the increase of the accumulation volume of the vapor cloud, the position of the vapor cloud slowly increased and moved towards #6 The top of the storage tank spreads gradually. Due to the high density of 5% vapor clouds, it is more susceptible to gravity to make the horizontal and vertical dispersion distances shorter than those of low-concentration vapor clouds. As shown in Fig. 8(e), when $t=260$s, It can be seen that the range of vapor cloud in the top area of #5 storage tank is small, and most of the vapor cloud is concentrated near the ground and the wall of #5 storage tank; a large number of vapor cloud is accumulated in the middle area of #5 and #6 storage tanks, and with the accumulation of vapor cloud, the vapor cloud is gradually lifted to the top area of #6 storage tank, and slowly diffuses to the top of #6 storage tank under the action of wind; and #6 storage tank is near the ground There's a vapor cloud nearby.

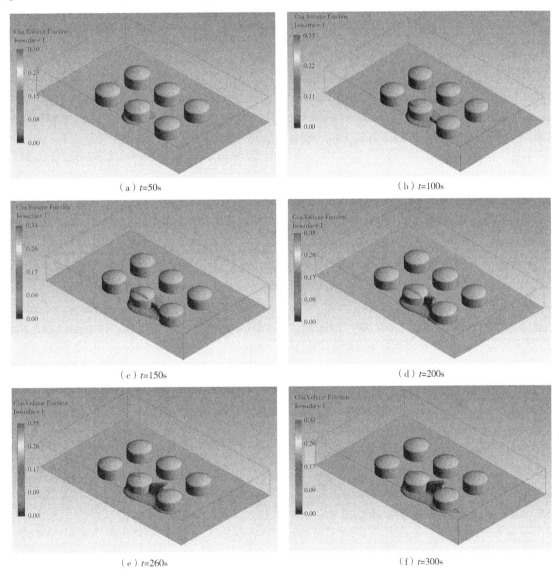

Fig. 8 Surface of LNG vapor cloud with a volume concentration of 5%

Fig. 9 shows the change trend of methane vapor cloud area with 5% concentration over time. With the extension of the leakage time, the area of the explosive vapor cloud continues to expand. At about 300s, the dispersion area of vapor cloud in the explosion limit is stable.

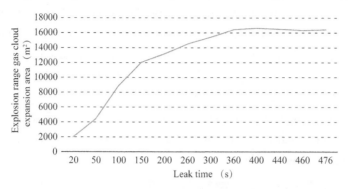

Fig. 9 Expansion area of LNG storage tank leaking and exploding vapor cloud

To study the distribution of vapor cloud concentration near the ground, Fig. 10(a—f) shows the distribution of methane vapor cloud near ground concentration in different time periods. The vapor cloud within the explosion limit is close to the ground, centered on the leakage, and continuously spread in the downwind direction. The concentration of methane gas at the center is darker and more concentrated, and the concentration gradually decreases toward the two sides. In the early stage of the leak, due to the low temperature and high density of the vapor cloud, gravity was the main driving force during the dispersion process. The gas cloud was dominated by lateral distribution, and the vapor cloud was flat and wide. With the mixing of air, the density of vapor cloud gradually decreases, and the atmospheric turbulence gradually replaces the sedimentation of gravity and becomes the main driving force of gas cloud dispersion. As shown in Fig. 10(f), when $t=300$s, the dispersion is stable, and it can be seen that the large area near the ground and between the two storage tanks of #5 and #6 tanks are within the explosion limit. If this area encounters an open fire, it will cause an explosion accident, and there are often frequent personnel activities and more process equipment near the ground. Therefore, the danger level in this area is high, and the personnel in this area should be given priority to evacuate as soon as possible in case of leakage.

As shown in Fig. 11, in the early stage of the leak, the vapor cloud within the explosion limit was mainly concentrated in the range of 0~40m from the ground. With the disturbance of the air, it began to spread in the downward wind direction. The tank wall surface of tank #5 upwind and tank #6 downwind near the ground and a large area between tanks #5 and #6 have been in the explosion limit range. Therefore, this area is a dangerous area that endangers people's lives and the safety of related equipment and facilities if the explosion occurs and will become the secondary hazard unit of the domino effect of the leakage accident.

4.3 Calculation results of pool fire

4.3.1 Flame development law of tank side leakage pool fire

As shown in Fig. 12(a), the LNG leaked to the ground to form a stable liquid pool and was sightedthe liquid pool was burning violently for about 5s. The flame became lenient with its dispersion, and the fire began to spread upward rapidly.

As shown in Fig. 12(b), the flame's height has exceeded the storage tank after burning for about 10s. The fire is a stage of rapid development at this moment. The blaze below the storage tank

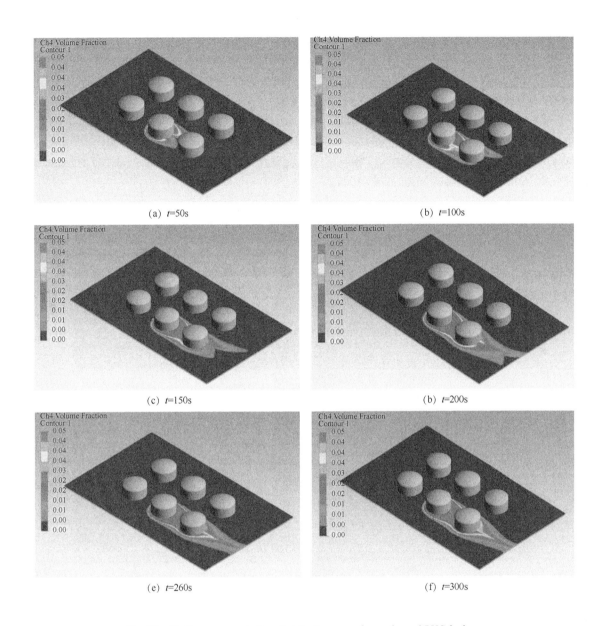

Fig. 10 Methane concentration distribution near the surface of LNG leakage

Fig. 11 Longitudinal distribution of gas cloud with 5% concentration
of LNG storage tank leaking 300s

height will no longer expand to the surrounding due to the obstruction of the tank wall, and the fire above the highness of the storage tank is blocked by no obstacles around it and is insufficient contact with the air. Under the airflow action, the methane diffuses to the surroundings to make the flame lenient, and the blaze is like a mushroom cloud.

As shown in Fig. 12(c—d), the combustion has beensteady in about 20s, forming a stable fire column, and the flame direction is upright. As a result of the combined action of the air generated by the flame, the flame is twisted, showing a clear S shape. There was no significant change in the fire intensity until about 100s.

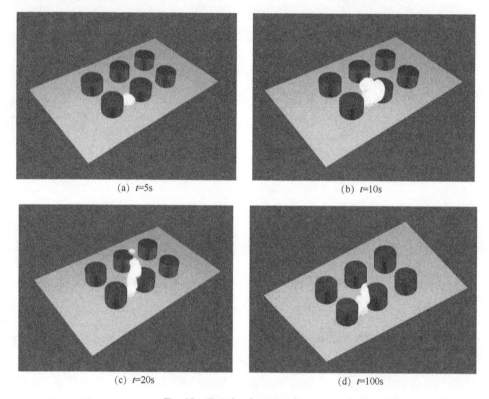

Fig. 12 Fire development diagram

4.3.2 Temperature field analysis of LNG tank fire

High temperature is one of theleading evaluation indicators to measure the scope of fire damage. After the fire, the temperature of the flame flow area rises rapidly, which may cause specific damage to the LNG reinforced concrete structure, other equipment, and related operators.

Fig. 13 shows the temperature field distribution when the fire is fully developed, and the maximum temperature of the fire-affected area is about 670℃. According to the flue gas flame temperature on reinforced concrete structure and human body impact criteria. At this temperature, the stucco layer of the LNG storage tank has all fallen off, causing cracks, the protective layer fell off, and the steel bar leaked out, and the steel lost all its strength. It can be known from the vertical section that the high-temperature area of the fire can reach up to about 150m in the vertical direction, and the high-temperature area of the fire can be known from the horizontal section concentrated in the lower half of the area between #4 and #5 storage tanks. The influence range of fire temperature near the bottom is mainly in the wall of #4 storage tank and #5 storage tank and its middle area. People who activities on the ground should stay away from this area as much as possible to avoid burns.

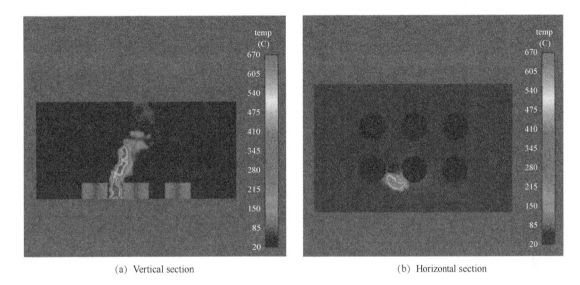

(a) Vertical section (b) Horizontal section

Fig. 13 Cross-sectional temperature field when the fire is fully developed

4.3.3 Thermal radiation analysis of LNG tank fire

Thermal radiation is a factor that has amore significant impact on personnel, equipment, and rescue. Analyzing the scope of radiation caused by fire can provide scientific suggestions for avoiding thermal radiation damage. Fig. 14 shows the change of heat radiation intensity of #4 and #5 tank walls affected by the fire with time. The radiation intensity fluctuates over time since the fire is in dynamic.

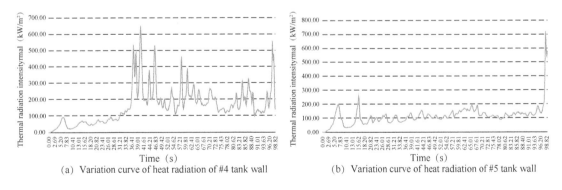

(a) Variation curve of heat radiation of #4 tank wall (b) Variation curve of heat radiation of #5 tank wall

Fig. 14 Time variation of thermal radiation from tank wall of LNG storage tank

(1) The influence of heat radiation on the wall of #4 storage tank:

Afterthe fire broke out, the fire kept expanding, and the thermal radiation flux of the tank wall near the fire source began to rise. About 7s after the fire, the maximum thermal radiation flux of the tank wall is about $90kW/m^2$. After that, the heat radiation flux changed with time at high frequency, reaching the maximum value of 624kW/m² in about 41s, and then began to fluctuate at high frequency with the general trend decreasing. At about 99s, the flame started to offset, so the heat radiation flux at this time suddenly oscillated and surged to 536.5kW/m². After that, the flame direction changed little, and the change of thermal radiation flux has been flattening out. According to the guidelines for the damage of fire thermal radiation to the human body and equipment, the radiation intensity will have a catastrophic effect on the storage tank. A long-term fire may cause the collapse of the #4 storage tank, which will increase the fire and cause. It may cause a fire in the surrounding storage tanks, cause a domino effect, and cause a large number of casualties, forming a catastrophic accident.

(2) The influence of heat radiation on the wall of #5 storage tank:

After the fire broke out, the fire continued to expand, and the thermal radiation flux of the tank wall near the fire source began to rise rapidly. The fire occurred for about 7s, and the maximum thermal radiation flux at the tank wall was about 194.5kW/m². Then there was a small peak at about 16s, and the thermal radiation flux was about 251.4kW/m². Then the change of heat radiation flux tends to be flat. After that, the heat radiation flux began to rapidly decrease to about kilowatts per square meter. After that, the heat radiation flux began to decrease rapidly to about 65kW/m², and the change of heat radiation flux tended to be flat. According to the damage criterion of heat radiation flux to the human body and equipment, steel materials such as the #5 storage tank and surrounding pipelines will be seriously damaged under this heat radiation intensity. If the fire lasts for more than 30min, it may cause the steel structure to collapse. People will die under the action of this heat radiation for more than 1min.

5 Conclusion

This paper builds a large-scale LNG storage tank leak chain catastrophe consequence assessment model, reveals the disaster mechanism of the LNG phase change dispersion cascade pool fire caused by the storage tank leakage, analyzes the characteristics of the LNG leakage chain disaster, obtains the impact of LNG leakage on personnel, equipment, and the impact of environmental safety. However, the purpose of this study is to present the cascading disaster due to LNG tank leak. The effect of environmental conditions on dispersion is not included in the scope of this study.

When leakage lasts for about 300s, the dispersion reaches a stable state. The flammable gas cloud coverage area is maintained at about 16250m². Fire accidents are prone to occur in the case of ignition sources, and a domino effect is formed. Damage to other storage tanks, related equipment, and operators; The LNG leaked to the ground to form a stable liquid pool and was ignited. The liquid pool burned vigorously in about 5s, and the flame spread to the surrounding area. The flame height was higher than the storage tank in about 10s, and the fire was mushroom-shaped. The combustion has been stable in about 20s, forming a stable pillar of fire; the flame covers the whole tank top area in about 5s. The maximum temperature of the fire-affected area reached about 670℃. The high-temperature area can reach a maximum of about 150m in the vertical direction. The maximum thermal radiation of fire to the #4 storage tank is 624kW/m², and that to the #5 storage tank is 251.4kW/m². The thermal radiation of #4 and #5 storage tank is greater than the maximum injury index. Long time action will cause storage tank collapse and cause a number of casualties.

Abbreviations

CFD	Computational fluid dynamics
FDS	Fire dynamics
LNG	Liquefied natural gas
LES	Large eddy simulation
LEL	Lower explosion limited
Q_L	The leakage mass flow
A	The hole area
C_d	The liquid leakage coefficient
ρ_1	The liquid density in the storage tank
h	The height of the leakage location from the fluid level in the tank
p_t	The pressure in the storagetank
p_0	The pressure of the external environment
r	The radius of the liquid pool
t	Time
Q	The mass flow rate of the liquid that leaks to the ground to form the liquid pool
ρ_2	The density of the liquid LNG
k	The thermal conductivity of the substance
B	The area of the liquid pool
T	The ambient temperature
T_0	The boiling point of the liquid under normal pressure
H	The latent heat of vaporization of the liquid
α	The thermal dispersion coefficient
ρ	The density
φ	The general variable
Γ	The dispersion coefficient
S	The source term
G_k	The turbulent energy term due to the existence of the velocity gradient
G_b	The turbulent energy item due to buoyancy
g_i	The component of gravity in the direction
β	The coefficient of thermal expansion
p_{γ_i}	The prandtl number
γ_i	The mass content of substance i
$\overline{J_j}$	The mass dispersion rate of the substance
T_{mix}	The unit temperature of the mixing zone
T_{sat}	The saturation temperature
$m_{l \to v}$	The phase change rate of the liquidphase into the gas phase

$m_{v \to l}$	The phase change rate of the gas phase into the liquid phase
b	A factor that controls the intensity of the phase transition
α_l	The surface tensions of the liquid phases
α_v	The surface tensions of the gas phases
ρ_l	The density of liquid phase
ρ_v	The density of gas phase
ρ_g	The gas density
u	The velocity vector
p	Pressure
f	The external force vector
τ_{ij}	The viscosity vector
h_e	The enthalpy value of leakage gas components
q	The radiant heat flux
φ	The dissipation rate
Y_i	The mass fraction of the i-th component of the leakage gas
D_i	The dispersion coefficient of the i-th component

Acknowledgements

The financial supports from the National "Thirteenth Five-Year Plan" national key research plan(2016YFC0801200).

References

[1] KOOPMAN R P, BAKER J, CEDERWALL R T, et al., LLNL/NWC 1980 LNG Spill Tests. Burro Series Data Report[R]. UCID-19075-Vol. 1, Lawrence Livermore National Lab., Livermore, CA, USA, 1982.

[2] HIRST W J S, EYRE J A. Maplin Sands experiments 1980: combustion of large LNG and refrigerated liquid propanespills on the sea[M]. Heavy Gas and Risk Assessment-II, Springer, Dordrecht, Netherlands, 1983: 211-224.

[3] BROWN T C, CEDERWALL R T, CHAN S T, et al. Falcon series data report: 1987 LNG vapor barrier verification field trials [R]. UCRL - CR - 104316; GRI - 89/0138, Lawrence Livermore National Lab, Livermore, CA, USA, 1990.

[4] KONIG-LANGLO G, SCHATZMANN M. Wind tunnel modeling of heavy gas dispersion[J]. Atmospheric Environment Part A. General Topics, 1991, 25(7): 1189-1198.

[5] NEFF D E, MERONEY R N. Behavior of LNG vapor clouds: wind-tunnel tests on the modeling of heavy plume dispersion[D]. Colorado State University, Fort Collins, CO, USA, 1982.

[6] CALAY R K, HOLDO A E. Modelling the dispersion offlashing jets using CFD[J]. Journal of Hazardous Materials, 2008, 154(1-3): 1198-1209.

[7] GUO D, ZHAO P, WANG R, et al. Numerical simulation studies of the effect of atmospheric stratification on the dispersion of LNG vapor released from the top of a storage tank[J]. Journal of Loss Prevention in the Process Industries, 2019: 275-286.

[8] LUO T, YU C, LIU R, et al. Numerical simulation of LNG release and dispersion using a multiphase CFD model[J]. Journal of Loss Prevention in the Process Industries, 2018, 56: 316-327.

[9] SALEEM A, FAROOQ S, KARIMI I A, et al. A CFD simulation study of boiling mechanism and BOG

generation in a full-scale LNG storage tank[J]. Computers & Chemical Engineering, 2018, 115: 112-120.

[10] ZHU D. Example of simulating analysis on LNG leakage and dispersion[J]. Procedia engineering, 2014, 71: 220-229.

[11] QI R, NG D, CORMIER B R, et al. Numerical simulations of LNG vapor dispersion in brayton fire training field tests with ANSYS CFX[J]. Journal of Hazardous Materials, 2010, 183(1-3): 51-61.

[12] ZHUANG X Q. Numerical Simulation for LNG Release & Dispersion from Large Scale Tank[D]. Wuhan: Wuhan University of Technology, 2012.

[13] ZHANG C. Numerical Simulation of Vertical Natural Gas Storage Tank Release and Dispersion[D]. Nanchang: Jiangxi University of Technology, 2015.

[14] JIANG F Q. Simulation Research on the Influence of Dike Dam on LNG Storage Tank Release and Dispersion [D]. Wuhan: Huazhong University of Science & Technology, 2016.

[15] ZHANG B, YU G F, WU W Q, et al. Numerical simulation on LNG spilling and dispersion[J]. Journal of Dalian Maritime University, 2013, 39(2): 99-102.

[16] ZHOU N, CHEN L, LV X F, et al. Analysis on influencing factors of LNG continuous release & dispersion process from large scale tank[J]. Chemical Industry and Engineering Progress, 2019, 38(10): 4423-4436.

[17] YANG Z J, HOU L, ZHU M. Research on leakage and diffusion of a large LNG storage tank and its influencing factors[J]. Natural gas and oil, 2020, 38(1): 47-53.

[18] LI S, CHEN B D, ZHANG Z T, et al. Numerical analysis and research about the thermal response of the LNG storage tank in the fire environment[J]. Journal of Petrochemical Universities, 2011: 24(1): 78-81.

[19] SUN B, GUO K, PAREEK V K. Computational fluid dynamics simulation of LNG pool fire radiation for hazard analysis[J]. Journal of Loss Prevention in the Process Industries, 2014, 29: 92-102.

[20] XIE F, SONG W, CHEN Z, et al. Study on pool fire model applied to the fire risk assessment of dichloropropane storage tank farm[J]. Acta Scientiarum Naturalium Universitatis Nankaiensis, 2012: 45(3): 100-106.

[21] DI J H, CHEN F Q. FDS software simulation for fire consequences of LNG storage tank leaks[J]. Oil & Gas Storage and Transportation, 2013, 32(1): 70-77.

[22] SUN B, GUO K, PAREEK V K. Dynamic simulation of hazard analysis of radiations from LNG pool fire [J]. Journal of Loss Prevention in the Process Industries, 2015, 35: 200-210.

[23] YU Z D, WU J L. Consequence analysis of leakage accident of large LNG storage tank[J]. Journal of Guangdong University of Petrochemical Technology, 2016, 4: 90-94.

[24] ZHANG W D, ZHANG Y X. Simulation of leakage accident of large LNG storage tank based on PHAST software[J]. Petrochemical Safety and Environmental Protection Technology, 2014, 30(5): 27-31.

[25] CHEN G H, CHENG S B. Consequence simulation on LNG leakage accidents and its quantitative risk assessment[J]. Natural Gas Industry, 2007, 27(6): 133-135.

[26] ZHOU D H, FENG H, LI W, et al. Application of fire and explosion effect analysis method in risk assessment for LNG storage tank [J]. Industrial Safety and Environmental Protection, 2015, 41(10): 54-56.

[27] BAALISAMPANG T, ABBASSI R, GARANIYA V, et al. Modelling an integrated impact of fire, explosion and combustion products during transitional 14 mathematical problems in engineering events caused by an accidental release of LNG[J]. Process Safety and Environmental Protection, 2019, 128: 259-272.

[28] PIO G, CARBONI M, IANNACCONE T, et la. Numerical simulation of small-scale pool fires of LNG [J]. Journal of Loss Prevention in the Process Industries, 2019, 61: 82-88.

[29] LV D, TAN W, LIU L, et al. Research on maximum explosion overpressure in LNG storage tank areas [J]. Journal of Loss Prevention in the Process Industries, 2017, 49: 162-170.

[30] JUJULY M, RAHMAN A, AHMED S, et al. LNG pool fire simulation for domino effect analysis[J]. Reliability Engineering & System Safety, 2015, 143: 19-29.
[31] WEI T T. Quantitative evaluation on the fire and explosion accident of LNG leakage[J]. Industrial Safety and Environmental Protection, 2013, 39(8): 56-59.

本论文原发表于《Mathematical Problems in Engineering》2021年第2021卷。

Risk Assessment of Large Crude Oil Tanks Based on Fuzzy Comprehensive Evaluation Method

Wu Gang[1]　Li Lifeng[1]　Zhou Huiping[2]　Jiang Jinxu[3]
Luo Jinheng[1]　Duan Qingquan[3]　Zhang Shuxin[1]

(1. State Key Laboratory of Performance and Structural Safety for Petroleum Tubular Goods and Equipment Materials, Tubular Goods Research Institute of CNPC;
2. Petrochina West Pipeline Company; 3. College of Safety and Ocean Engineering, China University of Petroleum(Beijing))

Abstract: The large scale of crude oil tank may face huge risks during operation due to the effect of dangerous and harmful factors. The Leakage and fire accidents of tanks are easy to happen which causes serious environmental pollution and even casualties. In order to ensure the safe operation of storage tanks, the analysis of storage tank accidents was conducted and the tank failure database was established based on the field survey results. The statistical results revealed that fire and explosion are the main causes of tank failure. The risk evaluation index system was derived by the statistical analysis of tank failure accidents. The risk assessment of large crude oil tanks based on fuzzy comprehensive evaluation method was performed. The evaluation result shows that the level of safety evaluation for the crude oil tank area is relatively safe. The failure statistics of crude oil tanks and risk assessment of tank area can provide certain reference for safe maintenance of oil tanks.

1　Introduction

With the rapid growth of oil demand, the need of storage devices has increased. As an important storage equipment, large scale oil tanks have been widely used in recent decades. Fracture and corrosion failure of the crude oil tank may occur due to long-term service, which seriously affects the safe operation of storage tanks. It is essential to carry out inspection and maintenance of large scale tanks. In order to determine the reasonable maintenance period and reduce maintenance costs, risk assessment of large crude oil tanks is important and necessary.

In recent decades, numerous investigations on risk assessment of large scale crude oil tanks have been carried out. Landucci et al.[1] adopted a simplified model for the estimation of the vessel

---

Corresponding author: Wu Gang, dqq@ cup. edu. cn.

time to failure with respect to the radiation intensity on the vessel shell to develop the quantitative assessment of the risk caused by escalation scenarios triggered by fire. Kang et al. [2] presents a new risk evaluation model for oil storage tank zones based on Fault Tree Analysis (FTA), Analytic Hierarchy Process (AHP) and multi-factor fuzzy evaluation matrix. Wei et al. [3] proposed a quantitative methodology for the risk assessment of direct lightning strike on external floating roof tank, in which the risk-attenuating factors are considered by three special sub-models. Guo et al. [4] proposes an improved SAM based FBN model to better deal with various types of uncertainty, which makes the prediction results of the storage tank accident more accurate and reliable. Sun et al. [5] conducted the risk assessment and early warning for large crude oil storage tanks based on extension theory. Zong et al. [6] explore the terminal risk evaluation method of quantitative analysis and develop the risk evaluation system. Proper suggestions for safety management were given based on the quantitative risk assessment results. Yang et al. [7] proposed a new method of risk assessment for the leakage accident based on BP neural network and conducted a risk assessment model of benzene tank based on the proposed method.

As mentioned above, documented investigations were mainly carried out based on risk assessment model. To a certain extent, the main failure mode and failure statistics of storage tanks may not detailed enough. The evaluation indicators of assessment models may not reflect the actual service conditions of storage tanks. In this study, the failure statistics of crude oil storage tanks was obtained. Failure database of tanks was established which can provide an effective way for the analysis of storage tank accidents. Based on site investigation data and statistical results of tank failure, fuzzy comprehensive evaluation model was developed for risk assessment of crude oil tanks. Research results of this paper can be referenced for safe maintenance of storage tank area.

2 Statistical analysis ofcrude oil tanks failure

2.1 Analysis of storage tank accidents and failure

Statistical analysis of storage tank failure accidents can help to effectively identify the hazard factors of tanks, which is the basis for the risk assessment of large scale tanks. In this study, accident causes and failure modes of storage tanks have been analysed via site investigation of Site investigation of Shanshan Petroleum Reserve Base and Wangjiagou Crude Oil Depot. As can be observed from Fig. 1, the corrosion of storage tank floating roof and reinforcement ring have occurred. Meanwhile, statistical analysis of 112 crude oil storage tank accidents reveal that the accident causes of large scale oil tanks may include the following eight categories, such as fire, explosion, oil spill, suffocation, poisoning, leakage, collapse and fall. Storage tank accidents are mostly general accidents. As shown in Fig. 2, the main cause of storage tank accidents is fire and explosion.

The common failure modes of crude oil storage tanks include fracture, instability and corrosion perforation. The fracture of tanks is mainly caused by the failure of material and third-party damage. The storage tank cracks can be easily found in welding seam of tank wall or bottom plate. Weld is the weak part of tanks. The stress corrosion may easily occur in weld of tanks due to the corrosiveness of crude oil and the stress state at the weld. Corrosion is one of the important causes of tank failure, which may occur in different parts of tanks due to the effect of the external environment and the medium in tank. The instability of tank shell is mainly caused by the overall inclined slippage due to external pressure. While, the instability of tank roof is mainly local collapse.

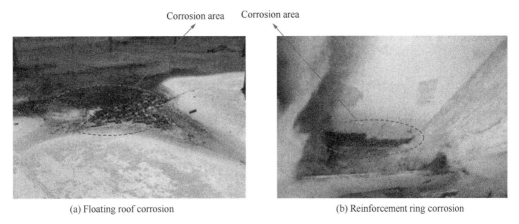

Fig. 1 The corrosion failure of crude oil tank

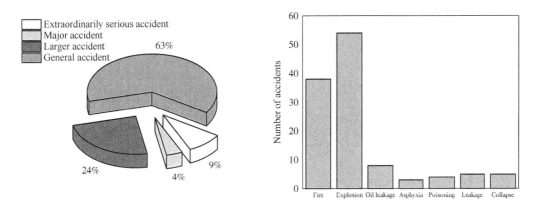

Fig. 2 Crude oil storage tank failure statistics

2.2 Establishment of crude oil tank failure database

In order to effectively carry out statistical analysis of storage tank accidents, the crude oil tank failure database was established in this study. Based on the statistics and analysis results of accidents of steel tanks in the previous 30 years, the name, location and time of accidents has been written into the database. Comparative analysis of various accidents of tanks can be conducted based on crude oil tank failure database. The development trend of storage tank accidents can be revealed which can be adopted to effectively predict the occurrence of accidents and achieve the purpose of reducing the probability of tank failure.

3 Fuzzy comprehensive evaluation model

Fuzzy comprehensive evaluation model is quantitative risk assessment method based on fuzzy mathematics and principle of maximum membership degree. The evaluation method can provide a better way for the vague and difficult to quantify issues. The safety status of crude oil storage tanks is affected by a variety of uncertain factors. Thus, fuzzy comprehensive evaluation model is an effective method for the risk assessment of tanks.

3.1 Establish the risk evaluation index system

Based on the statistical analysis of tank failure accidents, it is concluded that the main accident of crude oil tanks is fire and explosion. Fig. 3 presents the statistical results of the accident cause.

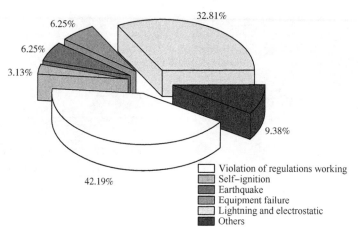

Fig. 3 Statistical results of tank fire and explosion accidents cause

The dangerous source of crude oil storage tanks mainly include physical and chemical properties of crude oil, operation and management and storage tank equipment and environment. From the results illustrated in the Table 1, several main risk factors for the failure of large crude oil storage tanks can be obtained.

Table 1 Accident cause and risk factors

Accident cause	Risk factors
Violation of regulations working	Management factors
Self-ignition	Environmental and management factors
Earthquake	Environmental factor
Equipment failure	Equipment and management factors
Lightning and electrostatic	Environmental and equipment factors

The risk evaluation index system was established in this section. As can be observed from Fig. 4, the first grade indexes of index system includes environment and surrounding conditions, design and auxiliary facilities of tanks, personnel organization and training, management and emergency response. Each first grade indexes includes the relevant second index.

3.2 Calculation of index weight

The Analytic Hierarchy Process (AHP) is usually adopted to achieve the weight distribution of multihierarchy evaluation, which can provide simple decision - making methods for complex decision-making problems with multiple goals and multiple criteria. In this study, the AHP system assigns the indexes into 3 layers: goal layer (objective level), criterion layer (first-level) and index layer (second-level). The results of risk evaluation index weight are shown in Table 2.

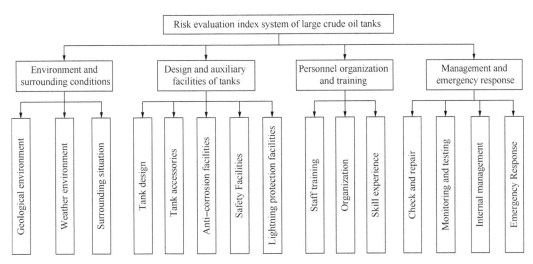

Fig. 4 Risk evaluation index system of crude oil tank

Table 2 Risk evaluation index weight

First grade indexes	Index weight	Second indexes	Index weight
Environment and surrounding conditions	0.5810	Geological environment	0.655
		Weather environment	0.250
		Surrounding situation	0.095
Design and auxiliary facilities of tanks	0.0560	Tank design	0.236
		Tank accessories	0.090
		Anti-corrosion facilities	0.146
		Safety facilities	0.382
		Lightning protection facilities	0.146
Personnel organization and training	0.1210	Staff training	0.500
		Organization	0.191
		Skill experience	0.309
Management and emergency response	0.2429	Check and repair	0.250
		Monitoring and testing	0.250
		Internal management	0.095
		Emergency response	0.405

3.3 Fuzzy comprehensive evaluation

In this study, the risk indicators of crude oil tank are divided into 5 types: very safe, slight security, comments security, slight insecurity, insecurity. There are five experts to score for 15 factors. The score values varies from 1 to 5. The index factor membership of storage tank is shown in Table 3. The index factor membership can be obtained by statistical analysis of expert scoring results.

$$V = \{\text{very safe, slight security, security, slight insecurity, insecurity}\}$$

Table 3 Index factor membership of storage tank

Serial number	Index	Comments				
		Very safe	Slight security	Security	Slight insecurity	Insecurity
U_{11}	Geological environment	0.2	0.6	0.2	0	0
U_{12}	Weather environment	0	0.2	0.2	0.6	0
U_{13}	Surrounding situation	0	0.2	0.6	0.2	0
U_{21}	Tank design	0.6	0.4	0	0	0
U_{22}	Tank accessories	0.2	0.6	0.2	0	0
U_{23}	Anti-corrosion facilities	0	0.2	0.2	0.6	0
U_{24}	Safety facilities	0	0	0.6	0.4	0
U_{25}	Lightning protection facilities	0.4	0.6	0	0	0
U_{31}	Staff training	0	0.6	0.4	0	0
U_{32}	Organization	0	0	0.6	0.4	0
U_{33}	Skill experience	0.4	0.6	0	0	0
U_{41}	Check and repair	0	0.2	0.6	0.4	0
U_{42}	Monitoring and testing	0	0.2	0.6	0.2	0
U_{43}	Internal management	0	0.2	0.4	0.4	0
U_{44}	Emergency response	0	0	0	0.4	0.6

Fuzzy comprehensive evaluation of environment and surrounding conditions was conducted. The index weight set W_1 can be obtained from Table 2. The fuzzy evaluation matrix R_1 can be determined based on Table 3.

$$W_1 = [0.6555 \quad 0.0796 \quad 0.2648]$$

$$R_1 = \begin{bmatrix} 0.2 & 0.6 & 0.2 & 0 & 0 \\ 0 & 0.2 & 0.2 & 0.6 & 0 \\ 0 & 0.2 & 0.6 & 0.2 & 0 \end{bmatrix}$$

The fuzzy comprehensive evaluation results of environment and surrounding factors are as follows:

$$B_1 = R_1 \times W_1 = [0.1311 \quad 0.4622 \quad 0.3059 \quad 0.1007 \quad 0]$$

Fuzzy comprehensive evaluation matrix of tank area can be decided by evaluation results of each single factor. The index weight set W of rule hierarchy can be obtained from Table 2.

$$W = [0.0560 \quad 0.5810 \quad 0.1201 \quad 0.2429]$$

$$R = \begin{bmatrix} 0.1311 & 0.4622 & 0.3059 & 0.1007 & 0 \\ 0.2669 & 0.2799 & 0.2402 & 0.2140 & 0 \\ 0.2533 & 0.5363 & 0.1424 & 0.0425 & 0 \\ 0 & 0.1068 & 0.3013 & 0.2400 & 0.2795 \end{bmatrix}$$

Then the fuzzy comprehensive evaluation result of the tank area is as follow:

$$B = W \times R = [0.1217 \quad 0.3750 \quad 0.2816 \quad 0.0796 \quad 0.0679]$$

The membership degrees of five risk level indicators are 0.1217, 0.3750, 0.2816, 0.0796, and 0.0679 respectively. Based on the result of fuzzy comprehensive evaluation method and principle of maximum membership degree, the following conclusions that the level of safety evaluation of the crude oil tank area is relatively safe can be obtained.

4　Summary and discussion

Statistical analysis of failure for large crude oil storage tanks was conducted based on the field investigation data. The failure database of crude oil tank was established. The analytic hierarchy process and fuzzy comprehensive evaluation were adopted to perform the quantitative risk assessment for the crude oil tanks. Results show that: (1) The main accidents of large crude oil storage tanks are general accidents. Failure accidents of storage tank are mainly caused by fire and explosion. (2) Crude oil tank failure database established in this study may provide an effective method for prevention and statistics of storage tank accidents. (3) The fuzzy comprehensive evaluation model can be adopted for risk assessment of oil tanks. The evaluation result shows that the level of safety evaluation for the crude oil tank area is relatively safe.

Acknowledgments

This study is supported by the National Key Research and Development Program of China (Grant Nos. 2017YFC0805804, 2017YFC0805801). We finished this study with the help of government agencies, school, colleagues, and students, and we are very grateful to these people for their generous assistance.

References

[1] LANDUCCI G, GUBINELLI G, ANTONIONI G, et al. The assessment of the damage probability of storage tanks in domino events triggered by fire[J]. Accident Analysis & Prevention, 2009, 41(6): 1206-1215.

[2] KANG J, LIANG W, ZHANG L B, et al. A new risk evaluation method for oil storage tank zones based on the theory of two types of hazards[J]. Journal of Loss Prevention in the Process Industries, 2014, 29: 267-276.

[3] WEI T T, QIAN X M, YUAN M Q. Quantitative risk assessment of direct lightning strike on external floating roof tank[J]. Journal of Loss Prevention in the Process Industries, 2018, 56: 191-203.

[4] GUO X X, JI J, KHAN F, et al. Fuzzy bayesian network based on an improved similarity aggregation method for risk assessment of storage tank accident[J]. Process Safety and Environmental Protection, 2020, 144: 242-252.

[5] SUN Y. Study on Early Warning Model for Petrochemical Wharf Storage Tank Zone Safety Based on Extension Theory[D]. Tianjin: Tianjin University of Technology, 2015.

[6] ZONG H. Research on Failure Probability and Risk Evaluation for Storages in Storage Firms[D]. Jiangsu: Southeast University, 2016.

[7] YANG J X, SHE X M, HUANG Y C, et al. Research on risk assessment model for leakage accident of benzene tank based on BP neural network[J]. Journal of Safety Science and Technology, 2019, 15(01): 157-162.

本论文原收录于 Sixth International Conference on Electromechanical Control Technology and Transportation(2022年)。

北溪管道用感应加热弯管的生产和质量水平

吉玲康[1]　董瑾[2]

(1. 中国石油集团石油管工程技术研究院，石油管材及装置材料服役行为与结构安全国家重点实验室；2. 西安石油大学期刊中心)

摘　要：北溪管道项目是迄今为止世界上最长的海上管道。大口径、高压力的特点要求管材必须进行严格的质量控制。本文介绍了北溪管道用38个大直径弯管的制造商 Salzgitter Mannesmann Grobblech GmbH 对制造装备的升级改造，以及弯管、弯管母管的材料设计、制造工艺和质量水平。生产结果表明，各项性能指标和尺寸公差均达到标准要求。

关键词：北溪管道；弯管；母管；质量；技术要求；机械性能

　　Nord Stream 管道(以下简称北溪管道)项目是从俄罗斯的维堡(Vyborg)经波罗的海到德国的格里夫斯瓦尔德(Greifswald)的天然气输送管道。其中北溪-Ⅰ管道已经建成，北溪-Ⅱ管道目前也接近完工。据报道[1]，自2011年北溪-Ⅰ管道竣工以来，截至2020年底，该管道已累计输送天然气约 $3820\times10^8m^3$。其中2020年，尽管受到COVID-19大流行的影响，该管道仍按照严格的行业标准进行定期维护和检查工作，确保了该管道安全、可靠和高效运行，并且向欧洲消费者输送了 $592\times10^8m^3$ 的天然气，其天然气运输量是自运营以来的历史最高水平。虽然北溪-Ⅰ管道项目建成已经过去了10年，但政治、经济等诸多方面的意义深远，而且在技术层面，它是迄今为止世界上最长的海底管道，在管线设计、管材准备、施工、完整性管理等各阶段都具有鲜明的特点，特别是其苛刻的输送条件对管材的质量提出了较高的要求，管材生产和质量控制方法等都值得研究和借鉴。

　　北溪-Ⅰ管道系统由两条直径为48in(外径1219mm，内径1153mm)的平行管道组成，每条管道长1223km。该管道系统年输气能力为 $550\times10^8m^3$ 的天然气，全线采用分段设计方法，设计工作压力从KP0到KP300为22MPa，从KP300到KP675为20 MPa，从KP675到KP1223为17.75MPa，节省了30多万吨钢材。线路钢管钢级为L485，壁厚为26.8~41.0mm。北溪-Ⅰ管道从2010年4月开工建设，第一条管道于2011年底完工，第二条于2012年底完工[2]。

　　大口径和高设计压力的结合意味着北溪-Ⅰ管道项目必须认真研究和规定管线、弯管、阀门、三通、隔离接头和清管器疏水阀所需的材料，并仔细选择供应商，对制造阶段监督，以确保达到高质量要求。

基金项目：国家重点研发项目"高应变海洋管线管研制"(编号：2018YFC0310300)；中国石油天然气集团公司科研课题"高应变海洋管道关键服役性能评估及环焊技术研究"(编号：2018D-5010-12)。

作者简介：吉玲康，男，1966年生，教授级高级工程师，1989年毕业于西安交通大学金属材料及热处理专业，工学博士，现主要从事油气输送管的应用技术研究。E-mail：jilk@cnpc.com.cn。

其中，感应加热弯管是除线路直管外重要的管材类型，经过认真考察和研究，只有少数公司通过了根据北溪-Ⅰ规范提供大直径弯管的资格预审。最终德国缪尔海姆的 Salzgitter Mannesmann Grobblech GmbH(MGB)中标，并赢得了向北溪-Ⅰ项目供应38个大直径弯管的合同，并向该项目提供了38个直径从28in到48in的弯管(表1)。生产结果表明，各项性能指标和尺寸公差均达到要求。

本文将对这些弯管的技术要求、生产情况和质量水平等进行详细介绍。

表1 北溪-Ⅰ项目38个大直径弯管的详细情况

弯管数量	规格(in)	弯管角度(°)	弯曲半径	弯管壁厚(mm)	钢级	单件重量(t)
16	28	90	3D	28.5	L485	2.67
3	28	60	3D	28.5	L485	1.98
3	48	93.33	5D	35.0	L450	12.59
7	48	90	5D	35.0	L450	12.59
9	38	45	3D	30.2	L485	2.75

1 感应加热弯管的技术要求

北溪-Ⅰ项目需要弯管材料等级为L485和L450的感应弯管，材料强度高，低温下延展性好。由于设计压力高，壁厚高，所有弯管均规定感应加热弯制后必须经过整体淬火和回火，并具有严格的几何尺寸公差。

德国接收站需要10个48in弯管。表1中列出的28in和38in弯管安装在俄罗斯海岸的压气站附近和德国一侧的接收站附近。由于它们应用条件有差异，必须遵守不同的规范标准和试验要求，如表1和表2所示。为满足低温试验要求，弯管由直缝埋弧焊(LSAW)母管制成，由 Eisenbau Krämer(EBK)制管厂生产，该公司还为北溪-Ⅰ项目提供48in×41mm 的止屈器。

表2 北溪-Ⅰ弯管母管尺寸和制造标准[3]

规格(in)	钢级	母管规格(内径×最小壁厚)(mm×mm)	制造标准	铺设位置
28	L485	654.0×35.0	ISO3183	俄罗斯
38	L485	904.6×35.4	EN10208-2	德国
48	L450	1153.0×38.0	DNV-OS-F101	德国

28in 和 38in 弯管采用L485级材料制造，与管线管相同等级。但对于48in弯管，要求降低到L450级，增加壁厚的原因是为了在弯曲后的整体淬火回火热处理期间保持适当的几何稳定性。其实，如果考虑干线和相邻管线管道尺寸的运行条件，由48in×34mm 壁厚的母管制成的L485级材料的48in弯管就足够了，但由于在感应加热的奥氏体化过程中，较大的径厚比会使弯管有变形和塌陷的风险，因此这些母管的壁厚最小增加到38.0mm，而其强度等级降低至L450级。

表3列出了北溪-Ⅰ项目不同规格弯管的基本力学性能要求。其中，冲击试验温度根据服役条件确定，例如俄罗斯一侧压气站28in弯管的测试温度为-38℃，德国一侧38in弯管的测试温度为-25℃。

表3 北溪-I项目弯管拉伸和韧性性能要求

弯管规格 (in)	拉伸性能				冲击性能				
	$R_{t0.5}$(MPa)	R_m(MPa)	Y(T)	A(%)	试验温度(℃)	母材横向		焊缝	
						最小值(J)	平均值(J)	最小值(J)	平均值(J)
28	≥485	≥570	≤0.93	≥18	-38	36	48	30	40
38	≥485	≥570	≤0.90	≥18	-25	49	65	30	40
48	≥450	≥535	≤0.92	≥18	-35	65	80	40	50

2 感应加热弯管成分设计、生产工艺及质量水平

2.1 成分设计

为了在感应加热并弯曲加工后达到所需的强度,选择碳当量CE_{IIW}为0.42、含0.09%C、1.5%Mn、Mo及Nb、V微合金化的热机械控制工艺(TMCP)生产的厚板作为弯管的原材料。

钒微合金化会在回火过程中产生析出强化作用。钒是非常有效的,特别是在热感应弯曲加工中,因为它在奥氏体化过程中在奥氏体中具有足够的溶解度,并且在回火过程中在纳米范围内形成细小的沉淀[4]。另一方面,铌在奥氏体中的溶解度较低,但在厚板材料的热机械轧制过程中起着重要作用,它降低了轧制道次之间可能再结晶的温度范围。在可能再结晶的温度范围内,奥氏体的变形会导致奥氏体晶粒的凝固和强化。这有助于在转变为铁素体或贝氏体的过程中细化组织[5]。北溪-I管道项目的不同尺寸弯管都是用这种化学成分制造的。

2.2 热处理工艺和显微组织

弯曲后的热处理包括在910℃下奥氏体化50~60min,然后在600~650℃下回火90min。根据弯管尺寸的不同,热处理参数也有微小差异。在热处理过程中,所有弯管都从内部用支撑固定,以避免压扁。由于直径48in的90°弯管的外形尺寸为10200 mm长,3200 mm宽,其尺寸已经超过目前已有的淬火可用托架,所以必须进行设备改造,同时还必须选择可用的热处理炉[3]。

图1显示了北溪项目48in弯管从天然气加热炉到淬火槽的转移过程。由于包括支柱在内的一个90°弯管的实际重量约为13t,加上用于转移至淬火池的C形钩和托盘的重量,必须对弯管厂的起重机的能力进行升级。

图1 北溪项目48in弯管从热处理炉转移到淬火槽过程(MGB弯管厂)

热处理对母材组织的影响如图2所示。母管的显微组织以铁素体为主,另外还有少量珠光体和贝氏体岛,这是典型的热机械加工和空冷的组织类型。然而,在淬火和回火后(调质处理),显微组织转变为均匀的贝氏体。

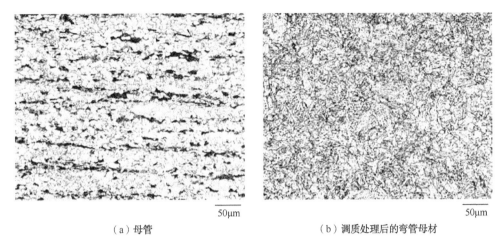

（a）母管　　　　　　　　　　　　　　（b）调质处理后的弯管母材

图2　母管和调质处理后弯管母材的显微组织

2.3　力学性能

2.3.1　拉伸性能

在最终热处理条件下，使用横向条状试样测量弯管的拉伸性能。弯管内弧、外弧、直管段和母管的拉伸试验结果对比情况见图3。结果表明，弯管的屈服强度在545～570MPa之间，抗拉强度在640～690MPa之间。使用该种成分设计可以达到强度要求，而且可以达到母管的强度水平甚至更高的强度水平。这主要是因为热处理后钒碳氮化物沉淀硬化的强化贡献比轧制条件高；另一个原因是淬火促进了细晶贝氏体组织的形成。48in弯管的屈服强度和抗拉强度与28in和38in弯管的屈服强度和抗拉强度处于同一水平，这表明即使强度要求较低，其也可以保持X70(L485)的强度水平。图4为弯管直管段、弯弧处的横向焊缝拉伸试验结果与母管焊缝拉伸试验结果的对比情况。结果表明，所有试验结果均达到了L485(X70)和L450(X65)的抗拉强度570MPa和535MPa的要求。

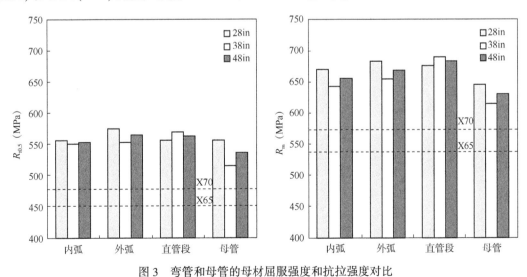

图3　弯管和母管的母材屈服强度和抗拉强度对比

2.3.2　冲击性能

弯管母材的韧性试验结果如图5所示，包括弯管内弧侧、外弧侧、直管段的母材冲击韧性均和母管处于同一水平，甚至略高，且平均值都在100J以上，远超出标准要求。

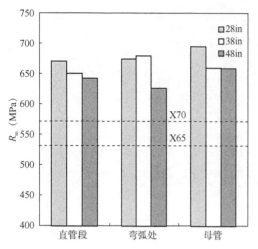

图4 弯管与母管的焊接接头抗拉强度对比

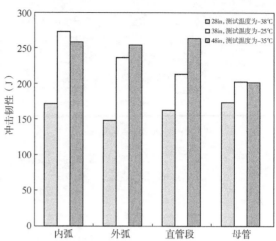

图5 弯管内弧、外弧和直管段的母材平均CVN冲击韧性和母管的对比

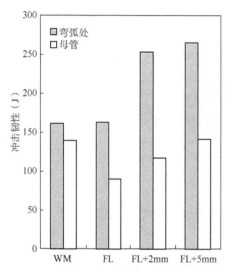

图6 48in弯管弯曲位置焊缝/热影响区-35℃时的平均CVN冲击能量与母管母材的对比

弯管的整体调质热处理使热影响区(HAZ)的韧性达到母材的水平[6-7],这对母管的HAZ韧性来说是非常有意义的。图6为48in试验弯管弯曲部位的焊缝金属(WM)、熔合线(FL)、FL+2和FL+5冲击试验结果,可见弯曲部位的焊缝、热影响区冲击韧性高于母管母材冲击韧性。

2.3.3 硬度

用维氏硬度计测定了母材和焊缝金属的硬度分布。母材和焊缝区的最大允许硬度分别为260HV10和270HV10。图7所示为48in弯管从弯曲位置获得的焊缝横截面的硬度分布情况,所有测量值都低于规定的最高水平。图8所示为48in弯管整体调质处理后多道次焊缝横截面的宏观形貌。

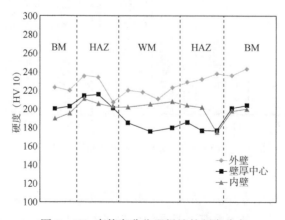

图7 48in弯管弯曲位置焊缝的硬度分布

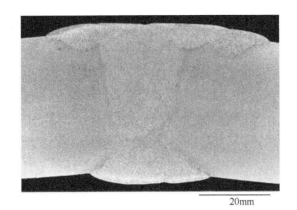

图 8 48in 弯管整体调质处理后多道次焊缝的宏观截面图

3 几个问题的思考

北溪-Ⅰ天然气管道成功进行了 38 个高钢级大口径弯管的设计、制造和使用,其中的一些做法值得思考和借鉴。

(1) 管道直径分别为 28in 和 38in 的弯管直接采用 L485 级材料交付,48in 弯管为保证热处理期间保持几何稳定性而采用了增大壁厚、降低钢级到 L450 的方法。也就是说,在同一管道工程中,根据不同的服役条件,可以采用不同钢级管材进行设计和制造,其目的是更好地满足管道的安全可靠性。中缅管线也采用了这种设计理念,中缅管线的主干线采用的是 X80 钢级钢管,但在地质灾害多发的基于应变设计地区,为了保证钢管具有更好的应变能力,采用了加大壁厚的 X70 大应变钢管。

(2) 该管道的三种规格弯管均采用相同成分体系的 TMCP 厚壁钢管作为原材料,采用感应加热后整体淬火+回火热处理的工艺路线,弯管的力学性能均满足订货标准要求。这种热处理方式可以很好地实现对弯管成分、组织、性能的控制,整体均匀性优异。但和中国大部分弯管厂采用的感应加热淬火方法相比,设备、制造过程成本更高,占地面积也更大。

(3) 该管道所使用的三个规格的弯管分别采用了 ISO3183、EN10208-2、DNV-OS-F101 三个不同的标准,其强度、韧性的要求和试验方法等都有所不同。这种做法可能是由于该管道是一条国际管道,不同国家的标准很难统一造成的。这显然会给管材的制造、使用、甚至将来的完整性评估等带来困难。

(4) 该管道采用 48in 大口径管道且在高设计压力下进行输送,因此对所有管材均提出了严格的要求,同时对各供应商进行资格预审,对生产过程进行监督制造,确保所有部件都达到高质量要求。

参 考 文 献

[1] https://www.nord-stream.com/press-info/press-releases/the-nord-stream-pipeline-transported-a-volume-of-592-billion-cubic-metres-of-natural-gas-in-2020-517/.

[2] PETTINELLI D, BERGOMI S P, BRUSCHI R, et al. Nord Stream Project – Segmented Pipeline System: Sizing vs Design for Operation [C]//Proceedings of the Twenty-second International Offshore and Polar Engineering Conference, Rhodes. Onepetro 2012.

[3] MUTHMANN E, GJEDREM T, STALLYBRASS C. Manufacturing of Large Steel Components for Nord Stream

Project[J].3R International, Special-Edition, 2010, 2：26-31.

[4] 樊朋煜. 钒及热处理工艺对高强塑积贝氏体钢组织与性能的影响[D]. 北京：北京交通大学, 2020.

[5] MALCOLM J.GRAY. 管线工程中的含铌钢[G]//中信微合金化技术中心编译. 铌·科学与技术. 北京：冶金工业出版社, 2003：557-568.

[6] MUTHMANN E, GRIMPE F. Fabrication of hot induction bends from LSAW large diameter pipes manufactured from TMCP plate[C]//International Symposium on Microalloyed Steels for the Oil & Gas Industry, Araxa, Brazil, 2006：573-587.

[7] 赵金兰, 刘腾跃, 王长安, 等. 热处理对X80钢弯管力学性能和组织的影响[J]. 金属热处理, 2017（6）：127-129.

本论文原发表于《石油管材与仪器》2021年第7卷第5期。

不同强度匹配 X80 钢环焊接头力学性能及变形能力

何小东[1,2] 高雄雄[1,2] David Han[2] 池 强[1,2]
霍春勇[1,2] José B. Bacalhau[3]

(1. 中国石油集团石油管工程技术研究院·石油管材及装备材料服役行为与结构安全国家重点实验室；2. 国际焊接研究中心；3. 巴西矿冶公司)

摘 要：环焊接头的强度匹配与韧性对管道服役安全至关重要。为了更为深入掌握环焊缝的失效机理，采用力学性能测试、微观分析等方法，测试了两种不同强度匹配的高铌 X80 环焊缝接头的组织和性能，并借助数字图像相关法(Digital Image Correlation, DIC)研究了焊接接头在拉伸载荷下的应变行为。结果表明：低强匹配与高强匹配的环焊接头均具有较好的冲击韧性，二者的夏比冲击吸收能量平均值相当。在拉伸载荷下，应变集中最先出现在环焊缝的根焊及热影响区部位。随着拉伸载荷的增加，高强匹配环焊缝接头的应变集中逐渐转移至母材，管道承受轴向载荷及变形的能力大于低强匹配环焊接头；低强匹配环焊接头虽具有较好的韧性，但因其在焊缝及热影响区存在塑性应变累积效应，易发生断裂。

关键词：高铌 X80 钢；强度匹配；环焊缝；应变累积；数字图像相关法

为提升输送效率，长距离输送管道材料多采用高强度管线钢。从 20 世纪 50 年代末开始，铌微合金化技术极大地推动了高强度管线钢的发展与工程应用[1-5]，在生产中常使用 C-Mn-Si-Mo-Nb 或 C-Mn-Si-Cr-Nb 合金设计的热机械控制轧制工艺(Thermo-mechanical control rolling process, TMCP)以满足 X70、X80 或更高钢级的组织及性能要求。可见，Nb 是高性能管线钢中重要的微合金化元素。

在早期高钢级管道建设中，由于管道焊接自动化水平相对较低，X70、X80 油气管道工程环焊缝常采用焊条电弧根焊与自保护药芯焊丝填充盖面的组合焊接工艺，管道运行一段时间后，常发生环焊缝断裂导致管道失效。尤其是，近年来国内外高钢级管道环焊缝断裂失效事故引起了社会广泛的关注[6]，研究者从设计、标准、材料成分、工艺、变形及断裂行为等方面开展了环焊缝质量控制、性能影响、应变能力及断裂行为等方面的研究，以期提升管道服役安全性[7-13]。王海涛等[14]研究认为环焊缝断裂失效主要与冲击韧性有关，其原因是 Nb 含量影响了管线钢粗晶区中 MA 组元的形态与韧性，在材料设计中应尽量降低 Nb 的含量[15]。但也有研究表明，在相同的焊接热输入条件下，由于未溶 Nb(C，N)颗粒抑制了奥

基金项目：巴西矿冶公司国际合作项目"X80 环焊接头强度匹配及 HAZ 软化行为研究"，2017-QT-52。
作者简介：何小东，男，1970 年生，正高级工程师，2004 年硕士毕业于西安交通大学材料加工专业。现主要从事管线钢焊接工艺、材料性能测试及表征方向的研究。地址：陕西省西安市长安区锦业二路 89 号，710100。电话：029-81887889。Email：xiaodonghe@126.com。

氏体晶粒长大，高 Nb 高温轧制（High Temperature Processing，HTP）管线钢粗晶热影响区（Heat affected zone，HAZ）的冲击韧性高于低 Nb 的 Mn-Mo 管线钢[16]。Nb 对环焊缝热影响区组织及性能的影响十分复杂，大量机制相互关联，同时与钢的化学成分及焊接参数有关。尽管存在争论，但油气管道失效案例分析表明，管道环焊接头断裂的主要原因是焊接质量控制不严格[17]，且以在低匹配环焊接头热影响区处发生拉伸断裂为主要失效形式[18]。由于环焊接头的焊缝与热影响区很窄，采用常规拉伸试验方法很难测得环焊接头不同区域的实际强度。因此，Midawi 等[19]采用仪器压痕法评估了 X80 管线钢焊接屈服强度的不匹配度，并使用数字图像相关法（Digital Image Correlation，DIC）获得的应变分布进一步详细说明了环焊接头局部特性与强度失配对失效位置的影响。

在此，采用焊条电弧焊（Shielded Metal Arc Welding，SMAW）与药芯焊丝自保护焊（Self-shielded flux cored arc welding，FCAW-S）组合焊接工艺，借助硬度云图与 DIC 研究了两种不同强度匹配的高铌 X80 环焊接头的性能及变形，以期促进对高钢级管道环焊缝失效机理的认识，有效控制管道环焊缝失效。

1 试验材料及方法

1.1 试验材料

试验材料为高铌 X80 直缝埋弧焊接钢管，管道直径为 762mm，壁厚为 14.1mm。试验材料的化学成分和纵向力学性能分别见表 1 至表 3，管体的显微组织为粒状贝氏体（Granular Bainite，GB）+准多边形铁素体（Quasi-polygonal Ferrite，QF）+珠光体（Pearlite，P），平均晶粒直径 11.5μm（图 1）。

表 1 试验用 X80 管线钢的化学成分表

成分	C	Si	Mn	P	S	Nb	Ti	Cr	Ni	Cu	Mo	N	V	Fe
质量分数	0.052%	0.13%	1.56%	0.012%	0.0033%	0.099%	0.012%	0.23%	0.14%	0.25%	0.0006%	0.0066%	0.0047%	余量

表 2 试验用 X80 管线钢纵向拉伸性能数据表

宽度×标距（mm×mm）	屈服强度（MPa）	抗拉强度（MPa）	断后伸长率
38.1×50	616	664	35.5%
	618	654	36.0%
	617	662	37.0%

表 3 试验用 X80 管线钢硬度及纵向夏比冲击吸收能量（-20℃）数据表

试样尺寸（mm×mm×mm）	夏比冲击吸收能量（J）		维氏硬度 HV10
	单个值	平均值	
10×10×55	298	304	227
	308		214
	307		225

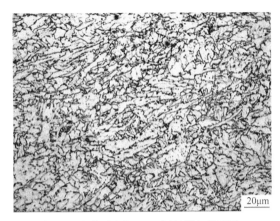

图 1 试验用高铌 X80 钢显微组织照片

1.2 环焊缝焊接及试验方法

环焊缝通过 SMAW 根焊与 FCAW-S 填充盖面的组合焊接工艺(图 2、表 4),采用 E7018+E81T8-Ni2J 与 E9016+E91T8-G 两种不同强度组合的焊接材料进行环焊缝焊接。根焊、填充及盖面焊接设备均为熊谷 MPS-500。预热温度为 100~150℃,层间温度为 60~100℃。

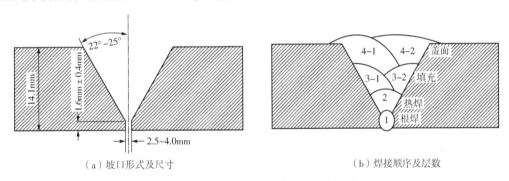

(a)坡口形式及尺寸   (b)焊接顺序及层数

图 2 焊接坡口形式及焊接顺序示意图

表 4 焊接工艺及参数表(1in=2.54cm)

焊接层数	焊接道数	焊接方法	焊接方向	电流(A)	电压(V)	送丝速度(in/min)	焊接速度(cm/min)
根焊	1	SMAW	上向	86~110	22~26		5~9
热焊	2	FCAW-S	下向	175~240	17~19	95~105	17~19
填充	3-1	FCAW-S	下向	160~230	18~20	95~105	17~19
填充	3-2	FCAW-S	下向	160~230	18~20	95~105	17~19
盖面	4-1	FCAW-S	下向	170~210	18~20	95~105	17~19
盖面	4-2	FCAW-S	下向	180~210	18~20	95~105	17~19

焊接完成后,使用 X 射线探伤仪对环焊缝进行无损检测,以检查是否存在可能影响后续力学测试结果的缺陷。而后分别从两种环焊缝上截取无焊接缺陷显示的力学试样与金相试样,其中拉伸及夏比冲击试样垂直环焊缝截取,条形拉伸试样尺寸(长度×宽度×厚度)为 300mm×25.4mm×14.1mm,且保留焊缝余高;圆棒拉伸试样尺寸(长度×直径)为 125mm×

10mm；夏比冲击试样的尺寸为10mm×10mm×55mm，缺口形式为V形，深度2mm，分别位于焊缝中心、熔合线（Fusion Line，FL，焊缝金属、热影响区各占50%处）、FL+1与FL+2（分别代表热影响区中距离熔合线1mm与2mm处）。依据ASTM A370—2017《钢制品力学性能试验的标准试验方法和定义》测试环焊缝接头的力学性能，在拉伸试验过程中利用DIC研究其变形、断裂过程，之后通过扫描电子显微镜（Scanning Electron Microscope，SEM）观察拉伸试样的断口形貌。环焊接头的微观组织形貌及硬度分布分别使用MEF4M金相显微镜与KB30BVZ-FA维氏硬度计进行观察及测试。

2 结果及讨论

2.1 强度及韧性

由不同强度匹配焊缝金属与母材纵向拉伸应力-应变曲线（图3）可知，采用E7018+E81T8-Ni2J焊接的环焊缝，其焊缝金属的实际强度低于母材纵向拉伸强度，为低强匹配环焊接头，记为UM；采用E9016+E91T8-G焊接的环焊缝为高强匹配接头，记为OM。低强匹配接头焊缝金属变形较大，母材变形相对较小，形变强化程度较低，因此其抗拉伸强度低于管体母材原有强度。

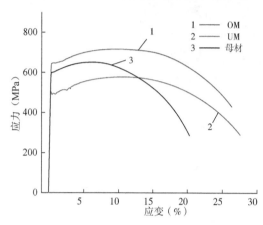

图3 不同强度匹配的焊缝金属与母材纵向拉伸应力—应变曲线

在制得的两个UM焊接接头与1个OM焊接接头上截取平行试样进行拉伸试验，由不同强度匹配环焊缝接头拉伸强度及断裂位置（图4）可见，在UM环焊接头试样中，保留焊缝余高的条形试样受拉伸载荷时，断裂多发生于母材（Base Metal，BM），仅有UM02断于焊缝（Welded Metal，WM）；去除焊缝余高的圆棒拉伸试样全部断于焊缝，断于母材的接头抗拉强度为635~652MPa。各OM环焊接头试样均断于母材，抗拉强度为655~667MPa，与管体母材的抗拉强度基本一致，且略大于UM环焊接头的抗拉强度，可见其主要体现为母材拉伸性能。

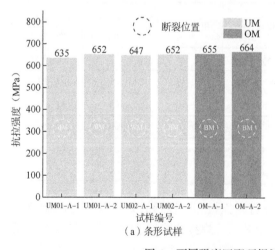

（a）条形试样

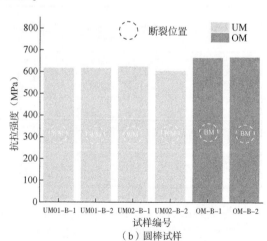

（b）圆棒试样

图4 不同强度匹配环焊缝接头拉伸强度及断裂位置图

由两种不同强度焊接材料环焊缝接头冲击韧性结果（图5）可见，两种强度匹配环焊接头各位置的夏比冲击吸收能量平均值相当。UM 环焊缝夏比冲击吸收能量较为分散，其值为 46~135J；OM 环焊缝夏比冲击吸收能量相对集中，其值为 72~128J。与 OM 匹配相比，受焊缝吸收能量离散性的影响，UM 环焊接头熔合线处的夏比冲击吸收能量较为离散。同时，FL+1 处夏比冲击试样的缺口大部分位于粗晶区，组织的不均匀性导致试样韧性较为分散，但其最低值达 141J，平均值高于 220J；FL+2 处夏比冲击试样的缺口位于细晶区，其冲击吸收能量数值相对集中，基本与母材的纵向冲击性能相当。

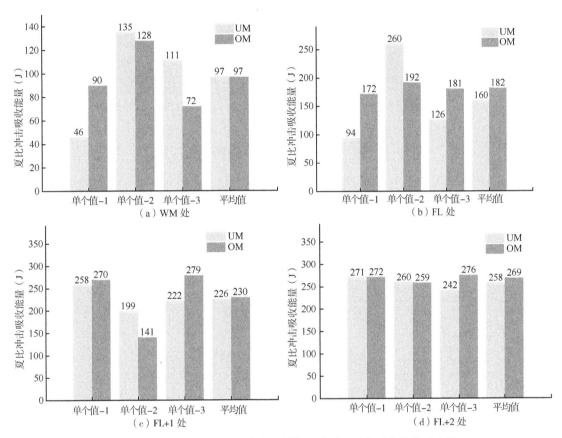

图 5 -20℃下 X80 钢环焊接头不同位置处夏比冲击吸收能量对比图

2.2 讨论

观察分别采用 E81T8-Ni2J、E91T8-G 焊丝填充、盖面所得焊缝与热影响区的显微组织形貌（图 6），可见两种焊缝的组织均为粒状贝氏体，其中采用 E91T8-G 填充、盖面焊接所得焊缝的组织晶粒更细小，强度较高，可见采用 E91T8-G 焊丝对 X80 钢管进行焊接更易实现等强或高强匹配的接头设计要求。采用 FCAWS 工艺焊接时，由于热输入小于制管焊缝所采用的多丝埋弧焊，其熔合线附近的微观组织为 GB+多边形铁素体（Polygonal Ferrite，PF），金相观察未发现高铌 X80 管线钢粗晶区及细晶区存在影响冲击韧性的 MA 组元（图 7）。

由不同强度匹配 FCAW-S 焊接接头硬度分布云图（图 8）可见，UM 环焊缝硬度明显低于两侧母材，而 OM 环焊缝硬度略高于两侧母材。两种强度匹配的焊缝、热影响区及母材之间硬度分布界限明显，且热影响区硬度均低于母材硬度，说明热影响区存在一定程度的软化，其宽度为 1.0~2.0mm。焊接热影响区的软化一般出现在 900~1000℃的两相临界区与细晶区的交界处[20]，软化宽度与程度不仅受焊接方法、焊接工艺参数影响，还与管体的碳当量、强度及轧制状态有直接关系。虽然试验所采用的 X80 钢有较高含量的 Nb 元素，但其碳当量相对较低，在采用热输入较大的 FCAW-S 焊接时，热影响区仍存在一定程度的软化。

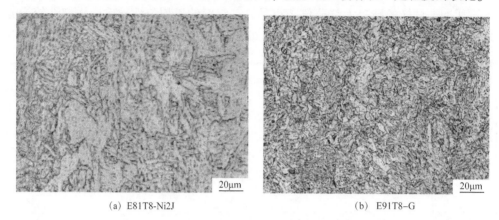

(a) E81T8-Ni2J　　(b) E91T8-G

图 6　不同焊丝填充、盖面所得焊缝的显微组织形貌图

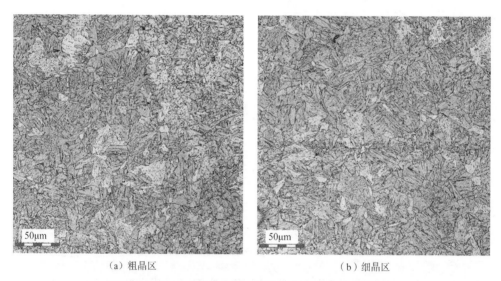

（a）粗晶区　　（b）细晶区

图 7　高铌 X80 钢 FCAW-S 热影响区的显微组织形貌图

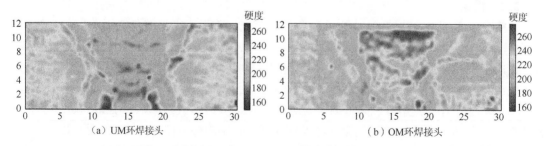

（a）UM环焊接头　　（b）OM环焊接头

图 8　不同强度匹配 FCAW-S 焊接接头硬度分布云图

由不同强度匹配环焊接头应变及断裂过程演化云图（图 9）可见，对于低强匹配环焊接头，在拉伸载荷作用下，由于焊缝盖面余高具有增强作用，应变集中优先在根焊及焊接热影响区产生。随着载荷增加，应变从热影响区扩展到焊缝，并在焊缝上产生应变集中，从而导

致焊缝或热影响区发生断裂。而对于高强匹配环焊接头,在拉伸载荷下,虽然根焊与热影响区最先出现应变集中,但是随着载荷增加,应变集中位置从根部焊缝与热影响区转移到母材上,并在母材上发生颈缩变形而断裂。UM 环焊接头断裂时,断口附近的应变约为 50%,试样断后伸长率为 22%;OM 环焊接头断裂时,断口附近的应变约为 76%,试样断后伸长率为 30%。可见,高强匹配环焊接头的变形能力大于低强匹配环焊接头的变形能力。过载时,高强匹配环焊接头的变形主要由管体母材提供,焊缝受到母材的保护,其变形较小。因此,在管道环焊接头强度匹配设计及焊材选择时,应采用等强或高强匹配设计,以避免地质灾害或土壤沉降等外部载荷造成应变累积所导致的管道环焊缝断裂失效。

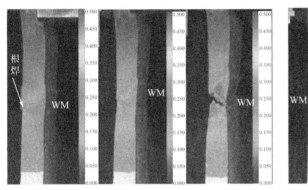
(a) 低强匹配

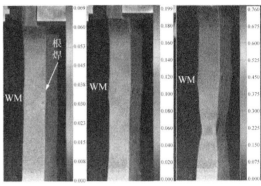
(b) 等强匹配

图 9 拉伸载荷下不同强度匹配环焊接头应变及断裂过程演化云图

有研究认为对于低强匹配环焊接头,应变最大位置是热影响区与焊缝交界的外表面[21],这与实际低强匹配环焊接头的 DIC 拉伸试验结果不一致。不同强度匹配环焊接头应变及断裂演化过程表明,无论何种强度匹配,由于根部焊缝宽度较窄,热影响区存在一定程度的软化,应变集中均最先出现于根焊与热影响区。众多环焊缝断裂失效事故的现象证实了这一点。随着载荷增加,在低强匹配环焊缝上不断产生应变累积,而等强或高强匹配接头的应变集中从根部焊缝和热影响区转移到母材。因此,即使低强匹配焊缝和热影响区具有较好的韧性,管道过载时也会因为塑性应变累计而发生断裂。

3 结论

(1) 采用 FCAW-S 工艺进行高铌 X80 环焊缝焊接时,低强匹配与高强匹配环焊接头的热影响区均具有较好的冲击韧性,其平均吸收能量相当。

(2) 无论是低强匹配还是等强或高强匹配,由于根部焊缝宽度较窄,热影响区存在一定程度的软化,应变集中最先出现在环焊缝的根焊与热影响区。但对于高强匹配环焊接头,随着载荷增加,应变集中逐渐从根焊与热影响区转移到母材,管道承受轴向载荷与变形的能力大于低强匹配环焊接头。

(3) 对于低强匹配环焊接头,当载荷过大时,即使焊缝具有较好的韧性,也会因塑性应变累积而发生断裂。而高强匹配环焊接头的变形主要由管体母材提供,降低了环焊缝应变累积,可有效防止管道环焊缝因过载而断裂。

参 考 文 献

[1] GRAY J M, SICILIANO F. High strength microalloyed linepipe: half a century of evolution[C]. Beijing:

Microalloyed Steel Institute, 2009: 20-45.

[2] STALHEIM D G. The use of high temperature processing(HTP) steel for high strength oil and gas transmission pipeline application[J]. Iron and Steel, 2005, 40(S): 699.

[3] REIP C P, SHANMUGAM S, MISRA R D K. High strength microalloyed CMn(V-Nb-Ti) and CMn(V-Nb) pipeline steels processed through CSP thin-slab technology: microstructure, precipitation and mechanical properties[J]. Materials Science and Engineering: A, 2006, 424(1/2): 307-317.

[4] 尚成嘉, 王晓香, 刘清友, 等. 低碳高铌X80管线钢焊接性及工程实践[J]. 焊管, 2012, 35(12): 11-18.

[5] 邓伟, 高秀华, 温志红, 等. 高铌X80管线钢的组织和性能[J]. 东北大学学报(自然科学版), 2009, 30(9): 1270-1273.

[6] 戴联双, 考青鹏, 杨辉, 等. 高强度钢管道环焊缝隐患治理措施研究[J]. 石油管材与仪器, 2020, 6(2): 32-37.

[7] 张振永. 高钢级大口径天然气管道环焊缝安全提升设计关键[J]. 油气储运, 2020, 39(7): 740-748.

[8] 李为卫, 何小东, 葛加林. 油气管道环缝焊接国外先进标准的启示和借鉴[J]. 石油管材与仪器, 2020, 6(2): 1-7.

[9] 陈延清, 牟淑坤, 刘宏, 等. 母材成分对X80管道环焊接头冲击性能的影响[J]. 电焊机, 2017, 47(7): 10-16.

[10] 沙胜义. 输油管道环焊缝缺陷疲劳寿命评估[J]. 管道技术与设备, 2017(2): 28-31.

[11] 任俊杰, 马卫锋, 惠文颖, 等. 高钢级管道环焊缝断裂行为研究现状及探讨[J]. 石油工程建设, 2019, 45(1): 1-5.

[12] 帅健, 孔令圳. 高钢级管道环焊缝应变能力评价[J]. 油气储运, 2017, 36(12): 1368-1373.

[13] 陈宏远, 张建勋, 池强, 等. 热影响区软化的X70管线环焊缝应变容量分析[J]. 焊接学报, 2018, 39(3): 47-51.

[14] 王海涛, 李仕力, 陈杉, 等. 高钢级天然气管道环焊缝断裂问题探讨[J]. 石油管材与仪器, 2020, 6(2): 49-52.

[15] 荆洪阳, 霍立兴, 张玉凤. 铌对高强钢焊接热影响区中马氏体-奥氏体组元形态的影响[J]. 焊接学报, 1997, 18(1): 37-42.

[16] 乔桂英, 郭宝峰, 陈小伟, 等. 热循环对高铌管线钢焊接热影响区冲击韧性的影响[J]. 金属热处理, 2009, 34(12): 32-35.

[17] 罗金恒, 杨锋平, 王珂, 等. 油气管道失效频率及失效案例分析[J]. 金属热处理, 2015, 40(增刊): 470-474.

[18] TAJIKA H, SAKIMOTO T, HANDA T, et al. Girth weld strength matching effect on tensile strain capacity of grade X70 high strain line pipe[C]. Calgary: 2018 12th International Pipeline Conference, 2018: IPC2018-78778, V002T06A005.

[19] MIDAWI A R H, SIMHA C H M, GERLICH A P. Assessment of yield strength mismatch in X80 pipeline steel welds using instrumented indentation[J]. International Journal of Pressure Vessels and Piping, 2018, 168: 258-268.

[20] 谷雨, 周小宇, 徐凯, 等. 高强X90管线钢焊接热影响区脆化及软化行为[J]. 金属热处理, 2018, 43(6): 74-78.

[21] 帅健, 张银辉. 高钢级管道环焊缝的应变集中特性研究[J]. 石油管材与仪器, 2020, 6(2): 8-14.

本论文原发表于《油气储运》2022年第41卷第1期。

高钢级天然气管道爆炸危害的影响

杨坤[1] 王磊[2] 高琦[2] 王琴[2] 池强[1]

(1. 中国石油集团石油管工程技术研究院石油管材及
装备材料服役行为与结构安全国家重点实验室；
2. 中石油管道有限责任公司西部分公司)

摘 要：冲击波、热辐射是天然气管道爆炸的主要危害形式。目前的研究手段主要是理论模型并结合小尺寸试验，缺乏全尺寸试验数据的验证，具有一定风险。本文基于高钢级管道的全尺寸气体爆破试验，对试验过程中的冲击波压力、热辐射通量进行了采集、分析和研究，并结合理论模型对试验结果进行分析。研究结果表明：随着与起爆点距离的增大，冲击波压力、热辐射通量迅速衰减；垂直方向相同位置上的热辐射通量值较高，而冲击波压力则差异较小；根据试验结果计算的热辐射安全距离为 124.9m，大于冲击波危害的安全距离；根据静态火球模型计算出的热辐射危害半径为 147.6m，与试验结果相差不大。

关键词：全尺寸气体爆破试验；高钢级管道；天然气管道爆炸危害；冲击波；热辐射；蒸气云爆炸；止裂韧性；危害半径

随着经济的发展和人民生活水平的日益提高，天然气的用量需求越来越大。管道输送作为最经济、安全的天然气运输方式，其也朝着高钢级、大管径、高输送压力的方向发展[1-3]。在埋地长输管道服役时，在腐蚀、外力损伤、地形变化等因素以及内压耦合作用下会发生管泄漏和断裂[4]，严重时可能发生管道爆炸，对临近管道、周围人或建筑造成危害，引发重大经济和安全损失[5-6]。

天然气管道泄漏后，由于管内高压使气体大量从泄漏点散出，管道在泄漏位置会发生裂纹扩展，加速气体泄漏过程。由于管道断裂以及天然气体积极速膨胀，会形成剧烈的物理爆炸，产生强烈的冲击波超压，造成危害[7-9]。此外，高压天然气管道泄漏后，由于泄漏点附近具有较高的压力以及天然气浓度(以甲烷为主)，并不会发生点火爆炸。随后，天然气体积迅速膨胀，并向空中进行扩散，并与大气中的氧气充分混合，形成预混蒸气云[10]。当满足气体的爆炸极限时，蒸气云会被点燃[11]。由于天然气与空气混合充分均匀，点燃过程极为剧烈，火焰前沿速度可达 50~100m/s，产生强烈的爆燃，并向周围释放冲击波及热辐射，

基金项目：国家重点研发计划课题"油气管道及储运设施损伤致灾机理与演化规律研究"(编号：2016YFC0802101)；中石油项目"高钢级管道可靠性设计及失效控制技术研究"(编号：2019B-3008)。

作者简介：杨坤，男，1985 年生，高级工程师，2013 年毕业于西北工业大学材料学专业，主要从事管道变形断裂控制、全尺寸试验技术、管道灾害失效分析等领域的研究。E-mail：kunyang073@cnpc.com.cn。

对人员、建筑和设置造成巨大伤害[12-13]，严重威胁管道服役安全，从而对管道铺设区域的公共安全带来重大威胁。

天然气管道爆炸既有管道断裂、气体泄漏膨胀的物理爆炸，又有与空气混合点燃后的化学爆炸。在管线设计及建设过程中必须针对冲击波及热辐射等信息进行综合考虑，设定管线建设、服役的安全半径，从而降低管道失效所带来的危害[14]。

天然气全尺寸气体爆破试验是管道服役安全及危害行为影响的重要研究手段和方法。利用该试验可模拟实际管道发生泄漏、断裂以及之后的整个爆炸过程，对管道断裂控制、爆炸危害评估方面具有重要的研究意义和价值[15]。本文通过开展 X80 OD1422mm 焊管的天然气全尺寸气体爆破试验对管道爆炸危害类型、范围、影响程度等内容进行了研究和分析，结合理论模型分析，估算出管道危害的安全距离，并与试验结果进行了对比分析。

1 爆破管道布置、数据采集方案及相关试验参数

1.1 全尺寸气体爆破试验管串布置及试验过程

爆破试验管串由 13 根 X80 钢级、外径（OD）1422mm、壁厚（T）21.4mm 直缝埋弧焊管组成（长度为130m），试验时管串内天然气压力为12.05MPa，管内气体温度为13.8℃，管道的回填深度为1.2m。管串中的钢管按止裂韧性高低进行排布（管串中间为止裂韧性较低的管道，两端为止裂韧性较高的管道）并焊接。管串沿轴向南北布置，起爆中心位于试验管串正中的管道上（启裂管）。试验中通过线性聚能切割器（切割器安装在启裂管中部垂直向上位置）引入贯穿裂纹，初始裂纹的长度为500mm。

试验时，通过聚能切割器将裂纹引入管串，在管道内压的作用下，裂纹由起裂位置向管串两侧发生扩展，同时管内天然气泄出并迅速膨胀，引起物理爆炸，产生冲击波超压。随着裂纹向两侧的扩展，钢管的止裂韧性升高，裂纹扩展阻力增大，裂纹扩展速度逐渐降低并发生止裂。

天然气泄出后，迅速向空中扩散，并与空气充分混合，达到爆炸极限。利用信号弹将其点燃，发生闪爆，形成天然气的蒸气云爆炸。

采用冲击波、热辐射等传感器及数据采集设备，在距起爆中心不同位置上进行传感器布置，对天然气管道爆炸过程中的冲击波、热辐射数据进行捕获分析。

1.2 传感器安装情况及参数

冲击波传感器分别安装于垂直管串、与管串成30°夹角两个方向上。两个方向上安装传感器的数量、距起爆中心的距离相同。距起爆位置100m内，每隔10m安装一个冲击波传感器，100m以外，每隔20m安装一个冲击波传感器，每个方向上的传感器数量为15个，如图1所示。

热辐射传感器分别安装于垂直管串以及与管串成30°夹角两个方向上。距起爆中心每隔50m安装一个热辐射传感器，管串垂直方向上安装了6个热辐射传感器，与管串呈30°夹角方向上安装了4个热辐射传感器。

冲击波传感器为针筒状，安装在距地面1.5m高的铁架上，用于测试试验过程中的冲击波压力[图2(a)]；热辐射传感器为方块状，安装在距地面1m高的铁架上，并呈一定角度（垂直于爆炸火球），用于捕获点燃后的燃烧热辐射，如图2(b)所示。

（a）全尺寸气体爆破试验管串

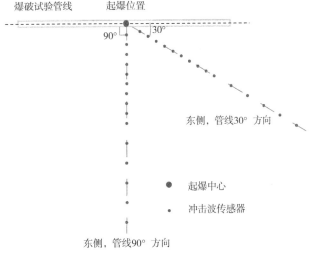
（b）冲击波传感器安装布置示意图

图1 X80 1422mm 全尺寸气体爆破试验管串及冲击波传感器安装布置示意图

（a）冲击波传感器

（b）热辐射传感器

图2 冲击波传感器和热辐射传感器安装效果

2 试验结果及分析

2.1 全尺寸气体爆破试验钢管测试结果

起爆后，裂纹由启裂管中心向南北两侧扩展，并在南1管（启裂管南侧相邻的第一根钢管）和北1管（启裂管北侧相邻的第一根钢管）止裂，如图3所示。在南侧，裂纹穿过起裂管后，在南1管内扩展9.10m后止裂，南侧裂纹扩展总距离13.965m，从起裂到止裂共耗时142.6ms。在北侧，裂纹穿过起裂管后，在北1管内扩展8.15m后止裂，北侧裂纹扩展总距离13.015m，从启裂到止裂共耗时133.3ms，如图3所示。

2.2 全尺寸气体爆破试验冲击波测试结果及分析

由于管道启裂和蒸气云点燃存在一定的时间间隔（信号弹点燃比切割器启裂晚0.5s），冲击波传感器仅采集到了由天然气在空中点燃所产生的化学爆炸冲击波数据，如图4所示。

（a）起爆点燃时刻　　　　　　　　　　　　（b）试验后的开裂钢管

图 3　OD1422mm/X80/21.4-12MPa 直缝埋弧焊管全尺寸气体爆破试验

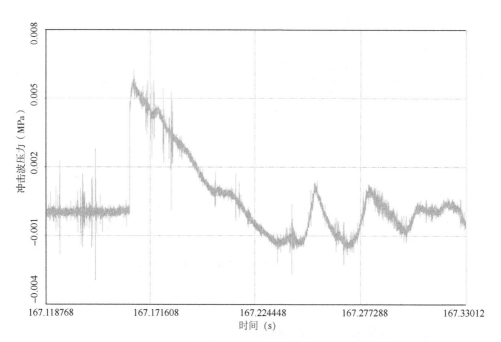

图 4　垂直方向距起爆位置 40m 位置上不同时刻的冲击波传压力

起爆管串放置在管沟中，并进行了回填处理，物理爆炸造成的冲击波受管沟及土壤回填的影响大大衰减。物理爆炸所产生的冲击波在向四周传播的过程中能量损失较大，因此远处传感器接收到的物理爆炸压力信号值较弱。

对两个方向上采集到的冲击波传感器数据进行分析，获得了试验过程中不同位置上产生的冲击波压力峰值，并对采集数据进行了滤波处理，获得了不同时刻上的冲击波压力变化曲线，如图 5 所示。从图中可见，不同位置上采集到的冲击波具有明显的时间间隔。距离除以时间，并求平均值，可计算出冲击波的传播速度为 357.1m/s，与声速(340m/s)接近。

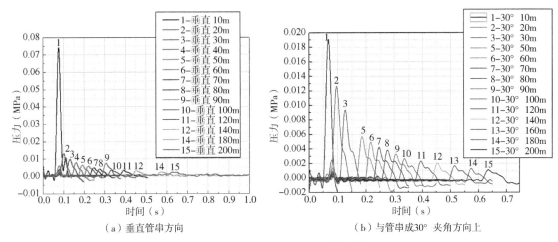

(a) 垂直管串方向　　　　　　　　　(b) 与管串成30°夹角方向上

图5　滤波处理后不同时刻的冲击波峰压力

图6为不同方向上冲击波峰压力随距离的变化规律。从图中可以看出，随着与起爆中心距离的增加，冲击波峰值迅速下降，冲击波能量衰减很快。距起爆中心80m以外的区域冲击波峰值相差不大。

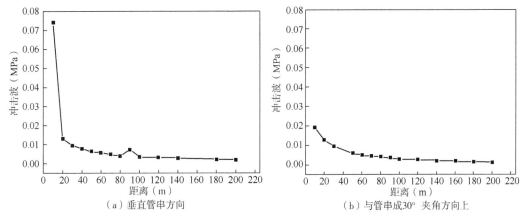

(a) 垂直管串方向　　　　　　　　　(b) 与管串成30°夹角方向上

图6　不同位置上的冲击波峰压力

表1为《安全评价方法应用指南》给出的冲击波压力对人的伤害。结合本次试验的冲击波超压峰值可以看出，只有在爆炸点附近(10m距离内)，爆炸冲击波才会对人造成致命杀伤。

表1　爆炸超压伤害分区标准[16]

伤害分区	超压(MPa)	备注
死亡区	≥0.1	大部分人员死亡
重伤区	0.05~0.10	内脏严重损伤或死亡
	0.03~0.05	听觉器官损伤或骨折
轻伤区	0.02~0.03	轻微损伤
	<0.02	安全

2.3 全尺寸气体爆破试验热辐射测试结果及分析

管道启裂后，天然气逸出并与空气充分混合，形成混合蒸气云。利用信号弹将其点燃，点燃高度距地面约150m。混合蒸气云发生闪爆，并向四周产生爆炸超压以及热辐射。对热辐射传感器采集到的热辐射通量峰值与传感器安装位置的关系利用公式进行拟合，可获得不同方向上的热辐射通量与距离的变化关系，如图7所示。从图中可以看出，随着距离的增加，热辐射通量峰值迅速衰减。相同距离位置上垂直方向上的热辐射值要高于30°方向上的热辐射值。

场地风速对天然气蒸气云爆炸有一定的影响。天然气泄漏后的混合蒸气云在风的作用下会发生一定的移动，引起爆炸位置的改变，从而导致地面相同位置不同方向上热辐射传感器数值的差异。

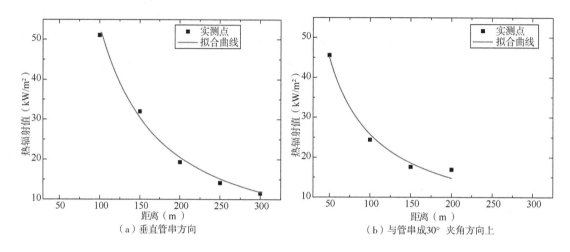

图7 不同位置上的热辐射通量

表2为根据热通量、热剂量准则对人员及设备的伤害情况，结合传感器测试结果，按照1%以下的致死率为安全距离来估计，当人暴露在热辐射小于10s时，安全距离为124.9m。综合考虑冲击波与热辐射的危害范围，可确定出此次爆破试验的安全距离为124.9m。

表2 热通量、热剂量准则的危害程度[16]

热通量 q (kW/m²)	热剂量 Q (kJ/m²)	对设备的损害程度	对人员的伤害程度	影响区域
37.5	375(10s)	操作设备全部破坏	10s，1%死亡；1min，100%死亡	死亡区
25	250(10s)	在无火焰、长时间的辐射下木材燃烧的最小能量	10s，重大伤亡；1min，100%死亡	重伤区
12.5	125(10s)	有火焰时，木材燃烧、塑料熔化的最小能量	10s，一度烧伤；1min，1%烧伤	轻伤区
4.0	40(10s)		20s以上感觉疼痛，未必起泡	
1.6	16(10s)		长期辐射无不适感	安全区

3 模型计算及与试验结果对比

本次试验中爆破段长度为130m,管内天然气体积为80000m³,天然气的热值为40.99~45.24MJ/m³,密度为0.7174kg/m³。这里考虑管道发生泄漏,根据泄漏口尺寸与管道尺寸的孔径比,假定泄漏百分比为3%,周围环境温度为-14~0℃,风速为4~7m/s,根据上述条件结合静态火球模型,可以估算出危害距离。

天然气质量与体积、密度存在以下关系:

$$m = \rho V \tag{1}$$

式(1)中 m 为天然气质量,ρ 为天然气密度,V 为天然气体积,可计算泄漏的天然气质量为57143kg。

天然气爆炸静态模型计算火球直径、火球高度可通过式(2)、式(3)分别计算:

$$D = 5.8 m^{1/3} \tag{2}$$

$$H = 4.35 m^{1/3} \tag{3}$$

上述公式中,D、H 分别为爆炸火球直径和高度,m 为天然气质量,根据公式可计算出火球的直径为223.4m,高度为167.5m。

根据天然气泄漏爆炸静态模型火球持续时间公式:

$$t = 0.25 m^{1/3} \tag{4}$$

可计算出火球的持续时间为9.6s。

热剂量与距离存在以下关系:

$$x = m^{1/3} e^{-0.541(10.3516 - \ln Q)} \tag{5}$$

式(5)中 x 为距起爆点的距离,Q 为热通量,m 为天然气质量,结合表2,可以计算出火球持续燃烧9.6s后,1%致死概率下的安全半径为147.6m,与试验结果相比较为保守。

4 结论

(1)长输管道泄漏爆炸后的主要危害因素为爆炸引起的冲击波以及爆燃时的热辐射。随着与起爆点距离的增大,冲击波压力迅速降低,冲击波传播的速度平均值为357.1m/s,不同方向(垂直管串、与管串成30°夹角方向)上,冲击压力随距离的变化规律相同。

(2)随着与起爆点距离的增大,热辐射通量迅速降低,垂直爆破管串方向上的热辐射通量值较高,结合热通量规范可计算出热辐射安全距离为124.9m,高于冲击波危害的安全距离。

(3)静态火球模型估算出的热辐射危害半径为147.6m,比试验结果更为保守、安全。

参 考 文 献

[1] MACIEJ W. Possibilities of using X80, X100, X120 high-strength steels for onshore gas transmission pipelines [J]. Journal of Natural Gas Science and Engineering, 2015, 27: 374-384.

[2] AIHUA Q. Present development situation and problem analysis of oil and gas pipeline in our country [J]. International Petroleum Economics, 2009, 17(12): 57-59.

[3] 杜伟,李鹤林,王海涛,等. 国内外高性能油气输送管的研发现状[J]. 油气储运, 2016, 36(9): 577-582.

[4] ZHU X K. State-of-the-art review of fracture control technology for modern and vintage gas transmission

pipelines[J]. Engineering Fracture Mechanics, 2015, 148: 260-280.

[5] 唐献述, 王树民, 龙源, 等. 爆炸空气冲击波对动物伤害效应试验研究[J]. 工程爆破, 2012, 18(2): 104-106.

[6] Pipeline Research Council International(PRCI). Line rupture and the spacing of parallel lines[R]. Chantilly, VA, United States: PRCI, 2002.

[7] 关丽, 刘德俊, 周志强. 天然气管道泄漏爆炸实验分析[J]. 中国安全生产科学技术, 2014, 10(12): 40-45.

[8] DONG Y H, GAO H L, ZHOU J E, et al. Mathematical modeling of gas release through holes in pipelines [J]. Chemical Engineering Journal, 2003, 92(1): 237-241.

[9] JO Y D, AHN B J. Analysis of hazard areas associated with high pressure natural-gas pipelines[J]. Journal of Loss Prevention in the Process Industries, 2002, 15(3): 179-188.

[10] 王若菌, 蒋军成. LPG 蒸气云爆炸风险评估中的参数不确定性分析[J]. 南京工业大学学报, 2005, 27(6): 12-15.

[11] 罗艾民, 李万春, 吴宗之, 等. 扁平圆环局限空间蒸气云爆炸数值模拟及其在事故调查中的应用[J]. 中国安全科学学报, 2008, 18(1): 95-100.

[12] 骆洪森, 王为民, 张月, 等. 高压天然气管道泄漏爆炸后危害分析[J]. 当代化工, 2014, 43(7): 1330-1332.

[13] 付小芳, 孙文栋, 方江敏. 高压天然气管道危险区域分析[J]. 油气储运, 2010, 29(2): 127-130.

[14] 梁瑞, 张春燕, 姜峰, 等. 天然气管道泄漏爆炸后果评价模型对比分析[J]. 中国安全科学学报, 2007, 17(8): 131-136.

[15] 霍春勇, 李鹤, 张伟卫, 等. X80 钢级 1422mm 大口径管道断裂控制技术[J]. 天然气工业, 2016, 36(6): 78-83.

[16] 刘铁民, 张兴凯, 刘功智. 安全评价方法应用指南[M]. 北京: 化学工业出版社, 2005.

本论文原发表于《石油管材与仪器》2021 年第 7 卷第 2 期。

国内外几种 GMAW 焊丝强韧性对比试验研究

李为卫[1]　杨耀彬[1]　何小东[1]　王正卿[2]

(1. 中油集团石油管工程技术研究院，石油管材及装备材料服役行为与结构安全国家重点实验室；2. 中国船级社质量认证公司四川分公司)

摘　要：为了掌握油气管道环焊缝用熔化极气体保护焊(GMAW)焊丝的性能，为现场焊接焊材选用提供指导，对国内外几种气保实心焊丝形成管道环焊缝的强度、韧性进行了试验研究。试验采用 50%Ar+50%CO_2 混合气体的 GMAW 方法，在 31.8mm 厚的 L485 管线钢板上进行。试验结果表明，国产的 GMAW 焊丝在合适的工艺条件下，具有较高的强度和韧性，可用于高强度管线钢管道环缝的焊接。

关键词：GMAW；焊丝；管道环缝；强韧性

经济发展对能源的需求促进了油气管道的大规模建设。由于历史原因，中国多年来的管道建设基本采用手工、半自动的焊接方法，由于管理和技术问题，长输天然气管道近年来出现了多次环焊缝失效事故，运行管道检测也发现大量焊缝存在质量问题。对环焊缝焊接技术和质量管控的要求提高，以往的半自动、手工焊等焊接工艺已不能完全满足工程需要[1-3]。

GMAW 是油气长输管道使用最广泛的焊接方法，国外大量用于油气管道的焊接[4]，中国近年来也正在大量推广这种方法。目前，国内管道环焊缝 GMAW 气保护实心焊丝大量使用的有美国 LINCOLN、意大利 FILLEUR、奥地利 BOHLER 等进口产品，少量使用大桥、金桥、大西洋、锦泰等国产焊丝。由于实心焊丝与母材熔合后共同形成的焊接接头，其强度受母材、工艺参数等因素影响很大，文献[5]研究表明，焊缝金属的强度比焊丝自身的熔敷金属强度高出 120~150MPa，合理选择和使用焊接材料对保证焊缝金属的优良力学性能至关重要。为了促进国产焊接材料的发展，本文对 3 种国产焊丝和 4 种进口焊丝形成的焊缝的拉伸性能和冲击韧性进行了试验研究。

1　试验材料和方法

试验用母材采用国产厚度 31.8mm 钢板，其纵向抗拉强度 610MPa，屈服强度 526MPa，纵向低温(-20℃)冲击韧性 352J。沿钢板横向边缘加工坡口(焊缝对应钢管环向)，焊接坡口形式见图 1。焊接方法为 GMAW，试板采用倾斜 45°、下向焊方式进行小热输入量的多层多道焊接，试件编号、焊接材料、工艺参数等见表 1，焊接过程如图 2 所示。

焊接试验完成后在焊缝中心、沿焊缝长度方向加工标距段直径为 6.25mm、长度为

基金项目：国家重点研发计划课题"L485 高应变海洋管道环焊材料及工艺技术"(2018YFC0310305)和中油股份有限公司课题"天然气管道环焊缝修复用 B 型套筒维抢修关键技术及换管焊接工艺优化研究"(2019E-23-0501)。

作者简介：李为卫，男，1965 年生，教授级高级工程师，1988 年毕业于西安交通大学焊接专业，现主要从事油气输送管道材料研究及标准化工作。E-mail：liweiwei001@cnpc.com.cn。

25mm 的圆棒试样进行常温拉伸性能试验。垂直焊缝方向取样，在焊缝中心加工截面尺寸 10mm×10mm 的夏比 V 形缺口试样，在-20℃下进行低温冲击韧性试验。

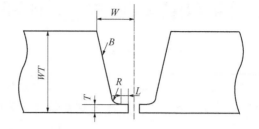

图 1 焊接坡口形式和尺寸

图 2 试件焊接试验过程

表 1 焊接试件编号、焊接材料及工艺参数

试件编号	焊丝	焊丝直径(mm)	保护气体	焊接热输入量(kJ/cm)
HS21	国产 A1	1.0	50%Ar+50%CO_2	4~6
HS22	国产 A2	1.0	50%Ar+50%CO_2	4~6
HS23	国产 A3	1.0	50%Ar+50%CO_2	4~6
HS24	进口 B1	1.0	50%Ar+50%CO_2	4~6
HS25	进口 B2	1.0	50%Ar+50%CO_2	4~6
HS26	进口 L1	1.0	50%Ar+50%CO_2	4~6
HS27	进口 L2	1.0	50%Ar+50%CO_2	4~6

2 试验结果及分析

采用 3 种国产焊丝和 4 种焊丝在试验条件下形成焊缝的拉伸和冲击试验结果见表 2。可以看出，在试验焊接条件下，国产焊丝 A1、A2、A3 具有较高的强度，同时具有良好的塑性和韧性，与国外进口焊丝 B1 相当，优于国外 B2、L1、L2 焊丝；国产焊丝 HS23 的强度更高，塑性良好，韧性更好，是试验焊丝中同时兼有高强度和高塑韧性的最好焊丝。

表2 不同焊丝焊缝金属的拉伸和冲击韧性试验结果

试件编号	全焊缝金属拉伸			−20℃冲击	
	抗拉强度 R_m(MPa)	屈服强度 $R_{t0.5}$(MPa)	伸长率 A(%)	总吸收功 A_{kv}(J)	剪切面积 SA(%)
HS21	704	661	27.0	131	95
HS22	719	677	25.3	139	98
HS23	792	765	24.8	161	100
HS24	694	661	25.5	161	97
HS25	838	808	15.0	79	90
HS26	752	709	22.8	117	99
HS27	759	725	25.5	109	96

X70、X80高强度管道环焊缝用气保护实心焊丝要求具备较高强度的同时具有良好的塑性和韧性。试验采用的国产焊丝，通过合金成分的合理设计，在现场环焊较低热输入量的焊接条件下，得到的焊缝组织以细小的针状铁素体为主，保证了焊缝金属具有高强度的同时具有高的塑性和韧性，实现了强度和塑韧性的良好匹配。图3是几种焊丝形成的焊缝中心的显微组织照片，其以细小的针状铁素体为主，先共析铁素体量很少。

对于焊接结构应采用等强、高强还是低强匹配，国内外的认识不一致[4,6-8]。但大多数研究认为，金属结构焊缝的高强匹配，对焊缝的抗断裂、变形更加有利，低强匹配焊缝容易产生应变集中，加上焊缝缺陷，容易使焊缝开裂失效[9]。油气管道焊缝强度匹配，国外著名油公司有不同的规定，中国目前已建成的X80钢级管道，许多环缝的实际强度与母材相比为低强匹配[4]。建议加强对油气管道失效的断裂机理研究，对重要天然气管道提出合理的强度匹配指标，为设计和制造焊材提供指导。

另外，实际生产中由于钢管和焊丝性能的不稳定性，加上环缝施工中的多种因素导致焊缝性能的分散性，影响了焊缝与母材强度的合理匹配，给焊缝的服役安全带来隐患。建议提高管道用钢管性能的稳定性，缩小屈服强度波动范围，在标准中提出更严格的控制指标，同时也要加强对焊材的性能复验和焊接工艺参数的控制，保证焊接质量的稳定性[9]。

3 结论

（1）试验用国产焊丝A1、A2、A3具有较高的强度，同时具有良好的塑性和韧性，与国外进口焊丝B1相当，优于国外B2、L1、L2焊丝；国产焊丝A3的强度更高，塑性良好，韧性更好，是试验焊丝中兼有高强度和高塑韧性的最好焊丝。

（2）试验采用的国产焊丝，通过焊丝合金成分的合理设计，实现以针状铁素体为主的焊缝组织控制，保证了焊丝具有高强度的同时具有高的塑性和韧性，实现了强度和塑韧性的良好匹配。

（3）建议对重要天然气管道提出屈服强度的合理匹配指标要求，进行焊接材料强度级别的选择，合理设计和制造焊丝。

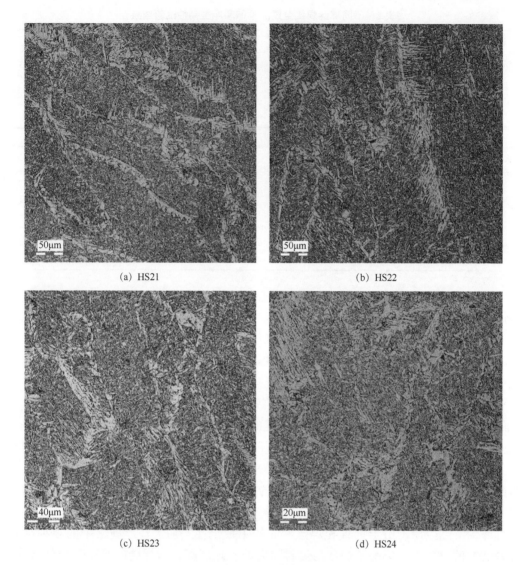

(a) HS21　　(b) HS22　　(c) HS23　　(d) HS24

图 3　焊缝中心显微组织照片

参 考 文 献

[1] 隋永莉,郭锐,张继成.管道环焊缝半自动焊与自动焊技术对比分析[J].焊管,2013,36(9):38-47.
[2] 张振永.高钢级大口径天然气管道环焊缝安全提升设计关键[J].油气储运,2020,39(7):740-748.
[3] 李少华,尹士科,刘奇凡.焊接接头强度匹配和焊缝韧性指标综述[J].焊接,2018,62(1):24-27.
[4] 李为卫,何小东,葛加林.油气管道环缝焊接国外先进标准的启示和借鉴[J].石油管材与仪器,2020,6(2):1-7.
[5] 陆阳,邵强,隋永莉,等.大管径、高钢级天然气管道环焊缝焊接技术[J].天然气工业,2020,40(9):114-122.
[6] 曹雷,孙谦,宗培.等强匹配焊接接头的特征及界定方法[J].焊接学报,2006,27(7):81-84.
[7] 庄传晶,冯耀荣,霍春勇.西气东输管道环焊缝强度匹配工艺探讨[J].机械工程材料,2005,29(8):32-34.

［8］STEPHEN L I U. Critical Concerns of Welding High Strength Steel Pipelines：X80 and Beyond Pipe［C］// Proceedings of the International Pipe Dreamer's Conference, Pacifico Yokohama, Janpan, 2002.

［9］李为卫，许晓锋，方伟.X80 及其以上高强度管线钢焊接的几个技术问题[J]. 石油管材与仪器，2016，2(2)：88-92.

本论文原发表于《石油管材与仪器》2021 年第 7 卷第 2 期。

热输入量对熔化极气体保护焊缝强韧性的影响

李为卫[1] 李嘉良[2] 梁明华[1] 何小东[1] 杨耀彬[1]

(1. 中油集团石油管工程技术研究院，石油管材及装备材料服役行为与结构安全国家重点实验室；2. 中移系统集成有限公司)

摘 要：为了掌握熔化极气体保护焊(GMAW)管道环焊缝的性能，为现场焊接焊材选用和工艺参数制订提供指导，对国内外几种气体保护实心焊丝在不同焊接热输入参数下焊缝的强度、韧性进行了试验研究。试验结果表明，热输入参数对焊缝金属的强度有很大影响，对韧性也有明显的影响，焊缝的强度较焊接材料熔敷金属有显著的升高。

关键词：GMAW；焊丝；管道环缝；强韧性

经济的发展对能源的需求，促进了油气管道大规模建设。由于历史原因，我国多年来的管道建设基本采用手工、半自动的焊接方法，焊接质量主要依赖焊工的技术水平。由于管理和技术问题，长输天然气管道近年来出现了多次环焊缝失效事故，运行管道检测过程也发现大量焊缝存在质量问题。对环焊缝焊接技术和质量管控的要求提高，以往的半自动、手工焊等焊接工艺已不能完全满足工程需要[1-3]。熔化极气体保护焊(GMAW)是油气长输管道最适用的焊接方法，一般采用自动或机械方式，效率高、质量优，国外将其大量用于油气管道的焊接[4]，我国近年来也正在大力推广这种方法。

管道环缝焊接工艺和选材一个重要的考虑就是焊缝的强度和韧性的匹配。焊接结构应采用高强匹配还是低强匹配，国内外有大量的文献，但认识不一致[4,5-7]。尽管国内的研究认识和标准规定不统一，但从大多数主流的观点来看，金属结构焊缝的高强匹配，对焊缝的抗断裂、变形更加有利，高强匹配的焊缝对缺陷的容限高，韧性要求低。低强匹配焊缝容易产生应变集中，加上焊缝缺陷，容易使焊缝开裂失效，给管道运行安全带来隐患[8]。

由于GMAW实心焊丝与母材熔合后共同形成的焊接接头，其强度受母材、工艺参数等因素影响很大，文献[9]研究表明，焊缝金属的强度比焊丝自身的熔敷金属强度高出120～150MPa，因此合理选择和使用焊接材料以及合适的焊接工艺参数对保证焊缝金属的优良力学性能至关重要。为了掌握焊接工艺参数对焊缝性能的影响规律，本文对2种国产焊丝和2种进口焊丝GMAW焊缝的拉伸性能和冲击韧性进行了试验研究和分析。

基金项目：国家重点研发计划课题"L485高应变海洋管道环焊材料及工艺技术"(2018YFC0310305)和中油股份有限公司课题"天然气管道环焊缝修复用B型套筒维抢修关键技术及换管焊接工艺优化研究"(2019E-23-0501)。

作者简介：李为卫，男，1965年生，教授级高级工程师，1988年毕业于西安交通大学焊接专业，现主要从事油气输送管道材料研究及标准化工作。E-mail：liweiwei001@cnpc.com.cn。

1 试验材料和方法

试验用母材采用的焊接材料为直径为1.0mm的实心焊丝,按焊材相关标准进行熔敷金属性能试验,见表1。试件采用钢级为L485M、厚度31.8mm钢板,其纵向抗拉强度610MPa,屈服强度526MPa,纵向低温(-20℃)冲击韧性352J。沿钢板横向边缘加工坡口(焊缝对应钢管环向),焊接坡口形式见图1。焊接方法为GMAW,试板采用倾斜45°、下向焊方式进行不同热输入量的多层多道焊接,保护气为50%Ar+50%CO_2,试件编号、焊接材料、工艺参数等见表2。

表1 焊丝熔敷金属拉伸和冲击韧性试验结果

焊丝	焊接电流(A)	电弧电压(V)	焊速(mm/s)	热输入量(kJ/cm)	屈服强度 σ_s(MPa)	抗拉强度 σ_b(MPa)	延伸率 A(%)	韧性 A_{kV}(J)	备注
国产A2	260-290	27-32	5.5±1.0	~14.8	500	590	26	-30℃:124	厂家试验数值
国产A3	200	26	4.67	~11.1	576	672	26	-40℃:157	试验批实测数据
进口B	260-290	27-31	5.5±0.5	~14.5	510	640	35	-40℃:75	厂家典型数值
进口L	260-290	27-31	5.5±0.5	~14.5	500	600	25	-40℃:81	厂家典型数值

表2 焊接试件编号、焊接材料及工艺参数

序号	试件编号	焊丝	焊接电流(A)	电弧电压(V)	焊速(cm/min)	热输入量(kJ/cm)	预热/层间温度(℃)
1	HS15	国产A2	150~190	25~28	18.1	15	60~120
2	HS22	国产A2	240~280	24~27	61.2	6.5	60~120
3	HS33	国产A3	250~270	26~28	35.1	12	60~120
4	HS23	国产A3	240~280	24~27	49.7	8.0	60~120
5	HS11	进口B	160~200	25~28	23.9	12	60~120
6	HS24	进口B	240~280	24~27	58.5	6.8	60~120
7	HS13	进口L	150~190	25~28	28.6	15	60~120
8	HS26	进口L	240~280	24~27	56.8	7.0	60~120

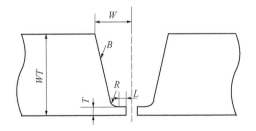

图1 试件焊接坡口形式和尺寸

焊接试验完成后在焊缝中心、沿焊缝长度方向加工标距段直径为6.25mm、标距长为25mm的圆棒试样进行常温拉伸性能试验。垂直焊缝方向取样,在焊缝中心加工截面尺寸10mm×10mm的夏比V形缺口试样,在-20℃下进行低温冲击韧性试验。

2 试验结果及分析

采用4种焊丝在不同热输入参数条件下形成焊缝的拉伸性能和冲击试验结果见表3。与表1焊丝熔敷金属拉伸和冲击韧性试验相比，由于试验条件不同，焊缝的屈服和抗拉强度比熔敷金属的强度有明显的提高，尤其是低热输入参数下有大幅度提高。例如，国产A3焊丝，在较高热输入下，焊缝的屈服和抗拉强度比焊丝熔敷金属的试验值分别升高128MPa、94MPa；在较低热输入下，焊缝的屈服和抗拉强度比焊丝熔敷金属的试验值分别升高189MPa、122MPa。进口B焊丝，在较高热输入下，焊缝的屈服和抗拉强度比焊丝熔敷金属的典型值分别升高105MPa、31MPa；在较低热输入下，焊缝的屈服和抗拉强度比焊丝熔敷金属的典型值分别升高151MPa、54MPa。

造成上述差异的原因主要因素有：(1)采用的试件材料不同。焊丝熔敷金属试验一般采用碳钢试件，本试验采用的为微合金管线钢试件；(2)焊接坡口形式和尺寸差别大；(3)焊接工艺参数的不同。由于这些因素的差异，造成母材金属过渡到焊缝金属的合金元素不同（母材过渡到试验焊缝中的合金元素高于熔敷金属试验），合金元素烧损以及焊缝冷却速度的快慢造成显微组织存在差异，因而造成焊缝金属与焊丝熔敷金属拉伸强度的较大变化。

从表3试验数据可以看出，与较高热输入参数相比，在较低的焊接热输入参数下，每种焊丝全焊缝金属的屈服和抗拉强度均升高，其中屈服强度升高幅度较抗拉强度更明显，屈服强度最大升高167MPa，抗拉强度最大升高131MPa。与较高的热输入参数相比，在较低热输入参数下，焊缝金属的伸长率有不同程度的下降，最大下降3.2%，但仍有较高的伸长率。较低热输入参数下，焊缝的韧性表现不一致，两种焊丝（国产A2、进口B焊丝）有明显的升高，另外两种焊丝保持基本不变。

表3 不同焊接热输入参数焊缝金属的拉伸性能和冲击韧性试验结果

序号	试件编号	焊丝	热输入量 (kJ/cm)	全焊缝金属拉伸			焊缝-20℃夏比吸收功平均值(J)
				屈服强度 $R_{t0.5}$(MPa)	抗拉强度 R_m(MPa)	伸长率 A(%)	
1	HS15	国产A2	1.5	510	588	28.5	78
2	HS22	国产A2	0.65	677	719	25.3	139
变化量（较低热输入-较高热输入）				167	131	-3.2	61
3	HS33	国产A3	1.2	704	766	25.8	162
4	HS23	国产A3	0.80	765	792	24.8	161
变化量（较低热输入-较高热输入）				61	26	-1	-1
5	HS11	进口B	1.2	616	671	27.5	115
6	HS24	进口B	0.68	661	694	25.5	161
变化量（较低热输入-较高热输入）				45	23	-2	46
7	HS13	进口L	1.5	595	667	24.5	123
8	HS26	进口L	0.70	709	752	23.5	117
变化量（较低热输入-较高热输入）				114	85	-1	-6

性能的差异归功于其显微组织的变化，焊接热输入既可改变焊缝金属一次结晶组织，又

可改变多层多道焊时焊缝金属的二次组织。图2为国产A2焊丝在两种不同的热输入参数下的填充焊缝典型的显微组织，可以看出，其均由针状铁素体和先共析铁素体为主组成，但是，与1.5kJ/cm较高的热输入参数相比，在0.65kJ/cm较低的热输入参数下先共析铁素体的量明显减少，针状铁素体变得更加细小，因而其强度和韧性提高更加明显。图3为进口B焊丝在两种不同的热输入参数下的填充焊缝典型的显微组织，可以看出，与图2相比，1.2kJ/cm较高热输入参数与0.68kJ/cm较低热输入参数的先共析铁素体的量和针状铁素体变化不明显，因而其强度和韧性提高程度相对较小。

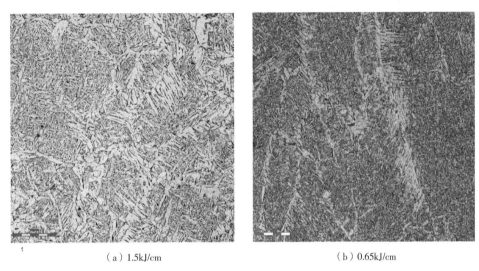

（a）1.5kJ/cm　　　　　　　　　　（b）0.65kJ/cm

图2　国产A2焊丝在不同热输入参数下的填充焊缝显微组织

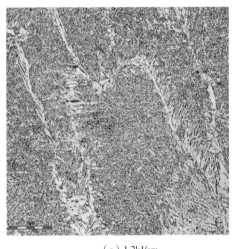 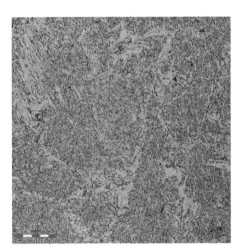

（a）1.2kJ/cm　　　　　　　　　　（b）0.68kJ/cm

图3　进口B焊丝在不同热输入参数下的填充焊缝显微组织

从理论上分析，在较高的热输入参数下，因为焊缝结晶冷却速度较慢，焊缝先共析铁素体量较多，针状铁素体较粗大，而在较低的热输入参数下，焊缝结晶冷却速度较快，焊缝先共析铁素体量较少，针状铁素体更细小。焊缝中大量针状铁素体和可显著提高微裂纹扩展抗力，增强焊缝金属的强韧性[9]。

焊缝金属的强度和韧性，一直是焊接结构关注的重点，尤其是高强度高压天然气管道环缝的焊接，由于受到不明确的外力，加上不可避免的缺陷，近年来成为行业关注和研究的重点。从焊接材料和工艺的角度，选用合适的焊材，在合适的工艺参数下形成强度合理匹配、韧性高的焊接接头，从而保证焊接结构的安全服役。从以上试验分析可以看出，工艺参数对GMAW实心焊丝形成焊缝的强度和韧性有很大的影响，应引起大家关注。

3 结论

通过试验研究和结果分析，可以得出以下结论和建议：

（1）GMAW实心焊丝的焊接接头在试验条件下，焊缝金属的屈服强度和抗拉强度较焊材熔敷金属的强度有明显的提高，尤其是在较低热输入参数下，有大幅的提高。

（2）焊接热输入参数对GMAW实心焊丝形成焊缝金属的显微组织、强度和韧性有很大影响。在较低的热输入参数下，填充焊缝的针状铁素体更加细小，先共析铁素体量更少，焊缝的强度更高，韧性更好。

（3）建议系统研究不同的焊接材料，结合具体的管道用钢材成分，在现场工艺参数下强韧性的变化规律，合理选用强度和韧性与母材匹配的焊接材料，并在与实际管道相近的焊接条件进行焊接材料的性能复验。

参 考 文 献

[1] 隋永莉,郭锐,张继成. 管道环焊缝半自动焊与自动焊技术对比分析[J]. 焊管,2013,36(9)：38-47.

[2] 张振永. 高钢级大口径天然气管道环焊缝安全提升设计关键[J]. 油气储运,2020,39(7)：740-748.

[3] 李少华,尹士科,刘奇凡. 焊接接头强度匹配和焊缝韧性指标综述[J]. 焊接,2018,62(1)：24-27.

[4] 李为卫,何小东,葛加林. 油气管道环缝焊接国外先进标准的启示和借鉴[J]. 石油管材与仪器,2020,6(2)：1-7.

[5] 陆阳,邵强,隋永莉,等. 大管径、高钢级天然气管道环焊缝焊接技术[J]. 天然气工业,2020,40(9)：114-122.

[6] 曹雷,孙谦,宗培,等. 强匹配焊接接头的特征及界定方法[J]. 焊接学报,2006,27(7)：81-84.

[7] 庄传晶,冯耀荣,霍春勇. 西气东输管道环焊缝强度匹配工艺探讨[J]. 机械工程材料,2005,29(8)：32-34.

[8] LIU S. Critical Concerns of Welding High Strength Steel Pipelines：X80 and Beyond Pipe[C]//Proceedings of the International Pipe Dreamer's Conference,7-8,November 2002,Yokohama,Janpan：91-107.

[9] 李为卫,许晓锋,方伟. X80及其以上高强度管线钢焊接的几个技术问题[J]. 石油管材与仪器,2016,2(2)：88-92.

[10] 毕宗岳,刘海璋,牛辉,等. X80管线钢环焊缝气体保护焊焊丝的研制[J]. 焊管,2012,35(10)：5-9.

本论文原发表于《焊管》2021年第49卷第4期。

某凝析油集输管线内腐蚀影响因素分析

李磊[1]　陈庆国[2]　袁军涛[1]　宋鹏迪[3]　白真权[1]

(1. 中国石油集团石油管工程技术研究院，石油管材及装备材料服役行为与结构安全国家重点实验室；2. 中国石油塔里木油田公司油气工程研究院；
3. 西安石油大学材料科学与工程学院)

摘要：某凝析油集输管线投产后内腐蚀穿孔频发，复杂的腐蚀影响因素导致难以制订有效的防腐措施。采用高压釜模拟管线内部腐蚀环境，结合光学显微镜、扫描电子显微镜和X射线衍射仪等对试样进行分析，以探讨微量硫化氢(H_2S)、二氧化碳(CO_2)和氯离子(Cl^-)对碳钢内腐蚀的作用规律。试验结果表明：该管线以CO_2腐蚀控制为主，腐蚀产物为$FeCO_3$，既有均匀腐蚀又有局部腐蚀，微量H_2S和Cl^-的存在促进局部腐蚀的发生；H_2S含量由50mg/L增加至100mg/L时，平均腐蚀速率增加约30%，局部腐蚀坑尺寸显著增大；Cl^-浓度由69600mg/L增加至120000mg/L时，平均腐蚀速率未见明显变化，但局部腐蚀坑增多。

关键词：集输管线；内腐蚀；CO_2腐蚀；H_2S腐蚀；局部腐蚀

随着油气田开发的不断深入，采出油气介质含水率不断上升，地面管线腐蚀问题越来越严重[1-2]，已显著影响油气田的生产效率。腐蚀不仅造成生产安全事故，带来严重经济损失，而且极易引发生态环境污染，影响人类生存环境，产生恶劣的社会影响[3]。

油气集输管道通常输送气、水、烃、固共存的多相流介质，采出水总矿化度较高，易产生水垢的离子多，还有溶解氧、二氧化碳、硫化物等腐蚀性介质和大量的SRB、TGB细菌以及泥沙，这导致油气集输管道结垢和腐蚀严重[4]。其中，硫化氢、二氧化碳和氯离子等对油气集输管道的腐蚀影响最大，国内外已有大量学者对此进行了研究[5-6]。然而，各油气田的腐蚀环境不同、管材类型各异，由此导致它们的腐蚀形态、腐蚀主要因素和腐蚀机理变得复杂多样。因此，针对具体的油气田腐蚀问题，开展相应的腐蚀与防护研究既十分必要，又意义重大。

本文针对某凝析油集输管线连续腐蚀穿孔问题，通过采用高压釜模拟管线实际服役工况，进而研究硫化氢(H_2S)、二氧化碳(CO_2)和氯离子(Cl^-)对钢管内腐蚀的影响规律，目的是为此类管线的腐蚀防护提供理论依据。

1　实验方法

试验材料取自油田现场所用钢管，钢级L245M，其化学成分检测结果见表1，符合标准GB/T 9711的要求，高压釜试片规格为50mm×10mm×3mm。试片逐级(240#、400#、600#和

作者简介：李磊，男，1987年生，2014年硕士毕业于西北工业大学材料专业，目前主要从事石油管材的腐蚀与防护工作。E-mail：lilei08@cnpc.com.cn。

800#)打磨后,清洗吹干,贮于干燥器中,放置1h后测量尺寸和称量(精确至0.1mg)。

试验溶液为模拟水,参照现场水样检测结果配制,溶液离子浓度和管线服役工况见表2。试验采用大连科贸生产的25MPa动态高压釜,以实际服役工况为基础,通过改变CO_2分压(0.02MPa和0.09MPa),H_2S含量(0mg/L和100mg/L)及Cl^-浓度(0mg/L和120000mg/L)中的单一参量来模拟管线内部不同的流动腐蚀环境。试验前通入高纯氮2h以除氧,随后将高压釜密封,升温升压,待温度压力达到试验参数后,开始旋转试样以达到试验要求,试验周期120h。试验结束后,取出试样,用化学法清除表面腐蚀产物。化学清洗液按照SY/T 5273中附录A进行配置,具体成分为100mL盐酸+10g六亚甲基四胺+去离子水(加至1000mL)。试样经化学清洗后,依次用饱和氢氧化钠水溶液、去离子水、丙酮和无水乙醇超声清洗,冷气吹干后称重并计算平均腐蚀速率。

经高压釜试验后,对试样的腐蚀产物膜进行扫描电子显微镜(SEM)和X射线衍射仪(XRD)分析,以及利用光学显微镜对除膜后试样表面进行腐蚀形貌观察。

表1 试验材料的化学成分检测结果   单位:%(质量分数)

元素	C	Si	Mn	P	S	Cr	Mo	Ni	Nb	V	Ti	Cu
含量	0.074	0.055	0.77	0.0084	0.0032	0.020	<0.005	0.0057	<0.005	<0.005	0.0017	0.014
标准	≤0.22	≤0.45	≤1.20	≤0.025	≤0.015	≤0.30	≤0.15	≤0.30	≤0.05	≤0.05	≤0.04	≤0.50

表2 试验溶液的离子浓度

项目	HCO_3^-	SO_4^{2-}	Cl^-	Ca^{2+}	Mg^{2+}	Na^+
含量(mg/L)	956	1216.2	69600	5170.2	193.9	49170

注:溶液的pH值5.5(通入气体之前),温度45℃,总压3MPa,流速0.22m/s,CO_2分压0.18MPa,H_2S含量50mg/L。

2 结果与讨论

2.1 不同H_2S含量对管线内腐蚀的影响

2.1.1 平均腐蚀速率

图1为不同H_2S含量下平均腐蚀速率的变化。从图1可以看出,三组试样的平均腐蚀速率都大于0.25mm/a,属极严重腐蚀[7];随H_2S含量增加,平均腐蚀速率先减小后增大,最大增幅达30%;当H_2S含量为50mg/L时,平均腐蚀速率最低。

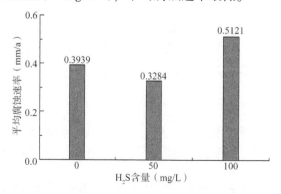

图1 不同H_2S含量下平均腐蚀速率的变化

据文献研究报道[5,8,9]，当 H_2S 含量小于某个临界含量时，腐蚀速率随 H_2S 含量的增加而减小，当 H_2S 的含量大于该临界含量时，腐蚀速率随 H_2S 含量的增加而增大。本试验条件中 H_2S 含量最高不超过 100mg/L，符合微量 H_2S 对 CO_2 腐蚀的影响条件，试验结果也与该研究吻合。通常环境温度较低(100℃以下)时，H_2S 通过加速腐蚀的阴极反应而加快腐蚀的进行，这是 H_2S 含量增大到 100mg/L 时平均腐蚀速率增大的原因；至于 H_2S 的含量为 50mg/L 时，平均腐蚀速率最低的原因应与腐蚀阴极反应受抑制有关，具体的机理比较复杂，仍需进一步研究。

2.1.2 腐蚀产物膜形貌和组成

利用 SEM 观察腐蚀产物的微观形貌，如图 2 所示。从图 2(a)可以看出，不含 H_2S 时，腐蚀产物膜较完整，表面凹凸不平；当 H_2S 含量为 50mg/L 时，试样的腐蚀产物膜完整性大大降低，呈破碎状[图 2(b)]；而当 H_2S 含量为 100mg/L 时，试样表面的腐蚀产物已基本脱落，仅有少量残留[图 2(c)]。这说明 H_2S 含量的增加破坏了腐蚀产物膜的完整性，降低了其与基体金属的附着力。图 3 为腐蚀产物的 XRD 分析图谱。从图 3 可以看出，三种情况下，腐蚀产物的组成均为 $FeCO_3$，而无 FeS，说明 H_2S 未与基体直接反应，依然属于 CO_2 腐蚀。

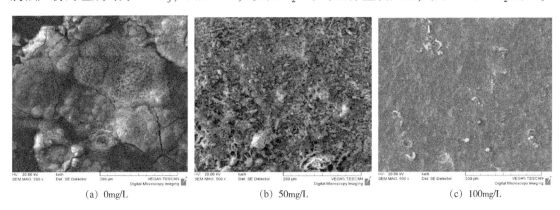

(a) 0mg/L　　　　　　　(b) 50mg/L　　　　　　　(c) 100mg/L

图 2　不同 H_2S 含量下腐蚀产物的微观形貌

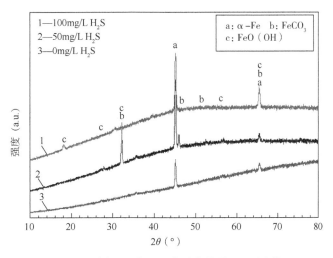

图 3　不同 H_2S 含量下腐蚀产物的 XRD 图谱

2.1.3 局部腐蚀

利用光学显微镜对除膜后试样的局部腐蚀形貌进行观察,如图4所示。从图4可知,不含H_2S时,试样表面基本无腐蚀坑;当H_2S含量为50mg/L时,试样表面腐蚀坑显著增多,其直径范围为5~35μm;而当H_2S含量增加到100mg/L时,试样表面腐蚀坑密度减小,但其尺寸变大,直径范围为25~80μm。由此可见,H_2S含量增加,局部腐蚀程度显著增强。这主要是因为H_2S破坏了腐蚀产物膜的完整性,导致形成大阴极小阳极的腐蚀形态,促进了局部腐蚀的发生。

(a) 0mg/L (b) 50mg/L (c) 100mg/L

图4 不同H_2S含量下试样除膜后的点蚀形貌

2.2 不同CO_2分压对管线内腐蚀的影响

2.2.1 平均腐蚀速率

图5为不同CO_2分压下平均腐蚀速率的变化。从图5可以看出,三组试样的平均腐蚀速率均大于0.25mm/a,属极严重腐蚀[7];CO_2分压由0.02MPa增加到0.18MPa时,试样的平均腐蚀速率变化未表现出单调递增或单调递减的规律,而且变化幅度较小。

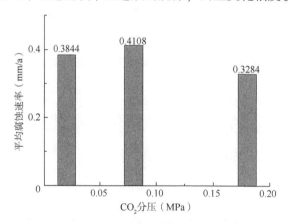

图5 不同CO_2分压下平均腐蚀速率的变化

有研究表明[6],在温度低于60℃,CO_2分压小于0.2MPa且介质为层流状态时,随着CO_2分压增加,腐蚀速率增大,而且CO_2腐蚀过程其实质是伴随着氢的去极化过程而进行的。这一过程是由溶液本身的水合离子来完成的,当CO_2分压高时,由于溶解的碳酸浓度高,从碳酸中分解氢离子浓度必然高,因而腐蚀被加速。但是,本试验中,平均腐蚀速率的变化不符合这一规律。这有三方面的原因:第一,在该腐蚀溶液中,还存在微量的H_2S和大量Cl^-等,它们对CO_2腐蚀会产生不同的影响;第二,在本试验中CO_2分压基本在中等腐蚀程度以内，而平均腐蚀速率均已达到极严重腐蚀，所以CO_2分压变化较小时，平均腐蚀速率差异不大；第三，受试样材质和设备系统误差的影响，也会使平均腐蚀速率出现偏差。综合这三方面的原因，所以试样的平均腐蚀速率无显著规律，且差异不大。

2.2.2 腐蚀产物膜形貌和组成

利用SEM观察腐蚀产物膜的微观形貌，如图6所示。由图6可知，当CO_2分压为0.02MPa和0.09MPa时，试样的腐蚀产物膜呈块状，比较完整，而当CO_2分压为0.18MPa时，腐蚀产物膜比较破碎。腐蚀产物XRD物相结分析结果见图7。从图7可以看出，三种CO_2分压下，腐蚀产物的组成均为$FeCO_3$，而无FeS，说明H_2S未与基体直接反应，属于CO_2腐蚀。

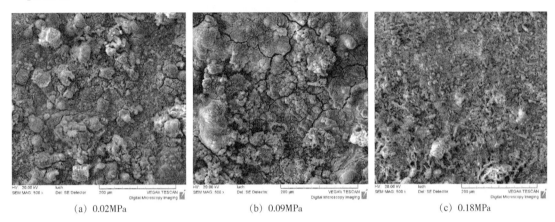

(a) 0.02MPa　　　　　　　　(b) 0.09MPa　　　　　　　　(c) 0.18MPa

图6　不同CO_2分压下腐蚀产物的微观形貌

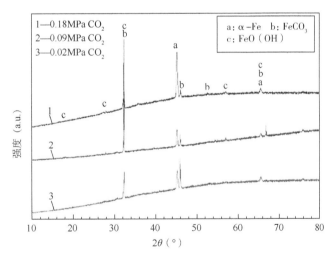

图7　不同CO_2分压下腐蚀产物的XRD图谱

2.2.3 局部腐蚀

利用光学显微镜对除膜后的试样进行观察，如图8所示。从图8可以看出，当CO_2分压为0.02MPa和0.09MPa时，试样表面基本无较大腐蚀坑；而当CO_2分压为0.18MPa时，试样表面腐蚀坑显著增多，其直径范围为5~35μm。这是因为CO_2分压增加后腐蚀产物膜增厚，但受H_2S的影响又使腐蚀产物膜遭受破坏，完整性降低，容易形成大阴极小阳极的腐蚀形态，从而会使点蚀程度增加。

图 8 不同 CO_2 分压下试样除膜后的点蚀形貌

2.3 不同 Cl^- 浓度对管线内腐蚀的影响

2.3.1 平均腐蚀速率

图 9 为不同 Cl^- 浓度下平均腐蚀速率的变化。从图 9 可以看出，Cl^- 浓度的变化对平均腐蚀速率的影响较小，Cl^- 浓度增加后，平均腐蚀速率的变化幅度在 8% 以内，且均为严重腐蚀。

研究表明，在 NaCl 水溶液中 CO_2 的溶解度随盐溶液的浓度增大而下降[10]，而当 CO_2 浓度降低后溶解于水中的碳酸浓度降低，从碳酸分离出的氢离子浓度也随之减少，从而导致氢的去极化反应减弱，最终使碳钢的腐蚀速率降低。然而，虽然 Cl^- 浓度增大使 CO_2 的溶解度降低，但是由前述 CO_2 对平均腐蚀速率的影响规律可知，CO_2 浓度降低后，平均腐蚀速率的变化较小，所以最终试样的平均腐蚀速率变化较小。

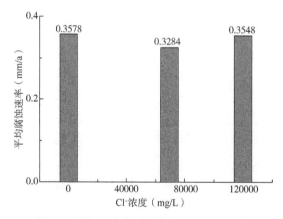

图 9 不同 Cl^- 浓度下平均腐蚀速率的变化

2.3.2 腐蚀产物膜形貌和组成

不同 Cl^- 浓度下腐蚀产物的微观形貌如图 10 所示。由图 10 可知，不含 Cl^- 时，腐蚀产物膜比较完整；当 Cl^- 浓度增加到 69600mg/L 时，腐蚀产物膜比较破碎；当 Cl^- 浓度增加到 120000mg/L 时，腐蚀产物膜凹凸不平，膜的完整性进一步降低。此外，腐蚀产物的 XRD 分析(图 11)可知，它们的组成均为 $FeCO_3$。

2.3.3 局部腐蚀

利用光学显微镜对除膜试样的局部腐蚀形貌进行观察，如图 12 所示。由图 12 可知，不含 Cl^- 时，试样表面有较大腐蚀坑，其直径约为 700μm；当 Cl^- 浓度增加时，试样表面腐蚀坑数量增加，但直径变小，这主要是因为 Cl^- 浓度增加后，更易穿透腐蚀产物膜，易于形成局部腐蚀，但又由于 Cl^- 浓度增加使 CO_2 溶解度降低，CO_2 的腐蚀产物较少，反应更倾向于均匀腐蚀，所以最终呈有较多小腐蚀坑的均匀腐蚀。

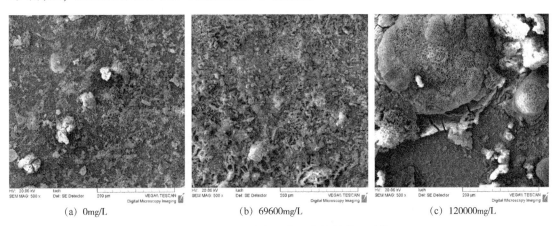

图 10 不同 Cl^- 浓度下腐蚀产物的微观形貌

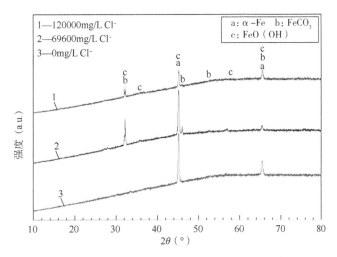

图 11 不同 Cl^- 浓度下腐蚀产物的 XRD 图谱

图 12 不同 Cl^- 浓度下试样除膜后的点蚀形貌

3 结论

(1) 该凝析油集输管线的内腐蚀以 CO_2 腐蚀控制为主，腐蚀产物为 $FeCO_3$，既有均匀腐蚀又有局部腐蚀，微量 H_2S 和 Cl^- 的存在促进局部腐蚀的发生。

(2) 微量 H_2S 的存在破坏了 CO_2 腐蚀产物膜的完整性，H_2S 含量由 50mg/L 增加至 100mg/L 后，平均腐蚀速率提高约 30%，局部腐蚀坑尺寸显著增大。

(3) Cl^- 浓度由 69600mg/L 增加至 120000mg/L 时，平均腐蚀速率未见明显变化，但局部腐蚀坑增多。

参 考 文 献

[1] 宋成立，方艳，陈庆国，等．集油管道失效原因分析[J]．石油管材与仪器，2019，5(4)：74-77．

[2] 马相阳，王义，杨晓辉，等．安塞油田集输管道内腐蚀防护技术研究与应用[J]．石油管材与仪器，2020，6(1)：57-59．

[3] 王春泉．雅克拉气田集输管材的 CO_2 腐蚀研究[D]．北京：中国石油大学(北京)，2009：7-8．

[4] 吴九虎．油气集输管线内防腐技术研究[J]．科技风，2012(3)：38．

[5] 陈卓元，张学元，王凤平，等．二氧化碳腐蚀机理及影响因素[J]．材料开发与应用，1998，13(5)：34-40．

[6] DWEAARD C, MILINATS D E, Carbon Acid Corrosion of steel[J]. Corrosion, 1975, 31(5): 177.

[7] NACE RP-0775-2005, Preparation, Installation, Analysis, and Interpretation of Corrosion Coupons in Oilfield Operations[S].

[8] 李春福，王斌，张颖，等．油气田开发中 CO_2 腐蚀研究进展[J]．西南石油学院学报，2004，26(2)：42-46．

[9] 张忠烨，郭金宝．CO_2 对油气管材的腐蚀规律及国内外研究进展[J]．宝钢技术，2000，(4)：54-58．

[10] 顾飞燕．加压下二氧化碳在氯化钠水溶液中的溶解度[J]．高校化学工程学报，1998，12(2)．

本论文原发表于《石油管材与仪器》2021年第7卷第4期。

某输油管道腐蚀穿孔原因

吉 楠[1] 廖 臻[2] 朱 辉[2] 李丽锋[1]

(1. 中国石油集团石油管工程技术研究院,石油管材及装备材料服役行为与结构安全国家重点实验室;2. 中国石油天然气股份有限公司新疆油田油气储运分公司)

摘要:某油田在巡线过程中发现输油管道有泄漏现象。通过宏观分析、理化性能检测、金相分析、腐蚀形貌观察和腐蚀产物分析等方法,并结合服役工况,分析了输油管道腐蚀穿孔的原因。结果表明:环焊缝防腐补口的密封失效而导致的外腐蚀是输油管道腐蚀穿孔的主要原因,同时土壤中的 Cl^- 加速了腐蚀穿孔的发生。

关键词:输油管道;泄漏;外腐蚀;防腐补口;穿孔

石油开采后往往需要异地输送,其距离可达数千千米。管道输送作为一种经济、高效而安全的物料输送手段,在我国国民经济的发展中起着十分重要的作用。已通过质量标准控制的管道投入运行后,随着使用年限的不断增长,其发生各种失效事故概率的可能性也在不断增加。由于输油管道多埋于地下,管控难度较大,一旦发生失效事故,造成原油泄漏,不仅会对油田造成巨大的经济损失,同时也会对周边环境造成污染[1-6]。因此针对输油管道失效泄漏事故开展分析研究,确保输油管道本质安全,显得尤为重要。

本文针对一起典型的输油管道泄漏事故,通过宏观分析、理化性能试验、金相分析、腐蚀表面微观形貌及腐蚀产物分析等手段,明确了管道泄漏原因,并为现场安全管理提供了决策依据。该发生泄漏的管线属埋地管线,规格为 $\phi114mm×4.5mm$ 无缝管,材质为20钢,输送介质为原油,设计压力为 4.0MPa。防腐保温结构采用聚乙烯保护层—硬质聚氨酯保温层—环氧煤沥青防腐层三层结构,防腐补口方式采用热收缩带防腐—浇注聚氨酯泡沫保温—热收缩套防水。该管线于2011年建成投运,至今一直运行稳定,未发生任何泄漏失效事故。

1 理化检验

1.1 宏观分析

将失效埋地管道开挖后取样,其宏观形貌如图1所示。由图1可知,在管段外壁的环焊缝补口区附近可见有1处明显的泄漏孔,位于管道底部6点钟的位置,孔径为20mm。泄漏孔附近的管道外壁被黑色和红褐色的物质所覆盖,同时在管道外壁防腐补口区以外,可见蓝色的环氧煤沥青防腐涂层。

将泄漏孔附近的管道外表面打磨处理后,其宏观形貌如图2(a)所示。由图2(a)可知,除刺漏孔外,管道外壁存还存在着大量大小、深浅不一的腐蚀坑,局部腐蚀坑尺寸较大,其

基金项目:国家重点研发计划(2017YFC0805804)。

作者简介:吉楠(1988—),男,工程师,主要从事石油管材失效分析与质量监督工作。E-mail:jinan003@cnpc.com.cn。

中最大的1处腐蚀坑直径约12mm。将管段沿轴对剖后，可观察到在其内壁覆盖有致密的黑色垢层[图2(b)]。去除垢层后，在内壁除刺漏孔外，未见明显腐蚀痕迹。

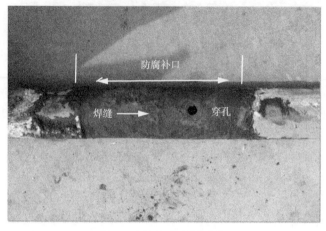

图1 腐蚀穿孔宏观形貌

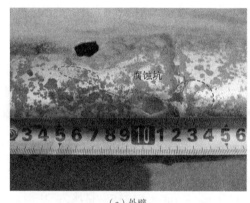
（a）外壁

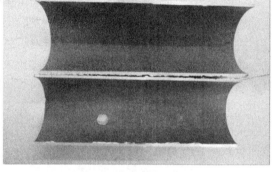
（b）内壁

图2 防腐补口区宏观形貌

1.2 几何尺寸测量

使用超声测厚仪对失效管道进行壁厚测量，测量结果显示：（1）环焊缝防腐补口区以外的壁厚最大测量值为5.08mm，最小测量值为4.29mm，壁厚未见明显减薄；（2）环焊缝防腐补口区以内的管道壁厚发生明显减薄，壁厚最小测量值出现在泄漏孔附近，测量结果为1.74mm，仅相当于公称壁厚的39%。

1.3 化学成分分析

从泄漏输油管道上截取块状样品，使用直读光谱仪进行化学成分分析，结果如表1所示。分析结果表明该失效输油管道的管体化学成分分析结果符合GB/T 8163—2018《输送流体用无缝钢管》对20钢的要求。

表1 泄漏输油管道的化学成分　　单位:%(质量分数)

元素	C	Si	Mn	P	S	Cr	Ni	Cu
实测值	0.17	0.21	0.51	0.010	<0.002	0.024	0.014	0.015
标准值	0.17~0.23	0.17~0.37	0.35~0.65	≤0.035	≤0.035	≤0.25	≤0.30	≤0.25

1.4 拉伸性能试验

在泄漏输油管道上沿纵向截取全壁厚板状拉伸试样,试样宽度20mm,标距50mm,进行室温拉伸试验,试验结果见表2。试验结果表明,该失效输油管道的拉伸性能试验符合GB/T 8163—2018标准的要求。

表2 泄漏输油管道拉伸性能试验结果

拉伸性能	抗拉强度(MPa)	屈服强度(MPa)	伸长率(%)
实测值	471	314	26.5
标准值	410~530	≥245	≥20

1.5 金相分析

在泄漏管道的管体未失效位置处取样,依据GB/T 13298—2015《金属显微组织检验方法》、GB/T 6394—2017《金属平均晶粒度测定方法》及GB/T 10561—2005《钢中非金属夹杂物含量的测定 标准评级图显微检验法》,对试样纵截面处的显微组织、晶粒度和非金属夹杂物进行检测分析。分析结果表明:管体未失效部位和泄漏孔处的金相组织均为铁素体和珠光体,晶粒度等级为9.5级,非金属夹杂物分别为硫化物、氧化铝、球状氧化物,如图3所示。管道外表面腐蚀坑形貌如图4所示。由图4可知,腐蚀坑内存在连续、较厚的腐蚀产物,腐蚀坑附近的组织与其他区域相同,均为铁素体和珠光体。

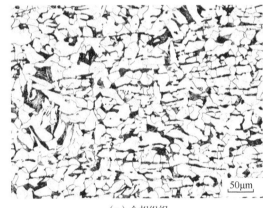

(a)金相组织　　　　　　　　　　(b)非金属夹杂物形貌

图3 泄漏输油管道未失效位置的显微组织形貌及非金属夹杂物形貌

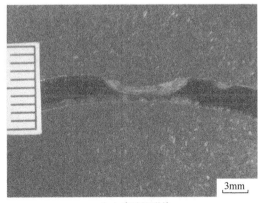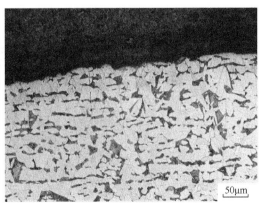

(a)腐蚀坑形貌　　　　　　　　　　(b)显微组织形貌

图4 泄漏输油管道外表面腐蚀坑形貌及显微组织形貌

1.6 微观分析

分别对泄漏孔及外壁腐蚀坑进行扫描电子显微镜形貌分析(SEM)及微区能谱分析(EDS)。分析结果表明:在泄漏孔处内、外表面可均见明显的腐蚀产物,微观形貌如图5和图6所示。分别在泄漏孔边沿(区域1)、泄漏孔内(区域2)及管体未失效部位(区域3)选取一定的区域,在管体未失效部位与发生腐蚀失效交界的边缘选取1点(区域4)进行微观能谱分析,分析结果见表3。由表3可知,泄漏孔处腐蚀产物以Fe和O元素为主,同时含有一定比例的Cl元素,越接近基体,Cl元素含量越高。管体外表面腐蚀坑底的腐蚀产物形貌及能谱分析结果如图7所示,由图7可知,外表面腐蚀坑底被腐蚀产物覆盖,且较为疏松,主要由Fe和O等元素构成,也含有一定量的Cl元素。

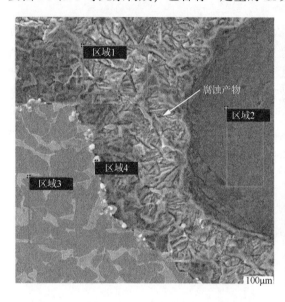

图5 泄漏输油管道穿孔处外表面SEM形貌　　图6 泄漏输油管道穿孔处内表面SEM形貌

表3 泄漏输油管道外表面的能谱分析结果　　单位:%(质量分数)

分析位置	C	O	Si	S	Cl	Ca	Fe
区域1		30.39	0.98		15.59		53.04
区域2	2.92	41.72		0.69	2.37	0.93	51.37
区域3	2.91						97.09
区域4		20.88			38.49		40.63

对防腐补口区表面所覆盖的黑色及褐色腐蚀产物进行XRD物相分析,XRD分析结果如图8所示,结果表明:外表面腐蚀产物主要是Fe_3O_4、$FeO(OH)$和SiO_2。

2 分析与讨论

2.1 材料理化性能分析

理化性能测试结果表明,失效样品材料的化学成分和拉伸性能符合GB/T 8163—2018标准中的相应要求,管体及腐蚀坑处的金相组织未见异常,所以材料性能不是造成此次失效事故的主要原因。

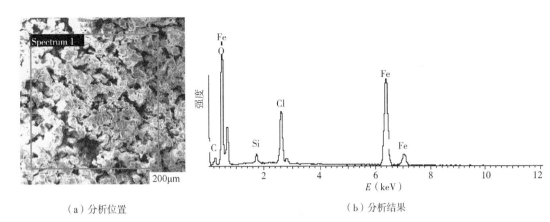

(a) 分析位置　　　　　　　　　　(b) 分析结果

图 7　泄漏输油管道外表面腐蚀坑底能谱分析位置及结果

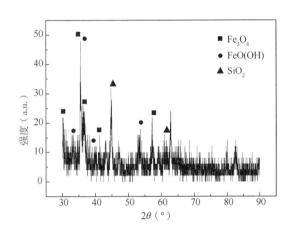

图 8　泄漏输油管道防腐补口腐蚀产物 XRD 分析结果

2.2　穿孔原因分析

从失效样品的宏观形貌分析来看,泄漏孔位于管道的防腐补口区,管体的外表面腐蚀较为严重,表面覆盖有大量的腐蚀产物,将腐蚀产物清理后可见大量深浅不一的腐蚀坑,而且此区域的壁厚减薄也较为明显,最小测量值仅为管道公称壁厚的 39%,管道内壁状况完好,未见有明显的腐蚀痕迹,由此可以判断,管道穿孔主要由外壁腐蚀引起。

通过观察发现,防腐补口区域以外的钢管外表面环氧粉末防腐层完好,无明显腐蚀特征,壁厚也未见有减薄现象,由此进一步判断,造成管道防腐补口区发生外腐蚀的原因主要在于防腐补口处的涂层发生漏点,这与防腐补口密封不严或防腐层破损有关。

从刺孔附近腐蚀产物能谱分析结果来看,失效样品表面腐蚀产物中含有大量的 O 和 Fe 元素,且在刺漏孔附近 Cl 元素含量也较高,所以推断腐蚀与 Cl^- 有关,同时结合 XRD 分析结果,管道外壁腐蚀产物主要为 Fe_3O_4、$FeO(OH)$ 和 SiO_2。

防腐补口处防腐层失效会导致外部土壤与管壁直接接触,土壤中的腐蚀物质会沿防腐层破损处渗入防腐层与管壁的空隙中,发生局部腐蚀。能谱及 XRD 的分析结果表明,管道外壁的腐蚀产物主要由 Fe_3O_4、$FeO(OH)$ 和 SiO_2 组成,其中 SiO_2 为与腐蚀产物结合在一起的土壤的主要成分,而 Fe_3O_4、$FeO(OH)$ 均为含水的铁的氧化物,说明管道外腐蚀的主要类型为氧的去极化腐蚀[7]。腐蚀机理如下[8-11]:

阳极反应:
$$Fe-2e^- \longrightarrow Fe^{2+} \tag{1}$$

阴极反应： $O_2+2H_2O+4e^- \longrightarrow 4OH^-$ (2)

总反应： $Fe+O_2+2H_2O+2e^- \longrightarrow Fe^{2+}+4OH^-$ (3)

Fe^{2+} 随后发生水解： $Fe^{2+}+2H_2O \longrightarrow Fe(OH)_2+2H^+$ (4)

Fe^{2+} 通常情况下很不稳定，容易被进一步氧化成 Fe^{3+}：

$$4Fe^{2+}+6H_2O+O_2 \longrightarrow 4FeO(OH)+8H^+ \quad (5)$$

FeO(OH)即为 $Fe_2O_3 \cdot H_2O$，通常处于腐蚀产物的外层，失水后形成红棕色的 Fe_2O_3，这与EDS及XRD分析结果相吻合。

外腐蚀发生后，最初生成的腐蚀产物会覆盖在金属表面，形成一层腐蚀产物膜，会对管道起到一定保护作用。与此同时，由于腐蚀产物膜较为疏松，在一定区域，土壤中的腐蚀介质还会不断渗入，与管道外壁金属接触，导致管道外壁的点蚀得以持续进行。同时，从泄漏孔处的能谱分析可知，其腐蚀产物中主要含有Fe和O元素，同时还有一定量的Cl元素。因为Cl^-的穿透性较强，对腐蚀产物膜有破坏作用，降低腐蚀产物膜对基体的保护能力，这就使得管道局部腐蚀程度不断加剧，最终导致穿孔[12-14]。

3 结论及建议

管道发生泄漏主要原因是外腐蚀引起穿孔所导致。环焊缝防腐补口的密封失效而导致的外腐蚀是造成管道发生腐蚀失效的主要原因，同时土壤中的Cl^-加速了腐蚀穿孔的发生。

在管道完整性管理中，应加强防腐补口失效风险控制，主要包括：(1)根据管道输送温度及所处环境，合理选择防腐补口方式及材质；(2)防腐补口材料应经国家计量认证的检测机构或国外第三方检测机构质量评定检验；(3)制定严格的防腐补口工艺，施工应由具有业主认定防腐资质的施工单位承担，补口操作人员应根据所使用产品的特点进行防腐施工培训并取得上岗证，方可进行补口施工操作；(4)在运行过程中，应定期开展防腐层直接检测评价，并及时对防腐层破损管段开挖修复；(5)对于具备开展内检测条件的管道，可根据内检测结果和适用性评价结果开挖验证，并对防腐层破损管段及时修复。

参 考 文 献

[1] 魏滨. 我国油气管道建设运行管理技术及发展展望[J]. 中国石油和化工标准与质量, 2018(15): 60-61.

[2] 马钢, 白瑞. 高强度油气长输管道腐蚀与防护研究进展[J]. 中外能源, 2018, 23(1): 55-62.

[3] 刘凯, 马丽敏, 陈志东, 等. 埋地管道的腐蚀与防护综述[J]. 管道技术与设备, 2007(4): 36-38, 42.

[4] 刘剑锋, 王文娟, 马健伟. 埋地管道腐蚀机理及应对措施[J]. 石油化工腐蚀与防护, 2006, 23(6): 20-22.

[5] 梁裕如, 姬丙寅. 某输油管道腐蚀泄漏失效原因分析[J]. 表面技术, 2016, 45(8): 68-73.

[6] 梁亚宁, 孟庆武, 毕凤琴. 大口径输油管道腐蚀及防腐层失效分析[J]. 科学技术与工程, 2010, 10(32): 8038-8041.

[7] 黄亮亮, 孟惠民, 黄晓林, 等. X60管线钢在盐碱性土壤中的腐蚀行为与机理[J]. 油气储运, 2013, 32(3): 257-262.

[8] 周波, 朱建雷, 李宁, 等. P110油管腐蚀穿孔原因分析[J]. 理化检验-物理分册, 2016, 52(5): 335-338, 344.

[9] 卢绮敏. 石油工业中的腐蚀与防护[J]. 北京: 化学工业出版社, 2001: 253.

[10] 蔡锐,吴鹏,赵金龙,等.某 L245 集输管道腐蚀失效原因分析[J].表面技术,2019,48(5):58-64.

[11] 王鸿膺,蒋涛,秦晓霞,等.川气东送管道土壤腐蚀埋片试验[J].油气储运,2010,29(10):769-771.

[12] 李健,汤荣,刘俊甫,等.长输原油管道的土壤腐蚀研究[J].装备制造技术,2012(9):31-32,85.

[13] 王叙乔.埋地油气管道外腐蚀成因及防腐技术分析[J].中国石油石化,2017(8):135-136.

[14] 聂向晖,王高峰,丰振军,等.X60M 钢管泄漏原因分析[J].理化检验-物理分册,2016,52(11):811-814.

本论文原发表于《理化检验(物理分册)》2021 年第 57 卷第 1 期。

二、油井管

Numerical Analysis of Casing Deformation under Cluster Well Spatial Fracturing

Wang Jianjun[1]   Jia Feipeng[1,2]   Yang Shangyu[1]   He Haijun[3]
Nan Zhao[4]   Le Zhang[1,2]

(1. CNPC Tubular Goods Research Institute, State Key Laboratory of Performance
and Structural Safety for Petroleum Tubular Goods and Equipment Materials;
2. College of Mechanical Engineering, Xi'an Shiyou University;
3. Daqing Oilfield Production Engineering & Research Institute;
4. Engineering Technology Research Institute of Xinjiang Oilfield Company)

Abstract: The large-displacement hydraulic fracturing technology widely used in the process of shale gas exploitation, which causes formation slip and serious casing damage under non-uniform load. It seriously restricts the exploitation and utilization of shale gas in China. In this paper, the writer employs the casing pipe with outer diameter of 139.7mm and grade N80 steel as an example and establishes the finite element model of stratum-cement ring-casing to analyze the influence rule of non-uniformity of stratum load, the difference of internal and external pressure and casing wall thickness to casing stress under non-uniform load. The analysis shows that: with the increase of the non-uniformity of formation load, the stress on the casing is gradually increased. Even the external load is far less than the casing's extrusion strength, but the non-uniformity of external load increases, the casing stress will reach its yield limit and lead to casing failure. Increasing the wall thickness is beneficial to reduce the stress value of casing under load. When the non-uniform coefficient of load decreases to 0.4, the increase of wall thickness has little effect on the casing's stress. When the formation non-uniform load is constant, the stress of casings decreases with the increase of the inside pressure of casings. Under the non-uniform load, reducing the pressure difference between the inside and outside of the casing is conducive to reduce the casing damage. The location where the casing failure occurs first is on the casing inner wall in the direction of the minimum horizontal stress. That is, the place where in the direction of the maximum horizontal in-situ stress on 90° is the casing stress risk area.

Keywords: Non-uniform load; Casing; Numerical analysis; Damage

---

Corresponding author: Wang Jianjun, wangjianjun005@cnpc.com.cn; Yang Shangyu, yangshangyu@cnpc.com.cn.

1 Introduction

China's shale gas is characterized by deep burial and great difficulty in exploitation. Used in shale gas horizontal multi-stage fracturing with large displacement, high pressure, effect time is long, etc, leads to formation fracturing construction of slip. As a result, an unusually high non-uniform load is generated on the outer wall of the casing, which causes serious casing damage and makes the well tools blocked. It severely restricts the exploitation and utilization of the shale gas in China.

At present, the speed of casing loss increases year by year, causing huge loss of manpower, material and financial resources. Casing damage has become one of the hot issues in petroleum engineering. It is an extraordinarily important theoretical and engineering significance to study casing stress under non-uniform load.

Dezhi Zeng has made use of the knowledge of elastic mechanics to solve and analyze the stress of thick wall casing under non-uniform load. His analysis suggests that the danger zone existed along the direction of minimum stress. By means of finite element analysis, Jun Fang finds that the ellipticity of casing weaken its strength and stiffness under non-uniform load.

Based on the existing researches, this paper analyzes the influence of non-uniform coefficient on the stress of casing, and the influence law of casing wall thickness and casing pressure difference on casing load under the action of non-uniform load.

2 Analysis of casing force under non-uniform load

According to statistics, 74% of casing damage of oil and water Wells in Daqing oilfield occurs at the bottom of the second floor, where is the interface between mudstone and other strata. One of the main causes of casing damage in Daqing oilfield is interlayer slip caused by mud and rock water absorption creep. In oil fields developed by water injection, when the pressure of water injection exceeds the fracture pressure of formation, micro-cracks will occur near the interface of mudstone layer. The injected water will penetrate into the mudstone layer along the micro-cracks, and the mudstone will absorb water, expand and soften, resulting in ground slip. At the same time, interfacial micro-cracks gather to form macroscopic interfacial cracks and begin to expand under the push of formation pressure difference. In the process of mudstone creep and crack propagation at stratum interface, the stress near the interface is highly concentrated. The deformation energy of the crack tip field is much higher than that of other parts of the formation. The outer load of oil and water well casing is concentrated near the formation interface, which leads to the occurrence of casing damage near the formation interface frequently. With the increase of water absorption of mudstone layer, the stiffness value and interface friction coefficient of mudstone layer gradually decrease. The non-uniform load at the interface of mudstone layer increases gradually and finally causes the formation sliding and the casing damage of shale gas well.

Under the action of non-uniform ground stress, according to the knowledge of elastic mechanics, the phenomenon of the maximum horizontal ground stress and the minimum horizontal ground stress acting on the outer wall of casing in the same horizontal direction shows an elliptic change, as shown in Fig. 1. The non-uniform load distribution model acting on the outer wall of casing can be expressed as follows:

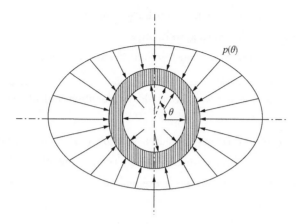

Fig. 1 Non-uniform load distribution on the casing

$$p(\theta) = p_0 + p_1 \cos(2\theta) \tag{1}$$

Where $p(\theta)$ is the radial load of acting on the casing, MPa; p_0 and p_1 is the equivalent external load of ground stress, MPa; θ is the angle between the radial load acting on the casing and the X-axis.

The inhomogeneity of load on casing is described by the ratio of maximum horizontal stress to minimum horizontal stress. The non-uniformity coefficient is calculated by the following equation:

$$n = p_{min} / p_{max} \tag{2}$$

Where p_{max} is the maximum horizontal stress, MPa; and p_{min} is the minimum horizontal stress, MPa.

Under the plane stress condition, the effective stress of the casing under heterogeneous load under the Mises stress criterion is:

$$\sigma_e = \sqrt{(1-\mu+\mu^2)(\sigma_1^2+\sigma_2^2) - \sigma_1\sigma_2(1-2\mu-2\mu^2)} \tag{3}$$

Where σ_1, σ_2 is the main stress of formation acting on casing, MPa; and μ is the poisson's ratio of stratigraphic rocks.

Under the action of non-uniform external load, taking $\sigma_1 = k\sigma_2$, the maximum Mises stress on the casing is:

$$\sigma_2 = \frac{(1-2\mu)\overline{\sigma}}{\sqrt{(n^2+1)(1-\mu+\mu^2) - n(1+2\mu-2\mu^2)}} \tag{4}$$

Where $\overline{\sigma}$ is the formation's uniform geostress under uniform loading, MPa.

3 Finite element modeling under non-uniform load

In order to eliminate the influence of the edge effect on the simulation results, the edge length of the formation modelis set as 3m. The stratum-cement ring-casing model as shown in Fig. 2 is established by using ABAQUS software. All the models use tetrahedral elements, add ZSYMM constraints to the model, and adopt sweeping mesh generation technology to generate CPS4R elements, and make the following assumptions in the model:

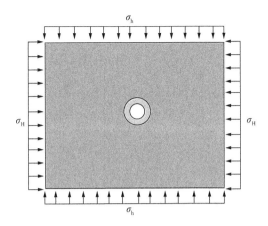

Fig. 2 Casing force model under non-uniform load

(1) Casing, cement ring and formation cementation are intact, there is no cement ring missing, casing eccentric phenomenon.

(2) The axial length of casing is much larger than its radial dimension. The longitudinal deformation of casing under the action of in-situ stress is ignored, and the problem is simplified to plane strain problem.

(3) Ignoring the initial defects such as casing ellipticity and uneven wall thickness, the casing end face is an ideal round end face.

4 Numerical calculation and analysis

According to the established finite element model, and the influence law of casing wall thickness and internal and external pressure difference on casing stress is simulated and calculated by changing the non-uniform load on the model and the parameters of the finite element model structure are changed.

Table 1 Calculation parameters of the casing model

Object	External diameter(mm)	Elasticity modulus(MPa)	Poisson's ratio	Wall thickness(mm)
Stratum		5×10^3	0.22	
Cement ring	249.7	1.0×10^4	0.18	55
Casing	139.7	2.06×10^5	0.3	9.17、7.72、10.64

4.1 The influence of different non-uniformity coefficients on the stress of casing

Under uniform load ($n=1$), the maximum stress of casing is 445.2MPa. With the increase of the degree of non-uniformity of formation non-uniform load, the maximum stress of casing gradually increases. When the coefficient of non-uniformity decreases to $n=0.4$, the maximum stress of casing is 633.1MPa, which increases by 42.2% compared with that of casing under uniform load. According to the API 5CT standard, the yield strength of N80 casing is 552MPa. That is, under the external load far less than the compressive strength of casing, when the non-uniform degree of external load increases, the casing will also reach its yield limit, leading to casing failure. The Mises stress cloud diagram of casing under different heterogeneity coefficients is shown in Fig. 3.

Fig. 3　Casing strain diagram with different non-uniform coefficients

4.2　The effect of casing wall thickness on casing stress under different non-uniform coefficients

According to the analysis in Fig. 4, when the casing wall thickness is smaller, the stress value of casing is larger. When the non-uniformity coefficient $n = 0.8$, the wall thickness is 7.72mm, 9.17mm and 10.64mm, and the stress on the casing is respectively 551.8MPa, 507.4MPa and 474.0MPa. Compared with the 7.72mm casing, the stress value of the casing with the wall thickness of 9.17mm is 7.9% lower and the casing with the wall thickness of 10.64mm is 16.3% lower. That is, improving the wall thickness is helpful to reduce the stress of casing load values under non-uniform load. With the increase of load non-uniform degree, the influence of wall thickness of casing stress value is reduced. When the coefficient of non-uniformity decreases to 0.4, the increase of wall thickness has little effect on casing stress under non-uniform load.

4.3　The influence of non-uniformity coefficient on casing stress under different pressure difference

According to the analysis in Fig. 5, with the increase of casing pressure under certain non-uniform load in formation, the stress of casing is reduced. That is, under the action of non-uniform load, reducing the pressure difference inside and outside casing is conducive to reduce the stress of casing.

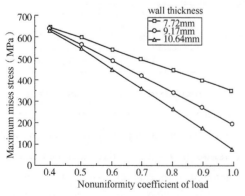

Fig. 4　Influence of non-uniform coefficient on casing stress under different wall thickness

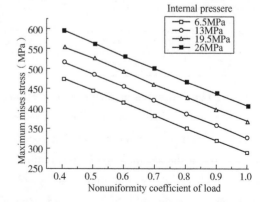

Fig. 5　Effect of casing non-uniformity coefficient on casing stress under different pressure difference

4.4 The stress distribution of casing under non-uniform load

Under non-uniform load, the annular stress of inner wall of casing (10.64mm), distance from outer wall of casing (7.08mm), distance from outer wall of casing (3.54mm) and outer wall of casing (0mm) were calculated respectively.

According to the analysis, under the action of non-uniform load, the stress of outer wall of casing is larger than that of inner wall of casing in the direction of larger horizontal stress, and that of outer wall of casing in the direction of smaller horizontal stress is larger than that of outer wall of casing. On the

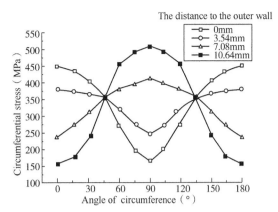

Fig. 6 Stress distribution map at different places away from casing outer wall

circumference of the casing, the first place where the casing failure occurs should be on the inner wall of the casing in the direction of the minimum horizontal in-situ stress, that is, where the maximum horizontal in-situ stress shows the position of 90° is the danger area of casing stress.

5 Conclusion

Through the above analysis, it can be concluded that the formation load non-uniformity coefficient has a significant impact on the casing stress:

(1) With the increase of non-uniformity of formation load, the local stress of casing is higher, that is, casing is easy to be damaged.

(2) When the non-uniform coefficient of load decreases to 0.4, the increase of wall thickness has little effect on casing stress under non-uniform load.

(3) On the inner wall of casing with the direction of the minimum horizontal ground stress. That is, the position of 90° with the maximum horizontal ground stress is the first place where the casing failure occurs.

In actual production, if conditions permit, the steel grade of casing material and its wall thickness can be increased to improve the extrusion resistance of casing and reduce the risk of casing damage.

Acknowledgments

The project is supported by National key Technologies R&D Program of China (2017ZX05009003-004), the Natural Science Foundation of China (U1762211), CNPC Basic Research Project (2019A-3911) and Shaanxi Outstanding Youth Fund (2018JC-030).

References

[1] LIAN Z H, YU H, LIN T J, et al. A study on casing deformation failure during multi-stage hydraulic fracturing for the stimulated reservoir volume of horizontal shale wells[J]. Journal of Natural Gas Science and

Engineering, 2015, 23: 538-546.
[2] CHEN Z M, LIAO X W, ZHAO X L, et al. Performance of horizontal wells with fracture networks in shale gas formation[J]. Journal of Petroleum Science and Engineering, 2015, 133: 646-664.
[3] OYARHOSSIEN M, DUSSEAULT M B. Risk Associated with Hydraulic Fracture Height Growth[C]. In 70th Canadian Geotechnical Conference, Ottawa, Ontario, 2017.
[4] XI Y, LI J, LIU G H, et al. Research review on casing deformation during multi-stage fracturing of shale gas horizontal Wells[J]. Special oil and gas reservoirs, 2019, 26(1): 1-6.
[5] LIU K, GAO D, WANG Y. Effect of local loads on shale gas well integrity during hydraulicfracturing process [J]. Journal of Natural Gas Science and Engineering, 2017, 37: 291-302.
[6] WANG J J, YANG S Y, JI H T, et al. Optimization of heavy oil thermal production well casing based on strain numerical analysis[J]. Petroleum pipes and instruments, 2019, 5(1): 42-45.
[7] ZENG D Z, LIN Y H, LI S G, et al. Analysis of compressive strength of thick wall casing under non-uniform load[J]. Natural gas industry, 2007(2): 60-62, 153.
[8] FANG J, GU Y H, MI F Z. Numerical analysis of casing extrusion failure under non-uniform load[J]. Petroleum machinery, 1999(7): 34-37, 59-60.
[9] LAST N, MUJICA S, PATTILLO P, et al. Evaluation, impact, and management of casing deformation caused by tectonic forces in the Andean Foothills, Colombia[R]. SPE 74560, 2002.
[10] YU H, LIAN Z H, XU X L, et al. Numerical simulation of casing failure during SRV fracturing in shale gas vertical Wells[J]. Petroleum machinery, 2015, 43(3): 73-77.
[11] ZHOU X, LIN G Q, ZHONG Y Y, et al. Analysis on the compressive resistance of oil casing under non-uniform load and internal pressure[J]. Prospecting engineering(geotechnical drilling engineering), 2017, 44(8): 76-80.

本论文原收录于 International Field Exploration and Development Conference(2020 年)。

Cu₂Se as Textured Adjuvant for Pb-Doped BiCuSeO Materials Leading to High Thermoelectric Performance

Jiang Long[1,2,3] Han Lihong[1,2,3,4] Lu Caihong[1,2,3] Yang Shangyu[1,2,3]
Liu Yaxu[1,2,3] Jiang Haoze[6] Yan Yonggao[5] Tang Xinfeng[5] Yang Dongwang[5]

(1. State Key Laboratory for Performance and Structure Safety of Petroleum Tubular Goods and Equipment Materials, CNPC Tubular Goods Research Institute;
2. Shaanxi Key Laboratory for Performance and Structure Safety of Petroleum Tubular Goods and Equipment Materials;
3. Key Laboratory of Petroleum Tubular Goods Engineering, CNPC;
4. School of Materials Science and Engineering, Chang'an University;
5. State Key Laboratory of Advanced Technology for Materials Synthesis and Processing, Wuhan University of Technology;
6. LWD Logging Center, China Petroleum Logging Co., Ltd.)

Abstract: Exploring the origin of intrinsic low thermal conductivity in BiCuSeO is of great significance for searching new oxide thermoelectric (TE) materials. In addition, from the perspective of material preparation, it is of great value to further develop the TE performance optimization strategy of BiCuSeO-based materials. In this work, the low-temperature TE transport properties of Pb-doped BiCuSeO-based materials are investigated. It is found that Pb doping can greatly optimize the carrier concentration, soften the lattice, and reduce the lattice thermal conductivity. The addition of Cu₂Se significantly enhanced the grain texture and then increased the interface concentration parallel to the pressure direction in the sintering process, which further reduced the lattice thermal conductivity of the material. Finally, the *ZT* value of $Bi_{0.96}Pb_{0.04}CuSeO-6mol\% Cu_2Se$ bulk material is as high as 0.85 at 840K. This provides important guidance to improve the properties of TE materials via interface engineering.

Keywords: BiCuSeO; Cu₂Se; Textured adjuvant; Texturization; Thermoelectric

1 Introduction

New energy materials and technology is an effective way to solve the energy crisis and

---

Corresponding author: Jiang Long, jianglong 003@ cnpc. com. cn; Han Lihong, hanlihong@ cnpc. com. cn; Yang Dongwang, ydongwang@ whut. edu. cn.

environmental pollution. Thermoelectric(TE) materials can directly convert heat into electricity in a solid-state means. As such, it is a very promising technology in deep space power supply, waste heat recovery, and industrial waste heat power generation[1]. The efficiency of a TE material is gauged by its dimensionless figure of merit ZT value, defined as $ZT=\alpha^2\sigma T/(\kappa_L+\kappa_e)$, where α, σ, κ_L, κ_e, and T are the Seebeck coefficient, electrical conductivity, lattice thermal conductivity, electronic thermal conductivity, and the absolute temperature, respectively[2,3]. Currently, a great amount of research efforts are mainly focused on the development of high-performance, thermally and mechanically stable, environmentally friendly, and low-cost thermoelectric materials[4-15] that can be synthesized rapidly in large quantities[16-23]. To optimize the electrical properties, various carrier engineering approaches, including band convergence engineering[24,25], resonant levels effect[26], band energy alignment engineering[27], spin-orbit engineering[28], magnetoelectric effect[29-31], and so on, have emerged to largely improve the power factor. A series of phonon engineering approaches have also been employed to soften lattice[32] or enhance phonon scattering by taking advantage of all scale hierarchical architectures[33] and then decrease κ_L.

BiCuSeO is one of the most superior oxide TE materials used in the mid-temperature range due to its ultralow lattice thermal conductivity ($\kappa_L \sim 0.4$ W·m^{-1}·K^{-1} for the pristine sample at high temperature), high Seebeck coefficient, and excellent thermochemical stabilities[34]. In recent years, the main performance optimization strategies of BiCuSeO compounds include carrier concentration engineering[8,35-37], modulation doping[38], chemical bond engineering[37,39], hierarchic structuring[40], and so on. On this basis, further development of new optimization strategies will greatly promote the commercial application of BiCuSeO-based materials.

It is particularly worth mentioning that the high-temperature TE properties of Pb-doped BiCuSeO compounds have been studied by different researchers[8,22,41-43]. Pb is considered to be an extremely effective doping element, which can not only significantly optimize the carrier concentration but also promote the self-propagating combustion synthesis reaction[43].

Recently, Yang et al. found that there is a special Schottky heterojunction at the interface of Cu$_2$Se host matrix and in situ formed BiCuSeO nanoparticles, which could regulate the behaviors of copper ions and electrons inside the composites[44].

As an extension of the work, what is the role of Cu$_2$Se in the BiCuSeO matrix?

In this work, we first measured the low-temperature TE properties of Pb-doped BiCuSeO compounds to explore its origin of the intrinsic low thermal conductivity. Based on this, we developed Cu$_2$Se as a textured adjuvant to achieve significant texture during the sintering process. The resulting strong interface scattering effect will have a significant impact on the phonon transport process and then optimize the TE performance.

2 Experimental section

Bi$_{1-x}$Pb$_x$CuSeO ($x=0$, 0.02, 0.04, 0.06, 0.08, 0.10) compounds were synthesized by the thermal explosion method[43]. Cu$_2$Se was prepared by the self-propagating high-temperature synthesis method[16]. $Bi_{0.96}Pb_{0.04}CuSeO$ and Cu_2Se powders were weighed according to the stoichiometric ratio of $Bi_{0.96}Pb_{0.04}CuSeO-xCu_2Se$ (x=0, 0.02, 0.04, 0.06). Then, the powders were uniformly mixed in an agate mortar and consolidated by spark plasma sintering (SPS) (SPS1050, Sumimoto, Japan) apparatus at 973K for 7min under a pressure of 40MPa in a vacuum level ≤20Pa. The resulting cylindrical ingots with $\phi15\times12mm^3$ sizes were cut into different shapes for the TE property measurements.

In the temperature range of 20~300K, the σ and α were measured simultaneously using a homemade measurement system similar to that in Clemson University[45], while the κ_{tol} was measured using a custom-designed steady-state technique[46], which was corrected for radiation losses above 200K according to the Stefan-Boltzmann power law[46].

More material synthesis, phase composition/microstructure characterization, and TE performance measurement details can be checked in our previous work[43,44].

3 Results and discussion

Fig. 1 shows the low-temperature electrical properties of $Bi_{1-x}Pb_xCuSeO$ compounds in the range of 20~300K. It clearly shows that pristine BiCuSeO exhibits semiconducting characteristics in the whole temperature region, whereas the Pb doping strongly changes the electrical transport behavior to metallic conduction[Fig. 1(a)]. Moreover, the electrical conductivity of pure BiCuSeO is very low, and the electrical conductivity increases by several orders of magnitude after Pb doping, which shows that Pb is a very effective dopant. The Seebeck coefficients of $Bi_{1-x}Pb_xCuSeO$ samples are positive over the whole temperature range, showing p-type conduction[Fig. 1(b)]. Pb doping has an obvious effect on the Seebeck coefficient, which decreased from $340\mu V \cdot K^{-1}$ for pure BiCuSeO to $94\mu V \cdot K^{-1}$ for $Bi_{0.90}Pb_{0.10}CuSeO$ at 300K. Finally, the significantly increased electrical conductivity compensates the moderately decreased Seebeck coefficient, which leads to highly improved power factors[Fig. 1(c)], and the 4mol% Pb-doped sample exhibits the highest value of $11.9\mu W \cdot cm^{-1} \cdot K^{-2}$ at around 300K.

To study the electrical transport mechanism at low temperature, the electrical conductivity and Hall coefficients of $Bi_{1-x}Pb_xCuSeO$ were measured between 2~300K through the physical properties measurement system(PPMS). It is worth mentioning that the electrical conductivities tested by two different measurement apparatuses are almost the same[Fig. 1(a) and Fig. S1], which shows that the test results are reliable. According to $p_H=1/R_He$ and $\mu_H=\sigma R_H$, the carrier concentration and mobility can be obtained. Fig. 2(a) shows the carrier concentration in the relationship with temperature. Around 300K, the hole concentration increases from $3.0\times10^{17}cm^{-3}$ to $1.35\times10^{21}cm^{-3}$ with 10mol% Pb doping in the Bi site. Because Pb has one less outermost electron than Bi, the material with Pb incorporation to the Bi site will significantly improve the hole concentration. Because a Pb atom provides a hole ideally, we found that the experimentally measured carrier concentration is consistent with that in the basic theory. All ionization after Pb doping into the Bi site can effectively improve the carrier concentration.

Combined with the low-temperature electrical conductivity, the curve of carrier mobility with the temperature change is shown in Fig. 2(b). The μ of x=0 sample increases with the increase in temperature in the range of 2~300K, which exhibits a $T^{3/2}$ dependence, indicating that ionized impurity scattering dominates hole conduction. After doping Pb, μ first exhibits T^0 dependence then switches to $T^{-3/2}$ dependence, indicating that the carrier scattering mechanisms transition to acoustic phonon scattering mainly by ionized impurity scattering.

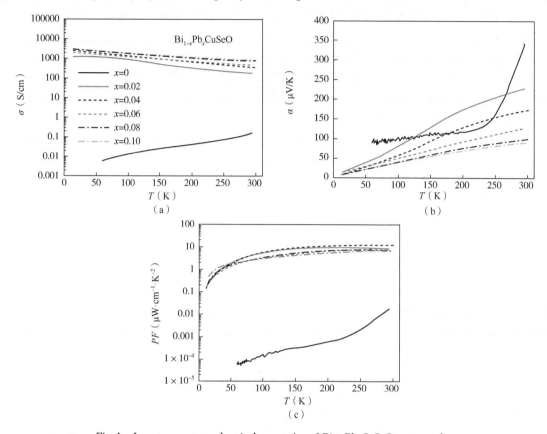

Fig. 1　Low-temperature electrical properties of $Bi_{1-x}Pb_xCuSeO$ compounds.
(a) electrical conductivity(σ), (b) Seebeck coefficient(α), and (c) power factor($\alpha^2\sigma$)

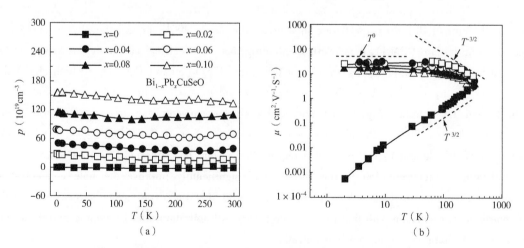

Fig. 2　Electrical transport mechanism of $Bi_{1-x}Pb_xCuSeO$ compounds at low temperature:
(a) carrier concentration (p) and (b) carrier mobility(μ)

Fig. 3(a) displays the temperature-dependent κ_{tot} of $Bi_{1-x}Pb_xCuSeO$ samples below 300K. The κ_{tot} of $x = 0$ sample decreases with the increasing temperature from $6.6 W \cdot m^{-1} \cdot K^{-1}$ at 30K to $1.4 W \cdot m^{-1} \cdot K^{-1}$ at 300K. Also, the κ_{tot} decreases with the contents of the Pb dopant before 143K but increases in the range of 143~300K. According to the Wiedemann-Franz relation, $\kappa_e = L\sigma T$, in which the Lorentz constant L can be obtained from fitting the corresponding Seebeck coefficient values with an estimate of the reduced Fermi level via using the single parabolic band model, assuming that the acoustic phonon scattering is predominant. Obviously, the electronic thermal conductivity increases significantly with the Pb content [Fig. 3(b)], which is consistent with the temperature-dependent trend of electrical conductivity. [Fig. 3(c)] displays the temperature dependence of κ_L, which can be calculated by directly subtracting κ_e from κ_{tot}. After the incorporation of heavy metals Pb, the κ_L decreases, which is consistent with our previous results[43]. Meantime, all $Bi_{1-x}Pb_xCuSeO$ samples have a very low lattice thermal conductivity over the entire temperature range, as low as $1.37~1.64 W \cdot m^{-1} \cdot K^{-1}$ at 300K. As a result, the synergy of anharmonicity, low-frequency vibrations, Pb-doping-induced lattice softening, mass fluctuations, stress field wave, and phonon scattering by point defects account for the observed low lattice thermal conductivity[34,43,47,48].

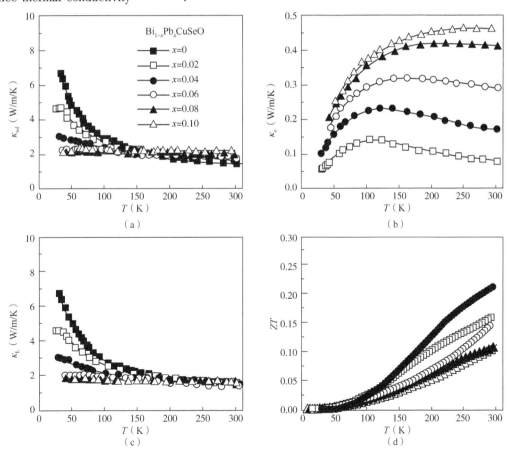

Fig. 3 Low-temperature thermal properties of $Bi_{1-x}Pb_xCuSeO$ compounds: (a) total thermal conductivity (κ_{tot}), (b) electronic thermal conductivity (κ_e), (c) lattice thermal conductivity (κ_L), and (d) figure of merit ZT

Fig. 3(d) plots the temperature dependence of ZT values, which increase first then decrease with the Pb content. The maximum $ZT = 0.21$ at 300K is achieved for the 4mol% Pb-doped sample. Based on this compound, we further adopt the texture strategy to optimize the TE properties.

Fig. 4(a) shows the X-ray diffraction(XRD) patterns of the $Bi_{0.96}Pb_{0.04}CuSeO-xCu_2Se$ ($x = 0$, 0.02, 0.04, 0.06) composites, in which all of the peak positions are consistent with the standard card of BiCuSeO (#98-015-9474), and no second phase was observed. Surprisingly, no phase transition peak of Cu_2Se was detected in the heat flux curves for all samples in the range of 293 ~ 473K[Fig. 4(b)]. Considering that during the material preparation process, the Cu vacancies in BiCuSeO are internally diffused into Cu_2Se, the phase transition temperature of Cu_2Se may be inferred to be below the room temperature. In fact, after broadening the test temperature range of the heat flux of the sample with $x = 0.06$ to 233 ~ 473K, no phase transition peaks below the room temperature were observed (Fig. S2). Moreover, the component contrast in $x = 0.06$ samples is almost the same (Fig. S3). Combined with the above XRD results, it can be seen that Cu_2Se no longer exists in the designed composites. Even if the design content of Cu_2Se increased to 8 ~ 10mol%, the XRD and differential scanning calorimetry(DSC) tests still failed to detect the Cu_2Se inside (Fig. S4 and Fig. S5). This shows that the solubility of Cu_2Se in BiCuSeO exceeds 10%.

It can be inferred that this special composite effect may also have a special impact on the microstructure of the material. Fig. 5 depicts the morphology of the fresh fracture surface of the $Bi_{0.96}Pb_{0.04}CuSeO-xCu_2Se$ ($x = 0$, 0.02, 0.04, 0.06) bulk material. It can be clearly seen that although the sintering pressure in the densification process is 40MPa, the orientation of the material gradually increases with the design content of Cu_2Se. Obviously, the introduction of Cu_2Se in the raw materials plays an important role in the microstructure formation process as a textured adjuvant.

To further detect the composite morphology of the material, more sophisticated electron microscopy observation techniques were used. The fine microstructure of the $Bi_{0.96}Pb_{0.04}CuSeO/4$ mol% Cu_2Se bulk material is depicted in Fig. 6. The high-angle annular dark-field scanning transmission electron microscopy(HAADF-STEM) image and its corresponding energy dispersive spectroscopy(EDS) elemental maps clearly show that Cu is enriched at the grain boundary and linked to the $[Cu_2Se_2]^{2-}$ layer in BiCuSeO. During the high-temperature sintering process, a large number of Cu vacancies in the BiCuSeO matrix flooded into the external Cu_2Se, which causes a large amount of Cu loss and instability of Cu_2Se. Finally, Cu_2Se dissolves into the $[Cu_2Se_2]^{2-}$ layer of BiCuSeO and extrudes excess Cu on the grain boundary. This process will contribute to the creeping of BiCuSeO grains at high temperatures and eventually forming a significant texture. Combined with the analysis of the aforementioned phase and heat flow curve, it can be known that this is the reason why Cu_2Se cannot be detected in the material matrix and the Cu thin layer can be observed on the grain boundary.

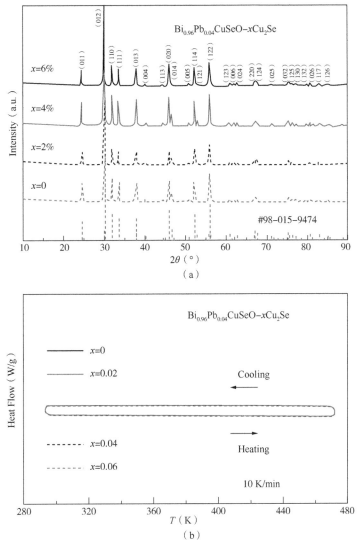

Fig. 4 $Bi_{0.96}Pb_{0.04}CuSeO-xCu_2Se$ ($x = 0$, 0.02, 0.04, 0.06) materials (a) XRD patterns and (b) heatflow. No second phase peak was observed in the XRD patterns and no phase transition peak of Cu_2Se was detected in the heat flow curves, which indicate that Cu_2Se dissolves in the BiCuSeO matrix

The orientation in the microstructure is likely to cause differences in thermoelectric transport properties, so the thermoelectric performance parallel to and perpendicular to the pressure sintering direction have been tested, and the curves are shown in Fig. 7. All samples exhibit metallic conduction behavior in both electrical conductivity [Fig. 7(a)] and Seebeck coefficient [Fig. 7(b)] in terms of magnitude and temperature dependence relation. Because the orientation has no direct effect on the carrier concentration, the Seebeck coefficient shows no obvious difference in orientation. On the contrary, the preferred orientation has a great influence on the carrier scattering process and leads to anisotropic electrical transport properties. The weaker scattering of carriers by the interface perpendicular to the sintering pressure direction causes the material to have higher electrical conductivity and power factor. According to the calculation results, the power factor of the

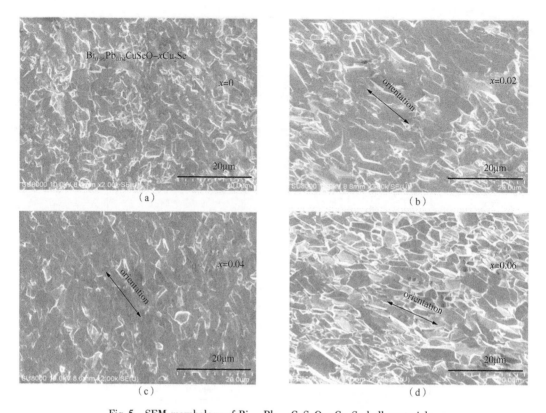

Fig. 5　SEM morphology of $Bi_{0.96}Pb_{0.04}CuSeO-xCu_2Se$ bulk materials:
(a)$x=0$, (b)$x=0.02$, (c)$x=0.04$, and(d)$x=0.06$. As the value of x increases, the orientation becomes stronger

$x=0.02$ sample is as high as $8.8\mu W \cdot cm^{-1} \cdot K^{-2}$, which is the most superior among all samples [Fig. 7(c)].

Fig. 7(d) displays the temperature-dependent κ_{tol} of $Bi_{0.96}Pb_{0.04}CuSeO-xCu_2Se(x=0, 0.02, 0.04, 0.06)$ bulk materials. Obviously, the κ_{tol} value is divided into two parts, which are perpendicular to and parallel to the sintering pressure direction, and the latter is significantly lower than the former. The temperature-dependent κ_L calculated by directly subtracting κ_e from κ_{tol} is displayed in Fig. 7(e), which is still clearly divided into two groups. In particular, in the group perpendicular to the sintering pressure direction, κ_L shows an upward trend with the increase of the Cu_2Se design content; however, in the group parallel to the sintering pressure direction, κ_L shows a downward trend with the increase of the Cu_2Se design content. This clearly shows that the interfacial concentration along the direction parallel to the sintering pressure in the bulk material gradually increases with the addition of Cu_2Se as a textured adjuvant and dominates the phonon scattering process, thus gradually reducing the lattice thermal conductivity. Especially, this component-induced preferred grain texture combined with the rich interface would scatter phonons hierarchically and lead to decreased lattice thermal conductivity throughout the temperature range. Finally, the sample parallel to the pressure direction showed a superior TE performance, with $x=0.06$ sample achieving the highest ZT value of 0.85 at 840K.

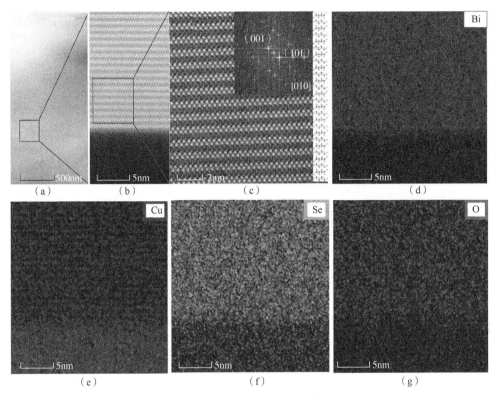

Fig. 6 Fine microstructure of the $Bi_{0.96}Pb_{0.04}CuSeO-4$ mol % Cu_2Se bulk material. (a) HAADF-STEM image at low magnification. (b) HAADFSTEM image at a high magnification of the blue marked region indicated in (a). (c) Enlarged HAADF-STEM micrograph with fast Fourier transform (FFT) patterns inside and crystal structure of BiCuSeO. (d-g) EDS elemental maps of the area in (b), Bi(blue), Cu(purple), Se(orange), and O(red).

Clearly, Cu is enriched at the grain boundary and linked to the $[Cu_2Se_2]^{2-}$ layer in BiCuSeO

4 Conclusions

We measured the low-temperature thermoelectric transport performance of Pb-doped BiCuSeO-based materials in detail. On the one hand, Pb cansignificantly increase the carrier concentration of the material and greatly improve the electrical transport properties; on the other hand, Pb-doping-induced mass fluctuations, stress field wave, and lattice softening effect account for the observed low lattice thermal conductivity. Taking $Bi_{0.96}Pb_{0.04}CuSeO$ as the matrix, the addition of Cu_2Se greatly enhanced the texture of the crystal grains, which significantly increased the concentration of the heterogeneous interface along the pressure direction in the sintering process and further reduced the lattice thermal conductivity. Finally, the ZT value of the $Bi_{0.96}Pb_{0.04}CuSeO-6mol\% Cu_2Se$ sample reached 0.85 at 840K.

Electrical conductivity measured by PPMS from 2 to 300K; heat flow of $Bi_{0.96}Pb_{0.04}CuSeO-6mol\%Cu_2Se$ bulk material from 233 to 473K; secondary electron image and back scattering image of a polished surface of $Bi_{0.96}Pb_{0.04}CuSeO-6$ mol%Cu_2Se bulk material; XRD patterns of $Bi_{0.96}Pb_{0.04}CuSeO-xCu_2Se$ ($x=0.08, 0.10$) materials; heat flow of $Bi_{0.96}Pb_{0.04}CuSeO-xCu_2Se$ (PDF).

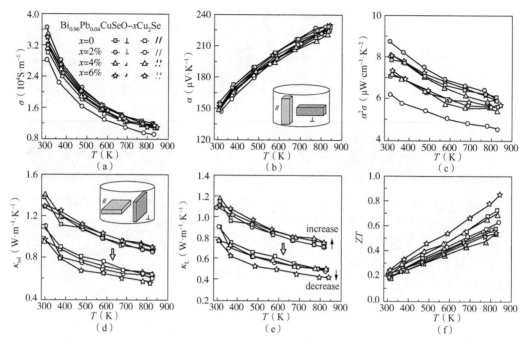

Fig. 7　Thermoelectric properties of $Bi_{0.96}Pb_{0.04}CuSeO-xCu_2Se$ bulk materials: (a) electrical conductivity; (b) Seebeck coefficient; (c) power factor; (d) total thermal conductivity; (e) lattice thermal conductivity; and (f) ZT

Acknowledgments

The study was sponsored by the National Key Research and Development Program of China (2019YFF0217504), the Basic Research and Strategic Reserve Technology Fund of CNPC (2019D-5008(2019Z-04)), and the Wuhan Frontier Project on Applied Research Foundation (2019010701011405).

References

[1] HE J, TRITT T M. Advances in Thermoelectric Materials Research: Looking Back and Moving Forward[J]. Science, 2017, 357(6358): eaak9997.

[2] ROWE D M. CRC Handbook of Thermoelectrics[M]. CRC Press, 2018: 1995.

[3] GOLDSMID H J. Introduction to Thermoelectricity[M]. Springer, 2010: 339-357.

[4] KIM S I, LEE K H, MUN H A, et al. Dense Dislocation Arrays Embedded in Grain Boundaries for High-performance Bulk Thermoelectrics[J]. Science, 2015, 348(6230): 109-114.

[5] SHI X, YANG J, SALVADOR J R, et al. Multiple-filled Skutterudites: High Thermoelectric Figure of Merit Through Separately Optimizing Electrical and Thermal Transports[J]. Journal of the American Chemical Society, 2011, 133(20): 7837-7846.

[6] ZHAO L D, LO S H, ZHANG Y, et al. Ultralow Thermal Conductivity and High Thermoelectric Figure of Merit in SnSe Crystals[J]. Nature, 2014, 508(7496): 373-377.

[7] ZHAO L D, TAN G, HAO S, et al. Ultrahigh Power Factor and Thermoelectric Performance in Hole-doped Single-crystal SnSe[J]. Science, 2016, 351(6269): 141-144.

[8] LAN J L, LIU Y C, ZHAN B, et al. Enhanced Thermoelectric Properties of Pb-doped BiCuSeO Ceramics

[J]. Advanced Materials, 2013, 25(36): 5086-5090.

[9] ZHU T, FU C, XIE H, et al. High Efficiency Half Heusler Thermoelectric Materials for Energy Harvesting [J]. Advanced Energy Materials, 2015, 5(19): 1500588-1500600.

[10] HU L, WU H, ZHU T, et al. Tuning Multiscale Microstructures to Enhance Thermoelectric Performance of n – type Bismuth – Telluride – based Solid Solutions [J]. Advanced Energy Materials, 2015, 5(17): 1500411-1500423.

[11] POUDEL B, HAO Q, MA Y, et al. High-thermoelectric Performance of Nanostructured Bismuth Antimony Telluride BulkAlloys[J]. Science, 2008, 320(5876): 634-638.

[12] BASU R, BHATTACHARYA S, BHATT R, et al. Improved Thermoelectric Performance of Hot Pressed Nanostructured n-type SiGe Bulk Alloys[J]. Journal of Materials Chemistry A, 2014, 2(19): 6922-6930.

[13] KOUMOTO K, WANG Y, ZHANG R, et al. Oxide Thermoelectric Materials: A Nanostructuring Approach [J]. Annual Review of Materials Research, 2010, 40: 363-394.

[14] OHTA H, SUGIURA K, KOUMOTO K. Recent Progress in Oxide Thermoelectric Materials: p – type $Ca_3Co_4O_9$ and n–type $SrTiO_3-\delta$[J]. Inorganic Chemistry, 2008, 47(19): 8429-8436.

[15] LIU W, YIN K, ZHANG Q, et al. Eco – friendly High – performance Silicide Thermoelectric Materials [J]. National Science Review, 2017, 4(4): 611-626.

[16] SU X, FU F, YAN Y, et al. Self-propagating High-temperature Synthesis for Compound Thermoelectrics and New Criterion for Combustion Processing[J]. Nature Communications, 2014, 5(1): 1-7.

[17] YANG D, SU X, MENG F, et al. Facile Room Temperature Solventless Synthesis of High Thermoelectric Performance Ag_2Se via A Dissociative Adsorption Reaction[J]. Journal of Materials Chemistry A, 2017, 5 (44): 23243-23251.

[18] SU X, WEI P, LI H, et al. Multi – Scale Microstructural Thermoelectric Materials: Transport Behavior, Non – Equilibrium Preparation, and Applications [J]. Advanced Materials, 2017, 29(20): 1602013-1602025.

[19] ZHENG G, SU X, LIANG T, et al. High Thermoelectric Performance of Mechanically Robust n – type $Bi_2Te_3-xSe_x$ Prepared by Combustion Synthesis [J]. Journal of Materials Chemistry A, 2015, 3(12): 6603-6613.

[20] ZHENG G, SU X, XIE H, et al. High Thermoelectric Performance of p–BiSbTe Compounds Prepared by Ultra-fast Thermally Induced Reaction[J]. Energy & Environmental Science, 2017, 10(12): 2638-2652.

[21] LIANG T, SU X, YAN Y, et al. Ultra-fast Synthesis and Thermoelectric Properties of Te Doped Skutterudites [J]. Journal of Materials Chemistry A, 2014, 2, 17914-17918.

[22] REN G K, LAN J L, BUTT S, et al. Enhanced Thermoelectric Properties in Pb – doped BiCuSeO Oxyselenides Prepared by Ultrafast Synthesis[J]. RSC Advanas, 2015, 5(85): 69878-69885.

[23] SAVARY E, GASCOIN F, MARINEL S. Fast Synthesis of Nanocrystalline Mg_2Si by Microwave Heating: A New Route to Nano – structured Thermoelectric Materials [J]. Dalton Transactions, 2010, 39(45): 11074-11080.

[24] PEI Y, SHI X, LALONDE A, et al. Convergence of Electronic Bands for High Performance Bulk Thermoelectrics[J]. Nature, 2011, 473(7345): 66-69.

[25] LIU W, TAN X, YIN K, et al. Convergence of Conduction Bands as a Means of Enhancing Thermoelectric Performance of n-Type $Mg_2Si_{1-x}Sn_x$ Solid Solutions[J]. Physical Review Letters, 2012, 108(16): 166601.

[26] HEREMANS J P, WIENDLOCHA B, CHAMOIRE A M. Resonant Levels in Bulk Thermoelectric Semiconductors[J]. Energy & Environmental Science, 2012, 5(2): 5510-5530.

[27] TAN G, SHI F, HAO S, et al. Non-equilibrium Processing Leads to Record High Thermoelectric Figure of Merit in PbTe-SrTe[J]. Nature Communications, 2016, 7(1): 1-9.

[28] WU L, YANG J, WANG S, et al. Two-dimensional Thermoelectrics with Rashba Spin-split Bands in Bulk BiTeI[J]. Physical Review B, 2014, 90(19).

[29] ZHAO W, LIU Z, SUN Z, et al. Superparamagnetic Enhancement of Thermoelectric Performance [J]. Nature, 2017, 549(7671): 247-251.

[30] ZHAO W, LIU Z, WEI P, et al. Magnetoelectric Interaction and Transport Behaviours in Magnetic Nanocomposite Thermoelectric Materials[J]. Nature Nanotechnology, 2017, 12(1): 55-60.

[31] WANG H, LUO X, CHEN W, et al. Magnetic-field Enhanced High-thermoelectric Performance in Topological Dirac Semimetal Cd_3As_2 Crystal[J]. Science Bulletin, 2018, 63(7): 411-418.

[32] HANUS R, AGNE M T, RETTIE A J E, et al. Lattice Softening Significantly Reduces Thermal Conductivity and Leads to High Thermoelectric Efficiency[J]. Advanced Materials, 2019, 31(21): 1900108-1900113.

[33] TAN G, ZHAO L D, KANATZIDIS M G. Rationally Designing High-Performance Bulk Thermoelectric Materials[J]. Chemical Reviews, 2016, 116(19): 12123-12149.

[34] ZHAO L D, HE J, BERARDAN D, et al. BiCuSeO Oxyselenides: New Promising Thermoelectric Materials [J]. Energy & Environmental Science, 2014, 7(9): 2900-2924.

[35] LI J, SUI J, PEI Y, et al. A High Thermoelectric Figure of Merit $ZT>1$ in Ba Heavily Doped BiCuSeO Oxyselenides[J]. Energy & Environmental Science, 2012, 5(9): 8543-8547.

[36] LI J, SUI J, PEI Y, et al. The Roles of Na Doping in BiCuSeO Oxyselenides as a Thermoelectric Material [J]. Journal of Materials Chemistry A, 2014, 2(14): 4903-4906.

[37] LI Z, XIAO C, FAN S, et al. Dual Vacancies: An Effective Strategy Realizing Synergistic Optimization of Thermoelectric Property in BiCuSeO [J]. Journal of the American Chemical Society, 2015, 137 (20): 6587-6593.

[38] PEI Y L, WU H, WU D, et al. High Thermoelectric Performance Realized in a BiCuSeO System by Improving Carrier Mobility Through 3D Modulation Doping[J]. Journal of the American Chemical Society, 2014, 136(39): 13902-13908.

[39] REN G K, WANG S, ZHOU Z, et al. Complex Electronic Structure and Compositing Effect in High Performance Thermoelectric BiCuSeO[J]. Nature Communications, 2019, 10(1): 1-9.

[40] REN G K, WANG S Y, ZHU Y C, et al. Enhancing Thermoelectric Performance in Hierarchically Structured BiCuSeO by Increasing Bond Covalency and Weakening Carrier-phonon Coupling[J]. Energy & Environmental Science, 2017, 10(7): 1590-1599.

[41] LIU Y C, LAN J L, ZHAN B, et al. Thermoelectric Properties of Pb-Doped BiCuSeO Ceramics[J]. Journal of the American Ceramic Society, 2013, 96(9): 2710-2713.

[42] PAN L, BE'RARDAN D, ZHAO L, et al. Influence of Pb Doping on the Electrical Transport Properties of BiCuSeO[J]. Applied Physics Letters, 2013, 102(2): 023902-023906.

[43] YANG D, SU X, YAN Y, et al. Manipulating the Combustion Wave during Self-Propagating Synthesis for High Thermoelectric Performance of Layered Oxychalcogenide $Bi_{1-x}Pb_xCuSeO$ [J]. Chemistry of Materials, 2016, 28(13): 4628-4640.

[44] YANG D, SU X, LI J, et al. Blocking Ion Migration Stabilizes the High Thermoelectric Performance in Cu_2Se Composites[J]. Advanced Materials, 2020, 32(40): 2003730-2003739.

[45] POPE A L, LITTLETON R T, TRITT T M. Apparatus for the Rapid Measurement of Electrical Transport Properties for Both "Needle-Like" and Bulk Materials[J]. Review of Scientific Instruments, 2001, 72(7):

3129-3131.

[46] POPE A L, ZAWILSKI B, TRITT T M. Description of Removable Sample Mount Apparatus for Rapid Thermal Conductivity Measurements[J]. Cryogenics, 2001, 41(10): 725-731.

[47] PEI Y L, HE J, LI J F, et al. High Thermoelectric Performance of Oxyselenides: Intrinsically Low Thermal Conductivity of Ca-doped BiCuSeO[J]. NPG Asia Materials, 2013, 5(5): e47.

[48] SAHA S K. Exploring the Origin of Ultralow Thermal Conductivity in Layered BiOCuSe[J]. Physical Review B, 2015, 92(4): 041202-041208.

本论文原发表于《ACS Applied Materials & Interfaces》2021年第13卷第10期。

Microstructure and Improved Thermal Shock Behavior of an In Situ Formed Metal-Enamel Interlocking Coating

Wang Hang[1,2] Zhang Chuan[3] Jiang Chengyang[4] Zhu Lijuan[1,2]
Han Lihong[1,2] Chen Minghui[4] Geng Shujiang[4] Wang Fuhui[4]

(1. Tubular Goods Research Institute of CNPC;
2. State Key Laboratory of Performance and Structural Safety for Petroleum Tubular Goods and Equipment Materials; 3. CNPC Bomco Drilling & Production Equipment Co. LTD;
4. Shenyang National Laboratory for Materials Science, Northeastern University)

Abstract: A novel metal-enamel interlocking coating was designed and prepared *in situ* by co-deposition of Ni-enamel composite layer and subsequent air spray of enamel with 10wt.% nanoscale Ni. During the firing process, the external enamel layer was melted and jointed with the enamel particles at the upper part of the Ni-plating layer to form the enamel pegs. Thermal shock tests of pure enamel, enamel with 10wt.% Ni composite and metal-enamel interlocking coatings were conducted at 600℃ in water and static air. The results indicated that the metal-enamel interlocking showed superior thermal shock resistance to both pure enamel and enamel with 10wt.% Ni composite coatings. The enhanced performance was mainly attributed to the advantageous effects of mechanical interlocking of the enamel pegs formed at the enamel/Ni-plating interface. Meanwhile, during thermal shock test, big clusters formed by nanoscale Ni agglomerations were oxidised to be a Ni/NiO core-shell structure while small single nanoscale Ni grains were oxidised completely, which both improved the thermal shock resistance of enamel coating significantly.

Keywords: Enamel; Interlocking; Thermal shock; Nanoscale Ni

1 Introduction

Ferritic-martensitic (FM) steels, e.g. P92, possessing a good combination of high thermal conductivity and excellent fatigue resistance, are considered as promising construction materials for high-temperature components in supercritical (SC)[1-2] and ultra-supercritical (USC) fossil fuel power plants[3]. However, early investigations[4-7] have shown that a non-protective oxide scale consisting of a porous Fe_3O_4 outer layer and a $FeCr_2O_4$ inner layer instead of an external chromia

Corresponding author: Chengyang Jiang, jiangchengyang@ mail. neu. edu. cn.

scale was formed on surface when exposed to high temperatures, which limits their long-term application in such harsh environments. FM steels used in tubes and boilers are prone to suffer from either steam oxidation or fireside corrosion at temperatures up to 650℃. Therefore, protective coatings are an effective alternative approach to gain superior resistance against oxidation/corrosion and achieve extended service life[8-9].

Traditional metallic coatings, like diffusional aluminide, exhibit remarkable oxidation resistance by forming a slowly growing, homogeneous and continuous alumina scale, protecting the underlying materials from attack in aggressive environments. However, early studies[10-11] have found that simple aluminide coatings cannot preserve the protection ability in harsh serving environment for a long period. Recently, Wollschläger et al.[12-13] prepared a micron-thick porous alumina layer by sol-gel technique on P92 steel against flue gas corrosion. Results have shown that deposition of a thin and porous alumina coating enabled the formation of a dense and protective chromium oxide scale at the interface between coating and steel, which significantly improved the flue gas corrosion resistance. But local breakaway oxidation still occurred on coated P92 steel due to a defective or incomplete chromia scale.

Enamel coatings possess superior oxidation/corrosion resistance against most corrosive media (salt[14], gas[15], acid[16], molten aluminium[17], etc.) due to their inertness and compactness with the underlying matrix. So they are good alternative to protect FM steels from the flue gas corrosion at fireside. However, as thermal shock is an ordinary state for most hot components in thermal power plants, the natural brittleness and crack sensitivity of enamel coatings, as well as their large thermal expansion coefficient (CTE) mismatch with substrate, will lead to a high susceptible to spallation, which limits their uses in thermal shock application. Therefore, designing and achieving a kind of enamel coating with superior thermal shock resistance becomes an important issue to be addressed.

A feasible way to improve thermal shock resistance of enamel coating is to introduce a second phase, such as nanonickel[18-19] and NiCrAlY particles[20]. The generally accepted toughening mechanism for ductile phases with large size is to obstruct cracks, such as crack bridging and deflection[21], while for the second phases with nano-size, one toughening mechanism proposed by Liao et al.[18] is that cracks initialised in the enamel coating can be self-healed due to the volume expansion by nano-sized nickel oxidation. However, no matter for the second phase with large size or nano-size, the toughening mechanisms are both from the perspective of inhibiting or repairing the existed crack to avoid crack propagation, rather than preventing the crack from initialising at the interface between enamel coating and substrate.

Mechanical interlocking, which contributed to interfacial adhesion, was first recognised by McBain[22]. Brockman also suggested that mechanical interlocking continued to operate even when chemical bonds failed[23]. Several studies have shown that coating/steel interfaces without mechanical interlocking fail early when exposed to corrosive conditions[24]. Another study argued that mechanical interlocking increased the stability of the coating/steel interface[25].

In the current work, therefore, a novel metal-enamel interlocking coating (named as NE10N hereafter) on P92 steel was designed. It was formed in situ during firing between the electroplated

Ni-enamel composite inner layer and the successive air-sprayed enamel outer layer with 10wt% Ni particles. The microstructure, as well as thermal shock behaviour, of the NE10N was investigated in contrast to pure enamel(PE) and enamel with 10wt% Ni composite coatings(E10N).

2 Experimental

2.1 Coating preparation

A ferritic-martensitic steel, P92(composition: C: 0.07~0.13, Mn: 0.3~0.6, Si: <0.5, Cr: 8.5~9.5, Ni: <0.4, Mo: 0.3~0.6, V: 0.15~0.25, W: 1.5~2 and balanced Fe, wt.%) was used as the substrate material. Test samples of approximate dimensions 15mm×10mm×1.5mm were cut from the pipe with 4μm in thickness using a spark discharge machine. The bare samples were ground with a final 400 mesh SiC sandpaper and then blasted with 200-mesh glass ball, then they were cleaned ultrasonically in acetone and ethanol for 30min, respectively.

The nominal compositions of enamel are presented in Table 1. The enamel frits were gained first by melting at high temperature, followed by quenching in water. After that, the enamel powder was obtained by milling the frits. The preparation method of enamel powder has been previously described in detail elsewhere[26].

Table 1 Nominal composition of enamel (wt.%)

SiO_2	B_2O_3	Al_2O_3	Na_2O	K_2O	CoO	CaF_2
54.62	12.32	5.96	12.32	5.96	2.46	6.36

A suspension to deposit the Ni-enamel composite layer(hereafter named as Ni-plating layer) was prepared by adding the obtained enamel powder(0.5~3μm) into the Niplating solution. The concentration of the suspension was 10g/100mL. A magnetic stirrer was used to ensure levitation of enamel particles at 10~20r/min. Details of the bath composition and electroplating parameters for depositing Ni are summarised in Table 2. After that, specimens coated with Ni-plating layer were conducted with homogenisation treatment at 650℃ for 2h in vacuum($<6\times10^{-3}$Pa).

Table 2 Parameters for Ni pre-deposition by electroplating technique

Parameters	Value	Parameters	Value
$NiSO_4 \cdot 6H_2O$	150~200g·L^{-1}	$C_{12}H_{25}NaSO_4$	0.1g·L^{-1}
H_3BO_3	20~30g·L^{-1}	pH value	5
Na_2SO_4	50~80g·L^{-1}	Temperature	55℃
NaCl	8~10g·L^{-1}	Current density	0.5~1A·dm^{-2}

Enamel, nanoscale Ni powder(approximately 100nm) and agate balls sealed and ball milled together in an agate container for mixing homogeneously. The mixed powder was then blended with ethanol in a ratio of 10g enamel powder to 150mL, forming a slurry. The external enamel layer with 10wt.% Ni incorporation(hereafter named as external enamel layer) was prepared by air spraying onto the annealed Ni-plating layer, after which it was baked at 70℃ for 10min and heated at 750℃ for 2min. The thickness of external enamel layer of PE, E10N and NE10N was all approximated to 40μm.

2.2 Thermal shock tests

Thermal shock tests in static air were carried out using an automated vertical furnace rig, where for each coating, three samples were used in the tests. During thermal exposure, the furnace was kept at 600℃, and the automated vertical rig lifted the sample into the furnace for heating and out of the furnace for cooling. Each thermal cycle consisted of maintaining the sample at 600℃ for 10min, followed by cooling it in static air for 5min, up to 200 cycles. Thermal shock tests in water were conducted using a muffle furnace with temperature maintained at 600℃. After heating for 10min, all the samples were transferred from the furnace directly into water (about 25℃) within 3 s. After cooling down for 10s in water, samples were taken out and dried in hot wind.

After the given number of cycles, the specimens were removed and weighed using an electronic balance (0.01mg precision, Sartorius BP211D, Germany), followed by the next round of cyclic oxidation tests.

2.3 Characterisation

Surface and cross-sectional morphologies were observed using a field-emission scanning electron microscope (SEM, Inspect F50, FEI Co., Hillsboro, OR) equipped with energy-dispersive X-ray spectrometer (EDAX, X-Max, Oxford Instruments Co., Oxford, U.K.). To observe surface morphology, a second electron (SE) mode was adopted, while the back-scattered electron (BSE) mode was utilised to characterise cross-sectional morphology of coating samples. The element mapping was carried out using an electron probe micro-analyser (EPMA, JXA-8530F, JEOL, J.P.). A transmission electron microscope (TEM, JEOL 2100F, Japan) equipped with a tracer energy-dispersive X-ray spectrometer (EDS) was used to identify the microstructure and chemical composition (in scanning transmission electron microscopy (STEM) mode) of the nanoscale Ni particles.

3 Results

3.1 Initial microstructure of the coatings

Fig.1(a) shows cross-sectional morphologies of the Ni-plating layer after homogenisation treatment. It can be observed clearly that irregular-shaped enamel particles ranging from 0.5 to 3μm in diameter dispersed mostly in the middle of the nickel layer, while some of them spread in the top. It should be noted that there was a thin NiO film formed at the surface after annealing. The formation of filmy NiO layer was proved to be beneficial to the interfacial adherence between Ni and enamel[27-28]. The cross-sectional morphology of as-deposited NE10N coating is presented in Fig. 1 (b). A bistratal-structured coating was formed on steel, consisting of an external enamel layer (~40μm in thickness) with homogeneous dispersion of bright Ni particles inside and the underlying Ni-plating layer (~10μm in thickness) with dark enamel particles spread. Substantial number of mortise-tenon-structured pegs was observed to form at interface between the Ni-plating and the external enamel layers.

Fig. 2 shows EPMA mapping at two interfaces of the as-deposited NE10N coating. Interfacial bonding strength is of critical importance for layered coatings especially when they are exposed to thermal cyclic environments. At the interface of steel/Ni-plating layer, a thin interdiffusion zone

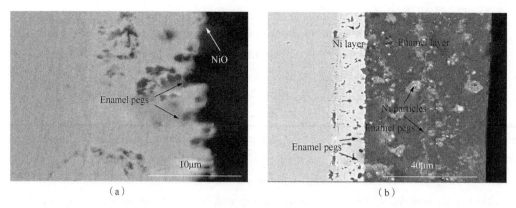

Fig. 1 Cross-sectional morphologies of Ni-plating layer after annealing(a), as-deposited NE10N coating(b)

was confirmed. Slight elements interdiffusion helped to form a strong bonding between them. The three main elements, Si, O and Ni, of the composite coating were seen clearly in both the inner Ni-plating and the external enamel layers. Several pegs embedded in the Ni-enamel composite layer were proved to be enamel due to enrichment of Si and O there. It is intriguing to note that some enamel pegs almost linked together with enamel particles embedded in the Ni-plating layer. This interlocking structure was expected to improve bonding strength between the Ni-plating layer and the external enamel one.

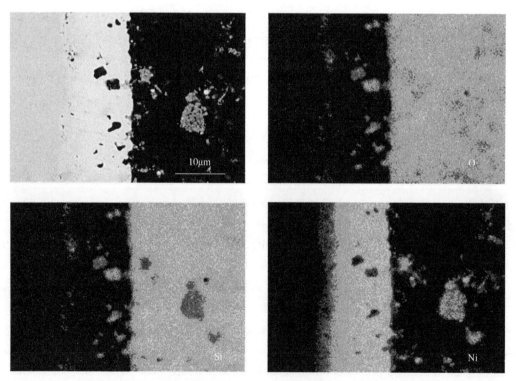

Fig. 2 EPMA mapping of the two interfaces of as-deposited NE10N coating

In comparison, the cross-sectional morphologies of the samples coated with PE and E10N are shown in Fig. 3. Both coatings consisted of a single enamel layer with approximate 40μm in thickness. The difference between them was only the dispersion of Ni particles in E10N coating while not in PE. Anyway, all the three enamel coatings were bonded well with the steel matrix or the Ni-plating layer with neither cracks nor spallation at interface.

Fig. 3　Initial cross-sectional morphologies of PE(a), E10N(b) coating samples

3.2　Thermal shock behaviour

3.2.1　Thermal shock in water

Fig. 4 shows kinetic curves of PE, E10N and NE10N coatings during thermal shock test at 600 ℃. For samples covered with PE coating, a sharp and continuous decrease in mass appeared from the beginning to the end, indicating its poorest thermal shock resistance among the three coatings. While for E10N coating, a steady mass change took place at the initial 40 cycles, then followed by a significant mass drop. On the contrary, the coating samples of NE10N remained steady mass change during the whole thermal shock test, indicating its best thermal shock resistance. It can be observed that the mass change of E10N and NE10N presented a little increase after thermal shock for certain times, this was mainly because Ni particles can be easily oxidised with the inward diffusional oxygen. After thermal shock for 60 times, the total mass changes were −4.79, −1.28 and +0.18 mg/cm² for PE, E10N and NE10N coatings, respectively.

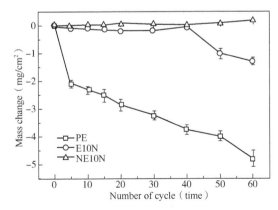

Fig. 4　Mass change curves of PE, E10N and NE10N coating samples after thermal shock in water at 600 ℃ for different times

Fig. 5 shows the surface and cross-sectional morphologies of the three coatings after thermal shock for 60 cycles. For the PE coating, it had almost all spalled off. Even within the remained zone, the enamel coating showed a large number of cracks and deep spallation pits, as shown in

Fig. 5(a). Correspondingly, a coherent and wide interfacial gap was formed between the substrate and coating, leading to most parts of coating spallation and severe oxidation of matrix[Fig. 5(b)]. While for E10N coating, cross cracks were developed as well at the coating surface[Fig. 5(c)]. These cross cracks divided the metal-enamel composite coating into several blocks, at central of which the white nickel clusters dwelled. The cross-sectional morphology of E10N coating was coincident with surface morphology[Fig. 5(d)]: a continuous interfacial crack appeared at the interface of the substrate and coating, and propagating cracks were observed within the enamel layer. In the case of NE10N coating, its surface was still flat and integrated after thermal shock, with some Ni agglomerates scattered[Fig. 5(e)]. Correspondingly, it can be observed from Fig. 5(f) that the external enamel layer with Ni particles was still bonded with the substrate well, without any visible crack or spallation.

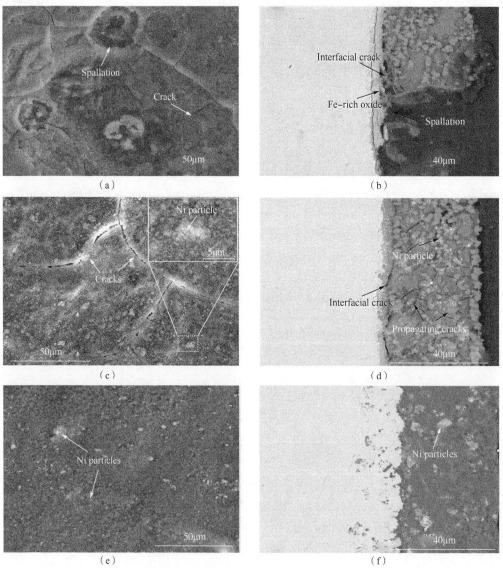

Fig. 5 Surface and cross-sectional morphologies of PE(a, b), E10N(c, d), NE10N(e, f) after thermal shock in water at 600℃ for 60 times

3.2.2 Thermal shock in air

Fig. 6 illustrates kinetic curves of three enamel coatings during thermal shock tests in air at 600℃. The sample coated with PE exhibited a rapid mass gain for the first 60 cycles, followed by a dramatic weight loss thereafter. The notable weight loss of PE coating after 60 cycles indicated that the coating began to suffer from severe spallation, which implied that the coating nearly lost its protective ability. In the case of E10N coating, the mass changed as a similar pattern with that of PE, while the turning point appeared at 100 cycles. For the samples coated with NE10N, the mass change curves presented a steady and slow increase during the entire thermal shock tests due to the oxidation of Ni particles.

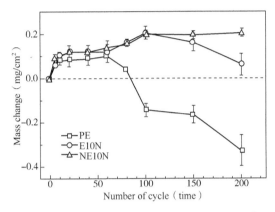

Fig.6 Mass change curves of PE, E10N, NE10N coating samples after thermal shock in static air at 600℃ for different times

Fig. 7 illustrates the cross-sectional and their enlarged morphologies of the three coating specimens after 200 cycles. As shown in Fig.7(a) and (b), interfacial crack and propagating cracks within the PE coating can be observed, and the thickness of enamel layer was less than 40μm, implying the occurrence of coating spallation. A similar failure phenomenon was formed in the E10N coating samples [Fig. 7(c) and (d)], cracks formed not only within the enamel layer but also at the enamel/steel interface. Their propagation induced the spallation of enamel coating layer-by-layer or peeling off along interface. Besides, the only difference was that in the E10N coating, some Ni particles stayed at the crack propagation path. These scattered nickel particles or clusters are able to blunt the propagated cracks as indicated by arrows in Fig. 7(d). By comparison, the enamel layer of the NE10N coating was still intact, where no cracks were observed. The thickness of external enamel layer was still approximately 40μm, implying no spallation happened [Fig. 7 (e)]. It can be observed from Fig. 7(f) that a large number of enamel pegs were pinning into the Ni-plating layer. Meanwhile, it is intriguing to note that some bright micron-sized Ni metallic particles were formed in the external enamel layer. EDS point results show that the core of particles was mainly composed of pure nickel (zone 1 in Table 3), while the grey phase decorated the core was NiO (zone 2 in Table 3). It is clear that the nickel particles are oxidised to form a Ni/NiO core-shell structure. By comparison, many nano-sized Ni particles that did not form clusters were grey mostly, implying their complete oxidation from inside to outside.

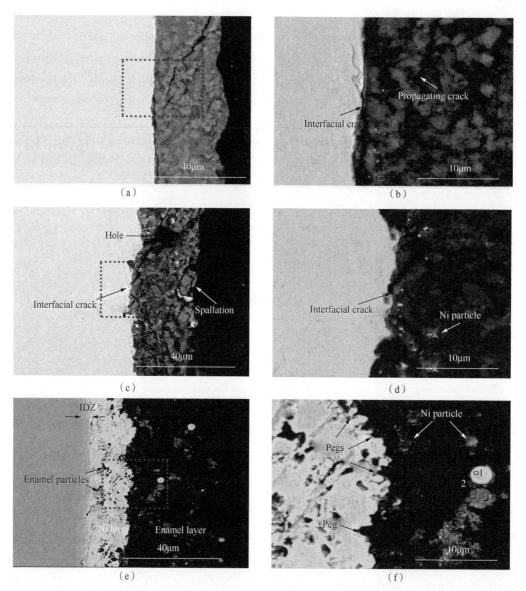

Fig. 7 Cross-sectional and enlarged morphologies of PE, E10N, NE10N coating samples after thermal shock in static air at 600℃ for 200 times(a and b for PE, c and d for E10N, e and f for NE10N)

Table 3　EDS results in Fig. 7(f)　　　　　　　　　　(at. %)

Zone	O	Na	Mg	Si	Ca	Fe	Co	Ni
Zone 1	5.06	—	—	0.66	0.64	0.69	—	92.96
Zone 2	49.77	2.28	7.32	3.27	1.73	1.70	2.35	31.57

In order to further shed light on the form of Ni particles, TEM analysis was conducted on the enamel layer of the NE10N coating after thermal shock tests in static air for 200 cycles. From the bright-field STEM cross-sectional morphology [Fig. 8(a)], some single nano-sized Ni grains and a big agglomerate were observed in this view. Based on the corresponding EDS mapping [Fig. 8(b)] and line scan profile [Fig. 8(d)], the agglomerate was confirmed to be Ni/NiO core-shell structure, while the single nano-sized Ni grain was confirmed to be NiO by the selected area electron diffraction(SAED) patterns, as shown in Fig.8(c).

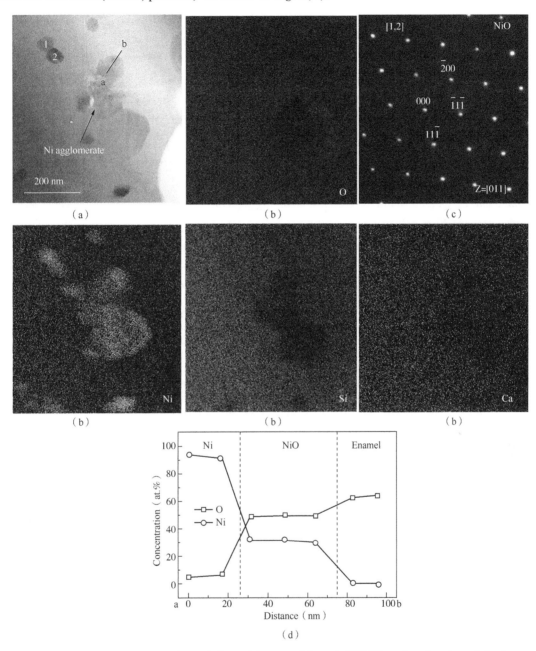

Fig. 8　BF-STEM cross-sectional morphology of the enamel layer in NE10N coating after thermal shock at 600℃ for 200 times(a), corresponding EDS mapping(b), SAED patterns of precipitate 1 and 2(c), corresponding line scan profile(d)

4　Discussion

As is clear, thermal shock resistance of the PE coating was improved by incorporating nano-sized Ni particles into the enamel layer, further enhanced by introducing a Ni-plating layer. The discussion starts from a consideration of the existing form of Ni particles, and their advantageous on thermal shock resistance. Then, the advantageous effect of Ni-plating layer on thermal shock resistance was discussed.

4.1 Effect of ni particles on thermal shock resistance

As can be seen from Fig. 8(a), the nano-sized Ni particles incorporated in this study are about 100 nm in diameter, it is easy for them to reunite to form larger clusters during the coating preparation stage due to the natural poor dispersion of nano-materials. It has been confirmed by TEM that big clusters formed by nano-Ni agglomerations were oxidised to be a Ni/NiO core-shell structure, while small single nanosized Ni particles were oxidised completely. According to the Ellingham diagram, Ni particles within the enamel should be oxidised at 600℃ to form NiO. For small nanosized Ni particles, the diameter is even smaller than the NiO thickness, the Ni/NiO interface can be ignored. While in the case of big agglomerates, once an outer NiO layer is formed on the surface of Ni metallic particles, the oxygen partial pressure at the interface of Ni/NiO drops significantly, which prevents Ni inside from further oxidation. Thus, Ni particles are decorated by a thin layer of NiO.

By comparing the thermal shock behaviour of PE and E10N coatings (Fig. 5 and Fig. 7), a small number of nano-sized Ni particles have little effect on interfacial crack initialisation but do inhibit the crack propagation within the enamel layer. The mechanisms of delaying crack progression by nickel agglomerates with a large size are mainly involved in crack bridging and deflection[21]. During the crack propagation, when the crack front reaches the Ni/NiO agglomerates, if the residual stress is high enough to pass through the agglomerates, the crack tends to expand in a straight manner, thus the agglomerates embedded in enamel are completely torn out. Otherwise, the crack will deflect, leading the crack path and the surface area of the fracture surface increasing. For nanoscale Ni particles with a small size, the crack self-healing mechanism was proposed by Liao et al.[18], the oxidation of nanoscale Ni particles leads to a volume expansion of ~65% at the crack surfaces. As the expanded (oxidised) nickel particles are chemically compatible to the [SiO_4] network of enamel, they are able to bridge the two crack surfaces, fill in the crack gaps, which combine with the high-temperature viscoelasticity of enamel to make the crack heal completely and avoid spallation.

In the present study, based on the experimental results of kinetics (Fig. 4 and Fig. 6) and morphologies after thermal shock tests (Fig. 5 and Fig. 7), whether cooling in water or static air, it is clear that the addition of nano-sized Ni particles had significantly improved thermal shock resistance of enamel coating. Though the reported studies[18-19] in the literature have explained well for the merits of utilising nanoscale Ni particles in enamel, the existing form of Ni particles in the enamel is observed directly in the current work. In the following part, the effort of this study would like to focus on the benefits of Ni-plating layer on thermal shock resistance.

4.2 Effect of Ni-plating layer and interlocking structure on thermal shock resistance

Generally, interfacial cracks are usually induced by the residual stress, which is caused by thermal stress due to the mismatches in the thermal expansion coefficient and growth stress because of the thickness increase in oxide scale. Once the accumulated stress exceeds limit of stress tolerance capacity, cracking would be the only outcome to release the stress, which eventually leads to delamination and failure of the whole coating sample[29].

In PE and E10N coatings, as no thermally grown oxide was formed at surface, the growth stress could be much smaller than the thermal stress, and the thermal stress is the dominant reason that is responsible for failure of coatings (spallation and cracks) by the rapid accumulation of stress during heating and cooling process. The thermal stress in the coating can be calculated by the following simplified equation[30]:

$$\sigma_e = \frac{-E_e \Delta T(\alpha_e - \alpha_s)}{1-\nu_e} \quad (1)$$

Where E_e is the Young's modulus of enamel, ν_e the Poisson's ratio of enamel, α_e is the CTE of materials, and ΔT represents the temperature change. The subscripts e and s refer to the enamel and the substrate.

From the above-mentioned equation, as the CTE mismatch between the substrate and coating is the largest, the thermal stress at the enamel/substrate interface thus reaches maximum, which leads to the crack occurring first at the enamel/substrate interface. Nonmetallic pure enamel ($9.3 \times 10^{-6}°C^{-1}$)[19] has a large mismatch with metallic substrate ($13 \times 10^{-6}°C^{-1}$)[31], while the incorporation of nano-sized Ni particles into the enamel layer slightly decreases the CTE mismatch between the substrate and coating, the effect is not significant. It was reported[19] that 5wt.% nano-Ni particles addition can only increase the CTE a bit, from $9.3 \times 10^{-6}°C^{-1}$ originally to $9.5 \times 10^{-6}°C^{-1}$. Therefore, although a small number of nano-sized Ni particles can inhibit the crack progression and extension, it has little effect on crack initialisation.

However, comparing Fig. 5(b) with Fig. 5(c) and Fig. 7(c) with Fig. 7(e), it is suggested that with the introduction of Ni-plating layer between the substrate and external enamel layer, the interfacial crack has been notably inhibited. Fig. 9(a) and b illustrates the formation of metal-enamel *in situ* interlocking coating schematically. First, as shown in Fig. 9(a), a Ni-plating layer with enamel particles homogeneous dispersion was developed by electroplating. A certain number of enamel particles were scattered at the top part of the layer, which acted as connecting pegs. The interdiffusion of Ni and Fe between the substrate and Ni-plating layer occurred once annealing, which strengthened the adhesion of substrate and Ni-plating layer. Then, the preplated samples were sprayed with enamel powder, heated to form the external enamel layer with Ni particles. During the heating process, the external enamel layer was melted and jointed with the enamel particles at the upper part of the Ni-plating layer to form the enamel pegs (see Fig. 9b). Thus, mechanical interlocking of the enamel pegs was formed at the enamel/Ni-plating interface. This interlocking structure at interface increases not only the compatibility between the enamel layer and the metallic one but also the bonding strength. It is thus the real reason why NE10N coating possesses the highest thermal shock resistance among the three enamel coatings.

Fig. 10(a) and b demonstrates the stress state within E10N and NE10N coatings during thermal shock tests. As mentioned above, both enamel layer and substrate suffered from cyclic thermal stress due to CTE mismatch between them. During the cooling stage in thermal shock, enamel would yield a compressive stress while substrate produces a tensile stress. However, during the heating process,

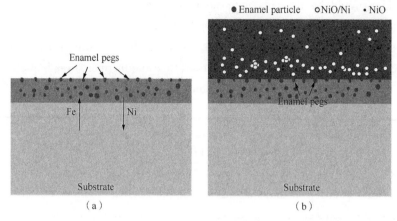

Fig. 9　Schematic illustration of NE10N coating formation

the stress status for two components is inverse. The interface between two components has to sustain a sliding shear. The enamel layer parallel to the interface pushes (or pulls) the substrate along the other, shearing the interface, which eventually generates the crack, as presented in Fig. 10(a). In the case of NE10N coatings [Fig. 10(b)], the thermal stress status is similar with that of E10N coating samples, but enamel pegs play a significant role of "pinning effect" at the interface. Many deep, vertical pits could provide high adhesion against the sliding shear. Meanwhile, the increase in contact area enhances the adhesion strength of the interface, even if the interfacial cracks initiate, a longer path is required for crack progression at the interface to bypass the enamel pegs.

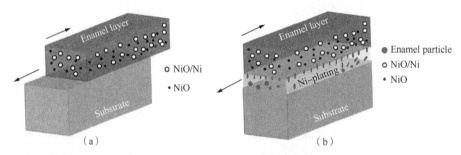

Fig. 10　Schematic illustration of stress state within E10N(a), NE10N(b) coatings during thermal shock tests

5　Conclusions

A novel metal-enamel *in situ* interlocking coating on P92 steel was designed. It was consisted of an electroplated Ni-enamel composite inner layer and the successive air-sprayed enamel outer layer with 10wt% Ni particles. Thermal shock behaviour of the coating was evaluated in water and static air at 600 ℃ in comparison with pure enamel and enamel with 10wt% Ni composite coatings. Based on the experimental results, the following conclusions can be drawn:

(1) With the incorporation of nano-sized Ni particles, the thermal shock resistance of enamel coating was notably improved.

(2) The introduction of Ni-plating layer with enamel particles incorporation further improved thermal shock resistance significantly, in which the enamel particles in the Ni-plating layer were

melted and jointed with the external enamel layer, mechanical interlocking of the enamel pegs was formed at the enamel/Ni-plating interface.

Acknowledgements

This project is financially supported by the Excellent Youth Foundation of Liaoning Province (No. 2019-YQ-03), the CNPC Science and Technology Development Project(Nos. 2019B-4013 and 2019A-3911), the National Key R&D Program of China (Nos. 2019YFF0217500 and 2016ZX05022-055), the Science Fund for Distinguished Young Scholars of Shaanxi Province and the Ministryof Industry and Information Technology Project(No. MJ-2017-J-99).

References

[1] HU L, WANG X, YIN X, et al. Influence of Inter-Pass Temperature on Residual Stress in Multi-Layer and Multi-Pass Butt-Welded 9% Cr Heat-Resistant Steel Pipes[J]. Acta Metall Sin, 2018, 54(12): 1767-1776.

[2] ZHONG X. Corrosion behaviors of nuclear-grade stainless steel and ferritic-martensitic steel in supercritical water[J]. Acta Metall Sin, 2011, 47(7): 932-938.

[3] YOSHIZAWA M, IGARASHI M, MORIGUCHI K, et al. Effect of precipitates on long-term creep deformation properties of P92 and P122 type advanced ferritic steels for USC power plants[J]. Materials Science and Engineering: A, 2009, 510: 162-168.

[4] GORMAN D M, FRY A T. Porosity connectivity within the spinel layer of the 9Cr steel grade 91 when exposed to high-temperature steam[J]. Oxidation of Metals, 2017, 88(3): 435-446.

[5] LUKASZEWICZ M, SIMMS N J, DUDZIAK T, et al. Effect of steam flow rate and sample orientation on steam oxidation of ferritic and austenitic steels at 650 and 700℃[J]. Oxidation of Metals, 2013, 79(5): 473-483.

[6] LAVERDE D, GOMEZ-ACEBO T, CASTRO F. Continuous and cyclic oxidation of T91 ferritic steel under steam[J]. Corrosion Science, 2004, 46(3): 613-631.

[7] TAKEDA M, KUSHIDA H, ONISHI T, et al. Influence ofoxidation temperature and Cr content on the adhesion and microstructure of scale on low Cr steels[J]. Oxidation of Metals, 2010, 73(1): 1-13.

[8] WU L K, WU J J, WU W Y, et al. Hot corrosion behavior of electrodeposited SiO_2 coating on TiAl alloy [J]. Corrosion Science, 2020, 174: 108827.

[9] WANG Q, WU W Y, JIANG M Y, et al. Improved oxidation performance of TiAl alloy by a novel Al-Si composite coating[J]. Surface and Coatings Technology, 2020, 381: 125126.

[10] JIANG C, YANG Y, ZHANG Z, et al. Preparation and enhanced hot corrosion resistance of Zr-Doped $PtAl_2$ + (Ni, Pt)Al dual-phase coating[J]. Acta Metall Sin, 2018, 54(4): 581-590.

[11] LI M J, SUN X F, GUAN H R, et al. High temperature hot-corrosion behavior of(Ni, Pd)Al coating [J]. Acta Metallrugica Sinica, 2004, 40(7): 773-778.

[12] WOLLSCHLÄGER N, NOFZ M, DÖRFEL I, et al. Exposition of sol-gel alumina-coated P92 steel to flue gas: Time-resolved microstructure evolution, defect tolerance, and repairing of the coating[J]. Materials and Corrosion, 2018, 69(4): 492-502.

[13] NOFZ M, DÖRFEL I, SOJREF R, et al. Thinsol-gel alumina coating as protection of a 9% Cr steel against flue gas corrosion at 650℃[J]. Oxidation of Metals, 2018, 89(3): 453-470.

[14] LIAO Y, FENG M, CHEN M, et al. Comparative study of hot corrosion behaviour of the enamel based composite coatings and the arclon plating NiCrAlY on TiAl alloy[J]. Acta Metall Sin, 2018, 55(2): 229-237.

[15] XIONG Y, ZHU S, WANG F. Synergistic corrosion behavior of coated Ti60 alloys with NaCl deposit in moist air at elevated temperature[J]. Corrosion Science, 2008, 50(1): 15-22.

[16] CHEN K, CHEN M, YU Z, et al. Simulating sulfuric acid dew point corrosion of enamels with different contents of silica[J]. Corrosion Science, 2017, 127: 201-212.

[17] YU Z, CHEN M, CHEN K, et al. Corrosion of enamel with and without CaF_2 in molten aluminum at 750℃ [J]. Corrosion Science, 2019, 148: 228-236.

[18] LIAO Y, ZHANG B, CHEN M, et al. Self-healing metal-enamel composite coating and its protection for TiAl alloy against oxidation under thermal shock in NaCl solution [J]. Corrosion Science, 2020, 167: 108526.

[19] FENG M, CHEN M, YU Z, et al. Comparative study of thermal shock behavior of the arc ion plating NiCrAlY and the enamel based composite coatings[J]. Acta Metall Sin, 2017, 53(12): 1636-1644.

[20] GUO C, CHEN M, LIAO Y, et al. Protection mechanism study of enamel-based composite coatings under the simulated combusting gas shock[J]. Acta Metall Sin, 2018, 54(12): 1825-1832.

[21] CHEN M, ZHU S, WANG F. Strengthening mechanisms and fracture surface characteristics of silicate glass matrix composites with inclusion of alumina particles of different particle sizes[J]. Physica B: Condensed Matter, 2013, 413: 15-20.

[22] MCBAIN J W. On adhesives and adhesive action[J]. The Journal of Physical Chemistry, 2002, 29(2): 188-204.

[23] BROCKMANN W. Durability of adhesion between metals and polymers[J]. The Journal of Adhesion, 1989, 29(1-4): 53-61.

[24] HAGEN C M H, HOGNESTAD A, KNUDSEN O O, et al. The effect of surface roughness on corrosion resistance of machined and epoxy coated steel[J]. Progress in Organic Coatings, 2019, 130: 17-23.

[25] VAN DAM J P B, ABRAHAMI S T, YILMAZ A, et al. Effect of surface roughness and chemistry on the adhesion and durability of a steel-epoxy adhesive interface [J]. International Journal of Adhesion and Adhesives, 2020, 96: 102450.

[26] CHEN K, CHEN M, WANG Q, et al. Micro-alloys precipitation in NiO-and CoO-bearing enamel coatings and their effect on adherence of enamel/steel[J]. International Journal of Applied Glass Science, 2018, 9(1): 70-84.

[27] DONALD I W, METCALFE B L, GERRARD L A. Interfacial reactions in glass-ceramic-to-metal seals [J]. Journal of the American Ceramic Society, 2008, 91(3): 715-720.

[28] SHIEU F S, LIN K C, WONG J C. Microstructure and adherence of porcelain enamel to low carbon steel [J]. Ceramics International, 1999, 25(1): 27-34.

[29] YANG J, WANG L, LI D, et al. Stress analysis and failure mechanisms of plasma-sprayed thermal barrier coatings[J]. Journal of Thermal Spray Technology, 2017, 26(5): 890-901.

[30] BULL S J. Modeling of residual stress in oxide scales[J]. Oxidation of Metals, 1998, 49(1): 1-17.

[31] AGÜERO A, MUELAS R, GUTIÉRREZ M, et al. Cyclic oxidation and mechanical behaviour of slurry aluminide coatings for steam turbine components[J]. Surface and Coatings Technology, 2007, 201(14): 6253-6260.

本论文原发表于《Acta Metallurgica Sinica(English Letters)》2021年第34卷第8期。

Evaluation of Glass Coatings with Various Silica Content Corrosion in a 0.5mg/L HCl Water Solution

Wang Hang[1,2]　Zhang Chuan[3]　Jiang Chengyang[4]　Zhu Lijuan[1,2]
Cui Jiakai[1,2]　Han Lihong[1,2]　Chen Minghui[4]　Geng Shujiang[4]　Wang Fuhui[4]

(1. Tubular Goods Research Institute of CNPC; 2. State Key Laboratory of Performance and Structural Safety for Petroleum Tubular Goods and Equipment Materials;
3. CNPC Bomco Drilling & Production Equipment Co. LTD.;
4. Shenyang National Laboratory for Materials Science, Northeastern University)

Abstract: Two enamel coatings with high and low silica content were prepared on 35CrMo steels via vacuum firing. Their corrosion behaviour in 0.5mg/L HCl solution, including corrosion kinetics, microstructures and electrochemistry performance were studied in comparison with uncoated steels. Results show that catastrophic corrosion occurred for uncoated steels, while enamel coatings significantly decreased corrosion rate. Enamel coatings with high silica content exhibited best corrosion resistance against hydrochloric acid due to their highly connected silicate network, which inhibited the leaching process of alkali metals in acid solution. Some white nano crystals were precipitated out from the parent glass. The corrosion inhibition efficiency for enamel containing high silica content reached maximum to 94.3%.

Keywords: Enamel; Nano crystal; Acid corrosion; Carbon steel; Electrochemistry

1 Introduction

Carbon steels have various applications in many kinds of fields due to their superior mechanical strength[1-2]. As some significant fields of their application are acid pickling, oil industry and petrochemical processes, during which strong acid solutions particularly the hydrochloric acid are always used[3-7], metallic components suffered from severe corrosion and their service lives are reduced significantly. Therefore, many attentions have been paid to looking for some efficient approaches to improve corrosion resistance of metallic components particularly the carbon steels in acidic medium.

Many of methods for protecting carbon steels against acid corrosion are available like surface coating techniques, e.g. metallic oxide film, corrosion inhibitors, electrochemical sacrificial

---

Corresponding author: Jiang Chengyang, jiangchengyang@mail.neu.edu.cn

anodic protection[8-10]. Among them, chemical inhibitor is always utilised to retard the acid corrosion of carbon steels effectively[11-12]. Work done by Saeed[11] reported that the corrosion resistance of carbon steels were significantly improved in sulfuric acid medium by bicyclic isoxazolidines inhibitor. Berrissoul et al.[13] used Lavandula mairei, a novel kind of corrosion inhibitor for mild steel in HCl solution, and indicated that its inhibitory effect can reach up to 92% maximally. Meanwhile, the metallic oxide coatings prepared by the sol–gel method, such as ZrO_2[14], SnO_2[15], SiO_2[16] also have shown superior acid corrosion resistance. Ates et al[15] reported that SnO_2 prepared by sol-gel method is a suitable coating that can protect the mild steel from corrosion in HCl solution. However, generally, regardless of the sol-gel solutions or most of the corrosion inhibitors, they are expensive synthetic chemicals and may contaminate the environment. Therefore, non–toxicity and ecofriendly aspects are always required by the environmental concerns.

Enamel coatings possess superior corrosion resistance against most corrosive media (salt[17], gas[18], molten aluminum[19], saline solution[20], etc.) owning to their inertness and compactness with the underlying matrix. In addition, the preparation process and raw material of enamel coatings are eco–friendly and cost–effective[21]. Some previous researches are available on the corrosion behaviour of enamel coating in acid solution, but most of them focused on the corrosion performance of titanium enamel for steel cookware in light acetic acid solution[22-25]. Few investigations have been conducted on the corrosion mechanism of enamel in strong hydrochloric acid.

Silica has been reported to have a significant effect on the microstructure and acid corrosion resistance of enamel coatings[26-28]. However, the relationship between silica content and corrosion resistance in strong acid has not been reported yet. Therefore, two enamel coatings with high and low silica content were designed and prepared on carbon steels. As carbon within the matrix may produce volatile gas and cause enamel coatings porous during the coating preparation, which led acid invading inward easily and coating degradation rapidly. So in this study enamel coatings were fired in vacuum. Corrosion behaviour, i.e. corrosion kinetics, microstructures and electrochemistry performance of the two enamels in 0.5mg/L HCl solution were investigated in comparison with that of the bare carbon steel.

2 Materials and methods

2.1 Sample preparation

A typical carbon steel, 35CrMo (composition: C: 0.32~0.40, Mn: 0.4~0.7, Si: 0.17~0.37, Cr: 0.8~1.1, Mo: 0.15~0.25, Ni: <0.03, Cu: <0.03 and balanced Fe, wt.%) was used as the substrate alloy. Test samples of approximate dimensions 15mm × 15mm × 1.5mm were cut from the pipe using a spark discharge machine. The bare samples were grounded with a final 400 mesh SiC sandpaper and then blasted with 200–mesh glass ball, then they were cleaned ultrasonically in acetone and ethanol for 30 min, respectively.

The nominal composition of enamel was presented in Table 1. The enamel containing high and low silica content was named as HS and LS, respectively hereafter. The enamel frits were gained first by melting at high temperature, followed by quenching in water. After that, the enamel powder was obtained by milling the frits. The preparation method of enamel powder has been previously described in detail elsewhere[29].

Table 1 Nominal composition of enamel (wt.%)

	SiO_2	B_2O_3	Al_2O_3	Na_2O	K_2O	CoO	CaF_2
LS	54.6	12.3	6.0	12.3	6.0	2.5	6.3
HS	64.6	9.6	4.7	9.6	4.6	1.9	5.0

Note: LS = low silica, HS = high silica.

The enamel powder was then blended with ethanol in a ratio of 10 g enamel powder to 150mL ethanol to form a slurry. The enamel coating was prepared by air spraying onto the prepared steel, after which it was baked at 70℃ for 10 minutes. Then the coated sample was taken into the furnace and heated in vacuum($<6\times10^{-2}$ Pa) at 750℃ to 2 minutes for LS and 880℃ to 10 minutes for HS, respectively. Both enamel coatings were approximated 100μm in thickness. Here, it should be noted that high content of silica will decrease the flowability of the enamel coating and increase the firing temperature. So, chemical composition of enamel coatings should not be adjusted within a large scale taking the firing process into consideration.

2.2 Corrosion test

To avoid hydrochloric acid invading inside the hanging hole in priority and leading to false mass change, only the bottom half of the test samples was immersed in 0.5mg/L HCl solution in a Teflon container at 30℃. Samples were taken out from the container after 24 h corrosion, flushed by deionized water for 1 min to remove the residual solution on their surface. The specimens then were dried using a blow drier and weighed using an electronic balance with sensitivity of 10^{-5} g(Sartorius BP211D). The above steps constitute a cycle of acid corrosion. The acid solution used in the former cycle was replaced by a fresh one before starting next cycle of the corrosion test. Three parallel samples of each coating were used to acquire the average value of mass change.

2.3 Characterisation

Surface and cross-sectional morphologies were observed using a field-emission scanning electron microscope (SEM, Inspect F50, FEI Co., Hillsboro, OR) equipped with energy dispersive X-ray spectrometer(EDAX, X-Max, Oxford Instruments Co., Oxford, U.K.). To observe surface morphology, a second electron(SE) mode was adopted, while the back-scattered electron(BSE) mode was utilised to characterise cross-sectional morphologies of coating samples.

2.4 Electrochemical measurements

The electrochemical measurements were carried out using a CHI660E Electrochemical Analyzer under computer control. An electrochemical cell with a three-electrode configuration consisting of the test sample electrode, a platinum sheet(with 2.25 cm^2 surface area) auxiliary electrode, and a saturated calomel electrode(SCE) as reference was used. All potentials were measured with respect to this reference electrode. The uncoated and enamel coated test samples were exposed to 0.5mg/L HCl solution at 30℃. The open circuit potential(OCP) was recorded as a functional of time up to 300 s. As a result, E_{ocp}, which corresponds to a steady-state OCP was obtained. Electrochemical impedance spectroscopy(EIS) and potentiodynamic polarization measurements were carried out after establishing a steady-state OCP. The polarization curves were potentio-dynamically obtained in the potential ranges from -300 mV to +600 mV based on Eocp with a scan rate of 1 mV/s. The EIS experiments were conducted in the frequency range from 100kHz to 0.01Hz at Eocp with a perturbation amplitude of 5 mV peak-to-peak.

3 Results

3.1 Vacuum firing

Fig. 1 shows the initial cross-sectional morphologies of HS firing in air and vacuum at 880℃. It can be observed from Fig. 1(a) that a large number of bubbles were scattered within the enamel coating, while the number of bubbles decreased dramatically when the enamel was fired in vacuum [Fig. 1(b)]. This is mainly because the low oxygen partial pressure in vacuum inhibited significantly the oxidation of carbon from the matrix. The content of carbon in 35CrMo steel ranged from 0.32 wt.% to 0.40 wt.%, belonging to medium carbon steel. During the coating preparation, when the steel was fired in air at 880℃, the carbon in the steel will be oxidised to produce gas, some of the gas could not escape out due to the coverage of enamel, and was trapped in the enamel. By contrast, the low oxygen partial pressure in vacuum suppressed the decarburization reaction, making the enamel layer almost free of bubbles.

It should be noted that vacuum firing provides a feasible method to applying enamel on steels that produce volatile gas at relative high temperatures, especially for high-carbon steels and cast iron. All the enamel coated samples for corrosion were fired in vacuum to avoid the acid invading through the bubbles and affecting the experimental validity.

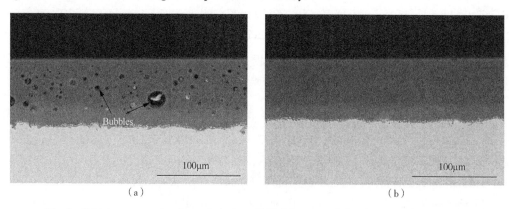

Fig. 1 Initial cross-sectional morphologies of HS firing in air(a) and vacuum(b) at 880℃

3.2 Macro morphology and corrosion kinetics

Fig. 2 shows macro morphologies of three test samples before and after corrosion in 0.5mg/L HCl solution for different days. Before corrosion (0 d), the uncoated sample displayed original metallic frosted color due to sand blasting. While both enamels were blue, but HS showed a little darker than LS due to the compositional difference. After one day corrosion, the red rust was formed at the surface of uncoated samples. LS lost its gloss and its color faded from blue into red. With the corrosion time increased to three days, the uncoated sample suffered from disastrous corrosion, the red rust grew to be looser and thicker. The original blue color of LS faded away completely, regardless of the immersed or non-immersed part. By contrast, HS maintained its gloss and only the color of immersed part faded a little during the entire corrosion test.

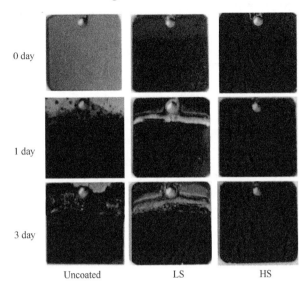

Fig. 2 Macro images of uncoated, LS and HS samples before and after corrosion in
0.5mg/L HCl solution at 30℃ for different days

Fig. 3 shows corrosion kinetics of three test samples in 0.5mg/L HCl solution. The mass loss of uncoated sample was catastrophic and much larger than that of the two enamel coated samples. It obeyed a linear relationship with the slope $k = -2.87$ mg/(cm² · d). LS lost its weight slowly following a linear relationship as well within the initial 1-day corrosion, but became faster with further increasing corrosion time. In the case of HS, its weight loss was the lowest, that it was hardly to definite clearly the corrosion law. After corrosion for three days, the total mass loss of uncoated, LS and HS samples was 8.38mg/cm², 4.30mg/cm² and 0.10mg/cm², respectively.

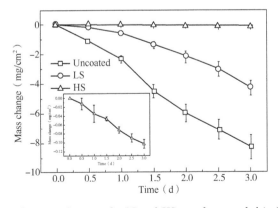

Fig. 3 Corrosion kinetics of uncoated, LS and HS samples corroded in 0.5mg/L HCl
solution at 30℃ (inset is the magnified corrosion kinetics of HS)

3.3 Surface and cross-sectional microstructures

Fig. 4 shows XRD patterns of three test samples after corrosion for three days. It is observed that the uncoated sample was mainly composed of matrix as most of the corrosion products have been

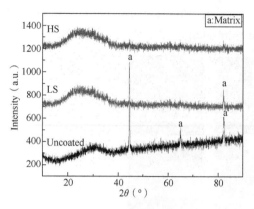

Fig. 4 XRD patterns of uncoated, LS and HS samples after corrosion in 0.5mg/L HCl solution at 30℃ for 3 days

dissolved into the acid. As for LS sample, a broad peak appearing around 25° indexed the existence of amorphous phase. As the outer enamel layer was corroded, the diffraction peaks corresponding to the matrix were detected at around 45° and 81°. In contrast, HS sample still consisted of the amorphous phase exclusively with little matrix phase detected. Fig. 5 shows surface morphologies of three test samples after corrosion in 0.5mg/L HCl solution for three days. It can be observed from Fig. 5(a) that a substantial number of corrosion holes formed at surface of the uncoated sample. The specimen surface was pretty rough and strongly damaged due to steel excessive dissolution in the presence of H^+ and Cl^- ions. Similarly, the surface of LS was also unsmoothed with a large number of spallation sites [Fig. 5(b)]. In the contrast, neither cracks nor spallation was observed at the surface of HS, which remained its superior protection against acid corrosion. In addition, some white nano crystals were precipitated out from the parent glass.

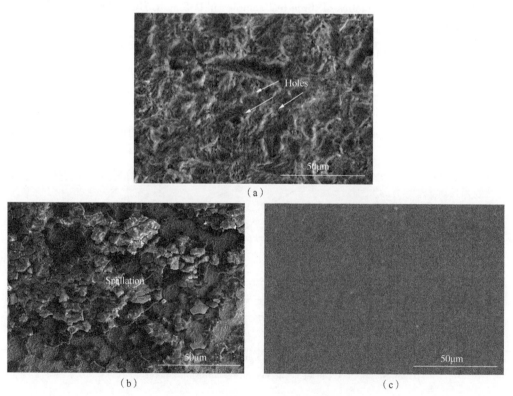

Fig. 5 Surface morphologies of uncoated(a), LS(b) and HS(c) samples after corrosion in 0.5mg/L HCl solution at 30℃ for 3 days

Fig. 6(a-c) shows corresponding cross-sectional morphologies of three test samples after corrosion in 0.5mg/L HCl solution for three days. A porous and unprotected Fe_2O_3 layer was observed to form at the surface of bare sample, as shown in Fig. 6(a). While for LS, the external enamel layer remained covered on the matrix, but there were some vertical and horizontal cracks within it [Fig. 6(b)], which has been reported and normally named as the leached layer (LL)[30-31]. The contrast difference between the external layer and inlayer was obvious. EDS analysis

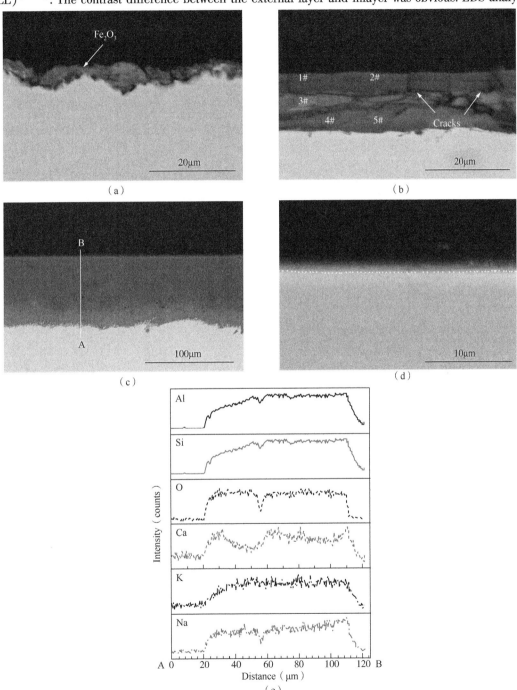

Fig. 6 Cross-sectional morphologies of uncoated(a), LS(b) and HS(c) samples after corrosion in 0.5mg/L HCl solution at 30℃ for 3 days, corresponding high-magnification view(d) and line scan profile(e) of image c

indicated that the external layer was in the absence of alkali metal, e.g Na, K, Ca, Mg, while Si was rich at the outer layer, as shown in Table 2. It is clear that elements leaching has broken the Si-O network of the enamel layer, which led to cracking and spalling. This process of leaching-cracking-spalling occurred successively till the LS was reduced to merely 20μm thick. In the case of HS, the enamel layer remained integrated and adhered well with the underlying matrix [Fig. 6 (c)]. Its thickness remained almost unchanged after three days immersion. From the high-magnification SEM images [Fig. 6(d)], a gray layer with approximate 1μm in thickness was observed at the outermost of the corroded HS. This gray layer was normally named as the gel layer (GL), which was observed typically at surface of other corroded glass system[26,32-34]. Meanwhile, it can be seen from the corresponding line profile [Fig. 6(e)] that there was little difference in the composition of the whole enamel, i.e. elements leaching from the Si-O network of HS was largely retarded.

Table 2 EDS results in Fig. 5(b) (at. %)

Zone	O	Na	Mg	Al	Si	K	Ca	Fe	Co
Zone 1	73.1	0.4	1.2	3.9	16.1	0.6	2.4	1.7	0.6
Zone 2	72.0	0.4	1.1	4.0	16.7	0.7	2.6	1.8	0.7
Zone 3	62.1	7.5	5.7	2.4	11.2	2.5	7.4	0.4	0.8
Zone 4	60.9	7.5	6.0	2.5	11.0	2.6	8.3	0.4	0.8
Zone 5	60.8	7.4	6.0	2.4	11.2	2.6	8.3	0.4	0.9

3.4 Electrochemical performance

3.4.1 Electrochemical impedance spectroscopy

Fig. 7 shows the Nyquist plots of uncoated, LS and HS samples in 0.5mg/L HCl solution. Generally, all the plots exhibited single capacitive semi-circles[35-36]. Based on the curves, the corrosion process was mainly charge transfer controlled. The total diameter of Nyquist plots, which determined the polarization resistance (R_p), considerably increased when the sample was coated by enamel, and with the increase of silica content. Correspondingly, polarization resistance was the lowest for uncoated sample, while it reached the highest value for HS, which indicated a great enhancement of corrosion resistance achieved by HS.

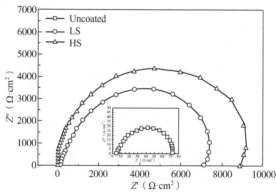

Fig. 7 The Nyquist plots of uncoated, LS and HS samples in 0.5mg/L HCl solution at 30℃ (inset is magnified Nyquist plots of uncoated sample)

Bode and phase angle plots for three test samples in 0.5mg/L HCl solution were presented in Fig. 8(a) and Fig. 8(b), respectively. The bode of HS reached the maximum value while that of uncoated sample was the minimum. Similarly, the phase angle, which referred to the formation of an effective protection layer on the surface[37], were also the highest for HS and lowest for uncoated sample, indicating that the most homogeneous surface with good corrosion resistance was formed at HS. These results are also consistent with Nyquist curves.

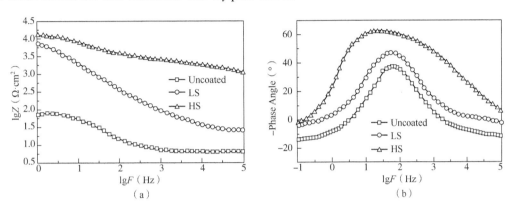

Fig. 8 The bode(a) and phase angle(b) plots of uncoated, LS and HS samples in 0.5mg/L HCl solution at 30℃

Equivalent circuits have been used to interpret the EIS data quantitatively (Fig. 9). The EIS spectrum of uncoated sample can be simulated and corresponded well by an impedance plot which has one time constant [Fig. 9(a)]. In the equivalent circuit, R_s represents the solution resistance, R_1 is the charge transfer resistance at the metal/electrolyte interface, and CPE1 is the constant phase element (CPE) which is used to replace the ideal double-layer capacitance. The impedance function of the CPE is as follows[38]:

$$Z_{CPE} = \frac{1}{Y_0(jw)^n} \quad (1)$$

Where Y_0 is admittance of CPE. $j^2 = -1$ defines as an imaginary number and ω is the angular frequency. n is a CPE exponent determining the phase shift which can be utilised as a gauge of roughness or heterogeneity of the surface ($0<n<1$).

In comparison, two time constants were used to simulate EIS data obtained from the enamel coated samples[39]. LS and HS were well fitted to the equivalent circuit, as shown in Fig. 9(b). As mentioned above, R_s represents the solution resistance, R_1 is the coating resistance, which is deeply influenced by defects (such as flaws and pores). CPE1 is the coating capacitance. R_2 and L correspond to the charge transfer resistance and the inductance at the coating/substrate interface, respectively.

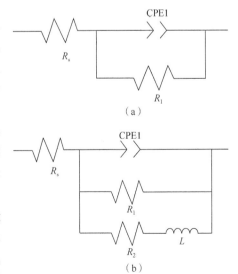

Fig. 9 The equivalent circuit used to fit the EIS results of uncoated(a), LS and HS(b) samples

The polarization resistance, as mentioned above, was defined as the difference between the intersection of the real axis and the impedance in the case of an infinitesimal and infinite frequency. The reciprocal of R_p was proportional to the corrosion rate, which was an important parameter. The disturbing alternating-current signal can be interpreted as the direct-current signal when the frequency became infinitesimal. In this case, the capacitance and the inductance can be regarded as an insulating component and a zero-resistance component, respectively. Therefore, for uncoated sample, R_p was equal to R_1, while for enamel coated samples, the reciprocal of R_p can be expressed by the following equations:

$$R_p = \frac{R_1 R_2}{R_1 + R_2} \tag{2}$$

Impedance parameters were presented in Table 3, the value of R_s, Y_0-CPE1, R_1, R_2, L and R_p, can be used to determine the corrosion resistance of three test samples. It can be seen that uncoated sample has a polarization resistance of $(66\pm1)\,\Omega\cdot cm^{-2}$. While for enamel coated samples, the polarization resistance improved significantly than that of uncoated samples, with $(960\pm5)\,\Omega\cdot cm^{-2}$ for LS and $(1178\pm23)\,\Omega\cdot cm^{-2}$ for HS, respectively, suggesting that the enamel coatings could markedly enhance the corrosion resistance of 35CrMo steel. In particular, HS showed the highest corrosion resistance among three test samples. Moreover, the L increased from $(2502\pm14)\,H\cdot cm^{-2}$ to $(3718\pm34)\,H\cdot cm^{-2}$ with the silica content increased from LS to HS. According to the EIS theory, the increase of L implied that the increase of the phase angle of inductive loop.

Table 3 Impedance parameters of uncoated, LS and HS samples in 0.5mg/L HCl solution at 30℃

Sample	$R_s(\Omega)$	$R_1(\Omega/cm^2)$	Y_0-CPE1 ($\mu S\cdot s^{-n}\cdot cm^{-2}$)	R_2 (Ω/cm^2)	L (H/cm^2)	R_p (Ω/cm^2)
Uncoated	6±1	66±1	286±5	—	—	66±1
LS	24±6	9304±7	22±9	1071±14	2502±14	960±5
HS	328±28	$(1583\pm6)\times10^7$	99±24	1178±23	3718±34	1178±23

3.4.2 Potentiodynamic polarization measurements

The polarization curves of three test samples in 0.5mg/L HCl solution were exhibited in Fig. 10. The study of these polarization measurements allowed having information related to cathodic and anodic reactions. Electrochemical parameters including the potential of corrosion (E_{corr}), corrosion current density (i_{corr}), anodic Tafel slope (β_a) and cathodic Tafel slope (β_c) were obtained and given in Table 4. Using polarization curves and electrochemical diagrams, the percentage of inhibition efficiency η_{Tafel} was obtained following Equation(3), as listed in Table 4:

$$\eta_{Tafel} = \left(\frac{i_{cor}^0 - i_{cor}}{i_{cor}^0}\right) \times 100 \tag{3}$$

where i_{cor}^0 and i_{cor} are, respectively, current densities without and with enamel coatings.

As can be seen from Fig. 10 and Table 4, it is clear that the value of i_{cor} was much smaller for enamel compared with uncoated samples, and decreased with increase of silica content. The most decreased corrosion current density belonged to HS with 0.007 mA/cm^2. Correspondingly, based on Equation(3), the inhibition efficiency raised up with increase in concentration of silica due to the decrease in corrosion current density, which reached the highest value (94.3%) for HS. The cathodic curves have given rise to parallel Tafel lines, meaning that the evolution of the hydrogen was activation-controlled[40]. A reduction in both anodic Tafel slopes was observed for the sample coated by enamel coatings, which indicated that the enamel decreased anodic metal dissolution. It should be noted that there was no difference in the cathodic Tafel slope between LS and HS, implying that the silica concentration did not affect the mechanism of the hydrogen evolution reaction. The decrease of the anodic Tafel slope can be interpreted as the prevention of anodic dissolution due to the formation of a protective enamel layer on the surface[41].

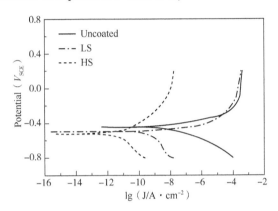

Fig. 10　The polarization curves of uncoated, LS and HS samples in 0.5mg/L HCl solution at 30℃

Table 4　Polarization data of uncoated, LS and HS samples in 0.5mg/L HCl solution at 30℃

Sample	$-E_{corr}$ (mV$_{SCE}$)	i_{corr} (mA/cm^2)	β_a (mV/dec)	$-\beta_c$ (mV/dec)	η_{Tafel}
Uncoated	447	0.122	13.98	8.53	—
LS	491	0.095	8.11	2.97	22.1
HS	517	0.007	5.78	2.94	94.3

4　Discussion

As is clear, the corrosion resistance of carbon steels against hydrochloric acid is notably improved by enamel coatings, and strongly related with silica content in the enamel. All the results, regardless of corrosion kinetics, microstructure evaluation or electrochemical performance, corroborate the idea that HS has the best inhibition efficiency against the attack of corrosive species like dissolved oxygen and chloride ions. Therefore, the discussion mainly focused on the effect of silica content within enamel on corrosion resistance.

As shown in Fig. 3, the total weight loss reduces dramatically by 97.6% with the Si content increases by 10 wt% from LS to HS. Combining with their microstructure evolutions and elemental distributions, the reason for the difference of corrosion resistance between LS and HS can be speculated to be the distinction in leachability of alkali metal in HCl solution, which is directly determined by the compactness of silica network. In order to shed light on the degree of network connectivity of LS and HS during the corrosion test, Raman spectra of LS and HS before and after corrosion for different days have been obtained and shown in Fig. 11. It can be seen from Fig. 11(a) that the characteristic peaks of LS are mainly at ~330 cm^{-1} and ~950 cm^{-1} before corrosion. While after corrosion, the characteristic peak of 950 cm^{-1} disappears gradually, but 330 cm^{-1} remains. In the case of HS, the characteristic peaks remain at ~1070 cm^{-1} before and after corrosion [Fig. 11 (b)]. The characteristic peaks of bridge oxygen band with Si ($Si-O_b-Si$) and non-bridge oxygen band with Si ($Si-O_{nb}$) have been reported previously[42-44] at ~1070 cm^{-1} and ~950 cm^{-1}, respectively, and the band at ~330 cm^{-1} belongs to CaF_2. Comparing the Raman spectra of HS with that of LS, obviously, the network connectivity of HS is higher than that of LS. The non-bridge oxygen band of LS is broken easily during the corrosion while bridge oxygen band of HS can be very stable. In addition, the intensity of 330 cm^{-1} peak weakens significantly with the decrease of CaF_2 content.

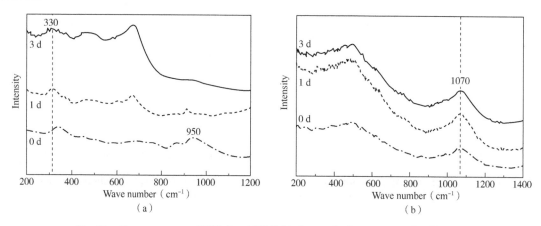

Fig. 11　Raman spectra of LS(a) and HS(b) before and after corrosion for different days in 0.5mg/L HCl solution at 30℃

Generally, three reactions can occur simultaneously during the leaching process in the aqueous or acid solution: (i) hydration, (ii) hydrolysis, and (iii) leaching. Hydration involves molecular water penetration into the glass network. Hydrolysis reaction involves the breakage of the M-O-M (M =Si, B, Al) network. Leaching is an ion-exchange reaction during which modifier cations like sodium and calcium diffuse out of the glass, while water or other hydrous species diffuse into the glass hydrating and hydroxylating the surface. Since leaching selectively depletes modifier cations, especially alkali ions, a surface layer resembling a silica-like gel forms which may extend up to a few hundred nanometers[45-47].

The corrosion process of LS is illustrated schematically in Fig. 12(a) and Fig. 12(b). Based on Raman spectra, for LS, since there are lots of non-bridging oxygen band and the network is loose, although a thin GL layer is formed at the surface of enamel coating initially, leaching with alkali metals ions, like Ca^{2+}, Na^+ and K^+ can easily occur, as shown in Fig. 12(a). Once alkali metals ions were leached out, the enamel will become to be fragile, and cracks can initialise and propagate easily within the enamel. Therefore, the dense GL will transform to a porous leached layer full of cracks in the enamel [Fig. 12(b)]. Meanwhile, the cracks formed in the leaching process will result in more acid solution invading deep inside the enamel and broke Si-O networks further. This process explains why the corrosion kinetics decreased slowly on the first day, then followed by drop dramatically later.

Fig. 12(c) and d shows the schematic diagram of the corrosion process of HS, in the case of HS, owning to its high content of silica, the content of non–bridging oxygen is less and the compactness of silica network is higher than LS. Thus, it is difficult to leach out the alkali metal ions after a dense filmy GL is formed at the surface of HS[Fig. 12(c)]. Hence, the enamel have good toughness and is free of cracks. At such a case, there is no crack formed within the enamel, and a dense gel-layer still remain and grow thicker at surface[Fig. 12(d)].

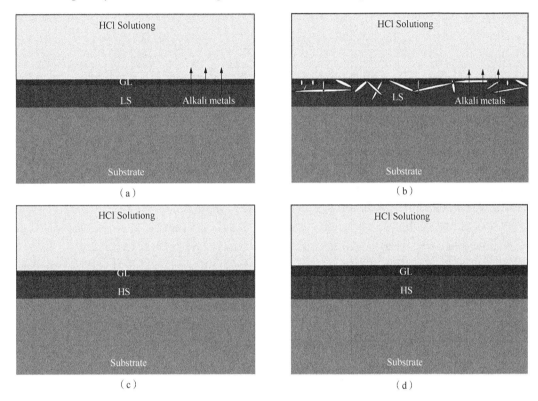

Fig. 12 Schematic diagram of the corrosion processes of LS(a, b) and HS(c, d)

5 Conclusions

Two enamel coatings with low and high silica content were prepared on 35CrMo steels via vacuum firing. Corrosion kinetics, microstructure evaluation and electrochemical performance of two enamel coatings in 0.5mg/L HCl solution were investigated, the following conclusions can be drawn:

(1) Compared with air firing, vacuum firing makes the enamel applied on carbon steels have few numbers of bubbles and pores inside.

(2) The corrosion resistance of carbon steels was improved notably by enamel coatings, the corrosion inhibition efficiency increased with the increase of silica concentration.

(3) HS exhibited superior corrosion resistance against hydrochloric acid than LS, because higher content of silica in the enamel causes the connected silicate tetrahedron network more

compact, leading the leaching process of alkali metals more difficult to happen in acid solution.

Acknowledgments

The authors wish to acknowledge the financial support by the Excellent Youth Foundation of Liaoning Province(No. 2019-YQ-03), the Ministry of Industry and Information Technology Project (No. MJ-2017-J-99), the CNPC Science and Technology Development Project(Nos. 2019B-4013 and 2019A-3911), the National Key Research and Development Program of China (Nos. 2019YFF0217500 and 2016ZX05022-055), and the Science Fund for Distinguished Young Scholars of Shanxi Province.

References

[1] SIRELI G K, BORA A S, TIMUR S. Evaluating the mechanical behavior of electrochemically borided low-carbon steel[J]. Surface and Coatings Technology, 2020, 381: 125177.

[2] CHEN J, LIU Z Y. The combination of strength and cryogenic impact toughness in low carbon 5Mn-5Ni steel [J]. Journal of Alloys and Compounds, 2020, 837: 155484.

[3] WANG H L, FAN H B, ZHENG H S. Corrosion inhibition of mild steel in hydrochloric acid solution by a mercapto-triazole compound[J]. Materials Chemistry and Physics, 2003, 77(3): 655-661.

[4] BENTISS F, LAGRENEE M, TRAISNEL M, et al. The corrosion inhibition of mild steel in acidic media by a new triazole derivative[J]. Corrosion Science, 1999, 41(4): 789-803.

[5] MOURYA P, BANERJEE S, SINGH M M. Corrosion inhibition of mild steel in acidic solution by Tagetes erecta(Marigold flower)extract as a green inhibitor[J]. Corrosion Science, 2014, 85: 352-363.

[6] ECH-CHIHBI E, NAHLÉ A, SALIM R, et al. Computational, MD simulation, SEM/EDX and experimental studies for understanding adsorption of benzimidazole derivatives as corrosion inhibitors in 1.0 M HCl solution [J]. Journal of Alloys and Compounds, 2020, 844: 155842.

[7] RBAA M, DOHARE P, BERISHA A, et al. New Epoxy sugar based glucose derivatives as eco-friendly corrosion inhibitors for the carbon steel in 1.0 M HCl: Experimental and theoretical investigations[J]. Journal of Alloys and Compounds, 2020, 833: 154949.

[8] ZHANG J, RAHMAN Z U, ZHENG Y, et al. Nanoflower like SnO_2-TiO_2 nanotubes composite photoelectrode for efficient photocathodic protection of 304 stainless steel [J]. Applied Surface Science, 2018, 457: 516-521.

[9] BRANZOI F, PAHOM Z, NECHIFOR G. Corrosion protection of new composite polymer coating for carbon steel in sulfuric acid medium by electrochemical methods[J]. Journal of Adhesion Science and Technology, 2018, 32(21): 2364-2380.

[10] ASHASSI-SORKHABI H, KAZEMPOUR A. Influence of fluid flow on the performance of polyethylene glycol as a green corrosion inhibitor[J]. Journal of Adhesion Science and Technology, 2020, 34(15): 1653-1663.

[11] SAEED M T. Corrosion inhibition of carbon steel in sulfuric acid by bicyclic isoxazolidines[J]. Anti-Corrosion Methods and Materials, 2004, 51(6): 389-398.

[12] RAHMAN S U, SAEED M T, ALI S A. Cyclic nitrones as novel organic corrosion inhibitors for carbon steel in acidic media[J]. Anti-Corrosion Methods and Materials, 2005, 52(3): 154-159.

[13] BERRISSOUL A, OUARHACH A, BENHIBA F, et al. Evaluation of Lavandula mairei extract as green inhibitor for mild steel corrosion in 1 M HCl solution. Experimental and theoretical approach[J]. Journal of Molecular Liquids, 2020, 313: 113493.

[14] LI H B, LIANG K M, MEI L F, et al. Oxidation protection of mild steel by zirconia sol-gel coatings

[J]. Materials Letters, 2001, 51(4): 320-324.

[15] ATES S, BARAN AYDIN E, YAZICI B. The corrosion behavior of the SnO_2-coated mild steel in HCl solution at different temperature[J]. Journal of Adhesion Science and Technology, 2021, 35(4): 419-435.

[16] STAMBOLOVA I, YORDANOV S, LAKOV L, et al. Preparation of sol-gel SiO_2 coatings on steel and their corrosion resistance[C]. MATEC Web of Conferences. EDP Sciences, 2018, 145: 05011.

[17] ZHENG D, ZHU S, WANG F. The influence of TiAlN and enamel coatings on the corrosion behavior of Ti6Al4V alloy in the presence of solid NaCl deposit and water vapor at 450℃[J]. Surface and Coatings Technology, 2007, 201(12): 5859-5864.

[18] XIONG Y, ZHU S, WANG F. Synergistic corrosion behavior of coated Ti60 alloys with NaCl deposit in moist air at elevated temperature[J]. Corrosion Science, 2008, 50(1): 15-22.

[19] YU Z, CHEN M, CHEN K, et al. Corrosion of enamel with and without CaF_2 in molten aluminum at 750℃[J]. Corrosion Science, 2019, 148: 228-236.

[20] CHEN K, CHEN M, YU Z, et al. Corrosion of SiO_2-B_2O_3-Al_2O_3-$CamF_2$-R_2O(R=Na and K)enamels with different content of ZrO_2 in H_2SO_4 and NaOH solutions[J]. Ceramics International, 2019, 45(12): 14958-14967.

[21] WU M Y, CHEN M H, ZHU S L, et al. Protection mechanism of enamel-alumina composite coatings on a Cr-rich nickel-based superalloy against high-temperature oxidation[J]. Surface and Coatings Technology, 2016, 285: 57-67.

[22] GOLEUS V I, NAGORNAYA T I, RUBANOVA O N, et al. Chemical stability of titanium enamel coatings [J]. Glass and Ceramics, 2012, 69(7-8): 274-275.

[23] RODTSEVICH S P, TAVGEN V V, MINKEVICH T S. Effect of alkali metal oxides on the properties of titanium containing glass enamels[J]. Glass and Ceramics, 2007, 64(7-8): 244-246.

[24] RODTSEVICH S P, ELISEEV S Y, TAVGEN V V. Low-melting chemically resistant enamel for steel kitchenware[J]. Glass and Ceramics, 2003, 60(1-2): 23-25.

[25] SVETLOV V A. Increasing corrosion resistance of titanium enamel coatings by modifying the slip[J]. Glass and Ceramics, 1989, 46: 293-295.

[26] GIN S, JOLLIVET P, FOURNIER M, et al. Origin and consequences of silicate glass passivation by surface layers[J]. Nature Communications. 2015, 6: 6360.

[27] CAILLETEAU C, ANGELI F, DEVREUX F, et al. Insight into silicate-glass corrosion mechanisms [J]. Nature Materials, 2008, 7(12): 978-983.

[28] HELLMANN R, COTTE S, CADEL E, et al. Nanometre-scale evidence for interfacial dissolution-reprecipitation control of silicate glass corrosion[J]. Nature Materials, 2015, 14(3): 307-311.

[29] CHEN K, CHEN M, WANG Q, et al. Micro-alloys precipitation in NiO-and CoO-bearing enamel coatings and their effect on adherence of enamel/steel[J]. International Journal of Applied Glass Science, 2018, 9 (1): 70-84.

[30] DOHMEN L, LENTING C, FONSECA R O C, et al. Pattern Formation in Silicate Glass Corrosion Zones [J]. International Journal of Applied Glass Science, 2013, 4(4): 357-370.

[31] GEISLER T, JANSSEN A, SCHEITER D, et al. Aqueous corrosion of borosilicate glass under acidic conditions: A new corrosion mechanism[J]. Journal of Non-Crystalline Solids, 2010, 356(28-30): 1458-1465.

[32] GIN S, RIBET I, COUILLARD M. Role and properties of the gel formed during nuclear glass alteration: importance of gel formation conditions[J]. Journal of Nuclear Materials, 2001, 298(1-2): 1-10.

[33] GIN S, GUITTONNEAU C, GODON N, et al. Nuclear glass durability: new insight into alteration layer properties[J]. The Journal of Physical Chemistry C, 2011, 115(38): 18696-18706.

[34] GIN S. Open scientific questions about nuclear glass corrosion[J]. Procedia Materials Science, 2014, 7: 163-171.

[35] BARAN E, CAKIR A, YAZICI B. Inhibitory effect of Gentiana olivieri extracts on the corrosion of mild steel in 0.5M HCl: Electrochemical and phytochemical evaluation[J]. Arabian Journal of Chemistry, 2019, 12(8): 4303-4319.

[36] BEHPOUR M, GHOREISHI S M, KHAYATKASHANI M, et al. Green approach to corrosion inhibition of mild steel in two acidic solutions by the extract of Punica granatum peel and main constituents[J]. Materials Chemistry and Physics, 2012, 131(3): 621-633.

[37] TANSUG G, TUKEN T, GIRAY E S, et al. A new corrosion inhibitor for copper protection[J]. Corrosion Science, 2014, 84: 21-29.

[38] HAGIHARA K, OKUBO M, YAMASAKI M, et al. Crystal-orientation-dependent corrosion behaviour of single crystals of a pure Mg and Mg-Al and Mg-Cu solid solutions[J]. Corrosion Science, 2016, 109: 68-85.

[39] JORCIN J B, ORAZEM M E, PEBERE N, et al. CPE analysis by local electrochemical impedance spectroscopy[J]. Electrochimica Acta, 2006, 51(8-9): 1473-1479.

[40] KHATTABI M, BENHIBA F, TABTI S, et al. Performance and computational studies of two soluble pyran derivatives as corrosion inhibitors for mild steel in HCl[J]. Journal of Molecular Structure, 2019, 1196: 231-244.

[41] AVCI, G. Corrosion inhibition of indole-3-acetic acid on mild steel in 0.5M HCl[J]. Colloids and Surface A: Physicochemical and Engineering Aspects, 2008. 317(1-3): 730-736.

[42] BRAWER S A, WHITE W B. Raman spectroscopic investigation of the structure of silicate glasses(II). Soda-alkaline earth-alumina ternary and quaternary glasses[J]. Journal of Non-Crystalline Solids, 1977, 23(2): 261-278.

[43] TSUDA H, JONGEBLOED W L, STOKROOS I, et al. Combined Raman and SEM study on CaF_2 formed on/in enamel by APF treatments[J]. Caries Research, 1993, 27(6): 445-54.

[44] BRAWER S A, WHITE W B. Raman spectroscopic investigation of the structure of silicate glasses. I. The binary alkali silicates[J]. The Journal of Chemical Physics, 1975, 63(6): 2421-2432.

[45] SHETH N, NGO D, BANERJEE J, et al. Probing hydrogen-bonding interactions of water molecules adsorbed on silica, sodium calcium silicate, and calcium aluminosilicate glasses[J]. The Journal of Physical Chemistry C, 2018, 122(31): 17792-17801.

[46] SHETH N, LUO J, BANERJEE J, et al. Characterization of surface structures of dealkalized soda lime silica glass using X-ray photoelectron, specular reflection infrared, attenuated total reflection infrared and sum frequency generation spectroscopies[J]. Journal of Non-Crystalline Solids, 2017, 474: 24-31.

[47] AMMA S, KIM S H, PANTANO C G. Analysis of water and hydroxyl species in soda lime glass surfaces using attenuated total reflection(ATR)-IR spectroscopy[J]. Journal of the American Ceramic Society, 2016, 99(1): 128-134.

本论文原发表于《Crystals》2021年第11卷第4期。

Research on Property of Borocarbide in High Boron Multi-Component Alloy with Different Mo Concentration

Ren Xiangyi[1] Han Lihong[1] Fu Hanguang[2] Wang Jianjun[1]

(1. State Key Laboratory of Performance and Structural Safety for Petroleum Tubular Goods and Equipment Materials, CNPC Tubular Goods Research Institute;
2. Research Institute of Advanced Materials Processing Technology, School of Materials Science and Engineering, Beijing University of Technology)

Abstract: In this work, the microstructure, alloying element distribution and borocarbide mechanical property of high boron multi-component alloy with Fe-2.0wt.%B-0.4wt.%C-6.0wt.%Cr-$x$wt.%Mo-1.0%Al-1.0wt.%Si-1.0wt.%V-0.5wt.%Mn (x = 0.0, 2.0, 4.0, 6.0, 8.0) is investigated. The theoretical calculating results and experiments indicate that microstructure of high boron multi-component alloy consists of ferrite, pearlite as matrix and borocarbide as hard phase. As a creative consideration, through the utilization of first-principles calculation, the comprehensive properties of borocarbide with different molybdenum concentration have been predicted. The calculation of energy, state density, electron density and elastic constant of Fe_2B crystal cell reveals that substitution of molybdenum atom in Fe_2B crystal cell can remarkably improve its thermodynamic stability, bond strength and covalent trend. For verifying the accuracy of this theoretical calculation, the nano-indentation testing is carried out, results of which indicate that the actual property of borocarbide presents favorable consistency with the theoretical calculation.

Keywords: High boron multi-component alloy; Borocarbide; Molybdenum; First-principles calculation; Nano-indentation.

1 Introduction

As a kind of cost-effectiveresource in China, boron is widely investigated in the past few years. Researchers indicated that matrix of ferrous alloys with slight boron solution performs excellent hardenability, tensile and bending strength. When boron content further increases, boride with high hardness and thermal stability is formed in the microstructure[1-8]. Besides, for wear-resistant ferrous alloys, reduction of production cost and simplification of manufacturing process can also be

---

Corresponding author: Ren Xiangyi, mmerenxiangyi@126.com; Han Lihong, hanlihong@cnpc.com.cn.

obtained with the addition of boron instead of expensive alloying element such as V, W, Ti, etc.[9-15]. High boron multi-component alloy is a kind of new-type wear-resistant material. Hardness of borocarbide in this alloy is 1500-1700HV. However, borocarbide in general high boron multi-component is unable to bear the abrasion because of its bad toughness.[14,16].

Molybdenum is a kind of widely usedalloying element on alloy steel, which possesses the effect of segregation eliminating, carbide formation promoting, hardenability and toughness improving. For boron-added ferrous alloy, it is significant to improve its toughness. Most of molybdenum in high boron multi-component alloy exists in borocarbide in the form of solid-solution, which possesses the effect of morphology and toughness improving[17].

According to the previous investigation, basic physical properties of solid materials depends on their internal electronic condition[18-22]. Thus the macro properties of materials can be theoretically predicted by the acquisition of electronic structure, energy and other information in the crystal. In recent years, with the development of computers, servers and other relevant equipment, the first-principles calculation which is based on density functional theory (DFT) is widely used to the calculation of electronic structure, stability and mechanical properties of condensed state materials[23-29]. According to the information above, it is inferred that the properties of borocarbide in high boron multi-component alloy can be predicted by first-principles calculation.

This work systematically investigate the microstructure, alloying element distribution and mechanical properties of high boron multi-component alloy with the composition of Fe-2.0wt.% B-0.4wt.% C-6.0wt.% Cr-$x$wt.% Mo-2.0% Al-1.0wt.% Si-1.0wt.% V-0.5wt.% Mn(x=0.0, 2.0, 4.0, 6.0, 8.0). Meanwhile, with the help of first-principles calculation of borocarbide lattice with various Mo atom addition, the effect of molybdenum on morphology, mechanical properties of borocarbide is studied.

2 Experimental procedure

2.1 Materials and specimens

Ferroboron, pig iron, steel scrap, industrial pure aluminum, ferromolybdenum, ferrochrome, ferromanganese, ferrosilicon and ferrovanadium were used as raw material for the manufacture of high boron multi-component alloy. Above materials were smelted in 8kg induction furnace under atmosphere and heated to 1600℃. Liquid steel was discharged into the ladle first and then poured into pro-heated Y-block dried sand mold at 1500~1550℃ in 2min. Part in the center of the ingot was cut into 15mm cubes as specimens for metallurgical analysis.

2.2 Characterization

X-rayflourescence analysis (XRF) was used to determine the chemical composition of the studied alloys, results of which are shown in Table 1. The optical microscope (OM), X-ray diffraction(XRD), scanning electron microscope(SEM) and electron probe microanalyzer(EPMA) were used to characterize the microstructure of metallurgical specimens. All of the metallurgical specimens were mechanical polished and then etched by 5% nital. Polished specimens for volume fraction calculation needed to be dyed using No. 1 Carlin corrosive.

The volume fraction reflects the amount of borocarbide[30-31], which can be directly calculated by image analyzing software ImageJ. According to the Stereological formula(1), volume fraction V_V could be replaced by area fraction A_A in microstructure image.

$$V_V = A_A \tag{1}$$

The TI950 nano-indentation tester was used to investigate the hardness, elasticity and plasticity of borocarbide in the studied alloys with different Mo contents. The pit caused from plastic deformation on one of the phases in the alloy caused by Berkovich indenter can be used to analyze the relationship between load and deformation, then the hardness, elastic modulus, plastic deformation resistance and other parameters of this phase can be obtained. The experimental parameters of this work are shown in Table 2.

Table 1 Chemical composition of the studied alloys   (wt. %)

	B	C	Cr	Mo	Al	Si	V	Mn	Fe
M0	1.8	0.4	5.5	0.0	0.7	1.0	0.9	0.6	Bal.
M1	1.8	0.4	5.5	2.1	0.8	1.0	1.0	0.6	Bal.
M2	2.0	0.4	5.9	3.9	0.9	0.9	0.9	0.5	Bal.
M3	1.8	0.4	5.4	6.2	0.7	1.0	1.0	0.6	Bal.
M4	1.8	0.4	5.4	8.3	0.7	1.0	1.0	0.6	Bal.

Table 2 Experimental parameter of nano-indentation

Parameter types	Value	Parameter types	Value
Max load(mN)	10	Loading rate(mN/s)	2
Loading time(s)	5		

3 Results and discussion

3.1 Microstructure of high boron multi-component alloy with different molybdenum contents

Fig. 1 shows the microstructure of high boron multi-component alloy with different molybdenum concentration. It is observed that matrix consists of ferrite and pearlite(dark area). The light reticular area in each figure is borocarbide in the microstructure. Reticular structure size of borocarbide in sample M0 is quite large. With the increase of molybdenum content, size of the reticular structure is reduced gradually, as shown in Fig. 1(b), 1(c), 1(d) and 1(e). Besides, some kind of extremely fine borocarbide appears and increases with the increase of Mo content. When Mo content reaches 6wt.%, it can be seen that quantity of tiny borocarbide is higher than that of normal borocarbide.

For certifying the phase types of high boron multi-component alloy, the XRD analysis of alloy M0, M2 and M4 is carried out, results of which are shown in Fig. 2. It is observed that α-Fe as matrix and borocarbide $M_2(B, C)$, $M_3(B, C)$ (M = Fe, Cr, Mo, V, Mn) are detected in the tested alloys. $M_2(B, C)$ is the resultant of eutectic reaction, which exists as reticular borocarbide in the microstructure. After solidification, boron atoms and other metal atoms diffuse into cementite in pearlite, resulting in the formation of boron-cementite $M_3(B, C)$. Besides, according to the difference of three patterns, it is inferred that variation of Mo content has no effect on the phase types of high boron multi-component alloy.

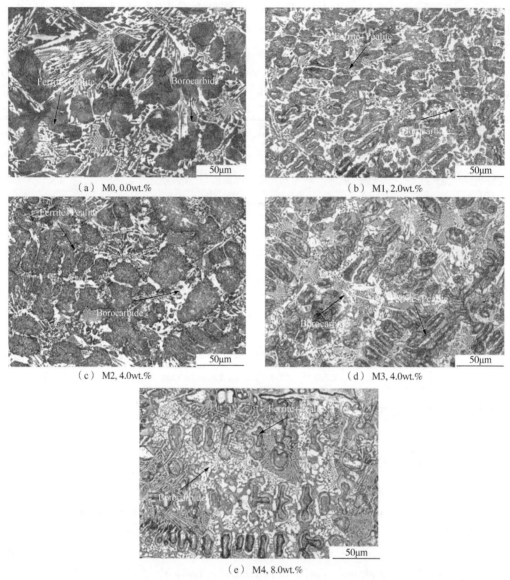

Fig. 1　Morphology of high boron multi-component alloy with different Mo contents

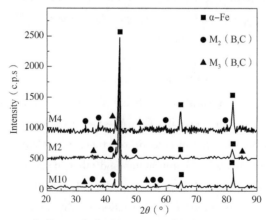

Fig. 2　XRD patterns of alloys with the Mo content of 0.0wt.%, 4.0wt.% and 8.0wt.%

The SEM BSE observation is carried out for investigating the effect of Mo content on element distribution of high boron multi-component alloy, results of which are shown in Fig. 3. The SEM BSE morphology reflects the inhomogeneous distribution of elements in borocarbide. Three kinds of borocarbide with different composition are observed. Table 3 shows the EPMA point scanning results of the studied alloys with different Mo contents. When Mo content is under 4.0wt.%, no variation of Mo content is detected in borocarbide at point 1, 4, 7 and 10. In this kind of borocarbide, Mo concentration rises only when Mo content is beyond 4.0wt.%. Borocarbide at point 2, 5, 8 and 11 possesses the similar composition and relatively high Cr content, in which Mo concentration keeps constant when Mo content in alloy changes. Composition of Mo-rich borocarbide is reflected at point 3, 6, 9 and 12. Moreover, Mo concentration in Mo-rich borocarbide rises when Mo content in alloy reaches 4.0wt.%.

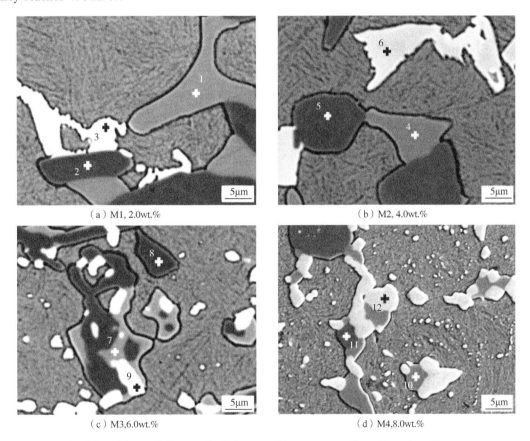

Fig. 3 SEM BSE morphologies of high boron multi-component alloy with various Mo contents

Table 3 EPMA point scanning results of high boron multi-component alloy with different Mo contents (wt.%)

Testing point	B	C	Cr	Mo	Al	Si	V	Mn	Fe
1	3.55	0.45	9.10	4.19	0.02	0.00	1.80	0.79	80.1
2	9.83	0.64	15.10	1.38	0.00	0.00	2.05	0.76	70.24
3	9.04	0.42	11.42	41.57	0.01	0.00	8.79	0.38	28.37

continued

Testing point	B	C	Cr	Mo	Al	Si	V	Mn	Fe
4	4.12	0.78	7.95	3.82	0.03	0.01	1.41	0.86	81.02
5	7.07	0.66	13.38	1.87	0.02	0.02	2.07	0.93	73.98
6	8.96	0.62	10.67	43.51	0.02	0.02	7.42	0.41	28.37
7	2.28	0.56	10.20	4.93	0.01	0.01	1.39	0.77	79.85
8	8.58	0.10	16.14	1.92	0.03	0.01	1.67	0.77	70.78
9	4.53	0.80	11.99	48.54	0.00	0.02	6.71	0.44	26.97
10	6.35	0.89	8.88	5.10	0.04	0.03	1.02	0.16	77.53
11	6.87	0.60	12.84	1.44	0.00	0.02	0.99	0.65	76.59
12	6.84	0.67	11.40	51.33	0.00	0.01	5.80	0.35	23.60

To certify whether the Mo content has effect on borocarbide quantity, the borocarbide volume fraction calculation is carried out, results of which are shown in Fig. 4. It can be seen that the variation of Mo content has no effect on the entire quantity of borocarbide. Volume fraction of borocarbide is 20% approximately in each alloy. However, quantity of borocarbide with different composition possesses obvious and regular changes. With the increase of Mo content, volume fraction of Mo-rich borocarbide ascends but of Cr-rich borocarbide reduces. This calculated results together with the morphology observing results of the studied alloys reveal that the main type of borocarbide is Cr-rich borocarbide when Mo content is relatively low, which changes into Mo-rich borocarbide when Mo content is higher than 4wt. %.

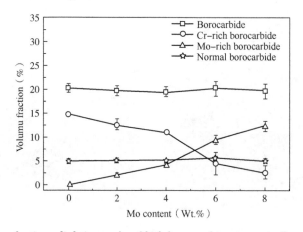

Fig. 4 Borocarbide volume fraction calculating results of high boron multi-component alloy with different Mo contents

3.2 First-principles calculation of borocarbide with different Mo concentration in high boron multi-component alloy

From the previous results, molybdenum exists in the borocarbide in the form of solid solution. According to the crystallographic principles, when one or more atoms are substituted by allochthonous atoms in crystal cell, the shape, electronic environment and other parameters of this cell will be changed. Thus it is inferred that borocarbide with different Mo content possesses various quantity of Mo atom substitution in its crystal cell. This work uses the CASTEP model of material calculation software Materials Studio in which the crystal cell of M_2B (M = Fe, Mo) with the substitution of 0, 1, 2 and 3 Mo atoms respectively can be created. These created cells can be used for structural and energy information calculation and mechanical property simulation. The calculating results can reveal the effect of Mo substitution in M_2B cell on general properties of borocarbide theoretically.

Before creating the M_2B crystal cell, the information of space group, lattice type and parameters of the cell need to be obtained from PDF card. Details of the information are shown in Table 4. For ensuring the conformance of stimulated results and actual situation, and considering the lattice distortion caused by Mo atom substitution, carrying out the geometric optimization after crystal cell creation is necessary. Parameter settings of the geometric optimization are shown in Table 5, which are based on the lattice parameters shown in Table 4. Besides, establishment of k-point range should be consistent with the corresponding reciprocal space of lattice parameter. Crystal cell after geometric optimization can be used for the calculation of population analysis, state density, energy information and elastic constant.

Table 4 Lattice parameters of M_2B crystal cell

Parameter	Type	Parameter	Type
Space group	1~4/mcm	Lattice constant a(nm)	0.511
Lattice	Tetragonal	Lattice constant c(nm)	0.424

Table 5 Parameter setting of geometric optimization in Castep model of Materials Studio

Parameter	Setting option
Optimizing mode	BFGS
Optimizing method	GGA-PBE
Cutting energy(eV)	300
SCF convergence energy(eV per atom)	10^{-5}
K-point range	5×5×6
Pseudo-potential type	Ultrasoft

Fig. 5 shows the ball-stick model of M_2B crystal cell with different numbers of Mo atom substitution created by Materials Studio software. It is inferred that the molecular formula of these cells are Fe_8B_4, Fe_7MoB_4, $Fe_6Mo_2B_4$ and $Fe_5Mo_3B_4$ respectively.

Table 6 shows the geometric optimizing results of the studied cells, which accurately reflect the structural information and distortion degree of these cells. From these data, cells with Mo substitution possess smaller values of a and b than that without Mo substitution. Values of a and b slightly changes with the variation of Mo atoms in cells. However, value of lattice parameter c ascends with the increase of Mo atom numbers in cells. Fe atom contains 26 extranuclear electron, radius of which is 0.127nm. Mo atom contains 42 extranuclear electron, and its radius is 0.140nm. When Fe atom in the crystal cell is replaced by Mo atom, part of the cell swells because of the substitution of Mo atom with larger radius. Moreover, atomic force of Mo atom with B and Fe atom increases. All of the variation mentioned before cause the change of Mo-included crystal cell reflected in Table 6.

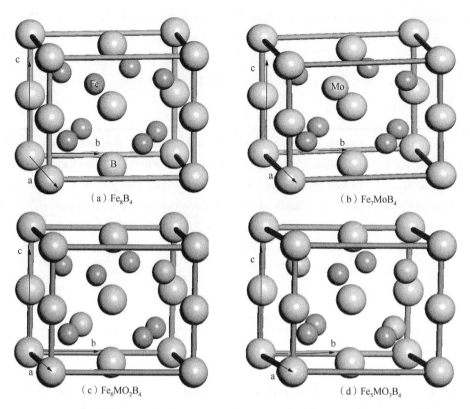

Fig. 5 Lattice structure of M_2B with different numbers of Mo atoms

Table 6 Results of geometric optimization of M_2B crystal cells with different Mo atom concentration

Crystal cells	Lattice parameters (nm)		
	a	b	c
Fe_8B_4	0.5880	0.5746	0.4249
Fe_7MoB_4	0.5490	0.5501	0.4694
$Fe_6Mo_2B_4$	0.5665	0.5667	0.4721
$Fe_5Mo_3B_4$	0.5558	0.5561	0.4737

The energy calculating results of M_2B crystal cell with different contents of Mo atoms are shown in Table 7. In the aspect of crystal cell volume, cells with 1 and 2 Mo atoms possess the smallest and largest value respectively. When the crystal cell contains 1 or 3 Mo atoms, distribution of Mo atoms in cell is inhomogeneous. For ensuring the balance of internal atomic force, volume of part without Mo atom or with 1 Mo atom reduces. Degree of this decrease is larger than the swelling degree of part with 1 (corresponding to no Mo atom) or 2 (corresponding to 1 Mo atom) Mo atoms. Thus cell Fe_7MoB_4 presents the smallest volume. Simultaneously, volume of cell $Fe_5Mo_3B_4$ is smaller than that of cell $Fe_6Mo_2B_4$. When 2 Mo atoms possess homogeneous distribution in the cell, the whole cell swells because of the rising of atom radius and atomic force. All of the cells possess the negative values of energy, which indicates that they can exist in the microstructure stably. Moreover, energy of cells descends with the increase of Mo atom numbers. Thus it is inferred that Mo atom can improve the property especially high temperature stability of borocarbide by improving the thermodynamic stability of crystal cell. Reference[32] suggests that brittleness of Fe_2B is caused by the low B-B bond energy along [002] crystal orientation. According to the results of B-B bond length in Table 7, when 1 or 3 Mo atoms are in the cell, length of B-B bond at [002] direction reduces. When the cell contains 2 Mo atoms, the length increases. Moreover, length of B-M bond (M = Mo, Fe) is lower than that of B-Fe bond. Descending of bond length results in the increasing of bond energy. Thus it is theoretically revealed that the substitution of Mo atom in Fe_2B crystal cell can effectively reinforce the B-B and B-M bond, so that the brittleness of borocarbide can be improved.

Table 7 Results of energy calculation of M_2B crystal cells with different Mo atom concentration

Crystal cells	Cell volume (nm^3)	Cell energy (eV)	B-B bond length (nm)	B-M bond length (nm)	M-M bond length (nm)
Fe_8B_4	14.356	-7217.12	0.2112	0.2428	0.2715
Fe_7MoB_4	14.176	-8291.65	0.1739	0.2373	0.2731
$Fe_6Mo_2B_4$	15.156	-9362.59	0.2359	0.2397	0.2822
$Fe_5Mo_3B_4$	14.641	-10436.52	0.1817	0.2395	0.2780

The calculation of state density for cells with different Mo atom concentration can reflect the stability of extranuclear electron and interactions of atoms. The closer to zero the state density at Fermi surface is, the more stable the extranuclear electrons are. Fig. 6 shows the calculated results of state density of M_2B crystal cell. It can be observed that cells with asymmetrical structure, such as Fe_7MoB_4 and $Fe_5Mo_3B_4$, possess quite low values of state density at Fermi surface. However, state density of cell $Fe_6Mo_2B_4$ at Fermi surface is higher than that of cell Fe_8B_4. These phenomena reveal that Mo concentration in the cell can improve the stability of electrons in the cell. When the stability of electron rises, property of bond is apt to the covalent bond which possesses higher strength than metallic bond. Consequently, it is further proved that Mo concentration in Fe_2B crystal cell can improve its property by increasing the bond strength.

As mentioned before, state density reflects the interacting type of electrons in the crystal cell. For revealing the degree of this interaction, calculation of electron density on (110) crystal plane is carried out, results of which are presented in Fig. 7. When no Mo atom is in the cell, level of electron density is relatively low. It is worth noting that electron density along [002] orientation is only 60e/Å^2 approximately. When one or more Fe atoms are substituted by Mo atoms, level of electron density in the whole cell remarkably rises. Moreover, density of boron atoms along [002] orientation also ascends obviously. Cell $Fe_5Mo_3B_4$ presents the highest electron density (approximately 140 e/Å^2). Consequently, bond strength especially B-B bond with lower strength along [002] orientation can be reinforced. In conclusion, substitution of Mo atom in Fe_2B crystal cell can effectively improve the thermodynamic stability and bond strength by increasing the electron stability, density and shortening the B-B bond length.

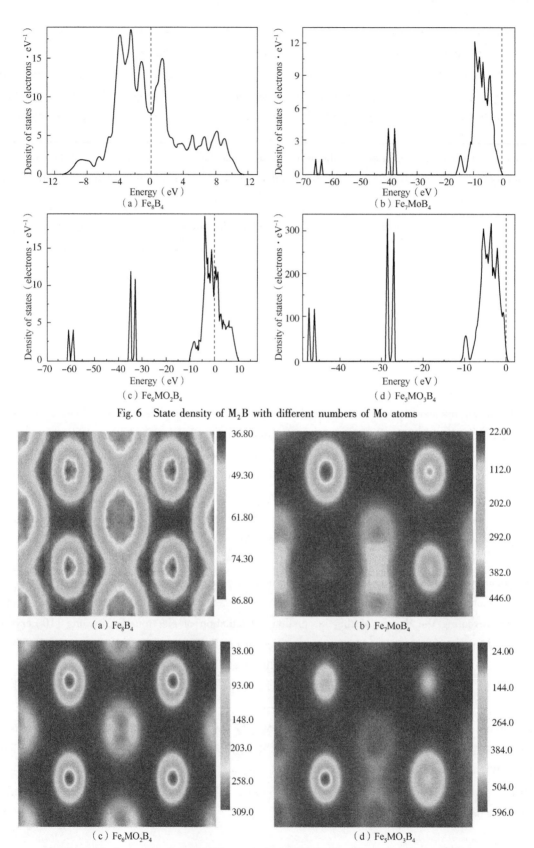

Fig. 6 State density of M_2B with different numbers of Mo atoms

Fig. 7 Total electron density on (110) crystal plane of M_2B with different numbers of Mo atoms

The utilization of first-principles calculation can not only predict the property of crystal cell by structural and energy calculation but also calculate the mechanical property directly. Results of elastic constant calculation of M_2B crystal cell with different Mo atom concentration are presented in Table 8. Ratio of bulk modulus and shearing modulus can reflect the toughness of material. The higher value of this ratio is, the better toughness of material possesses[33-34]. When the cell contains 1 Mo atom, toughness of which slightly rises. However, other parameters reduce. When 2 Mo atoms exist in the cell, all of the parameters increase. In cell with 3 Mo atoms, toughness ascends but elastic modulus reduces. Together with the structural and energy calculating results, it can be revealed that the substitution of Mo atom in Fe_2B crystal cell remarkably improves the thermodynamic stability, bond strength and mechanical property. Consequently, the improvement of general property of borocarbide in high boron multi-component alloy by molybdenum concentration can be theoretically proved.

Table 8 Results of elastic constant calculation of M_2B crystal cells with different Mo atom concentration

Crystal cells	Bulk modulus B	Shearing modulus G	Elastic modulus E	Poisson ratio ν	B/G
Fe_8B_4	124.84	57.48	149.50	0.30	2.17
Fe_7MoB_4	110.2	42.86	122.35	0.43	2.57
$Fe_6Mo_2B_4$	197.40	73.61	214.82	0.46	2.68
$Fe_5Mo_3B_4$	191.36	78.65	207.52	0.32	2.43

3.3 Experimental testing of mechanical property of borocarbide with different Mo concentration in high boron multi-component alloy

For verifying whether the predicted property by first-principles calculating results are accordance with the actual property of borocarbide, it is necessary to investigate the mechanical property of borocarbide through experiment. According to the previous results, size of Mo-rich borocarbide is quite small, on which regular hardness testing cannot be carried out. Consequently, property testing of Mo-rich borocarbide need to be implemented by the method of nano-indentation. Through this testing method, the corresponding relationship of borocarbide deformation and load on borocarbide surface during loading and unloading process will be obtained. This relationship can be used to reveal the hardness, elasticity and toughness of borocarbide. Fig. 8 shows the load-depth curve of Mo-rich borocarbide in sample M1, M2, M3 and M4. Curves on the left side reflect the loading process. On the right side of the curves reflect the unloading process. Different samples possess various extent of deformation which indicates that these samples have different hardness. Sample M2 presents the highest value. Gradient of the curves reflects the elastic modulus of borocarbide. High modulus corresponding to the high curve

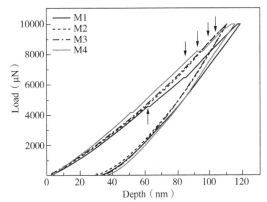

Fig. 8 Depth-load curve of Mo-rich borocarbide with different Mo contents

gradient. Through the elastic modulus calculation, it is indicated that elastic modulus of borocarbide ascends with proper Mo addition. Loading curve and unloading curve do not coincide, which indicates that all of the samples remain plastic deformation after unloading process. With the increase of Mo content in the alloy, plasticity of borocarbide descends first and then rises. Sample M2 presents the best plastic deformation resistance, because it shows the lowest pit depth after indentation process.

Table 9 shows the measuring results of hardness and elastic modulus of Mo-rich borocarbide in the studied alloys, which corresponding to the trend that Fig. 8 presents. Arrows in Fig. 8 point the steps in each curve. These steps reflect the critical point of elastic and plastic deformation of borocarbide. The higher load that corresponding to the step is, the better plastic deformation resistance borocabide possesses. It is worth noting that more than one step appears in the loading curve of alloy M1, which indicates that alloy M1 has no obvious critical point of elastic and plastic deformation. In the rest of the alloys, only one step appears in each curve. Moreover, when Mo content is higher than 6wt.%, plastic deformation resistance of the alloy reduces. To sum up, Proper Mo addition in borocarbide can remarkably improve its hardness, elastic modulus and plastic deformation resistance.

Table 9 Results of elastic constant calculation of M_2B crystal cells with different Mo atom concentration

Samples	Mechanical properties	
	Hardness H(GPa)	Elastic modulus E(GPa)
M1	20.28	209.19
M2	24.01	222.02
M3	20.08	221.53
M4	19.93	244.77

4 Conclusions

(1) Microstructure of high boron multi-component alloy consists of ferrite, pearlite and borocarbide. Alloying element in borocarbide possesses inhomogeneous distribution, resulting in the appearance of Mo-rich, Cr-rich and normal borocarbide. With the increase of Mo content, proportion of Mo-rich borocarbide gradually rises.

(2) The first-principles calculation executed by Materials Studio software is used to predict the comprehensive property of borocarbide theoretically. Through the calculation of energy, state density, electron density and elastic constant of crystal cell, it is theoretically revealed that substitution of Mo atom in Fe_2B crystal cell can effectively promote the thermodynamic stability, elastic modulus, bond strength and covalent trend. Consequently, hardness and toughness of borocarbide are obviously reinforced.

(3) Through the utilization of nano-indentation testing, the theoretical prediction of mechanical property of borocarbide by first-principles calculation is experimentally proved. With the addition of Mo in borocarbide, hardness, elastic modulus and plastic deformation resistance are remarkably improved.

Acknowledgements

This work was supported by the National key Technologies R&D Program of China(2017ZX05009-003, 2019YFF0217501, 2019YFF0217500), Natural Science Foundation of China(U1762211, 52005217) and CNPC Basic Research Project(2020B-4020, 2019B-4014, 2019E-2502).

References

[1] LIU S, CHENG Y, LONG R, et al. Research progress and prospect of wear-resistant Fe-B-C alloy[J]. Foundry Technology, 2007; 28(11): 1526-1530.

[2] HE L, LIU Y, LI J, et al. Effects of hot rolling and titanium content on the microstructure and mechanical properties of high boron Fe-B alloys[J]. Materials & Design, 2012, 36: 88-93.

[3] LV Z, FU H, XING J, et al. Microstructure and crystallography of borides and mechanical properties of Fe-B-C-Cr-Al alloys[J]. Journal of Alloys and Compounds, 2016, 662: 54-62.

[4] FU H, JIANG Z. A study of abrasion resistant cast Fe-B-C alloy[J]. Acta Met Sini, 2006, 42: 545-548.

[5] SONG X, LIU H, FU H, et al. Effect of Boron concentration on microstructures and properties of high-boron low-carbon Ferro-matrix alloy[J]. Foundry, 2008, 57: 498-503.

[6] ASTINI V, PRASETYO Y, BAEK E. Effect of Boron Addition on the Microstructure and Mechanical Properties of 6.5% V-5% W High Speed Steel[J]. Metals and Materials International, 2012, 18: 923-931.

[7] KIM J, KO K, NOH S, et al. The effect of boron on the abrasive wear behavior of austenitic Fe-based hardfacing alloys[J]. Wear, 2009, 267: 1415-1419.

[8] FU H. A study of microstructures and properties of cast Fe-B-C alloy[J]. Foundry, 2005, 54: 859-863.

[9] KIM J, HONG S, PARK H, et al. Improving plasticity and strength of Fe-Nd-B ultrafine eutectic composite [J]. Materials & Design, 2015, 76: 190-195.

[10] HUANG Z, XING J, GUO C. Improving fracture toughness and hardness of Fe_2B in high boron white cast iron by chromium addition[J]. Materials & Design, 2010, 31: 3084-3089.

[11] YUKSEL N, SAHIN S. Wear-behavior-hardness-microstructure relation of Fe-Cr-C and Fe-Cr-C-B based hardfacing alloys[J]. Materials & Design, 2014, 58(6): 491-498.

[12] ROTTGER A, LENTZ J, THEISEN W. Boron alloyed Fe-Cr-C-B tool steels-Thermodynamic calculations and experimental validation[J]. Materials & Design, 2015, 88: 420-429.

[13] SUNDEEV R, GLEZER A, MENUSHENKOV A, et al. Effect of high pressure torsion at different temperatures on the local atomic structure of amorphous Fe-Ni-B alloys[J]. Materials & Design, 2017; 135 (5): 77-83.

[14] GU J, ZHANG H, FU H, et al. Effect of Boron content on the structure and property of Fe-B-C alloy [J]. Foundry Technology, 2011, 32: 1376-1379.

[15] GUI Z, LIANG W, LIU Y, et al. Thermo-mechanical behavior of the Al-Si alloy coated hot stamping boron steel[J]. Materials & Design, 2014, 60(60): 26-33.

[16] REN X, FU H, XING J, et al. Effect of boron concentration on microstructures and properties of Fe-B-C alloy steel[J]. Journal of Materials Research, 2017, 32(16): 3078-3088.

[17] MA S, XING J, FU H, et al. Microstructure and crystallography of borides and secondary precipitation in 18wt.% Cr-4wt.% Ni-1wt.% Mo-3.5wt.% B-0.27wt.% C steel[J]. Acta Materialia, 2012, 60(3): 831-843.

[18] ZHOU C, XING J, XIAO B, et al. First principles study on the structural properties and electronic structure of X2B(X=Cr, Mn, Fe, Co, Ni, Mo and W)compounds[J]. Computational Materials Science, 2009, 44

(4): 1056-1064.

[19] TIAN H, ZHANG C, ZHAO J, et al. First-principle study of the structural, electronic, and magnetic properties of amorphous Fe-B alloys[J]. Physica B: Condensed Matter, 2012, 407(2): 250-257.

[20] LI L, WANG W, HU L, et al. First-principle calculations of structural, elastic and thermodynamic properties of Fe-B compounds[J]. Intermetallics, 2014, 46: 211-221.

[21] KITCHIN J, NØRSKOV JK J, BARTEAU M, et al. Trends in the chemical properties of early transition metal carbide surfaces: A density functional study[J]. Catalysis Today, 2005, 105(1): 66-73.

[22] SHEIN I, MEDVEDEVA N, IVANOVSKII A. Electronic and structural properties of cementite-type M3X(M = Fe, Co, Ni; X = C or B) by first principles calculations[J]. Physica B: Condensed Matter, 2006, 371(1): 126-132.

[23] QU Z, SU Y, SUN L, et al. Study of the structure, electronic and optical properties of $BiOI/Rutile-TiO_2$ heterojunction by the first-principle calculation[J]. Materials, 2020, 13(2): 323.

[24] GAO Y, QIAO L, WU D, et al. First principle calculation of the effect of Cr, Ti content on the properties of $VMoNbTaWM_x$(M = Cr, Ti) refractory high entropy alloy[J]. Vacuum, 2020, 179: 109459.

[25] GUO W, XUE H, WEI X, et al. Study on the interfacial properties of active Ti element/ZrO_2 by using first principle calculation[J]. International Journal of Applied Ceramic Technology, 2020, 17(3): 1286-1292.

[26] MAHMOOD A, SHI G, XIE X, et al. Adsorption mechanism of typical oxygen, sulfur, and chlorine containing VOCs on TiO_2(001) surface: first principle calculations[J]. Applied Surface Science, 2019, 471: 222-230.

[27] XU N, YAN H, JIAO X, et al. Effect of OH^- concentration on Fe_3O_4 nanoparticle morphologies supported by first principle calculation[J]. Journal of Crystal Growth, 2020, 547: 125780.

[28] LUO M, XU Y, SHEN Y. Magnetic properties of SnSe monolayer doped by transition-metal atoms: a first-principle calculation[J]. Results in Physics, 2020, 17: 103126.

[29] FENG C, CHEN Z, LI W, et al. First-principle calculation of the electronic structures and optical properties of the metallic and nonmetallic elements-doped ZnO on the basis of photocatalysis[J]. Physica B: Condensed Matter, 2019, 555: 23-60.

[30] REN X, FU H, XING J, et al. Effect of calcium modification on solidification, heat treatment microstructure and toughness of high boron high speed steel[J]. Materials Research Express, 2019, 6(016540): 1-13.

[31] REN X, FU H, XING J, et al. Effect of solidification rate on microstructure and toughness of Ca-Ti modified high boron high speed steel[J]. Materials Science and Engineering: A, 2019, 742: 617-627.

[32] JIAN Y, HUANG Z, XING J, et al. Effects of chromium addition on fracture toughness and hardness of oriented bulk Fe_2B crystals[J]. Materials Characterization, 2015, 110: 138-144.

[33] XIAO B, XING J, FENG J, et al. A comparative study of Cr_7C_3, Fe_3C and Fe_2B in cast iron both fromab initiocalculations and experiments[J]. Journal of Physics D: Applied Physics, 2009, 42(11): 115415.

[34] SUN L, GAO Y, XIAO B, et al. Anisotropic elastic and thermal properties of titanium borides by first-principles calculations[J]. Journal of Alloys and Compounds, 2013, 579: 457-467.

本论文原发表于《Materials》2021年第14卷第13期。

Effect of Titanium Modification on Microstructure and Impact Toughness of High-Boron Multi-Component Alloy

Ren Xiangyi[1]　Tang Shuli[2]　Fu Hanguang[3]　Xing Jiandong[4]

(1. State Key Laboratory of Performance and Structural Safety for Petroleum Tubular Goods and Equipment Materials, CNPC Tubular Goods Research Institute;
2. Xi'an Microelectronic Technology Institute;
3. Research Institute of Advanced Materials Processing Technology, School of Materials Science and Engineering, Beijing University of Technology;
4. State Key Laboratory for Mechanical Behavior of Materials, School of Materials Science and Engineering, Xi'an Jiaotong University)

Abstract: This work investigated the microstructure and mechanical property of high-boron multi-component alloy with Fe, B, C, Cr, Mo, Al, Si, V, Mn and different content of Ti. The results indicate that the as-cast metallurgical microstructure of high-boron multi-component alloy consists of ferrite, pearlite and borocarbide. In un-modified alloy, continuous reticular structure of borocarbide is observed. After titanium addition, structure of borocarbide changes into fine and isolated morphology. TiC is the existence form of titanium in the alloy, which acts as the heterogeneous nuclei for eutectic borocarbide. Moreover, impact toughness of the alloy is remarkably improved by titanium modification.

Keywords: High-boron multi-component alloy; Borocarbide; Titanium; Misfit degree; Heterogeneous nuclei

1 Introduction

In steel rolling industry, roll acts as an important component. Considerable consumption of rolls wastes resources and energy severly. For solving this problem, it is necessary to investigate roll materials with remarkable wear resistance. Previously, numerous metal materials with high hardness were successfully applied in roll making industry [1-4]. In a long time, high-chrome cast iron acts as the widest used wear-resistant roll material [5-8]. Factories in Europe started to use cast iron roll in the early 1990s. Subsequently, high-speed steel roll became the widely-used in the year of 1994[2]. Carbide in high-speed steel possesses excellent hardness, hardenability, toughness and

---

Corresponding author: Ren Xiangyi, mmerenxiangyi@126.com, +86-139-1996-4450.

wear resistance. Thus High-speed steel is the best material for roll manufacturing at that time. Soon afterwards, high-vanadium high-speed steel which contains more carbon and vanadium is invented in which numerous isolated MC carbide particles are found, bringing about considerable augmentation of thermal and mechanical property [9-14]. However, disadvantages are exposed with its wide utilization such as high cost and poor toughness and thermal fatigue resistance. [15]. Thus it is important for the investigation of new-type roll material with better properties.

China is a country with rich boron resource. In these years, numerous researches on boron-added alloy were carried out, which reveal that boron-added cast iron presents good mechanical property due to the hard and thermal-stable boride in the microstructure [16-23]. Furthermore, addition of boron instead of expensive alloying element reduces the production cost and simplifies the manufacturing process [24-30]. High-boron multi-component alloy is a new-type wear-resistant material. Hardness of borocarbide in this alloy is 1700HV approximately, which is lower than that of vanadium carbide VC (about 2000HV) [21, 31]. It is promising for high-boron multi-component alloy to be used widely as roll material. Howerver, in high-boron multi-component alloy, continuously distributed borocabide separates the matrix severely [29, 32], making it unable to support the borocarbide to resist the abrasion by means of its toughness. Some of the researchers investigated whether the structure of heat-treated borocarbide is isolated, but the results were unsatisfying [33, 34].

Modification treatment is a widely applied technology to refine the carbide in alloys [35-41]. According to the previous works, microstructure of high-speed steel can be changed by some kind of particular element so that property of which can also be improved. In previous investigations [42-51], high speed steel used surface-active elements (sodium, potassium and aluminum) and/or heterogeneous nuclei elements (cerium, yttrium and titanium) as modifying element. It was revealed that these elements effectively refine and/or spheroidize the carbide. For isolating the borocarbide in high-boron multi-component alloy, modifying elements are also used on refining or spheroidizing borocarbide in some researches. As a kind of effective modifying element as well as alloying element, titanium is widely used by investigators to improve the microstructure and property of cast iron and boron-added ferrous alloys. Chung et al. indicated that in high chromium cast iron with titanium added as alloying element, wear resistance of which is improved because of the refinement of eutectic carbides [52]. He et al. [17], Liu et al [36]. and Shi et al [40] suggested that boride in Fe-B alloy can be refined and sphereoidized by titanium modification. As a promising roll material, high-boron multi-component alloy should possess good property through ameliorating the microstructure. This research used titanium as modifying element and studied the effect of titanium on solidification structure and mechanical property of high-boron multi-component alloy with the composition of Fe-B-C-Cr-Mo-Al-Si-V-Mn-Ti.

2 Experimental process

2.1 Experimental materials

All of the materials for high-boron multi-component alloy casting were smelted to approximately 1600℃ under atmosphere in 8kg induction furnace. First of all, ladle in which liquid steel and FeTi70 Modifier particles in the diameter of 5mm was filled. 2min later, liquid steel with soluted modifier was discharged into pro-heated mold under 1500~1550℃. Fig. 1 presents the shape and size of the mold. The metallurgical analysis used 10mm cube specimensand 10mm× 10mm×55mm impact specimens (without notch). All of the specimens were cut from the center of the ingot. Table 1 shows the addition amount of modifier in each specimen.

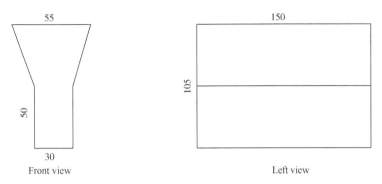

Fig. 1   Shape and size of mold cavity (mm)

2.2 Analysis and characterization

X-ray flourescence (XRF) and energy dispersion spectrum (EDS) analysis were used to determine the chemical composition of the studied alloys and FeTi70modifier. Results tested by XRF are shown in Table 2 and Table 3. microstructure of metallurgical specimens was analyzed by optical microscope (OM), X-ray diffraction (XRD), scanning electron microscopy (SEM) and electron probe microanalyzer (EPMA). The characterization of crystal structure and relationship of lattice orientation of phases was analyzed by transmission electron microscopy (TEM) and selecting area diffraction (SAD). Hardness of specimens was measured by Rockwell-hardness tester. Microhardness of different microstructure in each specimen can be measured by Vickers-hardness tester. Metallographic samples were polished by sandpaper and etched by 5% HNO_3 solution. For carrying out the quantitative analysis, specimens should be colored by Carlin corrosive. Metallurgical surfaces of alloy T2 and M were deeply etched by saturated $FeCl_3$ solution for XRD analysis. The heat treatment process is 1050℃, 2h, quenching+500℃, 1h tempering. Cooling rate of quenching is 40℃/s approximately. The JB30A impact testing device was used to measure the impact toughness tests of heat-treated specimens.

Table 1   Alloy symbols and the corresponding modifier contents in four specimens   (wt.%)

Alloy symbol	M	T1	T2	T3
Modifier content	0.0	0.3	0.6	0.9

Table 2   Chemical compositions of alloys   (wt.%)

Specimens	B	C	Cr	Mo	Al	Si	V	Mn	Ti	Fe
M	1.95	0.38	5.87	3.86	0.85	0.88	0.85	0.46	0.00	Bal.
T1	1.97	0.34	5.29	3.79	0.80	0.89	0.92	0.51	0.21	Bal.
T2	1.96	0.36	5.12	3.85	0.86	0.96	0.91	0.52	0.41	Bal.
T3	1.93	0.33	5.70	3.78	0.82	0.81	0.85	0.45	0.54	Bal.

Table 3 FeTi70 modifier composition (wt. %)

Si	Ti	Al	Mn	C	Fe
2.0	68.0	4.5	2.5	0.2	Bal.

The quantitative analysis includes volume fraction V_V, shape factor K and Feret diameter dF [53-54]. Software Image J can be used to realize the quantitative analysis. The Stereological formula (1) suggests that the volume fraction can be substituted by area fraction A_A.

$$V_V = A_A \tag{1}$$

The calculation of shape factor K is shown in Eq. (2) in which A refers to area (μm^2) and L refers to perimeter (μm). $K \in (0, 1)$. The larger K is, the closer to circle tested phase is.

$$K = \frac{4\pi A}{L^2} \tag{2}$$

The Feret diameter dF is the average width of a phase at not less than 40 different directions. Fig. 2 presents its calculating process.

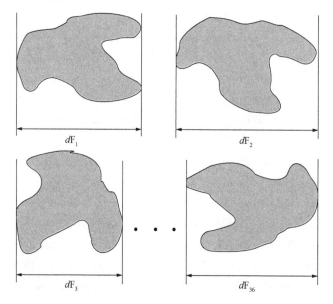

Fig. 2 Feret diameter calculating process

3 Results and discussion

3.1 Effect of titanium concentration on microstructure of high-boron multi-component alloy

Fig. 3 shows the as-cast microstructure of high-boron multi-component alloy which is unmodified and modified with different titanium concentration. It is observed that matrix in alloy consists of ferrite (light area) and pearlite (dark area). In Fig. 3(a), borocarbide (reticular area) in alloy M presents relatively large size. With titanium addition, borocarbide is obviously refined, as presented in Fig. 3(b), 3(c) and 3(d). Some of the previous works [36, 40, 42, 44, 45] have indicated that titanium in liquid steel forms plenty of intermetallic compound particles with other elements. These particles act as heterogeneous nuclei, providing more base for nucleation of borocarbide and significantly rise its nucleation rate. When amount of borocarbide is constant, the higher nucleation rate borocarbide has, the more isolated borocarbide will appear. Hence, the size of borocarbide is decreased. For certifying the formation of titanium-included compound in modified alloys and eliminating the interference of matrix, the deep etched alloy M and T2 are analyzed by XRD. Their results are shown in Fig. 4. Fig. 4(a) presents the morphology of deep etched alloy M, indicating that matrix is completely erased. In Fig. 4(b), three kinds of phases, α-Fe and borocarbide M_2(B, C), M_3(B, C) (M refers to metal elements in alloy except Al) are detected in both alloy M and T2. In the curve of alloy T2, obvious diffraction peaks of phase TiC can be observed. Formation of TiC proves that one of the conditions for titanium to take effect as a modifying element is required.

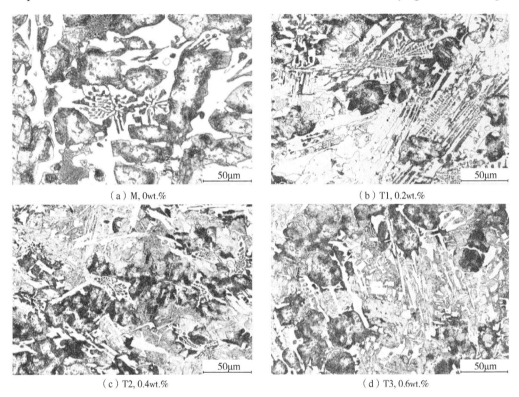

Fig. 3 As-cast morphology of high-boron multi-component alloy with different Ti contents

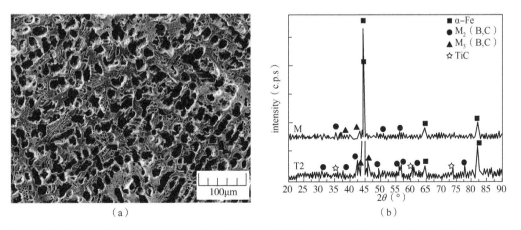

Fig. 4 Morphology deep etched as-cast alloy M(a) and XRD patterns of deep etched alloy T2 and M (b)

Fig. 5 shows the heat-treated morphologies of the alloys. Excellent hardenability is presented in boron-added matrix [32]. Consequently, the matrix completely transforms to martensite structure. In unmodified and modified alloys, angles of borocarbide are blunted and continuous borocarbide is disconnected. Size of borocarbide in modified alloys is smaller, borocarbide-matrix interface is more curved and less stable, so there are more locations that can be disconnected. In alloy M, borocarbide still keeps the continuous network, shape of which is quite similar to the as-cast structure. Boron possesses higher solubility in austenite than in ferrite [38], during martensite transformation, boron atoms near the interface diffuse into the matrix. Thus sharp angle of borocarbide is blunted, and narrow palce of borocarbide is broken.

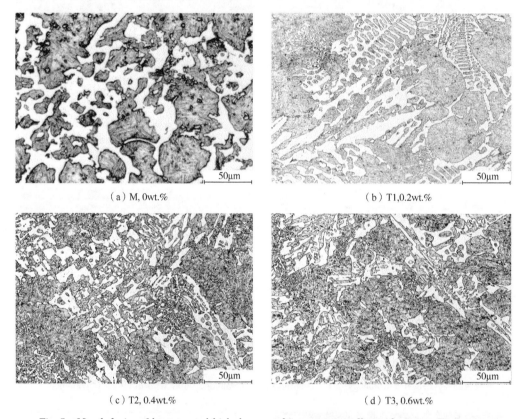

Fig. 5 Morphologies of heat-treated high-boron multi-component alloy with various Ti contents

The modifying effect of titanium on borocarbide morphology improving cannot be accurately characterized by normal observation. Thus it is necessary to use the quantitative analysis to reveal the varying law of borocarbide. Results of the quantitative analysis are presented in Fig. 6. In Fig. 6 (a), borocarbide volume fraction diminishes with the increase of titanium content in the as-cast alloys. From reference[55], after slight titanium addition in ferro-boron alloy, the amount of borocarbide descends when titanium content increases. Besides, borocarbide is coarsened after tempering treatment, thus the volume fraction increases[56]. Borocarbide is significantly spheroidized and refined after titanium modification, which can be seen in Fig. 6(b) and Fig. 6(c). Moreover, the heat-treated borocarbide is further spheroidized and refined.

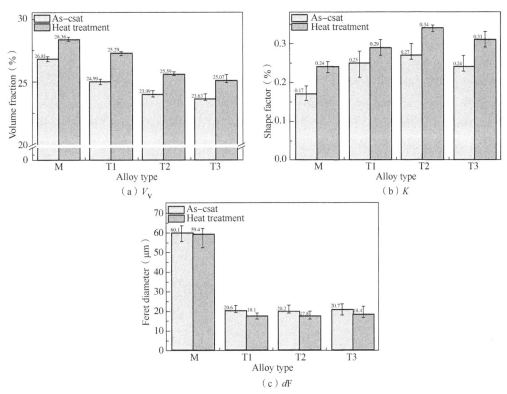

Fig. 6 Quantitative calculating results of borocarbide in high-boron multi-component alloy with various Ti contents

3.2 Analysis of titanium modifying mechanism in high-boron multi-component alloy

According to the results above, titanium distributes in the alloy in the form of TiC compounds. Three essential conditions are used to judge if TiC is effective heterogeneous nuclei: Precipitating priority, low misfit degree and dispersed distribution [47]. The thermodynamic estimation of TiC precipitating order is carried out. From the thermodynamic data of reference [57], the formation of TiC in liquid steel is as follows:

$$C(s) = [C] \quad \Delta_r G_1 = 22590 - 42.26T \text{ kJ/mol} \quad (3)$$

$$Ti(s) = [Ti] \quad \Delta_r G_2 = -25100 - 44.98T \text{ kJ/mol} \quad (4)$$

$$Ti(s) + C(s) = TiC(s) \quad \Delta_r G_3 = -186600 + 13.2T \text{ kJ/mol} \quad (5)$$

In liquid steel, forming reaction of TiC can be obtained by equation (5) - equation (4) - equation (3):

$$[Ti] + [C] = TiC(s) \quad \Delta_r G_4 = -182290 + 99.79T \text{ kJ/mol} \quad (6)$$

Reference [32] indicates that the melting point of high-boron multi-component alloy is 1602K. When T is 1602, it is calculated that value of $\Delta_r G_4$ -224.26kJ/mol. Hence, it is inferred that the formation of TiC under this temperature of is spontaneous, which is practicable for prior precipitation. Moreover, reference [47] suggests that under 1602K, content of TiC is higher than its solubility in the studied alloy. In conclusion, the prior formation of TiC is possible with the precipitation of Ti from liquid alloy before solidification.

For exploring the distribution of TiC in the alloy, the SEM BSE morphology of alloy T2 is observed and shown in Fig. 7. In Fig. 7(a), it is obviously seen that TiC distributes in the form of isolated particles dispersedly. Besides, there are no isolated TiC particles in the middle of matrix. Some of the particles are inside the borocarbide [Fig. 7(b)], others are at the interface of matrix and borocarbide [Fig. 7(c)]. Fig. 8(a) is the location of EPMA point scanning for analyzing the composition of TiC particle, results of which is shown in Table 4. It is observed that the atomic ratio of Ti and C is 1 : 1 approximately. Moreover, some other metallic elements are found such as V, Mo, Cr, Mn and Fe because of the solid solution. Fig. 8(b) presented the EPMA linear scanning results through TiC particle. By observing the distributing curves of each metallic element, content of V and Mo in TiC is higher than that in borocarbide. Besides, Mo mainly distributes near the interface of TiC and borocarbide, which leads to the different brightness of TiC under SEM BSE observation. It is because the diffusion is more difficult for Mo atom which has larger atomic radius.

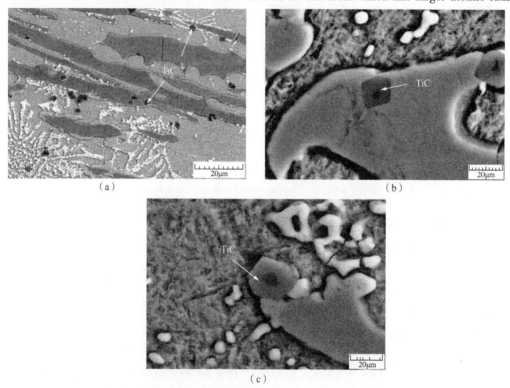

Fig. 7 Distribution of TiC particles in heat treated alloy T2 (a),
TiC particle inside the borocarbide (b) and particle at the interface of matrix and borocarbide (c)

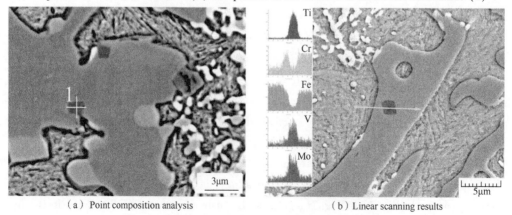

(a) Point composition analysis   (b) Linear scanning results

Fig. 8 Distribution of alloying elements in TiC particle

Table 4 EPMA results of point 1 (at. %)

Point numler	C	Cr	Mo	V	Mn	Ti	Fe
Point 1	54.95	0.27	0.92	2.79	0.05	40.13	0.89

The misfit degree is used to determine whether one kind of phase is the effective heterogeneous nuclei for another phase. Fig. 9 shows the lattice structures of TiC, austenite and boride Fe_2B. It is seen that structure of TiC, austenite is face-centered cube (fcc) and Fe_2B is body-centered tetragonal structure, lattice parameter of which is presented in Table 5. The misfit degree δ is calculated by the formula as follow [54]:

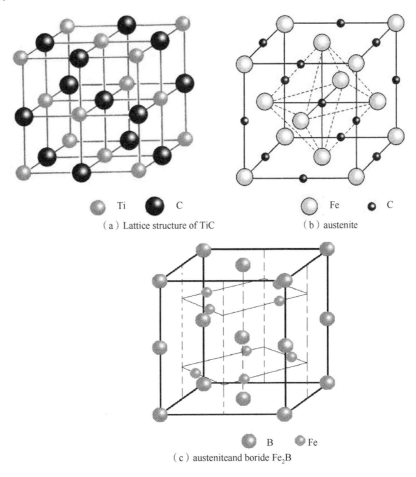

(a) Lattice structure of TiC (b) austenite

(c) austeniteand boride Fe_2B

Fig. 9 Lattice structure of TiC, austenite, and boride Fe_2B

$$\delta_{(hkl)_n}^{(hkl)_s} = \sum_{i=1}^{3} \frac{(\mid d[uvw]_s^i \cos\theta - d[uvw]_n^i \mid)/d[uvw]_n^i}{3} \times 100\% \qquad (7)$$

Here δ-misfit degree of two phases. (hkl) is the crystallographic plane index, $[uvw]$ is the crystal orientation index. d-the atomic length along $[uvw]$. s and n are the symbols of two different phases. θ-the included angle of two crystal orientations. The critical value of δ is under 12%. After calculation of various lattice combination of TiC with austenite and Fe_2B, it is found that the lowest misfit degree of TiC and austenite is 12.5%, whichis from $(100)_{TiC}//(110)_\gamma$. The lowest misfit degree of $M_2(B, C)$ and TiC is 9.3%, which from $(110)_{TiC}//(110)_{M_2(B,C)}$. From these results, it can be proved that TiC particle is the ineffective nuclei for primary austenite and effective for $M_2(B, C)$.

The TEM and SAD analysis is used to verify the concordance of theoretical calculation and actual condition, results of which is shown in Fig. 10. SAD patterns of region A and C reflect the lattice structure of TiC and $M_2(B, C)$, Data in Table 5 are details of them. Region C is the combined pattern of TiC and $M_2(B, C)$. Normally, in combined SAD pattern, when corresponding spots of two lattice and centric spot are detected alined, it can be proved that two crystallographic planes have orientation relationship. Spots $(110)_{TiC}$, $(110)_{M_2(B,C)}$ and the centric spot are detected alined obviously. This proves the orientation relationship $(110)_\gamma//(110)_{M_2(B,C)}$. These experimental results correspond to the results that areshown in Table 6.

Table 5 Lattice parameter of TiC, austenite and boride Fe_2B (nm)

Pase Types	Space group	a	b	c
TiC	Fm-3m	0.433	0.433	0.433
Austenite	Fm-3m	0.357	0.357	0.357
Fe_2B	I4/mcm	0.511	0.511	0.425

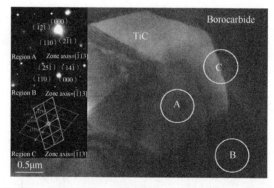

Fig. 10 TEM bright field of TiC and $M_2(B, C)$ with corresponding SAD patterns

Table 6 Calculated results of misfit degree

Possible relationship	δ(%)	Possible relationship	δ(%)
$(110)_{TiC}//(110)_{M_2(B,C)}$	9.3	$(111)_{TiC}//(001)_{M_2(B,C)}$	20.6
$(100)_{TiC}//(100)_{M_2(B,C)}$	36.2	$(100)_{TiC}//(100)_\gamma$	21.3
$(100)_{TiC}//(110)_{M_2(B,C)}$	23.8	$(110)_{TiC}//(110)_\gamma$	21.3
$(111)_{TiC}//(111)_{M_2(B,C)}$	24.5	$(111)_{TiC}//(111)_\gamma$	21.4
$(111)_{TiC}//(101)_{M_2(B,C)}$	29.1	$(100)_{TiC}//(110)_\gamma$	12.5

3.3 Effect of Ti content on mechanical properties of high-boron multi-component alloy

Hardness is one of the most important properties of materials for abrasion. The hardness of heat-treated high-boron multi-component alloy is measured (Table 7), which indicates that after quenching and tempering, high level of macrohardness is obtained. Moreover, Ti modification barely has effect on macrohardness. Titanium addition results in the solid solution strengthening, thus the microhardness rises slightly.

Table 7 Hardness of the studied alloys

Hardness Types	M	T1	T2	T3
Macrohardness after quenching/HRC	60.6	64.6	64.9	62.0
Macrohardness after quenching and tempering/HRC	58.2	58.2	59.8	57.6
Microhardness after quenching and tempering/HV	448.9	468.7	467.6	484.2

Modification is used to improve the toughness, so it is significant to test the impact toughness of the alloys. Fig. 11 shows the results of impact toughness testing of the alloys. Alloy T1, T2 and T3 possess much higher level of toughness than that of unmodified alloy M, which indicates that toughness of high – boron multi – component alloy can be remarkably increased by titanium modification. Besides, alloy T2 presents the smallest size and broken-reticular shape of borocarbide among modified alloys, but alloy which has the highest toughness is T3. It is because T3 has the hardest matrix which contributes to the enhance of toughness.

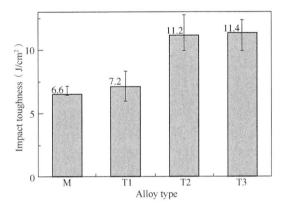

Fig. 11 Toughness results of high-boron multi-component alloy

The SEM SEI morphology observation of impact fractures is carried out, results of which are shown in Fig. 12. Cleavage fracture and dimples with divulsive arris are detected in each alloy. Thus it is revealed that the brittle-ductile composite fracture is the failuremechanism of high-boron multi-component alloy. In alloy M [Fig. 12(a)], this kind of large-size fractured borocarbide is the cause of bad toughness of unmodified alloy. In Fig. 12(b), fractured borocarbide is seen decrescent and more dimples are observed in T1. With the further ascending of titanium conent [Fig. 12(c), (d)], angles are barely observed in the fractures of alloy T2 and T3. According to all of the theoretical and experimental results, modification by titanium addition can be proved effectively for improving the mechanical property of high-boron multi-component alloy with the mechanism of spheroidization and refinement of borocarbide. These results make high – boron multi – component alloyfeasible for industrial application.

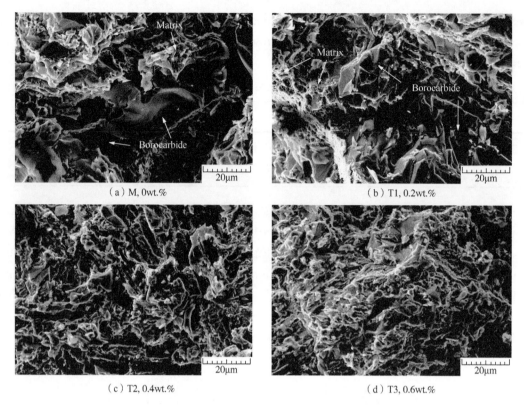

Fig. 12　Fractures of high-boron multi-component alloy with various titanium contents

4　Conclusions

(1) The as-cast microstructure of high-boron multi-component alloy is composed of ferrite, pearlite and $M_2(B, C)$ (M = Fe, V, Mo, Cr, Mn) (eutectic borocarbide). The un-modified borocarbide presents the structure of continuous network. In Ti-modified high-boron multi-component alloy, the borocarbide becomes isolated, fine and spherical.

(2) In liquid high-boron multi-component alloy with titanium addition, TiC particles precipitate first before solidification, which distribute dispersedlyin solidification microstructure. With the precipitation of TiC particles, eutectic borocarbide can precipitate from where TiC particles appear. The orientation relationship of borocarbide $M_2(B, C)$ and TiC is $(110)_{\gamma}//(110)_{M_2(B,C)}$.

(3) Matrix of modified high-boron multi-component alloy is strengthenedby titanium addition, hardness of which in modified alloy is higher than that of unmodified alloy. Toughness of high-boron multi-component alloy is significantly improved by titanium modification, which keeps rising when titanium content increases. Modified alloy T3 which contains 0.6wt.% titanium shows the best property. Fracture mechanism of high-boron multi-component alloy is mainly brittle with ductile fracture. It is the strengthening of matrix and improvement of shape, size of borocarbide that make the toughness increase.

Acknowledgments

This work was supported by the National key Technologies R&D Program of China

(2017ZX05009 – 003, 2019YFF0217501, 2019YFF0217500), Natural Science Foundation of China (U1762211) and CNPC Basic Research Project (2020B – 4020, 2019B – 4014, 2019E – 2502).

References

[1] ZAMRI W, KOSASHI P, TIEU A, et al. Variations in the microstructures and mechanical properties of the oxide layer on high speed steel hot rolling work rolls [J]. Journal of Materials Processing Technology, 2012, 212(12): 2597-2608.

[2] ANDERSSON M, FINNSTROM R, NYLÉN T. Introduction of enhanced indefinite chill andhigh speed steel rolls in European hot strip mills[J]. Ironmaking Steelmaking, 2004, 31 (5): 383-389.

[3] WU G, WANG Y. Fracture analysis of high chrominum cast iron roll on CSP mill [C]. Advanced Materials Research, 2012; 548: 538-543.

[4] SPUZIC S, STRAFFORD K, SUBRAMANIAN C, et al. Wear of hot rolling mill rolls: An overview [J]. Wear, 1994, 176 (2): 261-271.

[5] LIU D, LIU R, MEI Y, et al. Microstructure and wear properties of Fe – 15Cr – 2.5Ti – 2C – xB wt.% hardfacing alloys [J]. Applied Surface Science, 2013, 271: 253-259.

[6] KIM J, KO K, NOH S, et al. The effect of boron on the abrasive wear behavior of austenitic Fe – based hardfacing alloys [J]. Wear, 2009, 267: 1415-1419.

[7] ZHANG Y, WANG L. Study on the effect of RE modification treatment on the high chromium cast iron [J]. Journal of Harbin University of Science Technology, 2005, 10 (3): 7-10.

[8] MA G, GUO E, WANG L. Study on the effect of RE – Mg modification treatment on the high chromium cast iron [J]. Journal of Harbin University of Science Technology, 2005, 10 (4): 33-36.

[9] WEI S, ZHU J, XU L, et al. Effect of residual austenite on properties of high vanadium high speed steel [J]. Transcations of Materials and Heat Treatment, 2005, 26 (1): 44-47.

[10] XU L, XING J, WEI S, et al. Investigation on wear behaviors of high-vanadium high speed steel compared with high – chromium cast iron under rolling contact condition [J]. Materials Science and Engineering A, 2006, 434: 63-70.

[11] JI Y, LI Y, WEI S, et al. Influence of carbon content on properties of high – vanadium high – speed steel under dry sliding condition [J]. Journal of Harbin Institute of Technology, 2006, 38: 116-119.

[12] WANG Q, YANG D, LONG R, et al. Investigation of the damage in high vanadium high speed steel and high chrome cast iron roll under the laboratory condition [J]. Foundry, 2005, 54 (6): 570-574.

[13] HWANG K, LEE S, LEE H. Effects of alloying elements on microstructure and fracture properties of cast high speed steel rolls Part I: Microstructural analysis [J]. Materials Science and Engineering A, 1998, 254: 282-295.

[14] XU L, WEI S, LONG R, et al. Research on morphology and distribution of vanadium carbide in high vanadium high speed steel [J]. Foundry, 2003, 52 (11): 1069-1073.

[15] CAO Y, ZHANG J, YIN F, et al. Microstructure and properties and its failure mechanism of high speed steel roll [J]. Transactions of Materials and Heat Treatment, 2012, 33 (7): 50-54.

[16] LIU S, CHENG Y, LONG R, et al. Research progress and prospect of wear-resistant Fe-B-C alloy [J]. Foundry Technology, 2007, 28: 1526-1530.

[17] HE L, LIU Y, LI J, et al. Effects of hot rolling and titanium content on the microstructure and mechanical properties of high boron Fe-B alloys [J]. Materials & Design, 2012, 36: 88-93.

[18] LV Z, FU H, XING J, et al. Microstructure and crystallography of borides and mechanical properties of Fe-

B-C-Cr-Al alloys [J]. Journal of Alloys and Compounds, 2016, 662: 54-62.

[19] FU H, JIANG Z. A study of abrasion resistant cast Fe-B-C alloy [J]. Acta Metallurgica Sinica, 2006, 42: 545-548.

[20] SONG X, LIU X, FU H, et al. Effect of Boron concentration on microstructures and properties of high-Boron low-Carbon Ferro-matrix alloy [J]. Foundry, 2008, 57: 498-503.

[21] ASTINI V, PRASETYO Y, BAEK E. Effect of Boron Addition on the Microstructure and Mechanical Properties of 6.5% V-5% W High Speed Steel [J]. Metals and Materials Internation, 2012, 18: 923-931.

[22] KIM J, KO K, NOH S, et al. The effect of boron on the abrasive wear behavior of austenitic Fe-based hardfacing alloys [J]. Wear, 2009, 267: 1415-1419.

[23] FU, H. A study of microstructures and properties of cast Fe-B-C alloy [J]. Foundry, 2005, 54: 859-863.

[24] KIM J, HONG S, PARK H, et al. Improving plasticity and strength of Fe-Nd-B ultrafine eutectic composite [J]. Materials & Design, 2015, 76: 190-195.

[25] HUANG Z, XING J, GUO C. Improving fracture toughness and hardness of Fe_2B in high boron white cast iron by chromium addition [J]. Materials & Design, 2010, 31: 3084-3089.

[26] YUKSEL N, SAHIN S. Wear-behavior-hardness-microstructure relation of Fe-Cr-C and Fe-Cr-C-B based hardfacing alloys [J]. Materials & Design, 2014, 58 (6): 491-498.

[27] ROTTGER A, LENTZ J, THEISEN W. Boron alloyed Fe-Cr-C-B tool steels-Thermodynamic calculations and experimental validation [J]. Materials & Design, 2015, 88: 420-429.

[28] SUNDEEV R, GLEZER A, MENUSHENKOV A, et al. Effect of high pressure torsion at different temperatures on the local atomic structure of amorphous Fe-Ni-B alloys [J]. Materials & Design, 2017, 135 (5): 77-83.

[29] GU J, ZHANG H, FU H, et al. Effect of Boron content on the structure and property of Fe-B-C alloy [J]. Foundry Technology, 2011, 32: 1376-1379.

[30] GUI Z, LIANG W, LIU Y, et al. Thermo-mechanical behavior of the Al-Si alloy coated hot stamping boron steel [J]. Materials & Design, 2014, 60 (60): 26-33.

[31] SEN S, SEN U, BINDAL C. The growth kinetics of borides formed on boronized AISI 4140 steel [J]. Vacuum, 2005, 77: 195-202.

[32] REN X, FU H, XING J, et al. Effect of boron concentration on microstructures and properties of Fe-B-C alloy steel [J]. Journal of Materials Research, 2017, 32 (16): 3078-3088.

[33] LOU S, LIU G, FU H. Effect of heat treatment on structure and property of B-bearing LAHSS [J]. Foundry Technology, 2012, 33 (4): 408-411.

[34] YU Z, FU H, DU Z, et al. Effect of quenching treatment on microstructure and property of high boron high speed steel roll [J]. Transcations of Materials and Heat Treatment, 2013, 34 (4): 138-142.

[35] YI D, XING J, FU H, et al. Investigations on microstructures and three-body abrasive wear behaviors of Fe-B casting alloy containing cerium [J]. Tribology Letters, 2015, 58: 20-30.

[36] LIU Y, LI B, LI J, et al. Effect of titanium on the ductilization of Fe-B alloys with high boron content [J]. Materials Letters, 2010, 64: 1299-1301.

[37] YI D, XING J, MA S, et al. Effect of rare earth-aluminum additions on the microstructure of a semisolid low carbon Fe-B cast alloy [J]. Materials Science and Technology, 2011, 27 (10): 1518-1526.

[38] YI D, XING D, FU H, et al. Effect of RE-Al additions and austenitising time on structural variations of medium carbon Fe-B alloy [J]. Materials Science and Technology, 2010, 26 (7): 849-857.

[39] YI D, ZHANG Z, FU H, et al. A study of microstructures and toughness of Fe-B cast alloy containing rare earth [J]. Journal of Materials Engineering and Performance, 2015, 24 (2): 626-634.

[40] SHI X, JIANG Y, ZHOU R. Effects of rare earth, titanium, and magnesium additions on microstructures and

properties of high-boron medium-carbon alloy [J]. Journal of Iron and Steel Research International, 2016, 23 (11): 1226-1233.

[41] LI X, HOU J, QU Y, et al. A study of casting high-boron high-speed steel materials [J]. Material Swissenschaft and Werkstofftechnik, 2015, 46 (10): 1029-1038.

[42] FU H, LIU J, XING J. Study on microstructure and properties of high speed steel roll modified by RE-Mg-Ti [J]. Journal of Iron and Steel Research International, 2003, 15 (3): 39-43.

[43] ZHANG X, WEI S, NI F, et al. Effect of rare earth modification on vanadium carbide morphology of high vanadium high speed steel [J]. Foundry Technology, 2008, 29 (7): 869-872.

[44] FU H, JIANG Z, LI M, et al. Investigations on structure and performance of high speed steel roll modified by titanium additions [J]. Foundry, 2007, 56 (6): 590-593.

[45] ZHOU X, YIN X, FANG F, et al. Effect of calcium modification on the microstructures and properties of high speed steel [J]. Advanced Materials Research, 2011; 217-218: 457-462.

[46] QU Y, XING J, ZHI Z, et al. Effect of cerium on the as-cast microstructure of a hypereutectic high chromium cast iron [J]. Materials Letters, 2008, 62: 3024-3027.

[47] DENG X, YANG J. Effect of the multimodification by RE and Mg on the inclusions in SS400 steel [J]. Journal of Inner Mongolia University of Scieme Technology, 2008, 27(3): 227-231.

[48] FU H, XIAO Q, KUANG J, et al. Effect of rare earth and titanium additions on the microstructures and properties of low carbon Fe-B cast steel [J]. Materials Science and Engineering A, 2007, 466: 160-165.

[49] ZHI X, XING J, FU H, et al. Effect of niobium on the as-cast microstructure of hypereutectic high chromium cast iron [J]. Materials Letters, 2008, 62: 857-860.

[50] ABDELAZIZ S, MEGAHED G, MAHALLAWI E, et al. Control of calcium addition for improved cleanness of low C, Al killed steel [J]. Ironmaking Steelmaking, 2009, 36(6): 432-441.

[51] BEDOLLA-JACUINDE J, AGULLAR S, MALDONADO C. Eutectic modification in a low-chromium white cast iron by a mixture of titanium, rare earths, and bismuth [J]. Journal of Materials Engineering and Performance, 2005, 14(3): 301-306.

[52] CHUNG R, TANG X, LI D, et al. Effects of titanium addition on microstructure and wear resistance of hypereutectic high chromium cast iron Fe-25wt.%Cr-4wt.%C [J]. Wear, 2009, 267: 356-361.

[53] REN X, FU H, XING J, et al. Effect of calcium modification on solidification, heat treatment microstructure and toughness of high boron high speed steel [J]. Materials Research Express, 2019, 6(016540): 1-13.

[54] REN X, FU H, XING J, et al. Effect of solidification rate on microstructure and toughness of Ca-Ti modified high boron high speed steel [J]. Materials Science and Engineering A. 2019, 742: 617-627.

[55] ANTONI-ZDZIOBEK A, GOSPODINOVA M, BONNET F, et al. Experimental determination of solid-liquid equilibria with reactive components: example of Fe-Ti-B ternary system [J]. Journal of phase equilibria and diffusion, 2014, 35 (6): 701-710.

[56] FU H, WU Z, XING J. Investigation of quenching effect on mechanical property and abrasive wear behavior of high boron cast steel [J]. Materials Science and Technology, 2007, 23 (4): 460-466.

[57] YE D, HU J. Practicable manual of inorganic thermodynamic data [J]. Beijing: Metallurgical Industry Press, 1995.

本论文原发表于《Metals》2021年第11卷第2期。

Integrity Analysis of Casing Premium Connection under High Compression Load

Wang Peng[1] **Xie Junfeng**[2] **Zheng Youcheng**[3] **Hu Fangting**[2] **Ji Nan**[1]

(1. State Key Laboratory for Performance and Structure Safety of Petroleum Tubular Goods and Equipment Materials, CNPC Tubular Goods Research Institute;
2. Oil and Gas Research Institute, Tarim Oilfield Company of CNPC;
3. PetroChina Southwest Oil & Gas field Company)

Abstract: With the increasingly harsh conditions of complex oil and gas wells such as high-temperature and high-pressure deep wells and long-distance horizontal well, the integrity of casing string puts forward higher requirements for compression performance of premium thread connections. The requirements of high compression resistance of connection are complicated, including ensuring the integrity of structure and sealability for thread at the same time under high compression load being equal to the bearing capacity of casing body, and considering the structural fatigue, environmental fracture and seal failure caused by the weakening of thread bearing performance under cyclic load. Based on the failure cases of some casing connections, laboratory tests and finite element analysis results, this paper discusses the key technical points in this problem, and provides suggestions for the performance optimization of high - performance casing premium connections based on failure prevention.

Keywords: Casing; Premium connection; High compression; Integrity

1 Concept of connection compression performance

There are a large number of connections in the casing string, which are the weakest parts of casing string integrity system. The complex oil and gas wells such as high-temperature and high-pressure deep wells and long - distance horizontal well need the high performance of casing connections[1]. There are many kinds of premium connections from different factories with higher performance to contrast with the API connection[2-3], which mainly include high pressure gas sealability, high tension and compression resistance. High compression performance is particularly important because its implementation is more difficult. The casing premium connections with 100% compression efficiency are popular for complex oil and gas wells. The structural and seal integrity of connections is guaranteed at the same time under the same ultimate compression and combined load

Corresponding author: Wang Peng, wpengw123@163.com or wangpeng008@cnpc.com.cn.

as the casing body. Furthermore, high compression efficiency is of great significance to improve the fatigue and environmental fracture resistance of connections.

2 Integrity analysis of connection under high compression load

Structural integrity is the premise of safe service for premium connections. As for typical premium thread connections, structural integrity under high compression implies that the stress and deformation of the whole connection are in the reasonable range. It is indicated that the most of deformation is borne by the shoulder and the Mises stress extreme also locate here under high compression load from Fig. 1(a). The shoulder not only shares the load of the thread, but also provides a stable support for the radial sealing surface. Under large compression load, the plastic strain would preferentially appear at the shoulder of the coupling and the root of the first th read. It is generally considered that the plastic strain should be less than or equal to 0.1 within the allowable load conditions without obvious deformation, which is considered as structural safety [Fig. 1(b)]. In general, reasonable thread design can avoid structural damage in high compression process, but large deformation must be avoided, especially the deformation will lead to seal strength decline of metal to metal radial interference seal structure, resulting in leakage of connection.

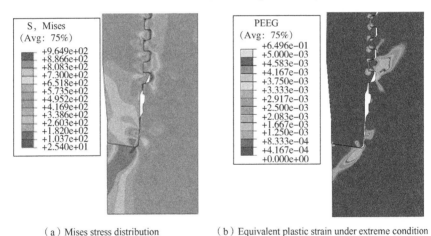

(a) Mises stress distribution   (b) Equivalent plastic strain under extreme condition

Fig. 1 Stess and strain distributions of connections under compression load

For the typical premium thread, the sealing performance evaluation of metal to metal sealing structure is a complex problem. The casing will bear different combined load at different time and position in downhole. Considering the limited case, the change of sealing contact pressure of threaded connections in four quadrants of combined load condition composed of internal pressure and external pressure, tension and compression is obtained by establishing load envelope according to axial and circumferential stress state of casing (Fig. 2). The worst sealing contact pressure does not occur in the compression load state of the connection. However, the contact pressure of the sealing surface gradually decreases with the increase of compression load under the internal pressure condition, which resulting in the sealing capacity is facing greater challenges under compression load. In addition, during the actual service of cyclic loading process, it is inevitable to weaken the mechanical properties of the material by Bauschinger effect, resulting in the increase of shoulder deformation and the decline of sealing strength of the connection. Therefore, it is very important to control the plastic deformation of the threaded connection (especially the shoulder part for typical premium thread) under the condition of high-compression condition to improve the compression resistance and sealing performance of the connection during long-term service.

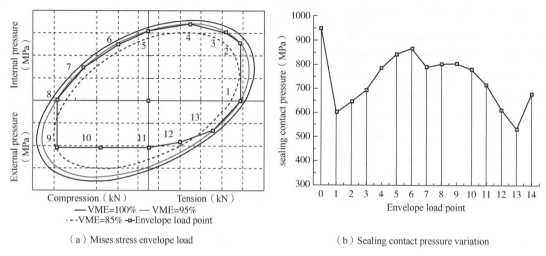

(a) Mises stress envelope load   (b) Sealing contact pressure variation

Fig. 2 Sealing contact pressure variation of connection for envelope load

3 Structural fatigue problem of connection bearing upon compression load

In fact, fatigue failure of casing in service is not common. The occurrence of fatigue failure depends on a certain degree of stress level and alternating cycle, especially for stress concentration effect. With the wide application of long-distance horizontal wells, casing fracturing, circulating multistage fracturing and casing completion, the above conditions can be met simultaneously under some conditions, such as rotary running casing, lifting up and down casing and casing multi-stage fracturing. It is founded that several proven casing fatigue failure problems, most of them occurred at the threaded connection due to the high local stress level and significant local stress concentration. As shown in Fig. 3, fatigue fracture occurs at the large end of the external thread, and the fatigue plateau area of the whole fracture section accounts for about 30% [Fig. 3(a)]. The fatigue begins at the root of external thread [Fig. 3(b)]. The fatigue striation can be seen from the microscopic photos [Fig. 3(c)]. These characteristics of low cycle fatigue are shown in Fig. 3.

 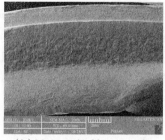 

(a) Thread fracture   (b) Macrograph of fracture surface   (c) Micrograph of fracture surface

Fig. 3 Fatigue analysis of connections

Most of the fracture surface is shear fracture morphology, which is a typical ductile fracture feature. These phenomena show that the instantaneous load is very large. This is very easy to be mistaken for the fracture caused by overload. As can be seen in Fig. 4, both the maximum tensile stress and the equivalent stress appear at the large end of the thread. The thread root where the thread disappears is a dangerous section. Thus the fracture shown in Fig. 3(a) is very reasonable. However, Fig. 3(b) and Fig. 3(c) clearly show that fatigue is the source of fracture, and the connection must bear a large range of tensile and compressive cyclic load. In fact, this occurs in the process of horizontal downhole running casing. The cycle times during the period is about $10^2 \sim 10^5$, and during the loading process, stress level is estimated to be more than 70% of the metal yield strength. After obtaining S-N curve through material fatigue test, the load boundary can be defined more accurately.

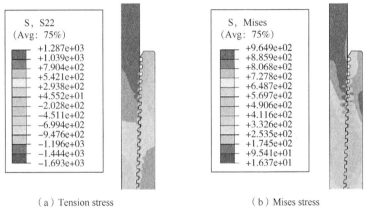

(a) Tension stress   (b) Mises stress

Fig. 4 Stess distributions of dangerous section for connections

Therefore, there is a correlation between the compression resistance and fatigue resistance of premium threaded connections. After the structural and sealing integrity under high-compression load is guaranteed, it is put forward for higher requirements that the control of local stress extreme value and uniformity of thread is particularly important to improve the fatigue resistance of connections.

4 Environmental fracture problem of connection bearing upon compression load

As we all know, the occurrence of environmental fracture needs three elements: material, stress and corrosion medium [4]. When the material is under service load, the corrosive medium will preferentially participate in the parts where the structure or stress level changes greatly [5]. Under the effect of more sensitive stress concentration, stress corrosion cracking (SCC) occurs at the dangerous section of thread. In general, there are two dangerous sections for premium threaded connections under axial tensile load, which are located at the large end of the external thread and the root of the first thread of the internal thread. As shown in Fig. 5 and Fig. 6, the stress corrosion cracking cases occurred at two dangerous sections. The failure at different parts depends on the structural characteristics and complex service state of the thread. In case 2, the bearing capacity of the coupling with small outer diameter is weaker, so it is better to understand that the failure occurs in the coupling.

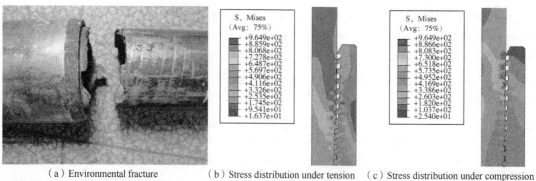

(a) Environmental fracture　　(b) Stress distribution under tension　　(c) Stress distribution under compression

Fig. 5　Enviromental fracture case 1 of connection

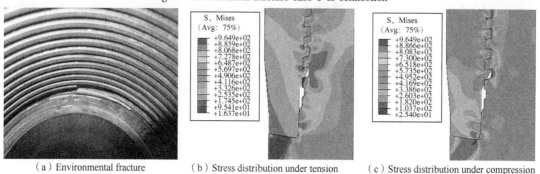

(a) Environmental fracture　　(b) Stress distribution under tension　　(c) Stress distribution under compression

Fig. 6　Enviromental fracture case 2 of connection

It can be seen from Fig. 7(a) that the typical crack morphology of stress corrosion cracking is similar to the shape of tree branchesor the cracking of ice block. The effect of corrosion medium on the stress concentration position causes serious embrittlement of the material. Fig. 5(b) and Fig. 6(b) show the larger tensile stress state at the failure site to help us better understand the occurrence of cracking. However, we do find some fatigue characteristics in some parts of the fracture [Fig. 7(b)]. If the failure property is SCC, it can be judged as corrosion fatigue. It is no doubt that the failure connection is subjected to tensile and compressive cyclic loads. [Fig. 5(c) and Fig. 6(c)] show that the stress at the fracture site under compression is also harsh. In fact, these failures do occur in the middle of the string (neither the place with the maximum tensile load near the wellhead nor the place with the maximum compressive load at the bottom hole). It can be seen that the compression resistance of threaded connections also plays a role in resisting environmental fracture.

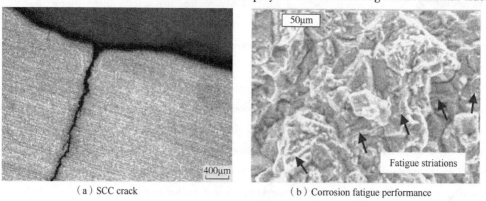

(a) SCC crack　　(b) Corrosion fatigue performance

Fig. 7　Micro analysis of environmental fracture for connections

5　Conclusions

The compression performance of thread needs to ensure the structural and sealing integrity of the connection under compression. Furthermore, the stress of the thread under compression has a great influence on the fatigue resistance and environmental fracture resistance of the connection. Therefore, the design of high-performance premium threaded connection can not only meet the requirements of joint strength and sealing capacity under compression state, but also consider the anti-fatigue performance and anti-SCC performance of the connection. The evaluation and selection of thread should also pay attention to these contents.

Acknowledgements

This paper is supported by key research and development project in Shaanxi Province (2018GY-172) and CNPC research and development project (2020B-4020).

References

[1] CARCAGNO G. The Design of Tubing and Casing Premium Connections for HTHP Wells [C]. Society of Petroleum Engineers, 2005.

[2] PENG Y, LI L. Analysis and Comparison of Sealing Performance of the Premium Connections [J]. January of Mechanical Engineering and Technology, 2016, 05(01): 31-37

[3] XU H, YANG B. A Quantitative Model to Calculate Gas Sealing Capacity and Design Sealing Parameters for Premium Connection [J]. Mathematical Problems in Engineering, 2020, 5: 1-7

[4] HU S Y. Manual of mechanical failure analysis [M]. Chengdu: Sichuan science and technology press, 1989.

[5] YANG C J, JIAN N. Corrosion & protection[M]. London: Springer London, 2004.

本论文原发表于《Materials Science Forum》第1035卷。

Coiled Tubing Plastic Strain and Fatigue Life Model

Wang Xinhu　Tian Tao

(State Key Laboratory for Performance and Structural Safety of Petroleum Tubular Goods and Equipment Materials, Tubular Goods Research Institute of China National Petroleum Corporation)

Abstract: Plastic strain calculation on multiaxial stress condition is the key to the fatigue life prediction of coiled tubing (CT). In this paper, the equivalent plastic strain estimate mathematical model of surface damaged coiled tubing at the joint action of bending and internal pressure was established and verified by fatigue testing in different pressure inside TC. The defects on the surface of CT are manufactured by Artificial method. The results showed that the increment of the plastic strain of CT caused by the fluid pressure inside CT will increase exponentially with the increase of the internal pressure, and the increment of the plastic strain caused by the surface defect will increase exponentially with the damage degree of the defect on the surface of CT, and the degree of defect damage can be characterized by the ratio of the parameters such as the depth, width and length of the defect to the geometric dimensions of the coiled tubing.

Keywords: Coiled tubing; Plastic strain; Fatigue life prediction

1　Introduction

The coiled tubing (TC) strings are used for acidification, fracturing, well repair, capacity modification, etc. Rely on specialized operating machines, the CT is straightened from the reel, then lowered into the oil well through the guide arch and injection head, and then wrapped back into the reel after the operation is completed. When bending on the reel and guide arch, the CT bending rang significantly exceeds the yield strength of the steel pipe. The strain fatigue of CT can occur due to repeated winding and straightening. It is necessary to predict the fatigue life of CT in order to operating safely. Because the plastic strain calculation of CT is difficult, the prediction result of fatigue life is not accurate. The standard API RP 5C8 (2017) *"Care, maintenance, and inspection of coiled tubing"* and SY/T 6698—2007 *"Recommended Practices for coiled tubing"* also lack practicable fatigue life prediction methods. This paperis to establish a simple and more accurate method.

2　Study progress review

Domestic and abroad experts predict the fatigue life of CT according to the classic Manson-

---

Corresponding author: Wang Xinhu, Wangxinhu002@ cnpc. com. cn.

Coffin fatigue life estimation equation $\frac{1}{2}\varepsilon = \frac{1}{2}\varepsilon_e + \frac{1}{2}\varepsilon_p = \frac{\sigma'_f}{E'}(2N_f)^b + \varepsilon'_f(2N_f)^c$, where fatigue life N_f is the number of stress or strain cycles to failure, E' is Cyclic elasticity modulus, or Yang's modulus; σ'_f is fatigue strength coefficient; b is fatigue strength exponent, ε'_f is fatigue ductility coefficient, c is fatigue ductility exponent, ε is cyclic strain rang, ε_e is cyclic elastic strain rang, ε_p is cyclic plastic strain rang. Unlike most mechanical equipment, which only works in the elastic strain range, a large plastic strain occurs when the coiled tubing (CT) operating, so the CT fatigue type belongs to strain fatigue, the effect of elastic strain on fatigue can be negligible, and the fatigue life calculation equation can only consider the plastic strain part. In the absence of internal pressure, the plastic strain range of CT is the bended plastic strain range. However, when the CT is operating, there is still a certain amount of pressure fluid inside tube, so how to calculate the plastic strain when bending and internal pressure work together is the key to predict the fatigue life of CT. Because the relationship between stress and strain is nonlinear in the condition of plastic deformation, it is difficult to calculate the strain in the multiaxial stress condition of CT by elastic mechanics.

Professor Teven M. Tipton in the University of Tulsa proposed a method[1] as below: $\varepsilon_{a,eff} = \varepsilon_{a,eq}(1+\varepsilon_{m,eq})^s$, where, $\varepsilon_{a,eff}$ is coiled tubing fatigue damage parameter, which can be understood as an effective plastic strain amplitude, is taken as $\frac{1}{2}\varepsilon$ or $\frac{1}{2}\varepsilon_p$ in Manson-Coffin fatigue equation, $\varepsilon_{a,eq}$ is equivalent Von Mises strain amplitude, which is computed based on half of the range of each strain component from a peak to a valley, $\varepsilon_{m,eq}$ is equivalent Von Mises peak strain, and S is coiled tubing exponent, which is showed as below: $S = Q\left(\frac{\sigma_{hh}}{s_y}\sqrt{\Delta\varepsilon_x}\right)^m$, where σ_h is Hoop Stress of CT, s_y is material yield strength, $\Delta\varepsilon_x$ is the bending strain range of CT, Q and m is constant, which can be obtained by fatigue test.

Ming Ruiqing et. al. analysed foreign development of CT software from the 1990[2], such as Fatigue Analysis for Coiled Tubing (FACT) and Tubing Analysis System (TAS) software developed by MEDCO (British Modelling Engineering and Development Company), Cerberus software developed by CTES, a subsidiary company of NOV (National Oilwell Varco), and CoolCADE software developed by Schlumberger. Cerberus software is widely used in the world, representing the highest level of CT software technology in the world. The core of the Cerberus software is the Achells 3.0model established by Professor Steven Tipton of Tulsa University as described above.

In the technical manual for the tubing analysis system (TAS) of the British Modeling engineering & development company limited (MEDCO)[3], the calculation model for the fatigue life of CT is $S = aN^\alpha + bN^\beta = bN^\beta$, where the aN^α is the elastic part of fatigue life, and the bN^β is the plastic part of fatigue life, and the index β value is -0.5. In essence, this model is the same as the Manson-Coffin fatigue life estimation equation. When there is fluid pressure inside CT, the stresses or strains of CT are calculated according to the following empirical equation $S = S_a + S_t^{1.895}$, where S_a is bending stress or strain as calculated according to the diameter of CT reel and guide arch, S_t is stress or strain caused by fluid pressure inside CT[2]. The technical manual provided the coefficient of hydrogen sulfide, acidification corrosion, welds, etc. on fatigue life, but it was not showed that the source of index 1.895 and the method of calculating the fatigue life of defected CT.

He Chunsheng et al. carried out the bend-straightening CT fatigue life experiment, and the equation of ellipse and wall thickness with the number of strain cycles was derived by mathematical regression method[4]. Zhao Le et al. used finite element software ABACUS to establish a series of CT models with volume defects, taking CT80 grade CT as an example, simulating the bending process of CT at the wellhead, and analyzing the impact of defect length, width and depth on CT[5]. It was found that the strain concentration coefficient of CT increases with the increase of defect depth and width, and decreases with length. The mathematical relationship between the strain concentration coefficient K_s and the defect parameter Q was obtained: $K_s = \dfrac{\varepsilon_b}{\varepsilon} = 1 + 35.7 Q^{3.13896}$, where ε_b was the strain amplitude of CT with volume defect and ε was the strain amplitude of defect-free CT [5]. Wang Zhenghan and Zhou Zhihong calculated the axial and circumference strain components of spheroid defects on CT in the process of winding and straightening under zero internal pressure, and then calculated the maximum shear strain range and Teven M. Tipton's effective strain amplitude, and the fatigue life of CT were estimated according to the Manson-Coffin equation. The results showed that the longer the length of the spheroid defect, the longer fatigue life was, and the wider and deeper the spheroid defect width, the shorter fatigue life was[6].

So the difficulty in predicting the fatigue life of CT was how to calculate the plastic strain when bending and internal pressure work together. Most of the research results focus on the fatigue life prediction of the defect-free CT, while fewer studies were conducted on the fatigue life prediction of the defected CT.

3 Mathematical model for predicting fatigue life

3.1 Basic scheme

The CT fatigue is strain fatigue, compared to a large plastic strain, the effect of elastic strain on fatigue can be negligible, so Manson-Coffin fatigue life estimate equation can only consider the plastic strain part. The fatigue life equation can be written as follow:

$$\frac{1}{2}\varepsilon = \varepsilon'_f (2N_f)^c \qquad (1)$$

Fatigue ductility coefficient ε'_f and fatigue ductility exponent c are constants related to material performance, there are a variety of methods to determine the constants, each method has advantages and disadvantages, sometimes the error is very large. In this paper, it was recommended the fatigue life are tested at different plastic strains, and then the two constants are calculated according to equation (1).

3.2 Plastic strain calculation model

Due fluid pressure working in CT during the operation, the CT is subjected to bending strain and circumference strain. Therefore, the strain range ε in the equation (1) is replaced by CT equivalent plastic strain range ε_{eq} at the joint action of bending and internal pressure, and the fatigue life equation changes as:

$$\frac{1}{2}\varepsilon_{eq} = \varepsilon'_f(2N_f)^c \tag{2}$$

If there is no fluid inside CT, the equivalent plastic strain range ε_{eq} is equal to the bending strain ε_b. If fluid is filled inside CT, assuming that the internal pressure produces a strain increment $\Delta\varepsilon_p$, then the equivalent plastic strain range ε_{eq} is given below:

$$\varepsilon_{eq} = \varepsilon_b + \Delta\varepsilon_p \tag{3}$$

where the ε_b is the bending strain range, and the calculation method is:

$$\varepsilon_b = \frac{D-t}{2R} \tag{4}$$

where D is the outside diameter of CT, t is the thickness of CT, R is the bending radius of CT when operating, Generally which is the radius of CT reel or the radius of the guide arch when CT enters the oil well. In equation (3), $\Delta\varepsilon_p$ is the increment of the plastic strain range caused by the internal fluid pressure, assuming that the internal pressure affects the bending strain of CT exponentially, and the calculation method is:

$$\Delta\varepsilon_p = (2\varepsilon_f - \varepsilon_b)k_p^m \tag{5}$$

Where ε_f is the true material strain, it is a constant reflecting the performance of the material, and it can be obtained by stretching test, m is the internal pressure strain equivalent index of CT, a constant that measures the effect of internal pressure on the bending strain of CT, which may be related to the performance of CT material, k_p is the internal pressure ratio of CT, i.e. the ratio of the operation pressure p to the yield pressure strength p_s of CT, which can be expressed as:

$$k_p = \frac{p}{p_s} = \frac{pD}{2\sigma_s t} \tag{6}$$

Where σ_s is the yield strength of CT material.

According to the above hypothesis, the CT equivalent plastic strain range ε_{eq} at the joint action of bending and internal pressure is calculated as follow:

$$\varepsilon_{eq} = \varepsilon_b + (2\varepsilon_f - \varepsilon_b)k_p^m \tag{7}$$

The CT fatigue life can be estimated by bringing the formula (7) into the equation (2).

3.3 Plastic strain calculation model for surface damage CT

The CT is not immune to damage in the course of operation, including scratches, corrosion pits, etc., which will cause stress and strain concentration, so the bending strain range $\varepsilon_{b,d}$ at the defect can be instead of the bending strain range ε_b. Then the equation (7) for calculating the plastic strain range of the defected CT becomes:

$$\varepsilon_{eq} = \varepsilon_{b,d} + (2\varepsilon_f - \varepsilon_{b,d})k_p^m \tag{8}$$

where the calculation method of the internal pressure ratio k_p is:

$$k_p = \frac{p}{p_s} = \frac{pD}{2\sigma_s t_d} \tag{9}$$

where the t_d is the effective wall thickness at the defect of CT, and the calculation method is:

$$t_d = t\left(1 - \frac{V_d}{V}\right) = t\left[1 - \left(\frac{h_d}{t}\right)\left(\frac{B_d}{\pi D}\right)\left(\frac{L_d}{\pi D}\right)\right] \tag{10}$$

where $\frac{V_d}{V}$ is the defect volume ratio, i.e., the ratio of defect volume V_d to the material volume

V of CT in a length range equal to one perimeter or more than one perimeter of CT, $\dfrac{h_d}{t}$ is the defect depth ratio, that is, the ratio of defect depth h_d to the wall thickness t of CT, $\dfrac{B_d}{\pi D}$ is the defect width ratio, i. e. the ratio of the defect width B_d on circumference of CT to the perimeter πD of CT, $\dfrac{L_d}{\pi D}$ is the defect length ratio, i. e. , the ratio of the defect length along the axial direction of CT in a length range equal to one perimeter of CT to the perimeter πD of CT.

Assuming the defect affects the plastic strain of CT in exponential form, then the plastic strain increment caused by the defect is:

$$\Delta\varepsilon_d = (2\varepsilon_f - \varepsilon_b)k_d^\mu \qquad (11)$$

where μ is the strain concentration index at the defect, which may be related to material performance, outside diameter and wall thickness of CT, k_d is the defect ratio, which is the parameter of the severity of the defect, and the calculation method is:

$$k_d = \left(\dfrac{h_d}{t}\right)\left(\dfrac{A_d}{A}\right)\left(\dfrac{h_d}{L_d}\right) \qquad (12)$$

where $\dfrac{h_d}{t}$ is the depth ratio of defect, i. e. the ratio of the defect depth h_d to wall thickness t of CT; $\dfrac{A_d}{A}$ is the cross-sectional area ratio of the defect, i. e. the ratio of the projected area of the defect on the cross-section of CT to the cross-sectional area of CT, $\dfrac{h_d}{L_d}$ is the depth-to-length ratio of the defect, i. e. the ratio of the depth h_d to the length L_d of the defect in the axial direction of CT. When actually evaluating, the defects with the deepest depth, the widest width, and the shortest axial length should beselected.

The bending strain range of CT at the defect is calculated as follow:

$$\varepsilon_{b,d} = \varepsilon_b + \Delta\varepsilon_d = \varepsilon_b + (2\varepsilon_f - \varepsilon_b)k_d^\mu \qquad (13)$$

The equivalent plastic strain range of the defective CT is calculated by substituting the equation (13) calculation result into the equation (8), and then the fatigue life of the defective CT can be estimated according to the equation (2).

4 Test and verify

4.1 Equivalent strain calculation model validation

The calculation equation (5) of the plastic strain range increment caused by internal fluid pressure can be written in logarithmic form:

$$\lg\Delta\varepsilon_p = \lg(2\varepsilon_f - \varepsilon_b) + m\lg k_p \qquad (14)$$

The equation (14) shows that if the equation (5) is established, the logarithm of plastic increments $\Delta\varepsilon_p$ is linearly related to the logarithm of the internal pressure ratio k_p.

The fatigue life N_f of can be obtained through CT fatigue test at different internal pressures, and the equivalent plastic strain range ε_{eq} of the corresponding internal pressure can be calculated according to the equation (2), then the plastic strain range increment $\Delta\varepsilon_p$ may becalculated as follow:

$$\Delta\varepsilon_p = \varepsilon_{eq} - \varepsilon_b = 2\varepsilon'_f(2N_f)^c - \varepsilon_b \quad (15)$$

Obviously, as long as CT fatigue test in different internal pressures, the linear relationship (14) can be verified.

The fatigue life test was carried out by using CT fatigue test machine developed by ourselves, and the bending fatigue test of CT at different curvature radius is realized by bending the mold. During the test process, the CT bends around a curved mold with a known radius, and then is straightened by a straight template, and the cylinder provides a two-way force to bend and straighten CT sample, and CT goes through a "bend and straight" for a cycle, with a cycle time of 60 seconds. The internal pressure of CT is provided by a pump inputing water and maintained at the desired pressure, when the internal pressure drops significantly, it is considered that CT fatigue damage occurs and the test device will automatically stop running.

CT specimens length was 2m, and the steel grade was CT90, and its outside diameter was 38.1mm, the wall thickness was 3.2mm, the material yield strength was 650MPa. In the case of approximately no internal pressure, CT fatigue life N_f was obtained through fatigue test at different bending strain ε, then the material fatigue toughness index and coefficient of CT were calculated according to the Manson-Coffin fatigue life equation (1), $c=-0.6$, $\varepsilon'_f=0.7$.

The radius of the curved mold was 48in (1219.2mm), and the internal pressures were 5, 20, 35 and 50MPa respectively, and the fatigue life test results were 980, 580, 265 and 90 cycles respectively. Then according to the equation (15), the plastic strain increment $\Delta\varepsilon_p$ caused by the internal pressure was calculated, according to the equation (6), the internal pressure ratio k_p was calculated. The results showed that the CT plastic strain increment $\Delta\varepsilon_p$ is linear with the internal pressure ratio k_p in logarithmic form as shown in Fig.2(a). It was proved that the CT plastic strain increment $\Delta\varepsilon_p$ increases exponentially with the increase of the internal pressure ratio k_p, and the equivalent plastic strain range ε_{eq} of the calculation equations (3) and (5) are established.

4.2 The bend strain calculation model validation for surface damage CT

Defected CT plastic strain increment equation (11) can be written as logarithmic form:

$$\log(\Delta\varepsilon_d) = \log(2\varepsilon_f - \varepsilon_b) + \mu\log k_d \quad (16)$$

If the equation (11) is established, defected CT plastic strain increment $\Delta\varepsilon_d$ is linear with defect ratio k_d in logarithmic form. Substitute equation (8) into equation (2):

$$2\varepsilon'_f(2N_f)^c = \varepsilon_{b,d} + (2\varepsilon_f - \varepsilon_{b,d})k_p^m$$

Then the bend strain range $\varepsilon_{b,d}$ is:

$$\varepsilon_{b,d} = \frac{2\varepsilon'_f(2N_f)^c - 2\varepsilon_f k_p^m}{1 - k_p^m}$$

Then the plasticity strain increment $\Delta\varepsilon_d$ is:

$$\Delta\varepsilon_d = \varepsilon_{b,d} - \varepsilon_b = \frac{2\varepsilon'_f(2N_f)^c - 2\varepsilon_f k_p^m}{1 - k_p^m} - \varepsilon_b \quad (17)$$

Obviously, the defective CT fatigue life N_f is obtained through the fatigue test, the plastic strain incremental $\Delta\varepsilon_d$ can be calculated according to equation (17).

As shown in Fig. 1, rectangular and circular damage with a depth of 1mm was manufactured on the outside surface of the steel grade CT90 tube (38.1mm outside diameter and 3.2mm in wall thickness), the diameter of the circular damage was 5mm, the width of rectangular damage was 3mm and the length was 10mm respectively, along the circumferential and longitudinal direction distribution of the tube.

The radius of the curved mold used in the testing was 48in (1219.2mm), and fatigue tests were carried out at 20, 35 and 50MPa pressure inner CT respectively to obtain fatigue life N_f. Then according to the equation (17), the plastic strain increment $\Delta\varepsilon_d$ caused by defects was calculated. The Results showed the relationship between the plastic strain increments $\Delta\varepsilon_d$ and the defect ratio k_d was linear at logarithmic form as shown in Fig. 2(b), which proving that the plastic strain increment $\Delta\varepsilon_d$ increase with the defect ratio k_d exponentially.

Fig. 1 Artificial defects on the surface of CT specimens

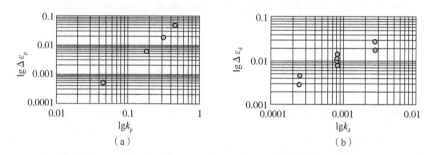

Fig. 2 The relationship between strain increment and (a) the internal pressure ratio or (b) defect ratio

5 Conclusion

Coiled tubing (CT) fatigue is strain fatigue, and the classical Manson–Coffin fatigue life calculation model is still the basis for estimating the fatigue life of CT, calculating the equivalent plastic strain in multiaxial stress condition is the key and difficulty to estimate fatigue life of CT. The study results showed that the increment of CT plastic strain caused by fluid pressure in the tube increases exponentially with the increase of internal pressure, and the increase of CT plastic strain caused by surface defects increases exponentially with the degree of defect damage. The degree of defect damage can be characterized by the ratio of defect depth, width, length to the geometry of CT.

Acknowledgments

The project is supported by China National Petroleum Corporation Science Foundation (2017B-4107).

References

[1] TIPTON S M, MITCHELL M R. Multiaxial Plasticity and Fatigue Life Prediction in Coiled Tubing [J]. ASTM special technical publication, 1996, 1: 283-304.

[2] MING R Q, HE H Q, TANG C J, et al. Study and Analysis on Domestic and Foreign Software Related with Coiled Tubing [J]. Oil Drilling & Production Technology, 2017, 39(6), 771-780, 794.

[3] Tubing Analysis System Technical Reference Manual, Courtenay House, Monument Way East, Woking, Surrey GU21 5LY, U.K.

[4] HE C S, LIU J B, YUE Q B, et al. Prediction of the Low-Cycle Fatigue Life of the Coiled Tubing Based on Ovality, and Wall Thickness Parameter[J]. Oil Drilling & Production Technology, 2013, 35(6), 15-18.

[5] ZHAO L, ZHANG H, DUAN Q Q. Influence Analysis of Geometrical Parameters of Volume Defects on Fatigue Life of Coiled Tubing[J]. Oil Field Equipment, 2016, 45(7), 1-5.

[6] WANG Z H, ZHOU Z H. Analysis of Fatigue Life of the Coiled Tubing with Ellipsoidal Defect[J]. China Petroleum Machinery, 2017, 45(5): 89-90.

本论文原收录于 International Field Exploration and Development Conference(2021 年)。

Comparison of CrN, AlN and TiN Diffusion Barriers on the Interdiffusion and Oxidation Behaviors of Ni+CrAlYSiN Nanocomposite Coatings

Zhu Lijuan[1,2] Feng Chun[1] Zhu Shenglong[2]
Wang Fuhui[2] Yuan Juntao[1,2] Wang Peng[1]

(1. State Key Laboratory of Tubular Goods Research Institute of CNPC,
Performance and Structural Safety for Petroleum Tubular Goods and Equipment Materials;
2. Institute of Metal research, Chinese Academy of Sciences)

Abstract: CrN, AlN, TiN layer were prepared as diffusion barrier between the K417 substrate and Ni + CrAlYSiN nano composite coatings via vacuum arc evaporation. Oxidation kinetics and microstructure evaluation of these nano coating systems at 1000℃ after 100h were studied. Results show that the AlN layer showed good thermodynamic stability, effectively inhibited the interdiffusion between the coating and the substrate, improved the oxidation resistance of Ni+CrAlYSiN nano composite coatings, and a single-layer Al_2O_3 film was formed on the coating. The CrN layer was decomposed, which did not block the diffusion of elements and had little effect on the oxidation resistance of the Ni+CrAlYSiN nano composite coating. The TiN layer effectively prevented the interdiffusion between the coating and the substrate. However, it deteriorated the oxidation resistance of the composite coating. Similar to the Ni+CrAlYSiN coating without a diffusion barrier, a double-layer oxide film structure with Al_2O_3 as the inner layer and $Ni(Al, Cr)_2O_4$ as the outer layer formed on the Ni+CrAlYSiN nano composite coatings with the CrN or TiN diffusion barrier.

Keywords: Oxidation; Nano; Diffusion barrier; K417; Interdiffusion

1 Introduction

High temperature protective coatings, such as MCrAlY (M is Ni, Co or NiCo) coatings, with outstanding high-temperature oxidation resistance, have been widely used to protect gas turbine components[1-5]. The content of Al in the temperature protective coating must be high enough to form a protective Al_2O_3 film and maintain its growth [6-8], due to the long-time exposure to the

---

Corresponding author: Feng Chun, fengchun003@cnpc.com.cn.

high-temperature environment. Considering the oxidation, corrosion resistance and consumption life of Al and Cr, the contents of Al and Cr should be increased as much as possible, but it is also subject to the embrittlement of MCrAlY coatings. According to the comprehensive requirements of oxidation resistance, heat corrosion resistance and coating plasticity, the Al content should range from 6wt.% to 12wt.%. There are two consumption modes of beneficial elements Al and Cr in high temperature protective coatings. One is the outward diffusion oxidation of Al and Cr to form a protective oxide film and maintain its growth [9-12]. The other is the diffusion of Al and Cr into the substrate, which reduces the content of beneficial elements in the coating and affects the service life of the coating [13-15].

Since interdiffusion is one of the key factors affecting the service life of coatings, diffusion barrier was introduced between the coating and substrate [16-17]. The basic requirements of diffusion barrier between the coating and substrate [16] are as follows: (1) Low diffusion coefficient of elements in the diffusion barrier; (2) No reaction between the coating and substrate with the diffusion barrier; (3) Good thermodynamic stability of the diffusion barrier at service temperature; (4) Good adhesion between diffusion barrier with the substrate and coating; (5) Low thermal stress at the interface between the substrate and coating. However, it is difficult to find a diffusion barrier that meets all the above conditions. Therefore, when the first condition is met, the diffusion barriers, such as TiN [18], CrN [19], Al_2O_3 [20], Al-O-N [21] or Cr_2O_3 [22], are selected by appropriately choosing other conditions.

In our previous work, a Ni + CrAlYSiN nano composite coating was prepared with high hardness, good wear resistance [23], good oxidation resistance [24] and thermal corrosion resistance [25], and the AlN diffusion barrier was introduced to effectively inhibit the interdiffusion between the coating and the substrate. " It is well known that the Gibbs free energies of CrN, AlN and TiN are -198.38kJ/mol, -378.95kJ/mol and -416.78kJ/mol respectively [26]. It is not clear whether TiN is more suitable for the diffusion barrier between Ni + CrAlYSiN nano composite coating and substrate. It has not been studied whether the TiN or AlN diffusion barrier can be formed in situ when a thin layer is deposited between the Ni + CrAlYSiN nanocomposite coating and substrate. Therefore, in the present work, TiN, CrN and AlN diffusion barriers were prepared, and the effects of the above diffusion barriers on the interdiffusion and oxidation behavior of K417/Ni+CrAlYSiN nano composite coating system were discussed.

2 Materials and methods

2.1 Sample Preparation

Nickel-base superalloy, K417 (nominal composition: 8.5~9.5wt.% Cr, 14.0~16.0wt.% Co, 4.8~5.7wt.% Al, 4.5~5.0wt.% Ti, 2.5~3.5wt.% Mo, 0.60~0.90wt.% V, minor C, balanced Ni) was used as the substrate alloy. Specimens of approximate dimensions 20mm×10mm× 2.5mm were ground to 2000-grit SiC paper and polished with diamond paste (1.5μm), then ultrasonically cleaned within ethanol and acetone for 20min. The AlN, TiN, CrN and Ni+CrAlYSiN nano composite coatings were prepared by vacuum arc evaporation. The chemical compositions of the cathode targets were Ni - 21Cr - 10Al - 0.5Si - 0.5Y (wt.%), Cr, Al and Ti with purity of

99.99wt.%, and typical deposition parameters are given in Table 1. During the coating deposition, the argon-nitrogen mixture with a total pressure of 0.2Pa was introduced, and the reactive gas (N_2, 99.99%) and inert gas (Ar, 99.99%) were introduced into the chamber using two independent mass-flow controllers. The detailed preparation process was described in our previous work [24]. Some specimens were coated with thick Ni+CrAlYSiN coatings only, while other specimens were coated with a CrN, AlN, or TiN thin film and then a Ni+CrAlYSiN coating via serial depositions using Cr, Al or Ti and then NiCrAlYSi targets respectively. In order to avoid of droplets that are commonly observed in the coatings deposited by vacuum arc evaporation, the CrN, AlN, and TiN thin films were prepared by filtered vacuum arc evaporation.

Table 1 Typical deposition parameters

Typical deposition	Arc-voltage (V)	Arc current (A)	Bias voltage (V)	Bias duty (%)	Temperature (℃)
CrN layer	19	70	−600	20	195~215
AlN layer	30	70	−500	20	195~215
TiN layer	19	70	−600	20	195~215
Ni+CrAlYSiN coating	19	70	−300	20	195~215

2.2 Oxidation Test

Cyclic oxidation tests of the specimens, uncoated K417, K417 coated with the composite coatings only, and K417 coated with the diffusion barrier layer and a composite coating, were conducted in static air in a muffle furnace for 5 cycles. In each cycle, the specimens were exposed at 1000℃ for 20h, taken out from the furnace, cooled down at room temperature, and then weighed using an electronic balance with sensitivity of 10^{-5}g.

2.3 Characterization

Surface and cross-sectional morphologies were observed using a field-emission scanning electron microscope (SEM, Inspect F50, FEI Co., Hillsboro, OR) equipped withan energy dispersive X-ray spectrometer (EDAX, X-Max, Oxford Instruments Co., Oxford, U.K.). To observe surface morphology, a second electron (SE) mode was adopted, while the back-scattered electron (BSE) mode was utilized to characterize cross-sectional morphologies of coating samples.

Surfaceand cross-sectional morphologies were gained by scanning electron microscopy (SEM) and energy dispersive X-ray spectrometer (EDX). The microstructures of the as-prepared diffusion barrier layers and the oxide scales were identified by X-ray diffraction (XRD), the reference codes of phases were checked from JCPDS. An electroless-nickel layer was applied on the surface of the cross-section specimens to prevent TGO (thermally-grown oxide) scales from spalling off in specimen preparation process.

3 Results

3.1 Coating morphology and microstructures

Fig. 1 shows the initial cross-sectional morphologies and elements depth profile of K417/CrN/

Ni+CrAlYSiN and K417/TiN/Ni+CrAlYSiN. There are grey phases mainly along with a few bright phases and dark voids in the Ni+CrAlYSiN nano coatings, consistent with our previous results [24]. The grey phases comprised mainly γ–Ni, fcc–AlN and fcc–CrN with average size less than 30nm [24]. The bright phases contained nanocrystalline solid solution γ–Ni(CrAlYSi) [23]. The almost un–nitride γ–Ni(CrAlYSi) was the splashed micro–droplet formed during vacuum arc evaporation. CrN, and TiN barrier layers, shown as a thick darken line in Fig. 1, are about 300nm and 500nm thick, respectively. The barrier layer was quite smooth as it was prepared by filtered vacuum arc vacuum arc evaporation. According to the EDX results, the atom ratio of Cr: N, and Ti: N of the barrier layers were about 1 : 1. K417/Ni+CrAlYSiN and K417/AlN/Ni+CrAlYSiN samples were also prepared for the comparative study. The thickness of AlN diffusion barrier is about 400nm.

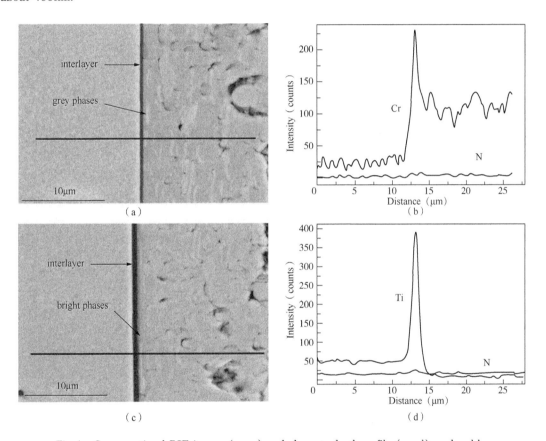

Fig. 1 Cross-sectional BSE images (a, c) and elements depth profile (c, d) analyzed by
EDX of the as-deposited K417/CrN/Ni+CrAlYSiN(a, b) and
K417/TiN/Ni+CrAlYSiN(c, d) nano composite coating systems

XRD analysis results showed that the structures of the three diffusion barriers are fcc–CrN and fcc–TiN with rock salt structure, hcp–AlN with braze zinc structure (Fig. 2), and their reference codes are 11–0065, 38–1420 and 25–1133 from JCPDS, respectively. In addition, the CrN diffusion barrier contains a small amount of Cr_2N (35–0803).

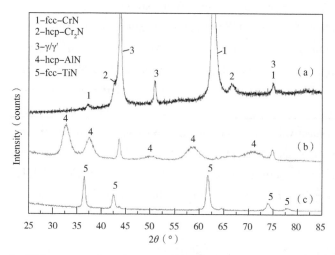

Fig. 2 XRD patterns of as-deposited (a) CrN, (b) AlN and (c) TiN film

3.2 Corrosion kinetics and corrosion products

Oxidation kinetics curves of K417 with and without coatings in air at 1000℃ after 100h are shown in Fig. 3. The coated specimens with a CrN barrier layer exhibited oxidation rates slightly higher than those without the barrier layer, while the coated specimens with an AlN barrier layer exhibited oxidation rates slightly lower than those without the barrier layer. The coated specimens with a TiN barrier layer exhibited oxidation rates slightly lower at the initial stage and then slightly higher than those with CrN barrier layer. Slight mass loss of the coated specimens with a TiN barrier layer after 80h, was detected, suggesting scale spallation occurrence on the surface. The mass gains of the K417/Ni+CrAlYSiN, K417/CrN/Ni+CrAlYSiN, K417/AlN/Ni+CrAlYSiN and K417/TiN/Ni+CrAlYSiN coating systems after 100h were 0.45, 0.51, 0.28 and 0.46mg/cm^2, respectively. That is, the nanocomposite coating with an AlN barrier layer exhibited the lowest oxidation rates.

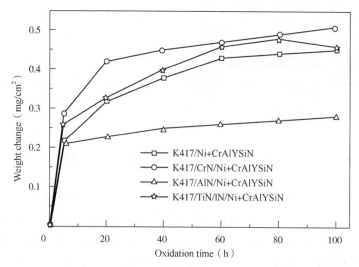

Fig. 3 Corrosion kinetics of the composite coating with or without a diffusion barrier after 100h oxidation at 1000℃ in static air

XRD patterns of the surfaces of specimens after 100h oxidation test are displayed in Fig. 4. The TGO scales on the nanocomposite coatings without an AlN diffusion barrier were composed of α-Al_2O_3 (50-1496) and $Ni(Al, Cr)_2O_4$ (23-1272) [Fig. 4(a)], those on the nanocomposite coatings with a CrN or TiN diffusion barrier were composed of Cr_2O_3 (38-1479), α-Al_2O_3 and $Ni(Al, Cr)_2O_4$ [Fig. 4(b) and Fig. 4(d)], while those on the nanocomposite coatings with an AlN diffusion barrier were exclusively alpha-alumina [Fig. 4(c)]. For the nanocomposite coatings, Cr_7Ni_3 (51-0637) was detected on the specimens without a diffusion barrier [Fig. 4(a)], while CrN was detected on those with the AlN and TiN barrier layer [Fig. 4(c) and Fig. 4(d)].

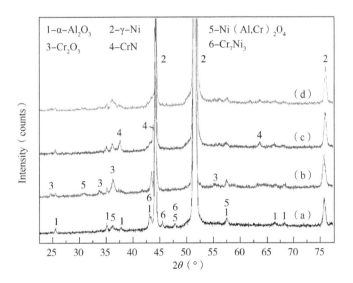

Fig. 4 XRD patterns of (a) K417/Ni+CrAlYSiN, (b) K417/CrN/Ni+CrAlYSiN, (c) K417/AlN/Ni+CrAlYSiN and (d) K417/TiN/Ni+CrAlYSiN coating systems after 100h oxidation at 1000℃ in air

3.3 Characterization by SEM and EDS

The morphologies and the element depth profiles of the K417/Ni+CrAlYSiN coating system after oxidation at 1000℃ for 100h were shown in Fig. 5. The oxidation and interdiffusion behavior of the K417/Ni+CrAlYSiN coating system generally consists with our previous investigation [24-25]. Neither cracking nor spallation of the TGO was detected on the nanocomposite coatings [Fig. 5(a)]. A TGO scale about 2.6μm was formed on the surface [Fig. 5(b)]. Combined with the results analyzed by XRD, the cross-section image demonstrates that a double-layered dense oxide scale, inner Al_2O_3 layer and outer $Ni(Al, Cr)_2O_4$ layer, formed on K417/Ni+CrAlYSiN coating system [Fig. 5(b)]. Within the composite coating, many dark hcp-AlN circles and particles were observed. Congregated hcp-AlN particles were also observed near the coating/substrate interface. Interdiffusion between the coating and the substrate was obvious. Many granular and even needle-like TiN phases were formed in the interdiffusion zone. Diffusion of N from the coating into the substrate and Co from the substrate into the coating was most evident [Fig. 5(c)].

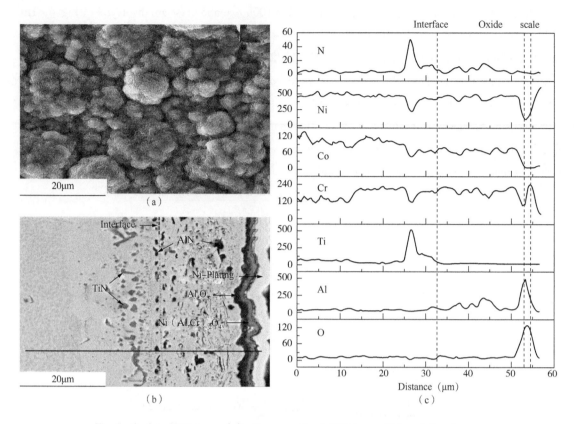

Fig. 5 Surface SEM image (a), Cross-sectional BSE image (b) and the element depth profiles [(c), analyzed by EDX] of K417/Ni+CrAlYSiN coating system after oxidation for 100h at 1000℃ in air

The morphologies and the element depth profiles of the K417/CrN/Ni + CrAlYSiN coating system after oxidation at 1000℃ for 100h were shown in Fig. 6. Large-sized particles were formed on the surface of the composite coatings [Fig. 6(a)]; however, neither cracking nor spallation of the TGO was detected. A thermally-grown oxide (TGO) scale about 2.9μm was formed on the surface [Fig. 6(b)]. Combined with the results analyzed by XRD, the cross-section image demonstrates that a double-layered dense oxide scale, inner Al_2O_3 layer and outer mixture layer mainly dominated by $Ni(Al, Cr)_2O_4$, formed on K417/CrN/Ni+CrAlYSiN coating system [Fig. 6 (b)]. Similar to the K417/ Ni+CrAlYSiN coating system, hcp-AlN particles were formed within the composite coating. However, no large-sized hcp-AlN particles near the coating/substrate interface was observed. The CrN diffusion barrier disappeared. Interdiffusion between the coating and the substrate was obvious. Many TiN particles were formed in the interdiffusion zone. Diffusion of N and Cr from the coating into the substrate, Co from the substrate into the coating was most evident [Fig. 6(c)].

In contrast, no such interdiffusion zone was observed on K417/AlN/Ni+CrAlYSiN (Fig. 7) and K417/TiN/Ni+CrAlYSiN coating systems (Fig. 8). The ©/©′ microstructure beneath the AlN and TiN barrier layers remained unchanged, the AlN and TiN diffusion barrier layer kept continuous and dense [Fig. 7(b) and Fig. 8(b)]. However, the thickness of the diffusion barrier for the K417/TiN/Ni+CrAlYSiN coating system was changed from 0.5μm to 1.1μm with a triple-layered structure after oxidation at 1000℃ for 100h. The center layer was rich in Ti and N, outer layer near the substrate was rich in Al, Ti and N, while outer layer near the coating was rich in Cr, Ti and N. Again, the phenomenon of AlN congregation within the coatings of K417/TiN/Ni+CrAlYSiN coating system occurred. However, the size of AlN particles was quite smaller than that in the K417/Ni+CrAlYSiN and K417/CrN/Ni+CrAlYSiN coating systems. A unique alumina scale about 1.6μm was formed on the surface of K417/AlN/Ni+CrAlYSiN coating systems [Fig. 7(b)] and neither cracking nor spallation of the TGO was detected [Fig. 7(b)]. In contrast, a double-layered dense oxide scale, inner Al_2O_3 layer and outer mixture layer mainly dominated by $Ni(Al, Cr)_2O_4$, was formed on K417/TiN/Ni+CrAlYSiN coating system [Fig. 8(b)]. The thickness of the oxide scale on K417/TiN/Ni+CrAlYSiN coating system ranged from about 2μm to more than 7μm in the bulged oxide zone [Fig. 8(b)].

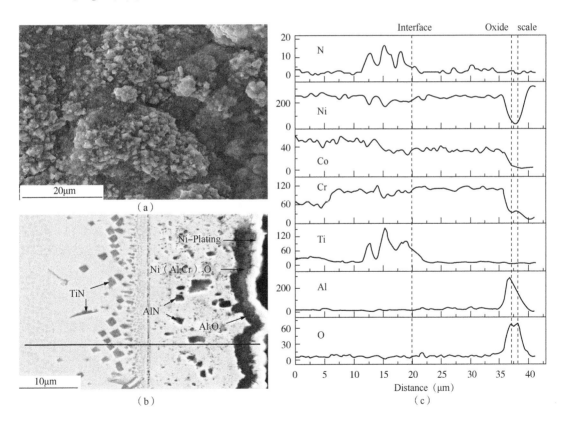

Fig. 6　Surface SEM image (a), Cross-sectional BSE image (b) and the element depth profiles [(c), analyzed by EDX] of K417/ CrN/Ni+CrAlYSiN coating system after oxidation for 100h at 1000℃ in air

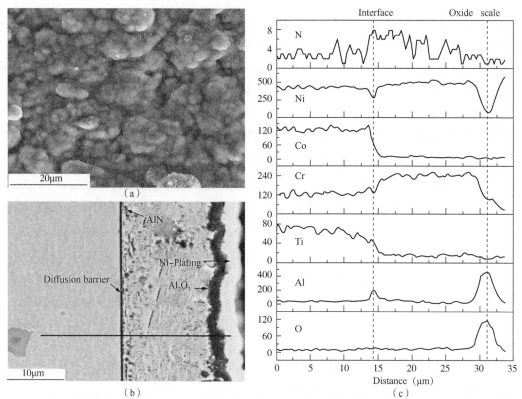

Fig. 7 Surface SEM image (a), Cross-sectional BSE image (b) and the element depth profiles [(c), analyzed by EDX] of K417/ AlN/Ni+CrAlYSiN coating system after oxidation for 100h at 1000℃ in air

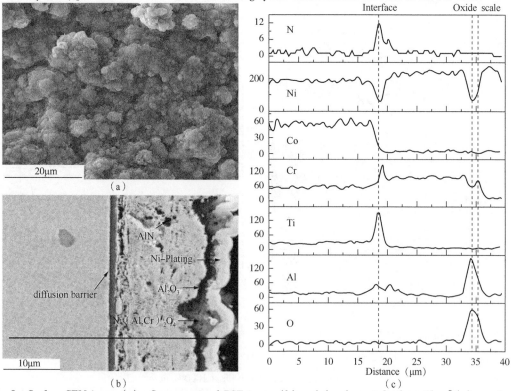

Fig. 8 Surface SEM image (a), Cross-sectional BSE image (b) and the element depth profiles [(c), analyzed by EDX] of K417/TiN/Ni+CrAlYSiN coating system after oxidation for 100h at 1000℃ in air

4 Discussion

4.1 Influence of diffusion barrier on interdiffusion

As is abundantly clear, the interdiffusion between the coating and the substrate will reduce the service life of the coating. Therefore, many compounds, such as TiN [18], CrN [19,27], Al_2O_3 [20,28], Al-O-N [21,29] or Cr_2O_3 [22], are used as diffusion barriers between MCrAlY coatings and superalloys.

In this study, CrN, AlN and TiN nitride were introduced as the diffusion barrier between the K417 matrix and Ni+CrAlYSiN composite coating. The standard formation Gibbs free energies of each nitride phase at 1000℃ is -198.38kJ/mol for CrN, -378.95kJ/mol for AlN and -416.78kJ/mol for TiN [26]. Therefore, from the perspective of thermodynamics, TiN diffusion barrier is the most stable.

The K417 substrate contains elements Ti, Al and Cr, Ni+CrAlYSiN composite coating contains elements Al and Cr. Therefore, for the K417/CrN/Ni+CrAlYSiN sample, the following reactions occurred near the interface between the substrate and coating:

$$Al+CrN \longrightarrow AlN+Cr \tag{1}$$

$$TTi+AlN \longrightarrow TiN+Al \tag{2}$$

Therefore, during the oxidation process, the CrN diffusion barrier was decomposed and granular and needle TiN was formed in the interdiffusion region.

If only from the perspective of thermodynamics, the TiN diffusion barrier should be more stable than the AlN diffusion barrier. The AlN diffusion barrier will eventually degenerate to TiN. However, after the K417/TiN/Ni+CrAlYSiN sample was oxidized at 1000℃ for 100h, a Cr rich nitride layer was formed near the coating side of the TiN diffusion barrier, and an Al rich nitride layer was formed near the substrate side. Therefore, the thickness of the nitride layer at the coating/substrate interface changes from 500nm as deposited to 1.1μm. In contrast, the AlN diffusion barrier is very stable after oxidation at 1000℃ for 100h.

Research shows that TiN is an ionic compound with FCC rock salt structure [26]. In general, the defect concentration of ionic compounds is relatively high. It can be seen from the Ti-N phase diagram that there are many compounds in TiN, such as Ti_2N, Ti_3N, etc. When the content of Al is less than 57%, the (Ti, Al)N can maintain the FCC rock salt structure. The Ti-Al-N ternary phase diagram [30] shows that there are two ternary phases Ti_3AlN and Ti_2AlN at 1000℃. The crystal structures of TiN phase and CrN phase are completely consistent. In addition, the lattice constants of TiN phase and CrN phase are 0.4242nm and 0.4148nm respectively, which are relatively close. Therefore, the positions of metal atoms in TiN phase and CrN phase can be replaced with each other to form (Ti, Cr)N films in the form of alloys. Therefore, when there are elements such as Al and Cr near the TiN diffusion barrier, the solid soluble Al and Cr in TiN can be transformed into TiCrAlN composite layer.

According to the element content in the matrix and the atomic size, from low to high content followed by Ti, Al and Cr, and from small to large size followed by Al, Cr, and Ti, respectively. Therefore, it is probably that the diffusion rate of Al in the matrix to the substrate and diffusion barrier interface is bigger than that of Ti and Cr, resulting in the formation of Al rich Ti nitride layer near the substrate side of TiN diffusion barrier. The content of Cr in the coating is higher than Al, and AlN is more stable than CrN. Therefore, a Cr rich nitride layer containing Ti was formed near the coating side of the TiN diffusion barrier. Due to the diffusion of beneficial elements Cr and Al from the coating to the interface where TiN is located, the effective content of Cr element in the coating was reduced, resulting in the formation of nodular oxide with large thickness on the surface of the coating and the peeling of the oxide film. Therefore, although TiN effectively prevented the interdiffusion between the coating and the substrate, it deteriorated the oxidation resistance of the composite coating. The structural transformation of TiN diffusion barrier in the oxidation process needs to be further studied. It should be pointed out that because the Ni content in the Ni+CrAlYSiN composite coating is equivalent to that in the matrix, it is impossible to judge whether TiN can prevent the diffusion of nickel in the present work. However, Cheng et al. [31] found that the TiN diffusion barrier could not prevent the Ni diffusion from the NiCrAlY coating into TiAl substrate during the oxidation experiment at 1000℃.

The standard formation Gibbs free energies of AlN and TiN at 1000℃ are −378.95kJ/mol and −416.78kJ/mol [26] respectively, and the driving force of the second reaction formula theoretically is −37.83kJ/mol. However, the AlN diffusion barrier is HCP braze zinc structure with covalent bonding, while TiN diffusion barrier is FCC rock salt structure with ionic bonding. Therefore, the second reaction formula is not a simple displacement reaction, but also involves the transformation of crystal structure, which probably increase the activation energy required for the second reaction. The conversion of hcp−AlN to fcc−AlN can be completed at a high pressure of 12GPa. As described in our previous work [24], there is no non stoichiometric AlNx. hcp−AlN with braze zinc structure is a broadband semiconductor with band gap width of 6.2 eV. hcp−AlN is covalently bonded, and the defect concentration in covalently bonded semiconductors is several orders of magnitude lower than that in ionic crystals [32]. In fact, the point defects in hcp−AlN are almost negligible [33]. Therefore, the AlN diffusion barrier with HCP braze zinc structure can remain stable and effectively inhibit the mutual diffusion between the coating and the substrate.

4.2 Influence of diffusion barrier on growth of oxide scale

In our previous work [24], the results showed that the K417 substrate suffered catastrophic corrosion after oxidation at 1000℃ for 100h, and the Ni + CrAlYSiN composite coating greatly improved the oxidation resistance of K417 substrate. In the present work, the experimental results show that the weight gain of K417/Ni+CrAlYSiN, K417/CrN/Ni+CrAlYSiN, K417/AlN/Ni+CrAlYSiN and K417/TiN/Ni+CrAlYSiN coating systems after oxidation for 100h is 0.45, 0.51, 0.28 and 0.46mg/cm^2 respectively. The decomposition of CrN diffusion barrier has little effect on the oxidation resistance of the Ni+CrAlYSiN composite coating, and the oxidation weight gain increases slightly. The AlN diffusion barrier is very stable in the oxidation process, which effectively inhibits the interdiffusion between the coating and the substrate. A single layer $\alpha-Al_2O_3$ film is formed on the coating surface, and the oxidation weight gain of the coating is reduced by about 50 percent. Therefore, the application of AlN diffusion barrier significantly improves the oxidation resistance of Ni+CrAlYSiN composite coatings. After TiN diffusion barrier is applied, the initial oxidation weight gain of the composite coating is smaller than that of the composite coating with the CrN diffusion barrier, and the oxidation weight gain is the same when it is oxidized to 60h. However, after 80h of oxidation, the composite coating with the TiN diffusion barrier showed a slight weight loss due to the peeling of oxide on the coating surface. The TiN diffusion barrier is relatively stable in the oxidation process, and effectively inhibits the mutual diffusion between the coating and the substrate. However, due to the enrichment of Al and Cr in the coating, the formation of nodular oxide with large thickness on the surface of the coating and the peeling of the oxide film occurred. Therefore, although the TiN diffusion barrier effectively prevented the interdiffusion between the coating and the substrate, it deteriorated the oxidation resistance of the composite coating.

5 Conclusions

CrN, AlN, TiN were prepared as diffusion barriers between the K417 substrate and Ni+CrAlYSiN composite coating via vacuum arc evaporation. Oxidation kinetics, microstructure evaluation and oxidation performance of these coating systems at 1000℃ after 100h were investigated, the following conclusions can be drawn:

(1) The CrN diffusion barrier decomposed, which did not block the diffusion of elements and had little effect on the oxidation resistance of Ni+CrAlYSiN composite coating. Similar to the Ni+CrAlYSiN coating without a diffusion barrier, a double-layer oxide film structure with Al_2O_3 as the inner layer and $Ni(Al, Cr)_2O_4$ as the outer layer formed on the coating.

(2) The AlN diffusion barrier showed good thermodynamic stability, effectively inhibited the interdiffusion between the coating and the substrate, and a single-layer Al_2O_3 film was formed on the coating. The AlN diffusion barrier significantly improved the oxidation resistance of Ni+CrAlYSiN composite coating.

(3) The TiN diffusion barrier effectively prevented the interdiffusion between the coating and the substrate. However, it deteriorated the oxidation resistance of the composite coating. Similar to the Ni+CrAlYSiN coating without diffusion barrier, a double-layer oxide film structure with Al_2O_3 as the inner layer and $Ni(Al, Cr)_2O_4$ as the outer layer formed on the coating.

Acknowledgments

The authors wish to acknowledge the financial support by the National Natural Science Foundation of China (NO. 51804335), the Youth Science and Technology Nova Plan of Shaanxi Province (No. 2020KJXX-064), Major science and technology projects of Inner Mongolia Autonomous Region (No. 2020ZD0019), and the CNPC Science and Technology Development Project (No. 2019E-25-05 and No. 2020B-4020).

References

[1] NICHOLLS J R, SIMMS N J, CHAN W Y, et al. Smart overlay coatings-concept and practice [J]. Surface and Coatings Technology, 2002, 149(2-3): 236-244.

［2］ GOWARD G W. Progress in coatings for gas turbine airfoils ［J］. Surface and Coatings Technology. 1998, 108-109, 73-79.

［3］ PODCHERNYAEVA I A, PANASYUK A D, TEPLENKO M A, et al. Protective Coatings on Heat-resistant Nickel Alloys (Review) ［J］. Powder Metallurgy and Metal Ceramics, 2000, 39(9): 434-444.

［4］ POMEROY M J. Coatings for gas turbine materials and long term stability issues ［J］. Materials & Design, 2005, 26(3): 223-231.

［5］ WANG J L, CHEN M H, YANG L L, et al. Nanocrystalline coatings on superalloys against high temperature oxidation and corrosion: A review ［J］. Corrosion Communications, 2021, 1: 58-69.

［6］ YANG S S, YANG L L, CHEN M H, et al, Understanding of failure mechanisms of the oxide scales formed on nanocrystalline coatings with different Al content during cyclic oxidation ［J］. Acta Materialia, 2021, 205: 116576.

［7］ BRANDL W, TOMA D, GRABKE H J. The characteristics of alumina scales formed on HVOF-sprayed MCrAlY coatings ［J］. Surface and Coatings Technology, 1998, 108: 10-15.

［8］ GODLEWSKI K, GODLEWSKA E. The effect of chromium on the corrosion resistance of aluminide coatings on nickel and nickel-based substrates ［J］. Materials Science and Engineering, 1987, 88: 103-109.

［9］ JOHNSON J B, NICHOLLS J R, HURST R C, et al. The Mechanical Properties of Surface Scales on Nickel-Base Superalloys-II. Contaminant Corrosion ［J］. Corrosion Science, 1978, 18(6): 543-553.

［10］ REN X, WANG F H, WANG X. High-temperature oxidation and hot corrosion behaviors of the NiCr-CrAl coating on a nickel-based superalloy ［J］. Surface and Coatings Technology, 2005, 198: 425-431.

［11］ CHEN G F, LOU H Y. Oxidation behavior of sputtered Ni-3Cr-20Al nanocrystalline coating ［J］. Materials Science and Engineering, 1999, 271(1-2): 360-365.

［12］ REN X, WANG F H. High-temperature oxidation and hot-corrosion behavior of a sputtered NiCrAlY coating with and without aluminizing ［J］. Surface and Coatings Technology, 2006, 201(1-2): 30-37.

［13］ ZHANG K, WANG Q M, SUN C, et al. Preparation and oxidation behavior of NiCrAlYSi coating on a cobalt-base superalloy K40S ［J］. Corrosion Science, 2008, 50(6): 1707-1715.

［14］ KNOTEK O, LUGSCHEIDER E, LÖFFLER F, et al. Diffusion barrier coatings with active bonding, designed for gas turbine blades ［J］. Surface and Coatings Technology, 1994, 68: 22-26.

［15］ LANG F Q, NARITA T. Improvement in oxidation resistance of a Ni_3Al-based superalloy IC6 by rhenium-based diffusion barrier coatings ［J］. Intermetallics, 2007, 15(4): 599-606.

［16］ NICOLET M. Diffusion barrier in thin films ［J］. Thin Solid Films, 1978, 52(3): 415-443.

［17］ MÜLLER J, SCHIERLING M, ZIMMERMANN E, et al. Chemical vapor deposition of smooth $\alpha-Al_2O_3$ films on nickel base superalloys as diffusion barriers ［J］. Surface and Coatings Technology, 1999, 120: 16-21.

［18］ LOU H, WANG F. Effect of Ta, Ti and TiN barriers on diffusion and oxidation kinetics of sputtered CoCrAlY coatings ［J］. Vacuum, 1992, 43(5-7): 757-761.

［19］ LI W Z, WANG Q M, GONG J, et al. Interdiffusion reaction in the CrN interlayer in the NiCrAlY/CrN/DSM11 system during thermal treatment ［J］. Applied Suface Science, 2009, 255(18): 8190-8193.

［20］ WANG Q M, ZHANG K, GONG J, et al. NiCoCrAlY coatings with and without an Al_2O_3/Al interlayer on an orthorhombic Ti_2AlNb-based alloy: Oxidation and interdiffusion behaviors ［J］. Acta Materialia, 2007, 55(4): 1427-1439.

［21］ WANG Q M, WU Y N, GUO M H, et al. Ion-plated Al-O-N and Cr-O-N films on Ni-base superalloys as diffusion barriers ［J］. Surface and Coatings Technology, 2005, 197(1): 68-76.

［22］ CHENG Y X, WANG W, ZHU S L, et al. Arc ion plated-Cr_2O_3 intermediate film as a diffusion barrier between NiCrAlY and γ-TiAl ［J］. Intermetallics, 2010, 18(4): 736-739.

［23］ ZHU L J, ZHU S L, WANG F H. Effects of bias voltage and nitrogen flow rate on the structure and properties of Ni+CrAlYSiN nanocrystalline composite coatings [J]. Chinese Journal of Materials Pesearch, 2013, 27 (1): 53-59.

[24] ZHU L J, ZHU S L, WANG F H. Preparation and oxidation behavior of nanocrystalline Ni+CrAlYSiN composite coating with AlN diffusion barrier on Ni-based superalloy K417 [J]. Corrosion Science, 2012, 60: 265-274.

[25] ZHU L J, ZHU S L, WANG F H. Hot corrosion behavior of a Ni+CrAlYSiN composite coating in Na_2SO_4-25 wt.% NaCl melt [J]. Pipplied Surface Science, 2013, 268: 103-110.

[26] LIANG Y J, CHE M C. Handbook of thermodynamic data for inorganic compounds, 1st ed [M]. Shenyang: Northeast University Press, 1993.

[27] LI W Z, YAO Y, WANG Q M, et al. Improvement of oxidation-resistance of NiCrAlY coatings by application of CrN or CrON interlayer [J]. Journal of Materials Research, 2008, 23(2): 341-352.

[28] MÜLLER J, NEUSCHÜTZ D. Efficiency of α-alumina as diffusion barrier between bond coat and bulk material of gas turbine blades [J]. Vacuum, 2003, 71(1-2): 247-251.

[29] CREMER R, WITTHAUT M, REICHERT K, et al. Thermal stability of Al-O-N PVD diffusion barriers [J]. Surface and Coatings Technology, 1998, 108: 48-58.

[30] ZHANG J, ZHAO Y H. Multi arc ion plating technology and Application [M]. Beijing: Metallurgical industry press, 2007: 90.

[31] CHENG Y X. Investigation on the diffusion barrier of γ-TiAl/NiCrAlY Coating Systems [D]. Shenyong: Groduate School of Chinese Academy of Science. 2009.

[32] MEHRER H. Diffusion in Solids: Fundamentals, Methods, Materials, Diffusion-Controlled Processes [M]. Berlin Heidelberg: Springer Science & Business Media, 2007: 155.

[33] CHEN H Y, STOCK H R, MAYR P. Plasma-assisted nitriding of aluminum [J]. Surface and Coatings Technology, 1994, 64(3). 139-147.

本论文原发表于《Crystals》2021 年第八卷第 11 期。

时效对石油钻杆用 Al-7.51Zn-2.37Mg-1.72Cu 合金力学及耐热性能的影响

冯 春¹ 张芳芳¹ 朱丽娟¹ 刘会群² 刘洪涛³

(1. 中国石油集团石油管工程技术研究院，石油管材及装备材料
服役行为与结构安全国家重点实验室；
2. 中南大学材料科学与工程学院；3. 中国石油塔里木油田公司)

摘　要：针对油气钻采用 Al-7.51Zn-2.37Mg-1.72Cu 铝合金钻杆材料，通过时效硬化行为、力学性能及微观组织表征等研究分析了时效处理对铝合金钻杆材料力学性能及耐热性能的影响。结果表明，时效温度为 120℃ 时，Al-7.51Zn-2.37Mg-1.72Cu 的最佳时效时间为 24h，其抗拉强度(R_m)、屈服强度($R_{0.2}$)和伸长率(A)分别由淬火态的 693MPa、566MPa 和 15%变化到 720MPa、692MPa 和 14%；200℃ 热暴露 500h 后，R_m、$R_{0.2}$ 相比时效态降低了 55.6%、67.3%，A 提高了 15.8%，基体内 GP 区和部分细小 η' 相回溶、尺寸较大稳定性较高的 η 相转变成 η' 相、部分 η 相聚集成粗大质点等是造成合金热暴露性能劣化的主要组织因素。

关键词：铝合金钻杆；时效；热暴露；力学性能；耐热性能

随着石油工业的发展，深井、超深井和水平井的数量不断增加[1]，未来石油勘探将逐步向深井钻井和海上钻井发展，这对油气井用钻杆材料提出了更高的要求。铝合金钻杆材料是超深井、大位移井及高含硫井等复杂井钻井作业的主要结构材料之一，在石油勘探钻井领域具有广泛的应用前景[2-4]。

当前，铝合金钻杆的商业化应用已积累了一定经验，合金体系主要有 Al-Cu-Mg 系、Al-Zn-Mg-Cu 系和 Al-Cu-Mg-Fe-Ni 系[5]，均为可热处理强化合金。冯春等[6]研究了时效时间对三种石油钻杆用 Al-Zn-Mg-Cu 系合金组织和力学性能的影响，结果表明，三种合金经 450℃/2h+470℃/1h 固溶处理及 120℃时效 12h 后，可获得较佳的强度与伸长率匹配；同时，通过自主设计 2xxx 系铝合金钻杆成分及配套热处理工艺结合实验室模拟及实物验证评价等方法系统进行了该合金强韧性、抗 H_2S 应力腐蚀性能、抗疲劳性能、组织演变规律及全尺寸实物承载性能等研究，2013 年由其开发的 460MPa 级铝合金钻杆管体材料在中国石油塔里木油田成功下井应用[7]。Fu Yanjun 等[8]在俄罗斯 1953 系列合金的基础上，通过添加

基金项目：国家重点研发计划"13Cr/110SS 产品服役全过程检测检验与质量管控技术"(2019YFF0217504)；国家科技重大专项-大型油气田及煤层气开发"深井超深井高效快速钻井技术及装备"(2016ZX05020-002)；塔里木油田公司科研项目"高强度铝合金套(筛)管室内试验及适用性评价研究"(T202118.01.01.03)。

作者简介：冯春，男，1980 年生，博士，正高级工程师，中国石油集团石油管工程技术研究院失效分析与智能仿真研究所，陕西 西安 710077，电话：18091805946，E-mail：fengchun003@cnpc.com.cn。

Cu 元素制备了 Al-6.2Zn-2.5Mg-1.6Cu 合金，结果表明，随着固溶温度的升高，抗拉强度 (R_m) 和屈服强度 ($R_{0.2}$) 降低，470℃/1h 固溶处理后，第二相很好地溶解在基体中，强度性能达到最佳。Zhao Juangang[9]设计了一种四阶段时效热处理方法来研究了时效处理对钻杆用 Al-Zn-Mg-Cu 合金热稳定性的影响，并在120℃下进行了 100~500h 的热暴露试验，结果表明，在 120℃下热暴露 500h 后，四阶段时效样品的抗拉强度仅下降 5.05%；自然时效加速 Mg 扩散，促进了在时效阶段 GP 区的沉淀，从而显著提高所研究合金的强度和热稳定性。

作为强度性能最高的 Al-Zn-Mg-Cu 系石油钻杆用铝合金，在深井超深井的钻井过程中，不可避免地需要在高温环境下服役，最高温度可达 200℃以上，其时效热处理工艺对其力学性能和耐热性能影响显著[10-11]。本文针对一种新近研发的铝合金钻杆用 Al-7.51Zn-2.37Mg-1.72Cu 合金，通过研究时效热处理对其力学及耐热性能的影响，以探索其室温和高温下性能演变及断裂行为规律，目的是为该合金钻杆材料在油气井管柱的设计和安全使用等提供实验室数据支撑。

1 实验材料与方法

1.1 实验材料

本实验设计了一种 Al-Zn-Mg-Cu 系合金，Zn 元素的添加有利于提高合金强度，Cu 元素的添加有利于提高合金伸长率。为了精确地测定所熔炼铸造的合金铸锭的实际成分，取少量合金粉末进行等离子光谱(ICP)分析，合金的化学成分见表1。

表1 Al-7.51Zn-2.37Mg-1.72Cu 合金的化学成分　　单位:%(质量分数)

Al	Cu	Mg	Zn	Mn	Fe	Ni	Ti	Si	Cr	Zr
Bal.	1.72	2.37	7.51	0.046	0.058	0.001	0.021	0.052	0.027	0.17

1.2 实验方法

合金的热处理工艺为时效处理，之后再经过热暴露服役试验。具体工艺为：在 120℃分别时效 4h、8h、12h、24h、36h、48h；随后选择分别在时效 4h、24h、48h 后再进行 200℃热暴露 500h 的热暴露试验。

时效合金的硬度值测试在 HV-5 型维式硬度计，其小载荷维式硬度计的加载载荷为 2kg，加载时间为 15s。实验开始前统一对样品进行抛光处理，实验测得的每个数据均测量了 7 个点，去掉最小最大的数值后取 5 个数据点的平均值。

拉伸试验在 CSS-44100 电子万能材料实验机上完成，拉伸试样的尺寸参照国标 GB/T 228.1—2021 制定，如图1所示。实验过程中，试样的拉伸速度为 2mm/min，每次试验取 2 个试样测试结果的算术平均值作为最终测试结果。

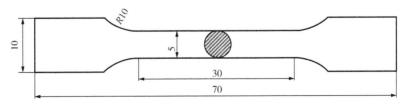

图1 拉伸试样的尺寸规格(mm)

将合金拉伸样的断口采用线切割的方式切割成统一的长度，然后浸泡到高纯乙醇中进行超声波清洗，目的是去除合金的表面和断口处的污渍和杂质，合金断口形貌采用 FEI Sirion

200场发射扫描电镜进行观察分析。

从时效12h合金管体上切割10mm×10mm×1mm的薄片，进行双喷减薄，获得透射电子显微镜样品。电解双喷在电解双喷减薄仪上进行，电压20~25V，保持电流在50~80mA之间，电解液为25%硝酸和75%甲醇溶液，用液氮将电解液温度保持在-30~-20℃。在TECNAIG220透射电镜上观察，加速电压为200kV。

2 实验结果

2.1 时效硬化曲线

图2为Al-7.51Zn-2.37Mg-1.72Cu合金120℃时效温度下不同时效时间的硬化曲线。

从图2中可以看出，铝合金钻杆硬度在时效开始后的4h内出现了明显上升的趋势，合金硬度达到了212HV，在时效4~12h之间缓慢上升，在时效12h时硬度达到了216HV。在12~24h内又缓慢下降到215HV，在时效开始后的24h后开始出现小幅上升趋势。此硬度变化趋势与析出相的析出速度、类型、尺寸和数量有关，依次析出GP区、η'相和η相，其中η'相的强韧性匹配最好[12-14]。

2.2 时效态合金力学性能

图3为Al-7.51Zn-2.37Mg-1.72Cu合金在120℃时效不同时间后的抗拉强度(R_m)、屈服强度($R_{0.2}$)和伸长率(A)等力学性能曲线。由图3可见，随时效时间的延长，在4h内合金的R_m、$R_{0.2}$均迅速升高，并在时效开始12h时分别达到741MPa、708MPa；在时效24h时，其R_m、$R_{0.2}$无明显变化，分别为720MPa、692MPa；在时效24~48h之间缓慢上升。A与R_m、$R_{0.2}$趋势不同，在时效后的4h内呈现下降趋势，在时效时间为4~24h的时间内缓慢上升到14%，但在时效24h后又开始出现下降趋势。合金淬火态的R_m、$R_{0.2}$和A分别693MPa、566MPa和15%，在120℃时效12h，合金具有较好的综合力学性能，其R_m、$R_{0.2}$和A分别为741MPa、708MPa和13%。

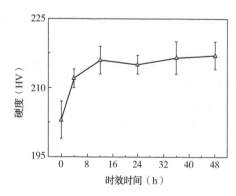

图2 Al-7.51Zn-2.37Mg-1.72Cu合金120℃时效硬化曲线

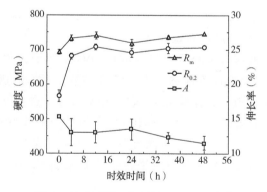

图3 Al-7.51Zn-2.37Mg-1.72Cu合金在不同时效状态下的力学性能曲线

2.3 热暴露后合金力学性能

根据合金时效后材料的综合力学性能，选择在4h、24h、36h的时效处理后在200℃进行500h的热暴露试验。图4为120℃时效不同时间的Al-7.51Zn-2.37Mg-1.72Cu合金在200℃热暴露500h后的力学性能。

由图4可知，合金在120℃时效4h、24h、36h后，经过200℃高温500h热暴露后合金的强度出现了大幅下降，其中120℃时效4h的试样在200℃热暴露500h后，合金的R_m和$R_{0.2}$相比时效态分别下降了54.4%和50.6%；时效24h的试样在200℃热暴露500h后，合金的R_m和$R_{0.2}$相比时效态分别下降了55.7%和67.6%；时效36h的试样在200℃热暴露500h后，合金的R_m和$R_{0.2}$相比时效态分别下降了56.7%和69.2%。而经过200℃高温500h热暴露后合金的A则稍有提升，相比时效4h和36h的合金，经过200℃热暴露500h的A分别增加了23.5%和14.3%，相比时效24h的合金，热暴露后A没有变化。

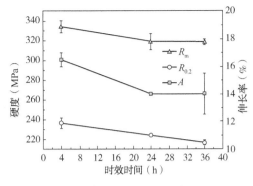

图4 120℃时效4h、24h、36h后Al-7.51Zn-2.37Mg-1.72Cu合金在200℃热暴露500h后的力学性能

2.4 时效态合金透射电镜微观组织

在时效过程中，GP区在时效初期析出，主要受Mg元素的控制。随着时效时间的延长，GP区中Zn元素的含量逐渐升高，Mg、Zn元素的扩散使其过饱和固溶体在时效过程中转变为GP区，再转变为η'相，而η'相的存在是Al-Zn-Mg-Cu系铝合金具有显著的硬化响应效应以及高强度性能的关键。图5为Al-7.51Zn-2.37Mg-1.72Cu合金在120℃下时效12h的TEM照片以及<112>和<100>方向的电子衍射花样。

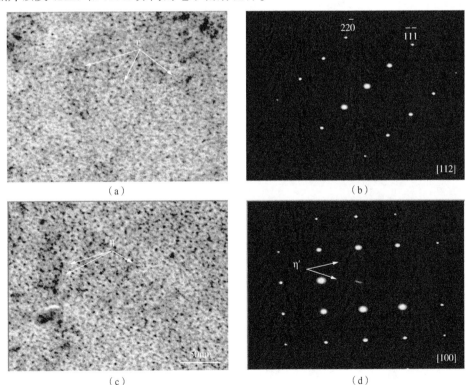

图5 120℃时效12h后Al-7.51Zn-2.37Mg-1.72Cu合金的TEM照片及<112>和<100>方向的选区电子衍射花样

在120℃时效初期，由于温度低，合金晶粒内部主要分布着高密度的GP区和η'相。从图5(a)和(c)中可以看出，时效开始12h时，晶内主要是较为细小的η'析出相，析出相平均尺寸

为 1~2nm，而且析出相密度非常高，此时合金的硬度值为 216HV，R_m 和 $R_{0.2}$ 分别达到 741MPa、708MPa，这与析出相硬化响应效应以及高强度理论相符合。由此可知，随着时效时间的继续延长，到达峰时效状态时，η′析出相的密度达到最大值，此时合金的强度最大，硬度最高。到达过时效状态时，η′相逐渐聚集成大的 η 相，导致强度逐渐减低，且硬度减小[15]。

2.5 合金室温拉伸断口形貌

在拉伸断裂行为中，粒子与基体界面有位错环被推过来，有些类似孔隙状的缺陷出现，随着力的作用裂纹会进行扩展，有的第二相留存，有的第二相脱离基体，此时就会我们在拉伸断口处观察到韧窝，韧窝的分布、形状尺寸、深浅都不同，这个过程不断地进行，当裂纹不断扩展，第二相不断与基体脱离，材料发生了拉伸断裂[16-17]。图 6 是淬火态和 120℃时效不同时间 Al-7.51Zn-2.37Mg-1.72Cu 合金的拉伸断口形貌图。

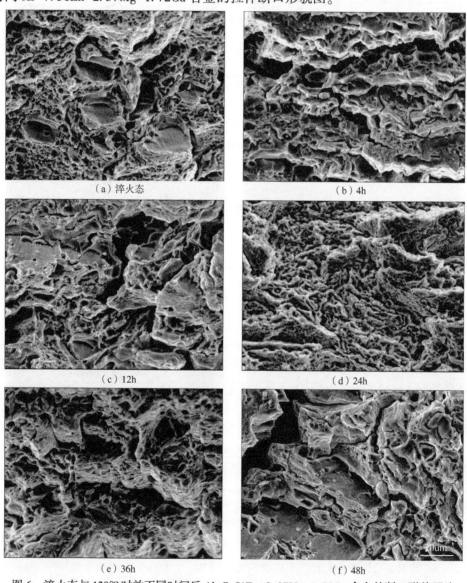

（a）淬火态　（b）4h　（c）12h　（d）24h　（e）36h　（f）48h

图 6　淬火态与 120℃时效不同时间后 Al-7.51Zn-2.37Mg-1.72Cu 合金的断口形貌照片

从图 6 合金在不同时效时间下的断口照片可以发现，淬火状态下，由于断口存在许多大小尺寸均匀且比较深的韧窝，因此其韧性较好，同时断口也存在类似解理面的台阶面，可知合金的断裂方式为穿晶韧断和沿晶断裂的混合断裂；合金在时效开始后的4~12h，合金有明显的脆断特征，分析认为：经过时效处理后，发生时效硬化行为，合金断口处的韧窝数量减少，并且大小不一，出现了较多的解理平面，还存在二次裂纹，导致合金的韧性降低；时效24h时，沿晶裂纹较之前减少且更细，相比时效4h和12h韧窝数量有所增加，韧性提高；合金在时效开始后的36~48h时，随着时效时间的延长，GP区溶解，η'相含量逐渐减少，η相含量增加，并且尺寸增加，相比淬火态A分别下降了19.7%和24.6%，但此时R_m和$R_{0.2}$相比淬火态有所提高，分别增加了5.8%和7.0%，这与图3中的结果相符。此时从图5中时效12h的断口形貌图可看出，此时沿晶裂纹粗大，韧窝分布不均匀且较小，说明局部力学性能并不好。这是由于在时效过程中，位错强化和析出强化同时决定材料强度大小。目前，析出强化和位错强化相互作用主要分为两种方式，一种是位错的相位机制，另一种是位错绕过机制[18-19]。

2.6 200℃热暴露后的合金拉伸断口形貌

图7为Al-7.51Zn-2.37Mg-1.72Cu合金120℃时效不同时间后在200℃热暴露500h的断口照片。

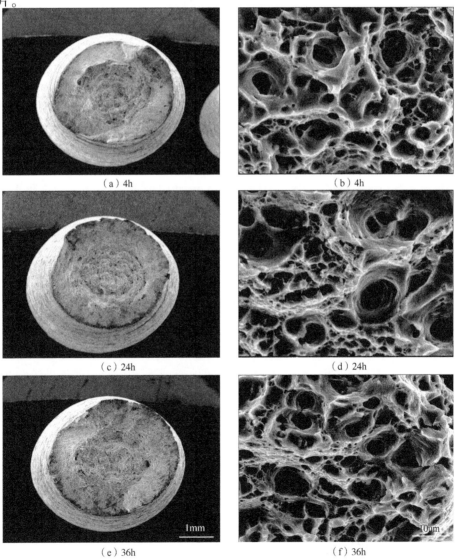

图7 Al-7.51Zn-2.37Mg-1.72Cu合金120℃时效不同时间后在200℃热暴露500h的合金断口形貌

从图 7 可以看出，120℃时效不同时间的合金经 200℃热暴露 500h 后，其宏观断口形貌图均有明显的颈缩现象，且时效 4h 和 36h 的宏观断口比时效 24h 的颈缩更加均匀，这与材料本身有关，组织和第二相分布不均匀等；其微观断口形貌图中所呈现的韧窝分布及大小正好对应了宏观断口现象，时效 24h 时，韧窝直径相差较大，最大约 10μm，最小约 0.5μm，时效 4h 和 36h 的韧窝数量相当，形状规则。此时，对比这 3 种拉伸断口形貌图，可知时效 4h 热暴露 500h 后的韧性较好，这与图 4 中的曲线结果一致。

3 分析和讨论

3.1 时效时间对合金力学性能的影响

7xxx 系合金时效过程中的沉淀析出顺序为[10-26]：α（过饱和固溶体）→ GP 区 → $\eta'(MgZn_2)$ → $\eta(MgZn_2)$。合金的强度在时效的初期阶段有所增加，表明合金具有较强的时效硬化能力。时效过程是合金基体中脱溶相不断析出和长大的过程，脱溶相的析出和长大共同决定了合金的整体强化效果[27-32]。

在时效初期-欠时效阶段（0~4h），铝合金强度迅速增加，伸长率下降。这是因为合金基体中析出尺寸很小且与基体共格的 GP 区，在这一过程中，GP 区的析出速度要远大于其长大速度，位错与 GP 区和合金基体有多种相互作用，当位错穿过基体时，需要更大的牵引力来切断 GP 区和合金基体，同时 GP 区形成时有较大的应变能产生，故合金的硬度在时效初期迅速增加。

在时效中期-峰时效阶段（12~24h），铝合金强度缓慢增加，伸长率维持稳定状态。这是由于随时效时间延长，GP 区逐渐转变成与基体半共格的过渡相 η' 相，η' 相在界面处形成的应变能可以有效地提高基体的强度。此外，峰时效阶段合金的 η' 相密度高于欠时效阶段的密度，尺寸小于过时效阶段基体的 η 相，对基体具有强烈的析出强化作用。且合金峰时效时的硬度值和抗拉强度均高于欠时效时的合金，表明 η' 相和 GP 区的复合强化作用比基体中仅有的 GP 区强。

在时效后期-过时效阶段（36~48h），铝合金强度稳定，伸长率下降。合金的粗 η 相大量分布在基体中，η' 相逐渐消失。随着 GP 区溶解和 η 相尺寸的增大，η 相的析出强化作用逐渐减弱，因此导致合金的时效硬化曲线呈现缓慢上升的趋势。可能随时效时间进一步延长，合金的硬度和强度都会出现下降，这是脱溶相已经转变为平衡相并发生了粗化引起的。

3.2 热暴露对合金力学性能的影响

在长时间热暴露时，GP 区转变为 η' 相的作用与 η 相的粗化作用共存，但最终 η 相的粗化作用大于 GP 区转变为 η' 相的作用。η 相的粗化作用使得位错阻力增大，从而合金的综合性能下降。根据实验结果可知，合金在 120℃时效后，再经过 200℃进行热暴露 500h，其 R_m、$R_{0.2}$ 相比时效态平均降低了 55.6%、67.3%，但 A 平均提高了 15.8%。这主要是因为合金在 200℃热暴露 500h 时，合金基体内的 GP 区和部分细小的且稳定性差的 η' 相开始发生回溶，而部分尺寸较大且稳定性较高的 η 相会转变成 η' 相，同时基体中仍然存在着聚集成粗大质点的 η 相。根据图 5 中热暴露 200℃的断口照片可知，部分韧窝底部具有很深的孔洞，这便是粗化后的第二相 η 相所引起的。因此，在 200℃热暴露 500h 后，这一系列的微观结构变化是导致合金强度下降、伸长率出现小幅上升的主要原因。

4 结论

（1）合金经 120℃时效 24h 后，可获得较佳的强度与伸长率匹配，R_m、$R_{0.2}$ 和 A 分别达

(2) 120℃时效不同时间后再经过 200℃热暴露 500h 的材料力学性能均大幅度变化，其 R_m、$R_{0.2}$ 相比时效态平均降低了 55.6%、67.3%，但 A 平均提高了 15.8%；

(3) 在 200℃热暴露后，合金拉伸断口形貌图中呈现出典型的韧性断裂特征，这与合金韧性提高的结果相一致；

(4) η′相是影响 Al-7.51Zn-2.37Mg-1.72Cu 合金强度和韧性匹配性的主要因素，稳定存在的 η′相对合金的强韧性提升效果明显。

参 考 文 献

[1] HOLDITCH S A, CHIANELI R R. Factore That Will Influence Oil and Gas Supply and Demand in the 21st Century[J]. Mrs Bulletin, 2008, 33(4): 317-323.

[2] 冯春, 寿文彬, 刘会群, 等. 石油钻杆用高强 Al-Zn-Mg-Cu 系铝合金的显微组织和力学性能(英文)[J]. Transactione of Nonferrous Metals Soeiety of China, 2015, 25(11): 3515-3522.

[3] 冯耀荣, 张冠军, 李鹤林. 石油管工程技术进展及展望[J]. 石油管材与仪器, 2017, 3(1): 1-8.

[4] 鄢泰宁, 薛维, 兰凯. 高可靠性铝合金钻杆乃其在超深井和水平井中的应用[J]. 地质科技情报, 2010, 29(01): 112-115.

[5] 王小红, 郭俊, 闫静, 等. 铝合金钻杆材料生产工艺及磨损研究进展[J]. 材料热处理学报, 2013, 34(S1): 1-6.

[6] 冯春, 张冠军, 韩礼红, 等. 热处理对超高强度铝合金钻杆用 Al-Zn-Mg-Cu 系合金力学性能和组织的影响[C]. 第十一次全国热处理大会论文集, 2015: 376-384.

[7] 冯春, 杨尚谕. 铝合金钻杆的特点及发展应用[J]. 石油管材与仪器, 2017, 3(4): 1-7.

[8] Fu Yanjun, Li Xiwu, Wen Kai, et al. Progress in Natural Science: Materials International [J]. 2019, 29(2): 217-223.

[9] WANG J, LIU Z, BAI S, et al. Microstructure evolution and mechanical properties of the electron-beam welded joints of cast Al-Cu-Mg-Ag alloy[JJ. Materials Science and Engineering: A, 2021, 801: 140363.

[10] Liang Jian, Sun Jianhua, Li Xinmiao, et al. Process Engineering [J]. 2014, 73: 84-90.

[11] 王一唱, 曹玲飞, 吴晓东, 等. 石油钻杆用 7xxx 系铝合金微观组织和性能的研究进展[J]. 材料导报, 2019, 33(07): 1190-1197.

[12] 顾伟, 李静媛, 王一德. 晶粒尺寸及 Taylor 因子对过时效态 7050 铝合金挤压型材横向力学性能的影响[J]. 金属学报, 2016, 52(1): 51-59.

[13] 邓运来, 李春明, 张劲, 等. 时效工艺对 Al-Zn-Mg-Cu 合金组织和力学性能的影响[J]. 中国有色金属学报, 2018, 28(9): 1711-1719.

[14] 张勇, 李红萍, 康唯, 等. 高强铝合金时效微结构演变与性能调控[J]. 中国有色金属学报, 2017, 27(7): 1323-1336.

[15] 李志辉, 熊柏青, 张永安, 等. 时效制度对 7B04 高强铝合金力学及腐蚀性能的影响[J]. 稀有金属, 2008, 32(6): 794-798.

[16] 兰滢. Al-Zn-Mg-Cu-Zr-Sc 合金的时效析出行为及力学性能[D]. 沈阳: 沈阳工业大学, 2015.

[17] 李海, 王芝秀, 郑子樵. 时效状态对 7000 系超高强铝合金微观组织和慢应变速率拉伸性能的影响[J]. 稀有金属材料与工程, 2007(9): 1634-1638.

[18] LIU Y, JIANG D, LI B, et al. Heating aging behavior of Al-8.35Zn-2.5Mg-2.25Cu alloy[J]. Materials & Design, 2014, 60: 116-124.

[19] SHERCLIFF H R, ASHBY M F. A process model for age hardening of aluminium alloys—I. The model[J]. Acta Metallurgica Et Materialia, 1990, 38(10): 1789-1802.

[20] MUKHOPADHYAY A K, YANG Q B, SINGH S R. The influence of zirconium on the early stages of aging of a ternary Al Zn Mg alloy[J]. Acta Metallurgica et Materialia, 1994, 42(9): 3083-3091.

[21] DESCHAMPS A, LIVET F, Y BRÉCHET. Influence of predeformation on ageing in an Al-Zn-Mg alloy—I. Microstructure evolution and mechanical properties[J]. Acta Materialia, 1998, 47(1): 281-292.

[22] JIANG X J, TAFTO J, NOBLE B, et al. Differential scanning calorimetry and electron diffraction investigation on low-temperature aging in Al-Zn-Mg alloys [J]. Metallurgical and Materials Transactions A, 2000, 31(2): 339-348.

[23] STILLER K, WARREN P J, HANSEN V, et al. Investigation of precipitation in an Al-Zn-Mg alloy after two-step ageing treatment at 100degree and 150degreeC [J]. MATERIALS SCIENCE AND ENGINEERING-LAUSANNE-A, 1999, A270: 55-63.

[24] GANG S, CEREZO A. Early-stage precipitation in Al-Zn-Mg-Cu alloy (7050)[J]. Acta Materialia, 2004, 52(15): 4503-4516.

[25] 李志辉, 熊柏青, 张永安, 等. 7B04铝合金的时效沉淀析出及强化行为[J]. 中国有色金属学报, 2007(02): 248-253.

[26] 张静, 杨亮, 左汝林. 固溶时效工艺对7055铝合金组织和力学性能的影响[J]. 稀有金属材料与工程, 2015, 44(4): 956-960.

[27] FENG C, LIU Z Y, NING A L, et al. Retrogression and re-aging treatment of Al-9.99%Zn-1.72%Cu-2.5%Mg-0.13%Zr aluminum alloy[J]. 中国有色金属学会会刊: 英文版, 2006, 16(5): 8.

[28] WANG S, LUO B, BAI Z, et al. Revealing the aging time on the precipitation process and stress corrosion properties of 7N01 aluminium alloy[J]. Vacuum, 2020, 176: 109311.

[29] ZANG J X, ZHANG K, DAI S L, et al. Precipitation behavior and properties of a new highstrength Al-Zn-Mg-Cu alloy[J]. Transactions of Nonferrous Metals Society of China, 2012, 22(11): 2638-2644.

[30] 陈旭. Al-Zn-Mg-Cu合金热处理工艺及组织性能研究[D]. 长沙: 中南大学, 2012.

[31] 夏卿坤, 刘志义, 李云涛, 等. 热暴露对欠时效态Al-Cu-Mg-Ag合金拉伸性能的影响[J]. 中国有色金属学报, 2009, 19(5): 808-815.

[32] Ren Jianping, Song Renguo. Rare Metal Materials and Engineering[J]. 2020, 49(04): 1159-1165.

本论文原发表于《稀有金属材料与工程》第49卷第11期。

文23储气库注采管柱接头密封性能指标研究

王建军[1,2]　孙建华[3]　李方坡[1,2]　鲍志强[3]　陈朝晖[3]　郭　浩[3]

(1. 中国石油集团石油管工程技术研究院；2. 石油管材及装备材料服役行为与结构安全国家重点实验室；3. 中石化中原储气库有限责任公司)

摘　要：本文结合文23储气库井注采工况，通过分析计算和注采交变工况下30周次气密封循环全尺寸实物模拟试验等方法，获得一种$\phi88.9mm \times 6.45mm$油管柱实际服役条件下的载荷谱、极限载荷及接头关键性能指标。结果表明，所研究注采管柱应选用拉伸效率100%、接头压缩效率不小于60%的气密封螺纹接头油管。上述关键性能指标对于保障文23储气库高效安全运行具有重要意义，相关研究方法可为其他储气库注采管柱接头的选用提供理论借鉴。

关键词：地下储气库；注采管柱；接头；拉伸；压缩

文23储气库地处中国中原腹地，设计注采井在百口以上，有效工作气量可达$45 \times 10^8 m^3$，以季节调峰和平衡管网压力为主，具有重要的战略地位。目前文3储气库气藏处于枯竭阶段，建成注气后地层压力会迅速提高至0.9，采气后又会逐步降低至0.7左右，而作为注采气关键通道的注采管柱，不仅承受着这种注采压力的交替变化，还承受着注采温度的交替变化，导致注采管柱在拉伸、压缩载荷下循环受载，加上服役年限超过30年，故对储气库注采管柱提出应选用气密封螺纹接头[1-2]。但众多气密封螺纹接头的密封性能存在差异[3-4]，尤其是气密封螺纹接头的抗压缩能力[5]。

笔者结合文23储气库注采管柱结构和服役工况，计算获取了文23储气库注采管柱载荷谱，进一步提出了文23储气库注采管柱气密封螺纹接头的最低技术指标要求，并通过全尺寸模拟试验进行验证。

1　文23储气库注采工况

依据文23储气库设计资料，知注采管柱主要使用$\phi88.9mm \times 6.45mm$油管柱，同时可获取到注采管柱主要结构参数和周围环境参数，详见表1。

基金项目：国家重点研发计划"非API石油专用管质量基础集成与服务共享体系"(2019YFF0217501)与国家科技重大专项课题"复杂结构井、丛式井设计与控制新技术"(2017ZX05009-003)资助。

作者简介：王建军(1979—)，男，山东单县人，教授级高级工程师，2015年毕业于中国石油大学(华东)安全技术及工程专业，博士研究生(工学博士)，现就职于中国石油集团石油管工程技术研究院，从事油气开发工程与管柱力学研究，发表论文近40篇，合作出版专著4部，多次获省部级科技奖励。地址：陕西西安市锦业二路89号(710077)，电话：029-81887677，E-mail：wg_j_jun@163.com。

表 1 文 23 储气库注采管柱结构和服役环境参数

油管外径（mm）	油管壁厚（mm）	油管钢级	油管下深（m）	封隔器坐封深度（m）	封隔器最小内径（mm）	封隔器最大外径（mm）	安全阀最大外径（mm）	安全阀最小内径（mm）
88.9	6.45	N80Q	2900	2850	76.20	149.23	99.99	71.45
注入气体相对密度	环空保护液密度（g/cm³）	环境温度（℃）	井口温度（℃）	地层温度（℃）	上限运行压力（MPa）	下限运行压力（MPa）	单井日注气（10⁴ m³/d）	单井日采气（10⁴ m³/d）
0.6	1.02	15	30	120	38.6	19	80	90

2 注采管柱载荷分析

2.1 注采管柱载荷计算模型

文 23 储气库注采管柱主要承受内压力、外压力和轴向力。内压力主要为注气压力和采气压力交变，外压力主要来自油套环空保护液静液柱压力，注采管柱轴向力主要来自油管下入、坐封等以及因温度效应、活塞效应、膨胀效应、屈曲效应而产生的轴向力[6-7]，另外在注气、采气过程中因气体流动产生的摩阻作用在管体内壁带来的轴向力。

因注采管柱接头性能标准参数主要为拉伸效率和压缩效率[8]，其选用是否安全与承受的轴向载荷密切相关。考虑到文 23 储气库注采管柱采用永久式封隔器，根据 SY/T 7370 标准[9]，可知注采管柱所受到的有效轴向力计算公式如下：

$$F_e = F_1 + F_2 + F_3 + F_4 + F_5 \tag{1}$$

当注采气过程中，坐封后管柱因应力松弛使原始轴向力丧失，即 $F_1 = 0$，此时有效轴向力为：

$$F_e = F_2 + F_3 + F_4 + F_5 \tag{2}$$

当注采气过程中，考虑最大压力降 $\Delta p_f = p_{smax} - p_{smin}$（纯压降），得最大摩阻力：

$$F'_5 = \pm \frac{\pi}{4} d^2 \Delta p_f = \pm \frac{\pi}{4} d^2 (p_{smax} - p_{smin}) \tag{3}$$

则，有效轴向力为：

$$F_e = F_1 + F_2 + F_3 + F_4 + F'_5 \tag{4}$$

当注采气过程中，$F_1 = 0$ 且考虑最大摩阻力 F'_5，此时有效轴向力为：

$$F_e = F_2 + F_3 + F_4 + F'_5 \tag{5}$$

以上各式中，F_e 为注采作业中有效轴向力，N；F_1 为注采作业中因管柱下入、坐封等产生的轴向力，N；F_2 为注采作业中因温度效应产生的轴向力，N；F_3 为注采作业中因鼓胀效应产生的轴向力，N；F_4 为注采作业中因活塞效应产生的轴向力，N；F_5 为注采作业中因摩阻效应产生的轴向力，N；F'_5 为注采作业中考虑纯压降产生的轴向力，N；Δp_f 为管柱内摩擦压力降，MPa；p_{smax} 为储气库上限运行压力，MPa；p_{smin} 为储气库下限运行压力，MPa；d 为油管名义内径，mm。

2.2 注采管柱载荷谱分析

按照表 1 中各项参数，根据 SY/T 7370 标准要求分三种注采工况（注气初始、采气运行、注气运行)。利用以上公式,可计算获得文23储气库注采管柱轴向载荷变化情况。因封隔器以下管柱自由(约束改变),以封隔器为界限,管柱受力方向出现拐点,但最大载荷均出现在封隔器以上管柱段,因此重点分析封隔器以上管柱受力情况。按照管柱有效轴向力[式(1)]、考虑应力松弛时有效轴向力[式(2)]、考虑纯压降时有效轴向力[式(4)]、考虑纯压降且应力松弛时有效轴向力[式(5)]等四种情况计算全井段注采管柱载荷变化,结果如图1所示。

从图1中发现,在注气过程中,管柱拉伸载荷随井深增加而增大;在采气过程中,管柱压缩载荷随井深增加而减小。若要获得管柱最大载荷变化,则需注气时不考虑应力松弛[式(1)],采气时考虑应力松弛[式(2)],同时考虑纯压降[注气为式(4),采气为式(5)]可获得管柱极限载荷变化。即各种情况下管柱承受的载荷,属考虑纯压降时的载荷最大。

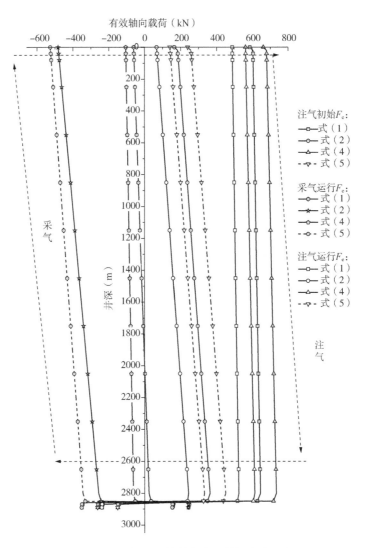

图1 注采管柱有效轴向力随井深变化规律

现把图1中各种注采工况下的最大载荷提取之后列入表2中,发现注采管柱接头最低要承受645.29kN拉伸载荷和534.05kN压缩载荷的能力;进一步把表2中的拉伸、压缩最大

载荷值与 $\phi 88.9mm\times 6.45mm$ N80Q 油管额定抗拉强度 921kN[10]相比较,可以获得实际载荷比(表2),说明注采管柱接头的拉伸效率不能低于 70.06%、压缩效率不能低于 57.92%。

结合 SY/T 7370 中要求接头抗拉伸安全系数 1.30、抗压缩安全系数 1.00,同时考虑到目前市场上主要气密封螺纹接头的拉伸效率均达到 100%,最终提出文 23 储气库注采管柱接头选用技术指标值为接头拉伸效率 100%、接头压缩效率不小于 60%。

表 2 注采作业过程管柱最大载荷变化

注采工况	有效轴向力计算类型	最大载荷(kN)		载荷比(%)	
		注气	采气	注气	采气
注气初始	有效轴向力[式(1)]	645.29		70.06	
	考虑应力松弛[式(2)]	363.25		39.44	
	考虑纯压降[式(4)]	615.32		66.70	
	考虑纯压降且应力松弛[式(5)]	451.67		49.04	
采气运行	有效轴向力[式(1)]		-61.24		-6.65
	考虑应力松弛[式(2)]		-482.85		-52.43
	考虑纯压降[式(4)]		-105.45		-11.45
	考虑纯压降且应力松弛[式(5)]		-534.05		-57.92
注气运行	有效轴向力[式(1)]	525.73		57.08	
	考虑应力松弛[式(2)]	243.69		26.46	
	考虑纯压降[式(4)]	614.15		66.68	
	考虑纯压降且应力松弛[式(5)]	332.11		36.06	

注:表中负号表示压缩。

3 试验验证

为了检验上述接头选用技术指标值的有效性,同时也为了保障文 23 储气库注采管柱安全,选用市场上已有的 $\phi 88.9mm\times 6.45mm$ N80Q T 型气密封螺纹油管进行全尺寸实物模拟试验评价,该油管接头拉伸效率为 100%、压缩效率为 60%。

若每年 1 个注采周期,则每年管柱载荷按照拉伸/压缩载荷交变 1 次(如图 1 中虚线所示,顺时针方向),30 年则有 30 次这种载荷交变。故在 ISO 13679 标准[9]基础上改进试验程序,分析拉压交变载荷下 $\phi 88.9mm\times 6.45mm$ N80Q T 型油管 30 周次气密封螺纹套管密封性能。

结合表 2 载荷值并考虑一定安全系数,试验过程中拉伸载荷施加至 782kN(85%额定抗拉强度),压缩载荷施加至 552kN(60%额定抗拉强度),内压施加至 50MPa。即该 T 型气密封螺纹油管在 50MPa 内压下经过 782kN 拉伸→552kN 压缩→782kN 拉伸→552kN 压缩→782kN 拉伸……如此往返试验加载 30 次循环后,管柱未发生泄漏,试验结果如图 2 所示。

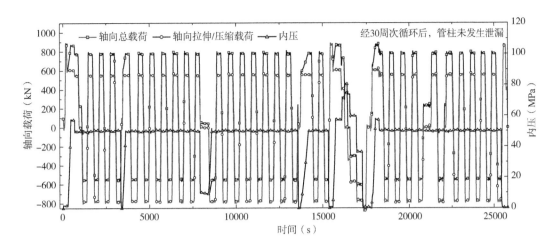

图 2 φ88.9mm×6.45mm N80Q T型油管多周次气密封循环试验结果

试验结果说明 2.2 节给出的文 23 储气库注采管柱接头技术指标值是合理可行的，而且试验载荷高于注采运行工况载荷，具有一定的安全余量。

4 结论

（1）文 23 储气库可按 SY/T 7370 标准载荷计算方法，获取注采管柱承受的最大拉伸载荷和最大压缩载荷，并据此选择相应性能的气密封螺纹接头。

（2）依据文 23 储气库注采管柱载荷变化，提出其注采管柱接头选用技术指标值应为接头拉伸效率 100%、接头压缩效率不小于 60%。

（3）对选用的气密封螺纹油管经 30 周次气密封循环试验证实文 23 储气库注采管柱接头选用技术指标的合理有效，且具有一定的安全余量。

（4）研究结果为文 23 储气库注采管柱设计提供了依据，也为其他储气库注采管柱接头选用提供了理论借鉴。

参 考 文 献

[1] 国家能源局. SY 6805—2017 油气藏型地下储气库安全技术规程[S]. 北京：石油工业出版社，2011：5-6.
[2] 国家能源局. SY/T 6848—2012 地下储气库设计规范[S]. 北京：石油工业出版社，2012：16-17.
[3] 朱强，杜鹏，王建军，等. 特殊螺纹套管接头柱面/球面密封结构有限元分析[J]. 郑州大学学报（工学版），2016，37(5)：82-85.
[4] 窦益华，于洋，曹银萍，等. 动载作用下特殊螺纹油管接头密封性对比分析[J]. 石油机械，2014，42(2)：63-65.
[5] 王建军，孙建华，高进伟，等. 拉压交变载荷下螺纹接头密封面接触压力分析[J]. 石油机械，2018，46(12)：111-116.
[6] 韩志勇. 液压环境下的油井管柱力学[M]. 北京：石油工业出版社，2011.
[7] 《海上油气田完井手册》编委会. 海上油气田完井手册[M]. 北京：石油工业出版社，1998.
[8] 王建军，孙建华，薛承文，等. 地下储气库注采管柱气密封螺纹接头优选[J]. 天然气工业，2017，37(5)：76-80.

[9] 国家能源局. SY/T 7370 地下储气库注采管柱选用与设计推荐做法[S]. 北京：石油工业出版社，2017：10-20.

[10] International Organization for Standardization. Petroleum and natural gas industries—Equations and calculations for the properties of casing, tubing, drill pipe and line pipe used as casing or tubing：ISO/TR10400-2018[S]. Geneva, Switzerland：International Organization for Standardization, 2018.

[11] International Organization for Standardization. Petroleum and natural gas industries — Procedures for testing casing and tubing connections：ISO13679—2002[S]. Geneva, Switzerland：International Organization for Standardization, 2002.

本论文原发表于《石油管材与仪器》2021 年第 7 卷第 1 期。

油套管用特殊螺纹连接密封完整性探讨

王建东[1]　李玉飞[2]　汪传磊[2]　陈禹含[2]　朱达江[2]　马　力[3]

(1. 中国石油集团石油管工程技术研究院；2. 中国石油西南油气田分公司；
3. 林州凤宝管业有限公司)

摘　要：开展了高温和室温环境下 A 系全包络线载荷试验评价和基于试验的密封性有限元模拟，依据气密封能判据分析了不同等效全包络线载荷 2 次循环的密封能和密封能倍数变化规律及危险载荷点，并对危险载荷点开展了应力松弛时间效应分析。研究结果表明：模拟工况载荷的试验评价是一种不考虑现场操作和井下载荷波动不确定性以及安全使用余量的方法；API RP 5C5 试验评价是一种依据材料强度性能计算出载荷的方法，未充分考虑材料高温应力松弛时间效应的影响。建议增加基于时间效应的试验评价和有限元密封性分析。

关键词：油套管；特殊螺纹；密封性能；应力松弛

油套管标准接箍式气密封特殊螺纹接头具有很好的抗拉伸/内压/外压性能，以及高抗压缩性能(压缩效率 60%~100%)，广泛应用于高温高压气井、高内压多段体积压裂非常规页岩气/油开发以及储气库注采管柱，ISO 13679《石油天然气工业用套管及油管螺纹连接试验程序》(即 API RP 5C5《套管和油管接头评价程序推荐作法》)是检验其性能是否优异的主要依据[1-7]。ISO 13679 标准由国际石油公司共同起草，是在充分吸取油田生产过程中螺纹连接失效教训的基础上，结合不断发展的钻井和完井工艺需要制定的，是国内外石油公司检验复杂苛刻环境工况用油套管螺纹连接性能的首要标准。但是通过 ISO 13679 标准验证评价的螺纹连接，在高温高压气井工况使用时仍然有气密封失效泄漏发生。本文通过分析 ISO 气密封特殊螺纹检验标准，提出对高温高压气井用气密封特殊螺纹还需要进一步开展基于高温材料应力松弛的密封时间效应验证和评价分析，以进一步提高管柱的密封完整性。

1　研究方法

开展螺纹接头的 API RP 5C5(2017 版)标准 CAL Ⅳ 四级密封完整性试验，采用有限元计算方法分析螺纹密封性能变化规律。鉴于材料在高温下会发生应力松弛，分析螺纹危

基金项目：国家科技重大专项"四川盆地大型碳酸盐岩气田开发示范工程"(2016ZX05052)，陕西省科技统筹创新工程计划项目"水平井用高性能气密封螺纹接头"(2015KTTSGY06-02-02)，中国石油天然气集团公司科学研究与技术开发项目"自动化与高效钻完井新装备新工具研制"(2019B-4013)，中国石油西南油气田分公司项目"川西超深大斜度水平井完井技术研究"(20200302-12)。

作者简介：王建东(1972—)，男，高级工程师，主要从事油井管与管柱力学研究。

险载荷点的密封性能随时间变化的规律，确定螺纹的适用范围，并提出补充试验评价方法。

目前主要有两种方法来评价气密封特殊螺纹的性能。一是试验评价方法，依据API RP 5C5（2017版）规定的密封准则气体泄漏量来检验螺纹泄漏量，若15min的气体泄漏量≤0.9cm³，则判定其密封性合格。虽然实物试验具有检测全面、真实有效的特点，但该试验只能观察接头是否发生泄漏，不能了解其微观受力。二是有限元计算分析方法能了解螺纹接头的微观受力，获得沿密封面轴向的接触压力和接触长度的分布规律，依据气密封性能判据分析螺纹在载荷状态下的密封性[8-12]。

目前，ISO/TR 10400：2018《石油和天然气工业套管、油管、钻杆和管线管性能公式与计算》是判断螺纹气密封性能的主要标准，该标准提出螺纹接头的最大接触压力应大于内压，但没有对密封准则泄漏速率作要求；文献[13]基于小试样试验，提出接触压力占主导地位的气密封性能判据，试验研究中没有考虑密封面直径变化对密封性能的影响，不适用于油套管螺纹连接规格多样化要求。

文献[14]通过实物试验形成了考虑密封直径D、密封接触压力σ、泄漏速率Q和表面涂层、粗糙度多因素修正系数的气密封判据，有限元分析密封能S_C见式（1），密封有效内/外压所需的密封能S_D见式（2），密封能倍数S_C/S_D见式（3），密封能倍数$b \geq 1$表明接头具有密封性。

$$S_C = \int_l \sigma^n(l)\,\mathrm{d}l \tag{1}$$

$$S_D = AD^K Q^m p \tag{2}$$

$$b = S_C/S_D \tag{3}$$

式中：l为密封接触长度，mm；n为密封能加权指数，取1.95；A为密封常数；K为表面处理影响系数，取0.8；m为表面粗糙度影响系数，取-0.033；p为密封有效压力，MPa。

1.1 API评价标准分析

气密封特殊螺纹的密封完整性评价一直都是国际石油界关注的热点和难点。中国石油塔里木油田公司高温高压气井用油套管的气密封失效分析表明，现用特殊螺纹接头油套管通过了API RP 5C5（1996版）标准CAL Ⅱ级+TGRC1+TGRC2试验评价（SY/T 6128—1995《油套管螺纹连接性能评价方法》中的TGRC1和TGRC2试验分别是85%VME内压恒定试验和外压挤毁试验，VME是指冯·米塞斯等效应力），但油田现场的多口井油管柱发生泄漏，套压升高。有些井刚开始投产时，套管的压力就出现异常上升情况。塔里木油田的DN2-8井，第一次下的完井管柱发生泄漏，套压高达78.7MPa。失效分析发现，DN2-8井所用的多根油管在接头现场端和工厂端发生泄漏，说明该特殊螺纹接头油管不适用于塔里木油田的工况。对DN2-8井所用油管进行模拟井况实物试验，该特殊螺纹接头油管全部通过了试验。因此，最终投产后的油套管柱是否泄漏是检验评价试验方案是否科学的唯一标准[15]。这说明基于工况载荷的模拟试验评价是一种不考虑现场操作和井下载荷波动不确定性以及安全使用余量的试验评价方法。API RP 5C5(1996版)针对气密封特殊螺纹接头的密封完整性检测试验评价只有封堵管端内压和热循环试验、室温下拉伸压缩+内压气密封试验，不符合井下油管柱设计三轴全包络线校核。

国际石油公司与全球知名油套管生产制造企业共同成立了API WG2工作组，制定了气密封螺纹连接检测评价方法。2017年美国石油学会发布了API RP 5C5(2017版)，采用实测材料屈服强度和几何尺寸计算的等管体性能极限包络线承载能力检验螺纹完整性，为管柱设计系数安全余量的确定提供了充分依据。该试验评价方法已成为各石油公司选择螺纹产品的首要标准。API全尺寸实物试验评价标准发展见表1，通过对比分析API全尺寸实物试验评价标准规定的最高评价等级可以反映检测的苛刻程度。

表1 API全尺寸实物试验评价标准发展

标准版本	发布年份	试验目的	试验原理	目标螺纹	试样数量(根)	试样选取
第1、2版	1990、1996	结构完整性①	单轴完整性验证	API螺纹及低压密封螺纹	27	极限公差配合6种
第3版	2002	结构和密封完整性②	单轴和三轴完整性验证	气密封特殊螺纹	8	极限公差配合4种
第4版	2017	结构和密封完整性，强化生产过程验证③	单轴和三轴完整性验证，温度载荷交变完整性验证	气密封特殊螺纹	5(4根气密封试验)	极限公差配合3种

标准版本	有效试样	高温气密封检验		室温气密封检验	极限失效试样(根)	单根试验时间(h)
第1、2版	每种1根	内压恒定的温度热循环试验(循环10次)		内压恒定密封试验④	18(每种失效3个试样)	33
第3版	每种2根	内压+拉伸载荷恒定的温度热循环试验(套管循环10次、油管循环100次)		包络线载荷试验(循环6次)	8(每种失效1个试样)	50
第4版	最差密封试样2根，其余各1根	高温恒定的载荷循环试验(4次包络线循环)，载荷和温度同时交变试验(5次)，内压+拉伸载荷恒定的温度热循环(10次)		包络线载荷试验(循环11次)	5(每种失效1个试样)	238

① 结构完整性是指螺纹在单轴载荷条件下不发生断裂、滑脱(拉伸)、结构失稳(压缩)、压溃(外压)、爆裂(内压)，如图1(a)所示；

② 密封完整性是指螺纹在复合载荷(轴向载荷拉伸/压缩+内压+弯曲)构成包络线循环下不发生密封失效泄漏，如图1(b)所示；

③ 高温环境生产过程载荷交变如图1(c)所示13c~22e压裂采气和10e~27e采气载荷波动的生产过程验证；

④ 螺纹在内压恒定、载荷变化时的密封能力，如图1(d)所示。

由表1可知：(1)螺纹完整性试验评价是伴随油气田生产勘探开发的需要而发展起来的，2000年以前世界石油勘探开发主要以油为主且多为低压浅井，主要采用API标准螺纹油套管；2000年以后随着深井、超深井及水平井等日益复杂苛刻工况和钻井新工艺的发展，需要更高性能的螺纹接头；（2）螺纹性能评价指标从单一的结构完整性向密封完整性发展，也为井下管柱三轴设计校核提供依据；（3）试验评价方法从单一的满足低压恒载荷密封验证向等管体密封性验证以及满足井下生产、高温环境下螺纹密封完整性验证评价发展。

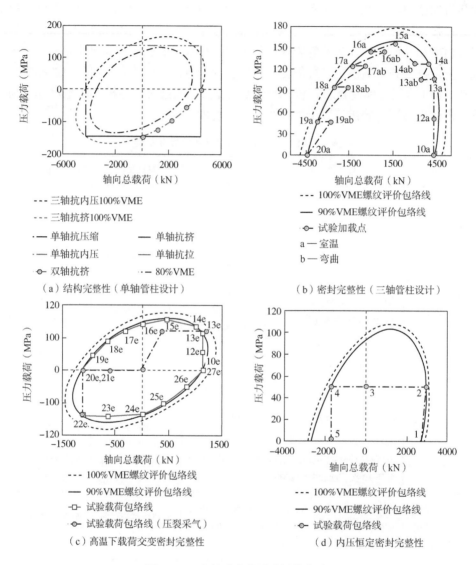

图1　API全尺寸实物试验评价方法

1.2　密封完整性的试验研究

依据API RP 5C5（2017版），评价ϕ88.9mm×6.45mm 110SS油管的标准接箍式气密封特殊螺纹。选取极限公差配合的1号试样来检验螺纹密封面的密封完整性，对1号试样进行A系包络线试验，1号试样的螺纹高过盈、密封低过盈，外螺纹台肩端面刻槽（破坏台肩密封作用）。螺纹A端1次上扣，B端10次上扣9次卸扣，未发生螺纹粘结现象。A系包络线试验载荷谱见表2。螺纹密封完整性试验评价结果如图1（c）、图2和图3所示。由试验结果可知，1号试样经A系包络线载荷温度循环未发生螺纹泄漏和结构失效。

表2 A系包络线试验载荷谱

项目	环境温度(℃)	循环次数	载荷(MPa)	载荷谱
高温载荷循环	180	2	90%VME	图1(c)
载荷温度循环	23,180	5	90%VME	图1(c)
室温载荷循环	23	2	90%VME	图2
室温载荷循环	23	2	90%VME	图3

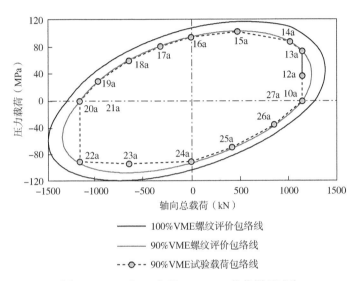

图2 CAL四级A室温90%VME载荷循环试验

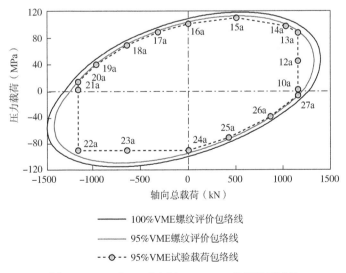

图3 CAL四级A系室温95%VME载荷循环试验

1.3 密封完整性的有限元分析

依据1号试样实测螺纹参数和管体几何尺寸以上扣扭矩对圈数曲线，进行有限元建模，采用轴对称模型四边形轴对称单元CAXA4，轴对称模型及载荷边界条件如图4所示。

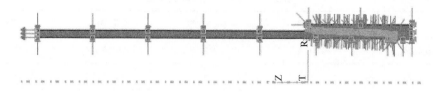

图4 轴对称模型及载荷边界条件

决定密封性能的主要因素是密封接触压力和密封接触长度,通过密封判据分析可知螺纹在不同载荷状态下的密封性。1号试样在A系包络线载荷循环时的密封能如图5至图8所示,在A系包络线载荷2次循环的密封能降低量如图9和图10所示。图6所示,载荷点13c的密封能 S_D 为 $67m \cdot MPa^{1.95}$,载荷点22e的密封能 S_D 为 $75m \cdot MPa^{1.95}$,数值接近。

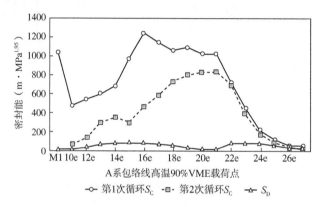

图5 1号试样在高温A系90%VME载荷循环时的密封能

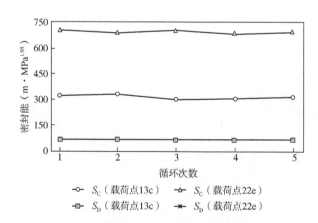

图6 1试样在A系13c~22e载荷温度循环时的密封能

(1)结合图1(c)、图5和图9可知,在高温环境90%VME载荷包络线第2次循环时,"拉伸+内压"载荷点(12e~15e)密封能显著降低,最大拉伸载荷点12e的密封能降低量最大(为75%),密封能倍数为4.3;随着内压的增加,载荷点13e的密封能降低量减少到50.3%,密封能倍数为4.8,相同拉伸载荷下,内压增大,外螺纹密封胀大提高了接触压力,表明在"高拉伸+低内压"工况下易发生泄漏,需要控制最大拉伸载荷。载荷点14e的密封能降低量为48%,密封能倍数4.8;最大内压载荷点15e的密封能降低量为70%,且密封能倍数最小(仅为3.4)。这表明高内压下降低拉伸载荷对密封没有显著影响,需要控制最大内压载荷。

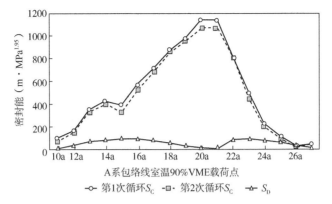

图7　1号试样在室温A系90%VME载荷循环时的密封能

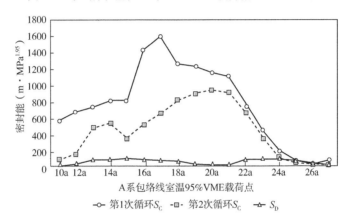

图8　1号试样在室温A系95%VME载荷循环时的密封能

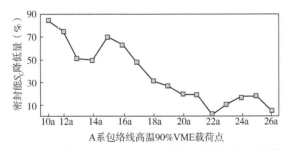

图9　1号试样在高温A系90%VME载荷2次循环的密封能降低量

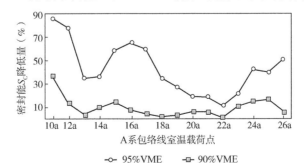

图10　1号试样在室温A系包络线载荷2次循环的密封能降低量

(2) 结合图2、图7和图10可知,在室温环境90%VME载荷包络线第2次循环时,密封能降低量均小于16.4%。"拉伸+内压"载荷点15a的密封能降低量最大(为14.1%),密封能倍数为3.5;"拉伸+外压"载荷点25a的密封能降低量最大(为16.4%),密封能倍数为1.38;载荷点26a的密封能倍数为0.66,存在外压泄漏的可能性,需要控制拉伸载荷使用范围;载荷点24a的密封能倍数为2.4。由此可知,1号试样在轴向压缩状态具有更好的外压密封性。

(3) 结合图3、图8和图10可知,室温环境95%VME载荷包络线第2次循环时,"拉伸+内压"载荷点12a的密封能降低量最大(为77.5%),密封能倍数为3.8;"拉伸+外压"载荷点25a和26a的密封能降低量分别为40%和51%,密封能倍数分别是0.60和0.11,存在外压泄漏的可能。外压试验时(介质液压油)未发生泄漏,其主要原因是采用"气密封泄漏量"作为密封判据时需要更大的密封能S_c。因此,通过密封能分析可以判定载荷点25a和26a是危险载荷点。

(4) 从图6可以看出,载荷点13c(室温)~22e(高温)的5次循环密封能无变化,载荷点13c密封能倍数为4.5,载荷点22e的密封能倍数为9.2。由此可知,气密封螺纹连接在室温90%VME"拉伸+内压"和高温90%VME"压缩+外压"工况下具有优异的密封性。

有限元分析结果表明,通过标准密封性试验评价的气密封特殊螺纹,在环境温度和包络线载荷循环状态下,其密封能显著降低,存在危险载荷点。因此,需要进一步研究危险载荷点密封性能随时间的衰减规律,为油套管螺纹在井下的安全使用提供依据。

1.4 螺纹在高温时的密封持久性分析

金属材料在承载状态下的总应变保持不变,但应力会随时间的延长而逐渐降低,这种现象叫应力松弛。材料在高温条件下会出现明显的应力松弛现象。随着时间的延长,一部分弹性变形转变为塑性变形,即弹性应变不断减小,所以材料中的应力相应地降低。蠕变与松弛在本质上差别不大,可以把松弛现象看作是应力不断降低时的多级蠕变,但用蠕变数据来估算松弛数据还是很困难的。因为应力松弛表达了材料在不同初始应力状态和一定温度下经规定时间后的剩余应力。蠕变是在恒定应力状态下材料的塑性应变不断增加直至发生断裂失效,不能反映应力的降低程度。因此,用蠕变分析密封接触压力的变化是不科学的。

对于采用金属对金属的气密封特殊螺纹,决定密封的主要因素是密封接触压力和密封接触长度。因此,材料在高温条件下经应力松弛后的剩余接触压力是决定密封持久性的关键因素之一。笔者对110SS油管材料进行了高温应力松弛试验,基于螺纹在载荷状态下密封面的平均塑性应变,施加初始应力进行高温下材料应力松弛试验。经高温180℃环境包络线载荷2次循环有限元分析可知:90%VME最大内压载荷点15e的密封能倍数最小,该载荷点是危险泄漏点。1号试样在上扣和15e载荷点的等效塑性应变如图11和图12所示,平均塑性应变为2.26%。密封面的平均塑性应变是影响密封性的重要因素;因此,对110SS油管材料进行了180℃高温环境下的应力松弛试验,试验结果如图13所示,应力松弛拟合方程见式(4)。

$$\sigma(t) = \sigma_s + \left[(x-1)\frac{yE}{\sigma_s x}t + (\sigma_0 - \sigma_s)^{1-x} \right]^{1/1-x} \quad (4)$$

110SS油管材料的应力松弛拟合方程是根据应力松弛试验结果和Cowper-Symonds法则[16]拟合出来的。其中,σ_s为稳态强度,MPa;x和y为应力松弛常数;E为弹性模量,MPa;t为保持时间,min;σ_0为初始应力,MPa。由式(4)可确定110SS油管材料的应力松弛参数，具体见表3。

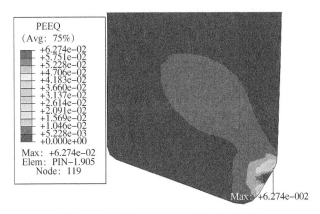

图11 1号试样在上扣的等效塑性应变

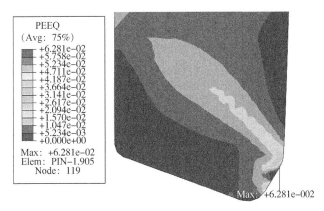

图12 1号试样在15e载荷点的等效塑性应变

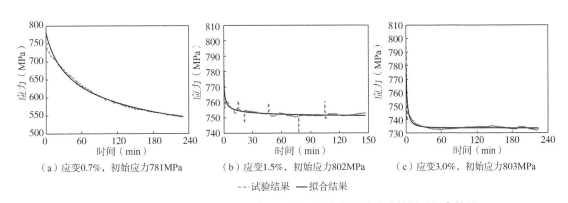

(a) 应变0.7%，初始应力781MPa (b) 应变1.5%，初始应力802MPa (c) 应变3.0%，初始应力803MPa

--- 试验结果　── 拟合结果

图13 110SS油管材料在180℃高温环境下的应力松弛试验结果及拟合结果

表3 110SS油管材料的应力松弛参数

保持时间(min)	应变(%)	σ_0(MPa)	y	x	σ_s(MPa)
230	0.7	781	5.92×10^{-6}	1.48852	545
150	1.5	802	4.93×10^{-6}	3.49797	749
210	3.0	803	3.06×10^{-4}	1.62476	733

由图13可知，110SS油管的初始塑性应变越大，初始阶段速率变化越显著，初始塑性应变3.0%时，试验30min后的初始应力从803MPa降到734MPa，下降速率为2.3MPa/min；初始塑性应变为0.7%时，试验230min后的初始应力从781MPa降到545MPa，下降速率为1.03MPa/min，后期应力松弛趋于缓慢。因此，对于气密封特殊螺纹密封设计，应控制密封面在载荷状态下的塑性应变小于材料高温屈服应变，使材料具有更大的应力松弛抗力和更低的松弛速率，提高密封稳定性。

基于110SS油管材料高温应力松弛试验结果拟合的应力松弛方程，采用有限元方法分析气密封螺纹在高温180℃载荷点15e的密封性随时间变化规律，结果如图14所示，密封性分析见表4。分析可知，密封能在保载初期10h变化最大（下降了16.0%），保载1个月后的降低量为22.0%，保载6个月后的降低量为22.4%，降低量随保载时间的增加而缓慢变化；保载6个月后的气密封螺纹密封能倍数为2.62，说明螺纹仍然具有密封性。如果管柱三轴设计系数提高30%，可以抵消高温环境长期载荷降低的密封性。气密封螺纹按照API RP 5C5（2017版）标准进行四级评价的保载时长238h是指在室温和高温环境B、C、A三个系列试验中保载时间总和，载荷点最长保载时间1h。经过上述分析可知，API标准密封性试验评价是基于材料强度验证螺纹密封性能的评价方法，没有充分考虑高温应力松弛时间效应的影响。为确保油套管螺纹在高温高压气井的长期安全使用，应增加基于时间效应的试验评价和有限元密封性分析。

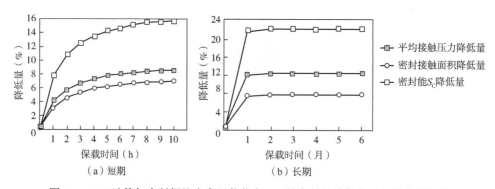

图14　110S油管气密封螺纹在高温载荷点15e的密封性随保载时间的变化规律

表4　110S油管气密封螺纹在高温载荷点15e的密封性随时间变化分析

保载时间(h)	密封能 S_C(m·MPa$^{1.95}$)		密封能 S_C 降低量(%)	密封能倍数 b
	载荷循环(第2次)	保载结束		
10	294	246	16.0	2.82
24×30	294	229	22.0	2.63
24×30×6	294	228	22.4	2.62

2　结论

（1）模拟工况载荷的试验评价是一种不考虑现场操作和井下载荷波动不确定性以及安全使用余量的试验评价方法。API RP 5C5标准密封性试验评价是一种依据材料强度验证螺纹密封性能的评价方法，没有充分考虑材料高温应力松弛时间效应的影响。为确保油套管螺纹在高温高压气井的长期安全使用，应增加基于时间效应的试验评价和有限元密封性分析。

（2）高温和室温环境下不同等效全包络线载荷2次循环的密封能倍数变化规律表明：气密封特殊螺纹连接密封性对高温（180℃）和室温95%VME包络线载荷循环极其敏感，密封能降低量较大；室温90%VME包络线载荷循环下密封能降低量较小。

（3）对于通过了API RP 5C5标准密封完整性试验评价的气密封特殊螺纹，还需采用有限元方法分析危险载荷点在高温环境下的材料应力松弛情况与密封性随时间变化的规律，从而为合理设计三轴安全系数提供理论依据。

参 考 文 献

[1] 王双来，彭娜，刘卜．高温高压井特殊螺纹接头的选用与评价试验[J]．钢管，2016，45(1)：64-71.

[2] 李渭亮，白松，陈晓华，等．基于ISO 13679的套管特殊螺纹极限承载能力仿真评价[J]．钢管，2017，46(3)：66-70.

[3] 朴龙华，张毅．我国特殊螺纹接头油套管生产使用现状及需要注意的几个问题[J]．钢管，2006，35(4)：45-48.

[4] 廖凌，崔顺贤，叶顶鹏，等．汉廷特殊螺纹接头油套管的技术特点与应用分析[J]．钢管，2009，38(4)：44-47.

[5] 吴稀勇，闫龙，陈涛，等．弯曲载荷下特殊螺纹接头密封性能的有限元分析[J]．钢管，2010，39(6)：70-73.

[6] 白鹤，党涛，何石磊，等．模拟工况下特殊螺纹密封特性的有限元分析[J]．钢管，2013，42(4)：60-63.

[7] 白鹤，何石磊，唐俊，等．台肩角度对特殊螺纹接头密封性能的影响[J]．钢管，2014，43(4)：53-56.

[8] 唐家睿，晁利宁，徐凯，等．油套管特殊螺纹连接密封性能试验研究[J]．钢管，2020，49(4)：64-69.

[9] 史彬，周晓锋．G3耐蚀合金油管特殊螺纹接头TP-G2的设计与开发[J]．钢管，2019，48(2)：63-67.

[10] 王怡，张建兵，聂艳，等．高温高压井用特殊螺纹接头的设计与评价现状[J]．钢管，2020，49(1)：72-76.

[11] 胡志立，李小兵，李建亮，等．钛合金特殊螺纹接头HSTI性能分析[J]．钢管，2020，49(3)：69-72.

[12] 李小兵，胡志立．油套管特殊螺纹的发展[J]．钢管，2020，49(5)：15-22.

[13] MURTAGIAN G R, FANELLI V, VILLASANTEJA, et al. Sealability of stationary metal-to-metal seals[J]. Journal of Tribology, 2004, 126(6)：591-596.

[14] XIE J R, MATTHEWS C. Experimental investigation of metal-to-metal seal behavior in premium casing connections for thermal wells[C]// SPE Canada Heavy Oil Technical Conference, 2017.

[15] 安文华，骆发前，吕拴录，等．塔里木油田特殊螺纹接头油套管评价试验及应用研究[J]．钻采工艺，2010，33(5)：84-88.

[16] FISCHER FD, REISNER G, WERNER E, et al. A new view on transformation induced plasticity (TRIP)[J]. International Journal of Plasticity, 2000(16)：723-748.

本论文原发表于《钢管》2021年第50卷第3期。

NQI 框架下的非 API 油井管质量管控技术

朱丽娟[1]　冯　春[1]　韩礼红[1]　谢军太[2]　徐　婷[1]　杨尚谕[1]　王　坤[1]　刘　震[2]　吴　峰[1]

(1. 中国石油集团石油管工程技术研究院，石油管材及装备材料服役行为与结构安全国家重点实验室；2. 西安交通大学，中国西部质量科学与技术研究院)

摘　要：介绍了国家质量基础(NQI)技术体系概念；论述了 NQI 技术体系是油气工业非 API 油井管质量管控的技术基础，NQI 框架下非 API 油井管产品的计量、标准、检验检测和认证认可技术共同组成了其质量管控技术体系；阐述了非 API 油井管质量管控技术面临选材评价技术不完善、质量管控认证体系缺失和现场检测与服役监控手段单一的技术现状；同时提出了在 NQI 框架下建立跨阶段集成产品质量管控技术体系、建立健全认证认可机制和工业物联网数据共享系统将促进我国形成先进适用的非 API 油井管质量管控技术体系。

关键词：国家质量基础；非 API；石油专用管；标准；认证

我国的 API 油井管系列产品从依赖进口到实现 80% 以上国产化率，历经了约 20 年[1]。目前，国产的油井管产品已经覆盖了 API Spec 5CT 与 API Spec 5DP 的全部钢级和规格，并针对油田苛刻环境开发生产了非 API 油井管系列产品。近年来，在满足国内需求的基础上，国产油井管产品的出口量持续增长。然而，我国目前出口的油井管大多属于低端产品，而进口的基本都是高技术含量的高端油井管产品，如高强韧性油井管、特殊螺纹接头油套管、镍基合金油套管等非 API 油井管。究其原因，有人认为是我国油井管发展起步晚，直至 20 世纪 80 年代，我国尚没有一套油井管标准体系[2]，严重制约了我国高端油井管产品的发展[3]；另一种观点认为是受我国冶金质量水平限制。但是，其更深层次的原因是我国缺乏健全的计量、标准、检验检测和认证认可技术共同组成的油井管质量管控技术体系。

国家质量基础设施(National Quality Infrastructure, NQI)是破解非 API 油井管技术性壁垒的关键因素，为非 API 油井管产品提供可参照的标准体系、检测检验、计量、认证认可服务。它有利于提升我国制管企业的核心竞争力和创新能力，缩短我国与国外知名企业生产的非 API 油井管产品的差距。因而，国家质量基础设施对非 API 油井管质量管控技术的发展具有重要影响，但鲜有文献将两者联系起来开展相关研究。本文将重点论述国家质量基础设施是油气工业非 API 油井管质量管控的技术基础的原因；基于 NQI 框架，分析了非 API 油井管质量管控技术现状，探讨了非 API 油井管质量管控技术的研究方向。

基金项目：国家重点研发计划项目"13Cr/110SS 产品服役全过程检测检验与质量管控技术"课题(编号：2019YFF0217504)。

作者简介：朱丽娟，女，1986 年 1 月，高级工程师，2013 年毕业于中国科学院金属研究所，腐蚀科学与防护专业，现在从事石油管工程领域研究工作。E-mail: zhulijuan1986@cnpc.com.cn。

1 NQI 技术体系的概念

NQI 的概念于 2002 年由德国联邦物理研究院提出，认为国家质量基础设施由计量、标准、检验检测、认证认可构成[4]。一直以来，关于 NQI 有国家质量基础、产业质量技术、国家质量技术等很多种不同表述，这些表述虽有差异，但内涵一致。经过 10 多年的研究，2017 年国际组织对质量基础设施提出了新的定义，确定质量基础设施是由支持与提升产品、服务和过程的质量、安全和环保性所需的公共和民间组织与政策、相关法律法规框架和实践构成的体系[5]。目前，在国家科技创新领域，国家质量基础设施技术体系即国家质量基础[6]。建立在现代科学技术基础上的计量、标准、检验检测和认证认可技术共同组成国家质量基础设施技术体系，包含关键技术和装备、技术机构、技术队伍、技术平台、信息资源、技术服务等内容；建设完善的 NQI 技术体系对促进国家各行业高质量发展具有重要的战略意义[3]。

2 NQI 技术体系是油气工业非 API 油井管质量管控的技术基础

API 油井管是根据美国石油学会的标准(简称 API 标准)生产制造的产品(简称 API 油井管)，其产品成熟可靠且具有良好的互换性，可以满足一般工况的油气井应用需求。我国是油井管产品制造大国，生产的 API 油井管产品质量已达到国际先进水平。然而，随着油气勘探开发力度加大，井深逐年增加(图 1)，随之带来油井管产品服役温度、压力上升，硫化氢、二氧化碳含量增加等更为苛刻的服役环境[7-8]。API 油井管产品的承载能力、密封特性与耐蚀性能已不能满足日益苛刻的油气开发环境需求。为此，近 20 年来，全球各生产厂开发了 300 余种非 API 油井管产品，按照功能用途可分为特殊钢种系列、特殊螺纹接头系列和特殊功能系列，约占油井管总产量 40% 左右[9-10]。

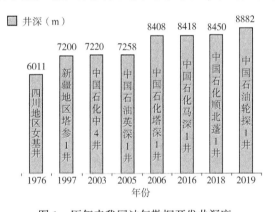

图 1 历年来我国油气勘探开发井深度

但是，我国生产的非 API 油井管产品存在质量参差不齐、附加值不高等问题。因非 API 油井管产品密封性和承载能力不足、耐蚀与耐磨性差等原因导致失效事件时有发生[11-17]，这不仅增加了油气田企业的生产成本，更严重影响了油田的安全生产，甚至引发环境污染、人员伤亡等严重社会问题。究其原因，除了非 API 油井管产品技术水平、人才队伍素质等因素外，NQI 技术能力薄弱也是重要的因素。伴随着经济全球化，NQI 逐渐成为国际竞合技术规则，NQI 技术体系的全球合作更为国际产业分工提供了坚实的技术保障，是关键性的产业技术基础[6]。因此，NQI 亦是油气工业非 API 油井管质量管控技术体系不可或缺的重要组成部分。

NQI 框架下的非 API 油井管产品的计量、标准、检验检测和认证认可技术共同组成其质量管控技术体系。非 API 油井管产品生产过程中每个环节的质量水平需要精准的计量来控制；高水平的非 API 油井管产品技术标准代表了产业技术发展水平，先进适用的标准能够带动从非 API 油井管产品基础材料、重大生产和作业装备、关键工艺到最终产品及配套使用技术的质量提高；检验检测是否科学、精准直接影响着非 API 油井管产品质量性能；严格的认证技术要求传递质量信任，是建立非 API 油井管产品加工制造业、油气行业良好生态循环的重要保障。因此，NQI 技术体系是油气工业非 API 油井管质量管控的技术基础。

3 非 API 油井管质量管控技术现状

非 API 油井管质量管控涵盖设计、制造、服役全过程，是包含材料冶金质量控制、管材制造工艺控制、关键性能指标监造与检验检测技术、关键装备校准与计量等关键技术、非 API 油井管产品及其全生命周期相关技术机构、技术队伍与技术平台的认证认可的非 API 油井管质量管控技术体系。目前，非 API 油井管质量管控存在以下难题：（1）无统一的国家（或行业）产品标准和验收规范，产品质量评价与分级结果不可靠；（2）选材评价试验缺乏系统性，评价结果与产品适用性匹配度不足；（3）无统一的特殊螺纹检测与计量标准；（4）服役状态监测手段单一，预测预警系统不健全；（5）产品质量与技术队伍认证管理体系缺失。具体从以下几方面论述。

3.1 非 API 油井管选材评价技术不完善

油井管选用与质量控制，是在明确其服役载荷工况与环境介质工况的基础上，依据 API 标准、国家标准、石油行业标准以及用户补充的订货技术协议等技术标准/规范，通过小尺寸样品室内评价、实物模拟工况和先导性现场试验等评价方法，优选适用于油田用户的不同等级的油井管。然而，API 标准基本上只能满足浅井和一般工况的油气井管柱选材，对于如高温、高压、高酸性、高腐蚀性的深井超深井、大位移井、火驱井、热采井等苛刻环境油气井用非 API 油井管选材并不适用。目前，典型的非 API 油井管产品代表有日本住友金属的 SM 系列、日本钢管公司的 NKK 系列、中国宝钢的 BG 系列、天钢的 TP 系列、以及德国的 V&M 公司的 VM 系列等。为此，针对非 API 油井管选材，各油井管生产企业通过系统的研究，针对 H_2S 应力腐蚀、CO_2 腐蚀、H_2S 与 CO_2 共存的全面腐蚀，以及温度变化的加速腐蚀作用等诸多因素，并结合产品性能，提出了选材依据；其中，最具有代表性的有 NKK 公司、住友公司、JFE 公司、欧洲的 V&M 公司，以及宝钢公司的选材图表/方案，这些选材方案为用户在非 API 油井管选用时提供了一定的参考依据。但这些选材图表/方案缺乏系统性，且各个厂家的选材方案通用性差，往往只适用于本公司的产品。此外，部分选材方案过于保守，工况应用范围有限。

目前，我国现有油井管标准共 91 项，其中国家标准 16 项，行业标准 46 项，中国石油企业标准 30 项。其中，国家和行业标准等同或修改采用 ISO/API 标准 26 项。石油专用管标准体系，按大类可分为通用基础标准、套管和油管标准、钻柱构件标准 3 类（图 2）。通用基础标准分为常用术语、性能公式与计算、加工与测量、评价与试验、驻厂监造、保护 6 类，套管和油管、钻柱构件标准可分为设计及选用、产品制造、检验与试验、保护、使用与维护、失效分析与完整性 6 类。非 API 石油专用管产品质量评估主要面临"标准分散、体系缺失、效能低下"的问题，无法对产品服役质量进行全过程综合评估。目前，非 API 油井管选材评价技术不完善，缺乏统一的非 API 油井管产品国家（或行业）产品标准和验收规范是制约非 API 石油专用管制造业高质量发展的首要因素。

3.2 非 API 油井管质量管控认证体系缺失

非 API 石油专用管服役质量状态是一个多因素影响和多场耦合的结果，非 API 油井管质量管控认证的对象涵盖了设计、制造、服役全过程，除了非 API 油井管产品认证，还包括其全生命周期相关技术机构、技术队伍与技术平台的认证认可。先进适用的非 API 油井管质量管控认证工作可系统、真实地反映制管企业的业务水平、第三方检测机构的专业能力与油田用户的质量管控成效。

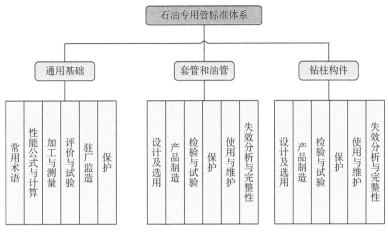

图 2 油井管标准体系

针对非 API 油井管产品的认证，可理解为具有资质的或者具有普遍公信力的第三方检测机构来证实非 API 油井管产品符合标准或订货技术协议的过程。目前，非 API 石油专用管产品质量评估主要存在"标准分散、体系缺失、效能低下"的问题，第三方监造与检验技术、认证管理体系缺失，第三方检测机构无法对非 API 油井管产品服役质量进行全过程综合评估，不能保证完整性。

3.3 非 API 油井管现场检测与服役监控手段单一

在钻完井和生产过程中，钻柱、油套管柱因外力作用或腐蚀损伤造成失稳是影响油气田安全生产的重大因素。因此，在油气生产过程中，往往采用现场检测或在线监测的方式探查管柱的服役状态。现有的井下管柱现场检测和腐蚀监测技术中，超声波等无损探伤技术、多臂井径测井等技术效率低，且无法达到实时监测的目的；超次声导波监测技术可实现实时监测，但成本较高且多用于管线等地面设备[18-20]；井下腐蚀挂片检测技术或挂环技术可用于油套管腐蚀检测[21-22]，但该方法存在无法连续腐蚀监测的问题；井下腐蚀挂片监测技术配套实验室模拟分析验证以及在役井油套管数据库等方法，可实现监测和预测相互验证的闭环管理模式[23]，但该法必须以在役井油套管数据库为前提。油套环空腐蚀探针可实时监测，但存在耐温耐压性能不足，且存在硫化氢应力腐蚀开裂风险。

因此，苛刻工况中服役的非 API 油井管，除了面临现场验收指标无统一规范标准要求的问题外，还存在油田现场高效检测手段缺乏，低效的无损探伤检测技术无法满足高效钻井需求，螺纹快速检测方法/装置缺失，服役状态监测手段单一，预测预警系统不健全等生产难题。

4 非 API 油井管质量管控技术研究方向探讨

为了更好地控制油井管的服役质量，笔者认为从以下几个方面实施质量控制会更加有效。

4.1 建立跨阶段集成产品质量管控技术体系

缺乏跨阶段集成产品质量管控技术体系是非 API 油井管产品质量参差不齐、附加值不高的重要原因。因此，在加大油气勘探开放力度的进程中，基于 NQI 技术体系框架，开展非 API 油井管产品质量基础设施研究，建设非 API 油井管产品制造和服役阶段数据共享平台，依托特色产品示范性研究，提升我国非 API 油井管核心技术能力，建立健全我国非 API 油井管服役质量技术体系(图3)，实现从全球各地采购非 API 油井管产品的兼容性、互换性和一致性，才能提高油气田企业分工协作的效率和质量，降低油气田用户的生产成本，促进构建高质高效的石油石化能源生态产业链。

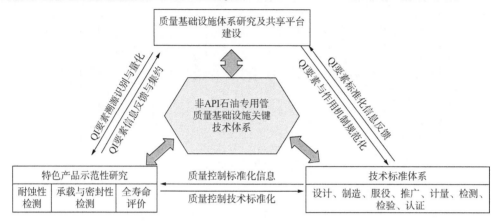

图3　NQI 框架下的非 API 油井管服役质量管控技术体系

此外，针对非 API 石油专用管产品制造和服役过程中产品质量评价具有多因素影响和多场耦合的特点，首先按照地质-工程一体化思路进行工况分级，基于制造与服役特征融合的产品质量集成评价进行产品等级划分，以保障非 API 石油专用管产品质量精准配置；其次，针对产品在宽范围变化工况中服役质量控制的实际需求，依据室内全尺寸试验评价结果，结合产品制造和服役工况大数据基础，开展基于多因素耦合的产品服役性动态监测与寿命评估，为非 API 石油专用管产品服役质量管控与认证管理提取关键支撑技术(图4)，促进非 API 石油专用管产品质量精准配置，以此形成良性的生态闭环。

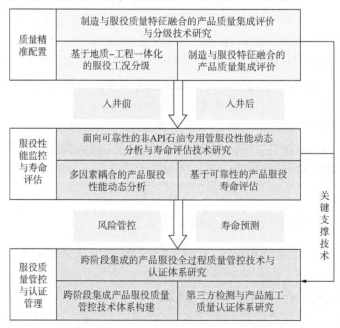

图4　非 API 石油专用管标准体系框架

最后，非API石油专用管产品的服役质量管控与认证管理相关技术，应在现有产品技术、第三方驻厂监造与质量监督、现场施工与检测监测、第三方检测与认证等标准体系的基础上，通过整合现有标准、查明缺失标准与方法，构建跨阶段集成产品质量管控标准体系及认证技术，以此形成多因素和多场耦合的跨阶段集成非API石油专用管产品质量管控技术体系(图5)，从而确保非API油井管产品质量控制有据可依，并为油气井的安全可靠运行提供技术保障。

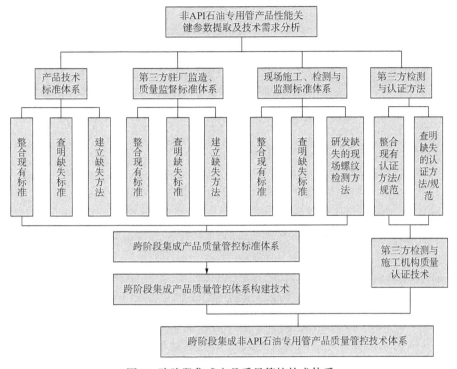

图5　跨阶段集成产品质量管控技术体系

4.2　建立健全认证认可机制

严格的认证技术要求传递质量信任，是建立非API油井管产品加工制造业、油气行业良好生态循环的重要保障。因此，应建立健全合格供应商、驻厂监造、第三方检测评价机构、现场施工与检测服务单位认证认可机制。

对制造企业实行入网许可，仅仅是物资采购实现对产品质量控制的基础。只有根据油套管订货技术要求，对制造厂的生产设备、技术水平、人员资质、质量管理体系等方面进行全面考核，满足用户订货技术协议要求并经过专业机构认证的制造企业才能获批为合格供应商。其次，非API油井管的生产制造过程包括冶炼、轧管、热处理、校直、螺纹加工、水压试验等多个关键工序，严格按照标准的要求进行生产、制造、检验是保证油套管质量的前提，而驻厂监造及第三方检测是落实标准执行的关键措施。随着油气勘探开发不断深入和井下、地面管网建设经验的积累，驻厂监造及第三方检测评价已成为石油专用管产品质量的控制方式。监造及第三方检测评价服务质量的优劣直接影响到石油专用管产品的实物质量、供货进度以及油田的安全生产。最后，现场施工质量与现场检测及监测技术服务成效则直接影响石油专用管的服役效果和使用年限。

因此，应在现有非API石油专用管服役质量评价的基础上，从第三方驻厂监造、质量监督抽查、质量认证工作数据中量化影响因素和主控因素，整合完善石油专用管产品制造过程中驻厂监造、第三方抽检、油田现场相关无损检测、螺纹计量、在线检测与监测技术。构建健全合格供应商、驻厂监造、第三方检测评价机构、现场施工与检测服务单位认证认可体系；结合非API石油专用管服役过程中多因素影响和多场耦合特点，构建基于已有产品质量检验检测结果的大数据性能预测模型和跨阶段集成产品质量管控认证模型，制定跨阶段集成产品质量管控认证规范(图6)，为提升石油专用管服役质量提供技术支撑。

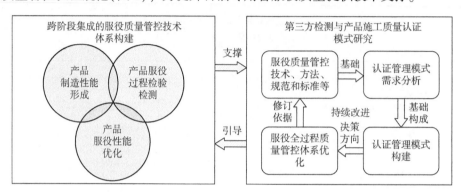

图6 石油专用管服役质量管控认证认可体系构建思路

4.3 建立工业物联网数据共享系统

非API油井管质量管控过程涉及冶金矿产、加工制造、储运、机械机电、石油化工、安全防护等多个行业，交叉融合材料、冶金、力学、腐蚀防护、石油工程，以及新兴的大数据、人工智能、云计算等多学科，数据信息分布点多、覆盖面广。通过数据采集的各种手段，采集非API油井管成分设计、材料制造、管材及接头加工、热处理及表面改性工艺、储运、现场施工、服役及失效故障处理等全生命周期过程中的静态与动态数据；将操作技术、数值化技术和通信技术深度融合，通过数据加密处理等措施，在保障信息安全的前提下，实现非API油井管生命网络互连互通，打破信息孤岛，建立工业物联网数据共享系统，实现数据全面统一；这将为实现非API油井管在全生命周期内各种运行场景的在线仿真、优化、预测预警、安全管控及科学决策提供技术支撑。

5 结语

（1）NQI技术体系是油气工业非API油井管质量管控的技术基础，NQI框架下的非API油井管产品的计量、标准、检验检测和认证认可技术共同组成其质量管控技术体系。

（2）非API油井管质量管控面临选材评价技术不完善、质量管控认证体系缺失和现场检测与服役监控手段单一的技术现状。在NQI框架下建立跨阶段集成产品质量管控技术体系、健全认证认可机制和工业物联网数据共享系统将促进我国形成先进适用的非API油井管质量管控技术体系。

参 考 文 献

[1] 李鹤林,张亚平,韩礼红.油井管发展动向及高性能油井管国产化(上)[J].钢管,2007(6):1-6.
[2] 方伟,许晓锋,徐婷.油井管标准化及非API油井管标准体系[J].石油工业技术监督,2010,26(6):20-23,42.
[3] 李为卫,方伟,冯耀荣.油井管标准的发展及选用[C].油气井管柱与管材国际会议(2014)论文集,

2014：428-432.

[4] 国家质检总局科技司. NQI：夯实质量强国战略引领经济社会发展[J]. 质量与认证，2016(4)：28-30.

[5] 肖建华. 认证认可服务国际贸易[J]. 上海质量-管理论坛，2019(2)：13-14.

[6] 徐成华. 国家质量基础设施技术体系建设的实践与思考[J]. 质量监管与发展，2020(1)：23-26.

[7] MIT-led Interdisciplinary Panel 'The Future of Geothermal Energy：Impact of Enhanced Geothermal Systems (EGS) on the United States', http：//geothermal.inel.gov/publications/future_of_geothermal_energy.pdf.

[8] R. Zeringue, HPHT Completion Challenges, SPE High Pressure/High Temperature Sour Well Design Applied Technology Workshop, May 17-19, 2005 (The Woodlands, TX), SPE 97589.

[9] 王双来. 非API油井管的发展及质量控制方式[J]. 石油工业技术监督，2019(9)：10-14.

[10] 李鹤林，韩礼红，张文利. 高性能油井管的需求与发展[J]. 钢管，2009，38(1)：1-9.

[11] 刘建勋，吕拴录，高运宗，等. 塔里木油田非API油、套管失效分析及预防[J]. 理化检验-物理分册，2013(49)：416-418.

[12] 吕拴录，张福祥，李元斌，等. 塔里木油气田非API油井管使用情况分析[J]. 石油矿场机械，2009，38(7)：70-74.

[13] 邵天翔，黎洪珍. 川东酸性环境中油管腐蚀分析及预防措施[J]. 中外能源，2013(18)：54-59.

[14] 吉楠，徐明军，周怀光，等. φ139.7mm加重钻杆内螺纹接头断裂失效分析[J]. 石油管材与仪器，2020，6(5)：66-73.

[15] 秦长毅，蒋家华，蒋存民，等. 钻杆失效分析与质量控制探讨[J]. 石油管材与仪器，2017，3(2)：1-4.

[16] 秦长毅，蒋家华，蒋存民，等. 高性能钻杆研发及应用进展[J]. 石油管材与仪器，2017，3(1)：9-13.

[17] 龚丹梅，余世杰，袁鹏斌，等. V150高强度钻杆断裂失效分析[J]. 金属热处理，2015，40(10)：205-210.

[18] 唐晓静. 一种有效的非侵入式在线腐蚀监测系统[J]. 石化技术，2019，26(9)：336-338.

[19] 高宝元，高诗惠，郭靖. 输油管线腐蚀泄漏在线监测系统研发及应用[J]. 石油化工自动化，2019，55(2)：54-56.

[20] 顾锡奎，何鹏，陈思锭. 高含硫长距离集输管道腐蚀监测技术研究[J]. 石油与天然气化工，2019，48(1)：68-73.

[21] 刘春斌，王琦，牛承东，等. 油套管腐蚀挂片监测技术[J]. 测井技术，2020，44(4)：418-421.

[22] 刘磊，何亚宁，张孝栋，等. 长北气田某气井油管腐蚀速率增大原因[J]. 理化检测-物理分册，2021，57(1)：50-57.

[23] 罗衡，元少平，杨超，等. 在役井腐蚀监测关键技术研究[J]. 腐蚀与防护，2020，38(4)：89-93.

本论文原发表于《石油管材与仪器》2021年第7卷第3期。

石墨烯改性环氧涂层在油田注水工况中的适用性研究

朱丽娟[1]　冯　春[1]　何　磊[1,2]　宋文文[1,2]　张芳芳[1,2]

(1. 中国石油天然气集团公司，石油管工程技术研究院，石油管材及装备材料服役行为与结构安全国家重点实验；2. 西安石油大学)

摘　要：中国石油集团石油管工程技术研究院(简称"管研院")联合天津宝坻紫荆创新研究院，共同开展了石墨烯技术在油管表面处理中的应用基础研究。研究结果表明，管研院牵头研制的油管用高性能石墨烯改性环氧涂层表面光滑、无漏点，且呈镜面效果，具有良好的结合力和耐磨性能；在高温高压酸性油气介质、碱性介质和模拟注水工况中具有优异的耐蚀性能。管研院联合长庆油田、辽河油田，国内首次完成了石墨烯改性涂层油管在石油石化领域的应用示范。此举标志着石墨烯材料在中国石油石化领域的应用基础研究取得重大进展，将有效提升油气井管材在高温、高压、高盐等苛刻钻采工况下的服役寿命和安全性。

关键词：石墨烯；涂层；腐蚀；油管；注水

石墨烯具有优异的光学、力学和电学等特性，被认为是21世纪一种革命性的材料，在材料学、微纳加工、能源和生物医学等领域具有广阔的应用前景。美国、英国、日本等发达国家已将这种先进材料的发展提升到战略高度，欧盟委员会也将其列为仅有的两个"未来新兴技术旗舰项目"之一。中国石墨烯产业技术创新战略联盟在2017年6月成立了石墨烯防腐应用推进工作组，快速推进石墨烯涂层材料产业发展，创造经济价值。

自2017年起，在中国石油集团基础研究和战略储备技术研究基金项目等支持下，中国石油集团石油管工程技术研究院(简称"管研院")联合天津宝坻紫荆创新研究院，共同开展了石墨烯技术在油管表面处理中的应用基础研究。科研人员采用先进工艺，在高性能环氧树脂、酚醛树脂、聚氨酯等基料中，将单层或多层石墨烯纳米材料有机分散或接枝改性，成功获得了致密、均匀的涂层结构[1-7]。相关试验评价表明，这种新型石墨烯改性涂层可显著提高油管的抗高温老化、耐磨损、耐氯离子环境损伤等性能，延长油气井管柱服役寿命30%至50%，并降低管材全生命周期成本30%以上。目前，管研院已经发布了中国首个石墨烯

基金项目：陕西省创新人才推进计划-青年科技新星项目"石墨烯改性环氧涂层强韧性及耐蚀性能研究"(项目编号：2020KJXX-064)；内蒙古自治区科技重大专项"石墨烯功能化涂层材料关键技术开发及应用"(项目编号：2020ZD0019)；中国石油基础研究和战略储备技术研究基金项目"石墨烯技术在油管表面处理中的应用基础研究"(项目编号：2017-Z04)；中国石油天然气集团有限公司科学研究与技术开发项目：煤炭地下气化关键技术研究与先导试验[项目编号：2019E-25(JT)，课题编号：2019E-2505]。

作者简介：朱丽娟，女，1986，2013年毕业于中国科学院金属研究所腐蚀科学与防护专业，现于中国石油集团石油管工程技术研究院从事石油管工程领域相关工作。E-mail：zhulijuan1986@cnpc.com.cn。

涂层CSTM选材设计标准[8]。

本文概述了管研院牵头研制的高性能石墨烯改性环氧涂层的研制过程和石墨烯改性环氧涂层的基本性能，探索研究了石墨烯改性环氧涂层在注水工况中的适用性。管研院联合长庆油田、辽河油田，国内首次开展了石墨烯改性涂层油管在石油石化领域的应用示范。

1 试验材料及方法

基体油管名义直径为73mm，壁厚为5.51mm。油管本体为N80钢级。采用Elcometer 456涂层测厚仪测量涂层厚度，采用漏点检测仪(XRD)对石墨烯改性涂层油管表面进行漏点检测；采用弯曲法和拉伸法测试涂层结合力；采用高温高压反应釜评估石墨烯改性涂层在酸性油气介质、碱性介质、注水工况中的耐腐蚀性能。

2 结果与讨论

2.1 涂层基本性能

2.1.1 外观与漏点

石墨烯改性涂层油管的宏观形貌如图1所示。从图1中可以看出，石墨烯改性涂层的内表面十分光滑，不存在橘皮、流淌、褶皱等缺陷，油管涂层内表面呈镜面效果。涂层油管内表面均未检测到漏点。

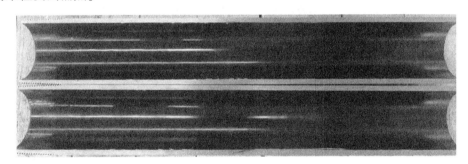

图1 石墨烯改性涂层油管宏观形貌

2.1.2 涂层厚度与耐磨性

在涂层油管内表面沿周分A、B、C、D四个象限，沿径向每隔10cm测量一次，每个象限测量25个点。测试结果表明涂层厚度为152~181μm，平均厚度为169μm。采用落砂实验法测试涂层的耐磨性，添加石墨烯后，涂层的耐磨性能提升了80%以上。

2.1.3 结合力

采用拉伸和弯曲试验测试石墨烯改性涂层与管材基体的结合力，测试后的宏观形貌如图2所示。从图2中可以看出，涂层与基体界面结合力良好，拉伸加载直至样品拉断，涂层除紧缩段以外，未发现涂层剥落现象。样品弯曲30°，涂层依然未见开裂和剥落情况。

2.2 涂层耐蚀性能

2.2.1 高温高压碱性介质

在148℃、70MPa下，pH值为12.5的NaOH溶液中，进行24h测试后，石墨烯改性涂层和普通环氧涂层的宏观形貌如图3所示。从图3中可以看出，与普通环氧涂层相比，石墨烯改性涂层未发生起泡、开裂或剥落现象，具有优异的耐蚀性能。

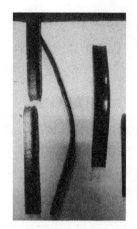  

图 2 石墨烯改性涂层结合力测试后的宏观形貌

(a) 石墨烯改性涂层　　　　　　(b) 普通环氧涂层

图 3 深层耐高温高压碱性介质测试后宏观形貌

2.2.2 高温高压酸性介质

在 107℃、35MPa 下，水、甲苯和煤油的混合液中，CO_2 气体加压的工况下，进行 16h 测试后，石墨烯改性涂层和普通环氧涂层的宏观形貌如图 4 所示。从图 4 中可以看出，与普通环氧涂层相比，石墨烯改性涂层表面未发现起泡、开裂、结垢等现象，具有优异的耐蚀性能。

(a) 石墨烯改性涂层　　　　　　(b) 普通环氧涂层

图 4 涂层耐高温高压酸性介质测试后宏观形貌

2.2.3 模拟油田注水工况

采用矿化度约为 $4×10^4 mg/L$ 的某油田注水井回注的污水,在16MPa、150℃下测试60h,测试后涂层形貌如图5所示。从图5可以看出,与普通环氧涂层相比,石墨烯改性涂层未发生起泡或剥落现象。

2.3 应用示范

2020年8月,由管研院牵头研制的TG110型石墨烯改性涂层油管,在长庆油田采油二厂西峰油区圆满完成首次下井试验(图6);同年10月份,TG110型石墨烯改性涂层油管在辽河油田曙光采油厂完成下井试验。此举标志着石墨烯材料在中国石油石化领域的应用基础研究取得重大进展,将有效提升油气井管材在高温、高压、高盐等苛刻钻采工况下的服役寿命和安全性。

(a)石墨烯改性涂层　(b)普通环氧涂层
图5　涂层耐高温高压模拟工况测试后宏观形貌

图6　TG110石墨烯改性涂层油管应用示范与新闻报道

3 结论

管研院牵头研制的高性能石墨烯改性环氧涂层表面光滑、呈镜面效果、无漏点,具有良好的结合力和耐磨性能;在高温高压酸性油气介质、碱性介质和模拟注水工况中具有优异的耐蚀性能。管研院联合长庆油田、辽河油田,国内首次完成了石墨烯改性涂层油管在石油石化领域的应用示范。

参 考 文 献

[1] ZHU L J, CHUN F, CAO Y Q. Corrosion behavior of epoxy composite coatings reinforced with reduced graphene oxide nanosheets in the high salinity environments[J]. Applied Surface Science, 2019, 493: 889-896.

[2] FENG C, ZHU L J, CAO Y Q. Performance of Coating Based on APTMS/GO/Epoxy Composites for the Corrosion Protection of Steel[J]. International Journal of Electrochemical Science, 2018, 13: 8827-8837.

[3] FENG C, ZHU L J, CAO Y Q. Performance of Coating Based on β-CD-g-GO/epoxy Composites for the Corrosion Protection of Steel[J]. International Journal of Electrochemical Science, 2019, 14: 1855-1868.

[4] FENG C, CAO Y Q, ZHU L J. Corrosion Behavior of Reduced-Graphene-Oxide Modified Epoxy Coatings on N80 Steel in 10.0 wt% NaCl Solution[J]. International Journal of Electrochemical Science, 2020, 15: 8265-8276.

[5] ZHU L J, FENG C, GE H J. Research Progress on Anti-Corrosive Properties of Graphene Modified Coatings[J]. Materials Science Forum, 2020, 993: 1140-1147.

[6] ZHU L J, FENG C, SONG Y C. Comparison of Corrosion Behavior of an Epoxy Coating and a RGO Modified Epoxy Coating on N80 Tubing Steel in 10.0 wt% NaCl Solution at Different Temperatures[J]. Materials Science Forum, 2021, 1035: 554-561.
[7] LU C H, ZHU L J, FENG C. Graphene modified epoxy coating tubing applied in the water injection wells [J]. Spring: Proceedings, 2017: 1075-1082.
[8] T/CSTM 00242—2021 油管石墨烯改性涂层质量要求及检验[S].

本论文原收录于2021年石油管及装备材料国际会议(2021年)。

三层环氧涂层油管在注水井中的应用研究

朱丽娟[1]　冯　春[1]　李长亮[1]　何　磊[1,3]　张芳芳[1,2]　宋文文[1,3]

(1. 中国石油天然气集团公司，石油管工程技术研究院，
石油管材及装备材料服役行为与结构安全国家重点实验；
2. 大庆石油管理局有限公司装备制造分公司；3. 西安石油大学)

摘　要：为了研究三层环氧涂层油管在注水井中的应用效果，分析了环氧涂层油管在注水井中使用 19~80 个月的服役行为。采用 SEM、EDS 和 XRD 等表征手段对腐蚀产物的形貌、成分和结构进行了分析，对环氧涂层和油管钢基体的失效原因和腐蚀机理进行了探讨。结果表明，三分之一的防腐油管已不符合用户继续使用的技术协议要求，然而，其中一根油管却仍具有优异的耐腐蚀性。油管前处理工艺不达标和环氧涂层耐高温腐蚀性差分别是导致环氧涂层修复油管和高温使用的环氧涂层油管失效的主要原因。涂层失效后，P110 油管基体发生溶解氧腐蚀、二氧化碳腐蚀和垢下腐蚀。

关键词：环氧涂层；腐蚀；油管；注水井；失效

随着油田开发进入中后期，注入水质工况日益苛刻，注水井管柱损坏问题日益严重[1-4]。为了提高注水效果，性价比较高的环氧涂层因其经济效益而被广泛用于提高注水管的耐腐蚀性[1-2]。然而，近年来，由于油田使用的涂层不合适和施工不当而导致的油管失效频繁发生。

截至 2017 年底，在我国 J 油田的注水井中，超过 36% 的 P110 油管采用环氧树脂涂层进行保护，油管的使用寿命显著提高。然而，部分防腐管的使用寿命不到两年。为研究环氧涂层的失效原因和 P110 油管的腐蚀机理，本研究团队对 J 油田注水井服役用 19~80 个月的环氧涂层油管进行了应用分析，以期为油田注水开发工程防腐工艺的选择和改进提供一定的参考依据。

1　试验材料及方法

以现场取出的环氧涂层油管、同批次未下井的环氧涂层油管为研究对象，基体油管名义

基金项目：陕西省创新人才推进计划-青年科技新星项目"石墨烯改性环氧涂层强韧性及耐蚀性能研究"（项目编号：2020KJXX-064）；内蒙古自治区科技重大专项"石墨烯功能化涂层材料关键技术开发及应用"（项目编号：2020ZD0019）；中国石油天然气集团有限公司科学研究与技术开发项目：煤炭地下气化关键技术研究与先导试验[项目编号：2019E-25(JT)，课题编号：2019E-2505]。

作者简介：朱丽娟，女，1986，2013 年毕业于中国科学院金属研究所腐蚀科学与防护专业，现于中国石油集团石油管工程技术研究院从事石油管工程领域相关工作。E-mail：zhulijuan1986@cnpc.com.cn

直径为73mm，壁厚为5.51mm。油管本体材质为P110钢级，在注水井中使用19~80个月，并对同一批未使用的环氧涂层油管的性能进行了对比分析。采用超声波测厚仪测量剩余壁厚。采用X射线衍射仪（XRD）对防腐油管表面的腐蚀产物进行物相分析；采用带能谱（EDS）的扫描电子显微镜（SEM）分析防腐油管截面形貌和腐蚀产物的成分，进而确定引起腐蚀的原因。采用高温高压反应釜评估未使用油管涂层的耐腐蚀性，采用J油田注入水作为试验溶液，其pH值为6.5（成分：9912.4mg·L^{-1} Cl$^-$、104.2mg·L^{-1} Mg^{2+}、110.8mg·L^{-1} Ca^{2+}、1107.2mg·L^{-1} HCO$_3^-$、4.9mg·L^{-1} SO$_4^{2-}$、其余为Na$^+$和K$^+$）。

2 结果与讨论

2.1 宏观分析

在J油田注水井中使用了19~80个月的9根环氧涂层油管和同批次的2个未使用的环氧涂层油管（10#和11#），基本情况如表1所示。从表中可以看出，服役后的环氧涂层油管受到不同程度的损坏。1/3油管的最小剩余壁厚小于4.82mm，不符合API SPEC 5CT标准[5]和用户订货技术协议要求。然而，其他6根使用过的油管却仍具有优异的耐腐蚀性，只有少量涂层损坏；在这些油管上未观察到腐蚀产物，只是在取样过程中出现了少量涂层损坏。值得注意的是2#、3#和9#油管是经修复在利用的油管，即在使用一段时间后重新涂覆用于井下应用。这意味着涂层下方的2#、3#和9#油管的基体已被腐蚀。

表1中所述部分油管样品的宏观形态如图1所示。2#修复油管内表面环氧涂层出现严重起泡和剥落；在涂层剥落处有红棕色和黑色腐蚀产物，很可能分别为氧化铁和碳酸亚铁；油管表面有大面积的沉积物，这将导致水垢腐蚀。5#油管上未出现起泡和剥落，然而，在9#油管上观察到了大面积的涂层起泡和剥落。

表1 在J油田注水井中服役的环氧涂层油管基本情况

序号	服役时间（30d）	井深（m）	温度（℃）	最小壁厚（mm）	涂层损伤区域（%）
1	19	500	20	5.45	1
2	22	500	20	3.7	100
3	23	1000	36	4.7	50
4	34	1000	36	5.5	0
5	34	3000	82	5.5	0
6	58	1000	36	5.5	1
7	58	2000	60	5.5	1
8	80	1000	36	5.51	0.5
9	80	3900	102	4.2	100

2.2 截面形貌与腐蚀产物

从表1所述油管上未起泡区取样分析，其截面形貌如图2所示。2#和5#油管样品上的环氧涂层致密，涂层中未观察到裂纹，然而，在9#油管样品上的涂层中有明显的裂纹，并且环氧涂层内部的结构疏松。因此，9#油管上的环氧涂层已无防腐性能。这很可能与改井中的使用温度高达102℃有关。

另外，在2#修复油管和9#油管样品的涂层/基体界面处检测到腐蚀产物。EDS分析结果表明，2#修复油管试样的腐蚀产物主要为Fe^{2+}、O和C，未检测到Ca^{2+}。值得注意的是，在2#修复油管样品涂层/基体界面下方的P110碳钢基体中也可以观察到腐蚀产物。由于2#修复油管样品上的环氧涂层仍然很致密，这表明在油管重新涂覆之前，在涂层与基体界面处和P110基体中已经存在这些腐蚀产物，这会降低环氧涂层的附着力。因此，2#修复油管重涂前的表面预处理工艺不达标。EDS和XRD物相分析结果表明，9#油管试样上的腐蚀产物主要由$FeO(OH)$、$FeCO_3$、FeO、Fe_3O_4、Fe_2O_3和$CaCO_3$组成(图3)。因此，环氧涂层失效后，9#管样P110管主要发生了溶解氧腐蚀、二氧化碳腐蚀和垢下腐蚀。

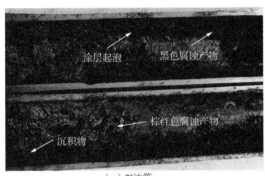

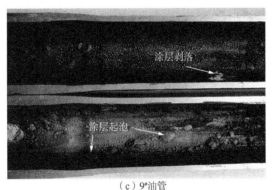

（c）9#油管

图1 2#、5#、9#环氧涂层油管宏观形貌

2.3 高温高压腐蚀试验

为了验证表面预处理及服役温度对环氧涂层附着力与耐腐蚀性的影响。采用油田注入水质为试验溶液，在90℃和120℃的温度下对环氧涂层油管样品进行了15d的高温高压模拟工况试验。在90℃高温高压腐蚀试验后，未使用的新油管和修复油管样品的表面形貌如图4所示。在90℃下进行试验后，新油管样品的环氧涂层未观察到任何变化；然而，由于重新涂覆的环氧涂层附着力较差，在修复油管样品的环氧涂层上观察到了大规模的涂层脆化。因此，油管重涂前表面预处理不达标是2#和3#环氧涂层修复油管在不到两年内失效的主要原因。

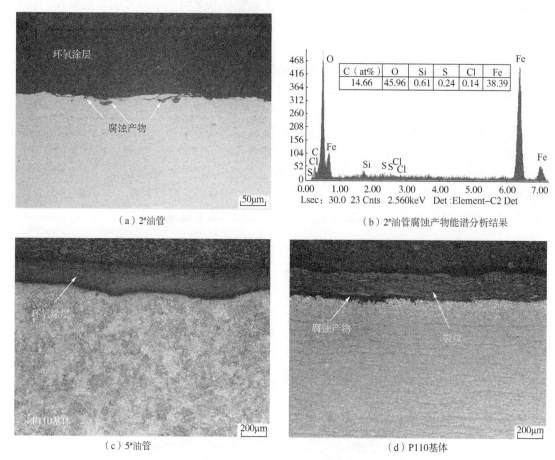

图 2 2#、5#、9# 环氧涂层(图片未标注)油管样品的截面形貌以及 2# 油管涂层/基体界面 P110 基体中腐蚀产物能谱分析

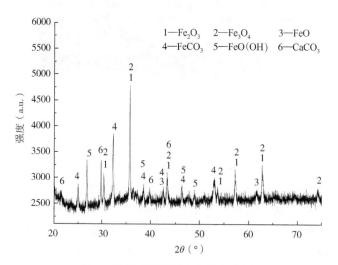

图 3 油管表面腐蚀产物物相分析

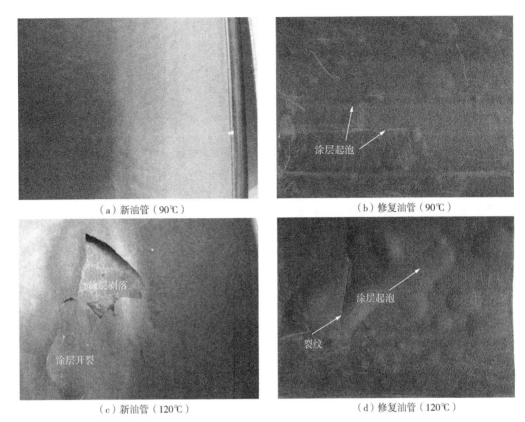

图4 新油管和修复油管在90℃和120℃高温高压模拟工况试验后的宏观形貌

另外,由于环氧涂层的耐高温腐蚀性差,在120℃下进行试验后,新油管和修复油管样品上都出现了大面积的涂层起泡(图4),这与图1中所述的在注水井中服役后的环氧涂层油管的失效特征一致。众所周知,有机聚合物材料的服役行为受温度影响较大[6-8]。在低于玻璃化转变温度时,聚合物非晶态材料发生自发变化,这归因于聚合物主链的构象变化,并与物理老化有关。涂层松弛应力能力丧失是物理老化和化学降解的结果。随着黏结强度的降低,接近局部黏结应力的临界值,出现黏结损伤[7]。因此,当裂纹被电解液填充以及金属基底的局部分层时,因在涂层中形成导电通路,涂层性能退化[9-11]。当达到累积损伤的特定极限时,涂层发生宏观失效。因此,环氧涂层耐高温腐蚀性差是9#环氧涂层油管失效的主要原因。环氧涂层失效后,腐蚀介质直接与P110油管基体接触,导致油管腐蚀。腐蚀机理为溶解氧腐蚀、二氧化碳腐蚀和垢下腐蚀。

3 结论

(1)在J油田取样分析的涂层油管中,30%以上的管材最小剩余壁厚小于4.82mm,不符合API SPEC 5CT标准和用户订货技术协议要求,其他的油管仍具有优异的耐腐蚀性。

(2)修复油管重涂前的表面预处理不达标是导致环氧涂层修复油管在不到两年的时间内失效的主要原因,环氧涂层耐高温腐蚀性能差是J油田高温深井中环氧涂层油管失效的主要原因。环氧涂层失效后,P110油管基体发生溶解氧腐蚀、二氧化碳腐蚀和垢下腐蚀。

参 考 文 献

[1] 柳言国. 胜利油田注水井管柱腐蚀防护技术应用效果分析[J]. 腐蚀与防护, 2003, 24 (8): 361-362.
[2] 黄斐, 胡建修, 李文通, 等. 油田注水井腐蚀及其防护[J]. 石油化工腐蚀与防护, 2010, 27 (4): 48-50.
[3] 赵凤兰, 鄢捷年, 胡海红. 注水系统腐蚀规律与防腐技术[J]. 油气地面工程, 2002, 21(6): 19-20.
[4] 李荣强. 胜利油田注水管柱腐蚀机理与防治技术[J]. 石油钻探技术, 2008, 36 (4): 64-66.
[5] API 5CT, Specification for Casing and Tubing (Washington, DC: API).
[6] HODGE I M, Enthalpy relaxation and recovery in amorphous materials [J]. Journal of Non-Crystalline Solids. 1994, 169(3): 211-266.
[7] MISZCZYK A, DAROWICKI K. Effect of environmental temperature variations on protective properties of organic coatings[J]. Progress in Organic Coatings, 2003, 46: 49-54.
[8] 朱丽娟, 田涛, 范晓东, 等. 一种环氧涂层防腐油管的室内模拟工况评价与实际服役性能对比研究[J]. 石油管材与仪器, 2017, 3(5): 36-39.
[9] LI X, HRISTOV H A, YEE A F, et al. Influence of cyclic fatigue on the mechanical properties of amorphous polycarbonate [J]. Polymer, 1995, 36(4): 759-765.
[10] NICHOLS M E, DARR C A, SMITH C A, et al. Fracture energy of automotive clearcoats-I. Experimental methods and mechanics [J]. Polymer Degradation and Stability, 1998, 60(2-3): 291-299.
[11] LEE H, KRISHNASWAMY S. Quasi-static propagation of sub interfacial cracks [J]. Journal of Applied Mechanics-transactions of the ASME, 2000, 67(3): 444-452.

本论文原收录于2021年石油管材及装备材料国际会议(2021年)。

三、腐蚀与非金属

Effects of Trace Cl^-, Cu^{2+} and Fe^{3+} Ions on the Corrosion Behaviour of AA6063 in Ethylene Glycol and Water Solutions

Fan Lei[1] Zhang Juantao[1] Wang Hao[2] Liu Yang[2] Cui Yu[3]
Wang Cheng[3] Liu Rui[3,4] Xu Dongxiao[2]

(1. State Key Laboratory for Performance and Structure Safety of Petroleum Tubular Goods and Equipment Materials, CNPC Tubular Goods Research Institute;
2. NO. 1 Gas Production Plant of Changqing Oilfield Company; 3. Shi-Changxu Innovation Center for Advanced Materials, Institute of Metal Research, Chinese Academy of Sciences;
4. Shenyang National Laboratory for Materials Science, School of Materials Science and Engineering, Northeastern University)

Abstract: The effects of Cl^-, Cu^{2+} and Fe^{3+} ions and their combinations on the corrosion behaviour of aluminium alloy 6063 (AA6063) in ethylene glycol and water solutions at 50℃ were investigated by electrochemical and immersion methods. Cl^- ions resulted in pitting corrosion of the alloy. In Cl^- free solutions, Fe^{3+} ions were prone to accelerate uniform corrosion, while Cu^{2+} ions tended to accelerate pitting corrosion. Serious pitting corrosion of AA6063 was observed in the cases of Cl^- combined with Cu^{2+} or Fe^{3+} ioins, especially in the case of Cl^- combined with Cu^{2+} and Fe^{3+} ions.

Keywords: Aluminium; Polarization; Weight loss; Pitting corrosion.

1 Introduction

Coolant is an essential cooling medium in some industrial fields such as automobile, railway motor car, steamship, solar collector system, electronic instrumentation, and diesel and gasoline engines. The most popular coolant systems are produced on the base of ethylene glycol (EG) and water mixtures. The EG-water system has exceptional anti-flame property, high boiling point (more than 130℃), high stability and excellent heat absorption capacity [1]. In general, a cooling system includes aluminium alloy, steel, copper, stainless steel and other metal or nonmetal materials. The use of aluminium alloy in these systems can significantly reduce the weight of an engine and its fuel consumption, and thus decrease environmental pollution. Moreover, the thermal conductivity of aluminium alloy is about 200W/(m·K), higher than those of steel about 50W/(m·K) and other

---

Corresponding author: Cui Yu, ycui@ imr. ac. cn; Wang cheng, wangcheng@ imr. ac. cn.

nonmetal materials.

Currently, the main composition of the conventional EG-based coolant is 30~70 vol. % ethylene glycol and added inhibitors normally include molybdate, phosphate, borate, nitrate, nitrite, tolyltriazole, benzoate and silicate [2]. Accidents resulted from engine failures are mainly caused by the corrosion of construction material by aggressive contaminants in its cooling system. In water medium, the major corrosion type of aluminium alloy is pitting corrosion, and considerable research has been carried out in this field [3-10], but the corrosion behaviour of alloy in cooling system has not been systematically investigated. Wong et al. [11] reported that the corrosion of pure aluminium, AA1100, AA 3003 and AA3304 alloys are accelerated in the uninhibited glycol-water solutions containing 200mg/L Cl^-, Cu^{2+} and Fe^{3+} ions. It is also reported that severe pitting corrosion occurred for 2S and 3SR aluminium alloy in EG-water solution containing 0.01mg/L combination of Cl^-, Cu^{2+} and Fe^{3+} ions, and resulted in a considerable mass loss of the aluminium alloys [12]. Starosvetsky et al. [13] reported that extensive perforation of some aluminium parts caused by severe corrosion was found in the cooling systems of 250-HP (186-kW) diesel engines during long-term storage. The failure analysis revealed that a major cause of the failure is pitting corrosion initiated by chloride ions present in tap water used for the coolant preparation. The corrosion behaviour of aluminium depends on the types of ions and their concentration in the solutions [14]. Chloride ions are often found in the coolant, cupric and ferric ions may likewise present in the cooling systems, and they are the resultant of the corrosion of copper and steel respectively in the system [15-16]. In practice, the concentrations of contaminates are as low as several ppm, especially the concentration of cupric and ferric ions may be less than 10mg/L.

This paper investigated the corrosion behaviour of AA6063 aluminium alloy in EG and water mixtures of solely containing Cl^-, Cu^{2+} and Fe^{3+} ions and in the combination of $Cl^- + Cu^{2+}$, $Cl^- + Fe^{3+}$ and $Cl^- + Cu^{2+} + Fe^{3+}$ ions by electrochemical and immersion methods.

The aim of this study is to provide a better understanding of the corrosion performance of AA6063 aluminium alloy in EG and water solutions containing contaminats such as Cl^-, Cu^{2+} and Fe^{3+} ions, and to reveal the main factors influencing the corrosion of aluminium alloy. It is also hoped that the results will provide essential knowledge for the design of novel ethylene glycol based coolants and routine maintenance of a cooling system.

2 Experimental procedures

2.1 Sample preparation

The material used in this study was AA6063 plate with thickness of 3mm. The nominal composition of the AA6063 (wt. %) is 0.55% Mg, 0.45% Si, 0.097% Cu, 0.17% Fe, 0.02% Cr, 0.06% Mn, 0.01% Ti, 0.02% Zn and the balanced Al. Specimens were machined into 1cm× 1cm coupons for electrochemical and immersion experiments. Before each experiment, the working surfaces of samples were wet ground with emery paper up to 800 grit and then degreased with ethanol in an ultrasonic cleaner, cleaned with distilled water, and finally dried in air. In order to avoid prolonged exposure to the open atmosphere, the specimens were stored in a desiccator after drying by air stream.

2.2 Solutions

All solutions were prepared from distilled water with a conductivity about 18 MΩ · cm and analytical grade reagents. The basic solution was a mixture of ethylene glycol (EG) and distilled water (volume ratio of 57:43 with an ice point about −45℃). NaCl, $CuSO_4 \cdot 5H_2O$ and $Fe_2(SO_4)_3$ were added to the basic solution in requisite amounts to make solutions containing trace amout of Cl^-, Cu^{2+} and Fe^{3+} ions or their combinations. The testing solutions are listed in Table 1.

Table 1 Testing solutions with different contaminates of Cl^-, Cu^{2+} and Fe^{3+} ions (10^{-5} mg/L)

Solutions	Cl^-	Cu^{2+}	Fe^{3+}
B[a]	—	—	—
B+Cl^- ions	30	—	—
B+Cl^- ions	60	—	—
B+Cl^- ions	150	—	—
B+Cl^- ions	300	—	—
B+Cu^{2+} ions	—	16.5	—
B+Cu^{2+} ions	—	33.0	—
B+Cu^{2+} ions	—	82.5	—
B+Cu^{2+} ions	—	165	—
B+Fe^{3+} ions	—	—	18.7
B+Fe^{3+} ions	—	—	37.4
B+Fe^{3+} ions	—	—	93.5
B+Fe^{3+} ions	—	—	187
B+Cl^-+Cu^{2+} ions	30	16.5	—
B+Cl^-+Cu^{2+} ions	60	16.5	—
B+Cl^-+Cu^{2+} ions	150	16.5	—
B+Cl^-+Cu^{2+} ions	300	16.5	—
B+Cl^-+Fe^{3+} ions	30	—	18.7
B+Cl^-+Fe^{3+} ions	60	—	18.7
B+Cl^-+Fe^{3+} ions	150	—	18.7
B+Cl^-+Fe^{3+} ions	300	—	18.7
B+Cl^-+Cu^{2+}+Fe^{3+} ions	30	16.5	18.7
B+Cl^-+Cu^{2+}+Fe^{3+} ions	60	16.5	18.7
B+Cl^-+Cu^{2+}+Fe^{3+} ions	150	16.5	18.7
B+Cl^-+Cu^{2+}+Fe^{3+} ions	300	16.5	18.7

Note:[a] B denotes a solution of EG and water with a volume ratio of 57:43.

2.3 Electrochemical measurements

Electrochemical measurements were carried out in a conventional three-electrode electrochemical glass cell with a platinum counter electrode and a saturated calomel electrode (SCE) as reference. The surface area of the working electrodes was 1cm^2, five faces of the specimens were sealed with epoxy resin, leaving one face for tests. All the potentials quoted in this paper are on the SCE scale. The polarization curves were recorded by changing the electrode potential automatically with a potentiostat type EG&G 273 controlled by PAR M352 electrochemical software at a scan rate of 0.3mV/s. Alloy AA6063 samples were immersed in a solution and allowed to attain a stable open circuit potential (OCP) before starting the polarization scan. The potentiodynamic polarization curves were plotted by starting scanning electrode potential from an initial potential of 200mV below the OCP up to the breakdown potential.

2.4 Mass change testing

Immersion tests were conducted in 200mL glass beakers containing basic solution with a requisite trace amount of Cl^-, Cu^{2+} and Fe^{3+} ions or their combinations at 50℃. The average mass change for each three identical experiment was conducted and expressed in mg/cm^2. The weights and sizes of the specimens were measured prior to tests. To determine themass change of the samples, the specimens were taken out and weighted after 385 hours immersion in testing solution. The corrosion rate of AA6063 was calculated as follows:

$$v = [(W_0 - W_t) \times 1000]/S \tag{1}$$

where, v in mg/cm^2 was corrosion rate, W_0 and W_t both in g were the mass of AA6063 before and after immersion for 385h, respectively. S in cm^2 was the total surface area of aluminium alloy sheet. All the experiments were conductedin static EG-water solutions at 50℃±1℃.

2.5 Corrosion morphologies of AA6063 alloy

The morphologies of the tested samples wereobserved using SEM (FEI-Inspect F) and EDS (INCA, X-Max). The phase compositions of the corrosion products were analysed using XRD (X' Pert Pro Panalytica Co.).

3 Results and discussion

3.1 The effects of Cl$^-$ ions

Fig.1 shows the potentiodynamic polarization cures of AA6063 alloy in basic solution containing Cl$^-$ ions varying from 0 to 300×10^{-5} mg/L at 50℃. The potentials of the alloy shifted to negative values when Cl$^-$ added in the basic solution as seen in Fig.2, which was attributed to the adsorption of negative Cl$^-$ ions on the surface of the alloy [17], and the potential decrease was similar to that of aluminium alloy in aqueous solution [18].

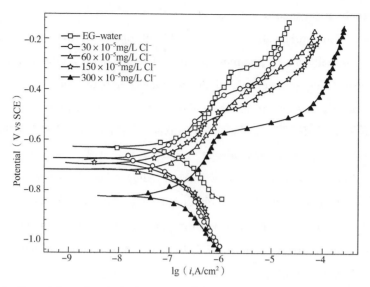

Fig. 1 Potentiodynamic polarization curves of AA6063 in basic solution containing Cl$^-$

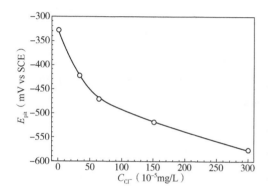

Fig. 2 E_{pit} for AA6063 alloy in basic solution containing Cl$^-$

The anodic current density of AA6063 increased slowly in the range from corrosion potential E_{corr} to breakdown or pitting potential E_{pit}, above which the current density increased suddenly. The E_{pit} decreased when the Cl$^-$ ions were added to the basic solutions, and the greater the concentration of Cl$^-$ ions, the more obvious the decrease of pitting potential, as shown in Fig. 2. In the basic solution containing Cl$^-$, the cathodic branches of the AA6063 alloy shifted to a smaller current density region compared with that in the basic solution, indicating that Cl$^-$ ions inhibited the cathodic reaction of the alloy. The cathodic current density increased gradually with increasing concentration of Cl$^-$. It is evident that the effect of Cl$^-$ on the anodic reaction of AA6063 in the basic solution is greater than that on the cathodic reaction, as shown in Fig. 1. The pitting susceptibility of passive alloys was evaluated by the difference between E_{corr} and E_{pit}[19-20], as described by Eq. (2). The higer the ΔE value, the lower the pitting susceptibility of the alloy.

$$\Delta E = E_{pit} - E_{corr} \qquad (2)$$

where, ΔE is the difference between E_{pit} and E_{corr}, E_{pit} is pitting potential or breakdown potential, and E_{corr} is open circuit potential.

The relationship between ΔE and the concentration of Cl$^-$ in basic solutions is shown in Fig. 3(a). ΔE decreased almost linearly with increasing Cl$^-$ concentration, indicating that pitting susceptibility increased with increasing the amount of Cl$^-$. The corrosion current densities (i_{corr}) of AA6063 in the basic solution containing Cl⁻ are shown in Fig.3(b). The i_{corr} increased with increasing concentration before it reached to $60×10^{-5}$ mg/L. After that, the i_{corr} decreased slowly.

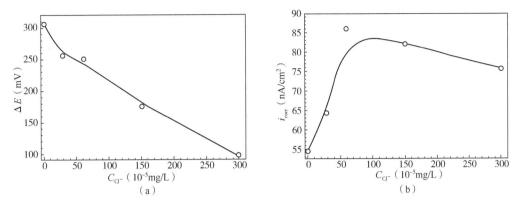

Fig. 3 Fitted results of AA6063 in basic solution containing Cl⁻ ions: (a) ΔE values; (b) i_{corr}

For aluminium alloys, pitting occurs in the media containing Cl⁻ [21-23]. The anodic reaction is the dissolution of Al, and the cathodic reaction is the reduction of dissolved oxygen in the solutions. The anodic and cathodic reactions can be described as Eqs. (3) and (4).

$$Al \longrightarrow Al^{3+} + 3e^- \tag{3}$$

$$O_2 + 2H_2O + 4e^- \longrightarrow 4OH^- \tag{4}$$

In the media containing Cl⁻, oxygen and Cl⁻ will adsorb competitively on the surface of aluminium alloy according to pitting formation and growth mechanism [24-25]. The high concentration of Cl⁻ reduces the solubility of oxygen [26], and Cl⁻ preferentially absorbs on the alloy surface, leading to the passive film rupture. The adsorption of Cl⁻ on the alloy surface decreases the adsorption of oxygen. Thus, the cathodic reaction rate depicted in Eq. (4) is reduced, and the cathodic reaction is inhibited. In Cl⁻ containing media, Eq. (3) could be written as Eqs. (5), (6) and (7) [27].

$$Al^{3+} + H_2O \longrightarrow H^+ + Al(OH)^{2+} \tag{5}$$

$$Al(OH)^{2+} + Cl^- \longrightarrow Al(OH)Cl^+ \tag{6}$$

$$Al(OH)Cl^+ + H_2O \longrightarrow Al(OH)_2Cl + H^+ \tag{7}$$

The increased concentration of Cl⁻ in the basic solution promotes the reaction rate of reaction (6). The intermediate product $Al(OH)^{2+}$ is consumed continuously. Reactions (5) and (7) are also promoted according to the law of mass action. Thus, the anodic reaction of the aluminium alloy is accelerated and the passive film dissolves, resulting in pitting corrosion ultimately.

Fig. 4 exhibits the surface morphologies of AA6063 after immersion for 385 h in the basic solution with Cl⁻ varying from 0 to $300×10^{-5}$ mg/L. Some white spots covered the alloy surface after immersion. EDS revealed that these white spots' compositions were principally composed of Al, O and a slight amount of Si and Zn (marked A in Fig. 4). The corrosion products may consist of amorphous Al_2O_3 or $Al(OH)_3$ as shown in Fig. 5(a). The amount of corrosion products increased when Cl⁻ was added into the basic solution, as indicated in Fig. 4(b). Large and dense corrosion products appeared when the concentration of Cl⁻ increased to $60×10^{-5} \sim 300×10^{-5}$ mg/L. Meanwhile, the number of little white spots decreased, indicating that inhomogeneous corrosion occurred in these environments. In these cases, except for Al and O, a small amount of Fe and Si elementswas detected in the corrosion products (marked B in Fig. 4). It is speculated that the corrosion preferably occurred in the second phase (Mg_2Si) region. Wang et al. [28] and Birbilis et al. [29] reported that pitting corrosion occurs in or around intermetallic compound particles present in the alloy as a result of the difference in potential energy between the matrix and the intermetallic compound. The potential of Mg_2Si is more negative than that of aluminium substrate in NaCl solution. Thus, the second phases dissolve preferentially and Si remain un-attacked. This phenomenon resembles the pitting corrosion of aluminium in aqueous solutions [30-32], and the corrosion is accompanied by the adsorption of EG on the alloy surface. The obvious passive film rupture was observed when the concentration of Cl^- ions increased to $300×10^{-5}$ mg/L [Fig. 4(e)].

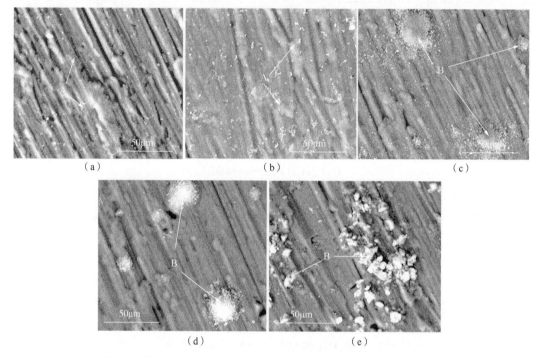

Fig. 4 SEM morphologies of AA6063 in basic solution containing Cl^-:
(a) 0; (b) $30×10^{-5}$ mg/L; (c) $60×10^{-5}$ mg/L; (d) $150×10^{-5}$ mg/L; (e) $300×10^{-5}$ mg/L

3.2 Effects of Cu^{2+} ions

The potentiodynamic polarization curves of AA6063 in thebasic solution solely containing Cu^{2+} from 0 to $165×10^{-5}$ mg/L are shown in Fig. 6. The corrosion potential of AA6063 in the basic solution containing Cu^{2+} ions shifted to a more positive region. The elevation of E_{corr} was caused by the deposition ofCu^{2+} ions on AA6063 surface in the solution, which is similar to that in aqueous solutions [33]. The elevated potentials made the alloy easily approach its pitting corrosion potential and increased the pitting corrosion susceptibility of the alloy. The deposited copper acted as cathodic electrodes. The dissolution rate of aluminium around copper was accelerated due to the galvanic corrosion effect, as in the case of cathodic protection systems with sacrificial anodes [34]. However, the potential of AA6063 decreased when the concentration of Cu^{2+} increased to $165×10^{-5}$ mg/L (still more positive than that in the basic solution). Accordingly, there are more fresh aluminium surfaces exposed to the solution. The passive characteristic of the alloy disappeared when the basic solution contained Cu^{2+}, although its concentration was as low as 16.5×10^{-5} mg/L. The cathodic current density of the alloy shifted to a larger current density region compared with that in the Cu^{2+} ions free basic solution, indicating an accelerating effect on the cathodic reaction of AA6063, especially in the presence of 165×10^{-5} mg/L Cu^{2+}.

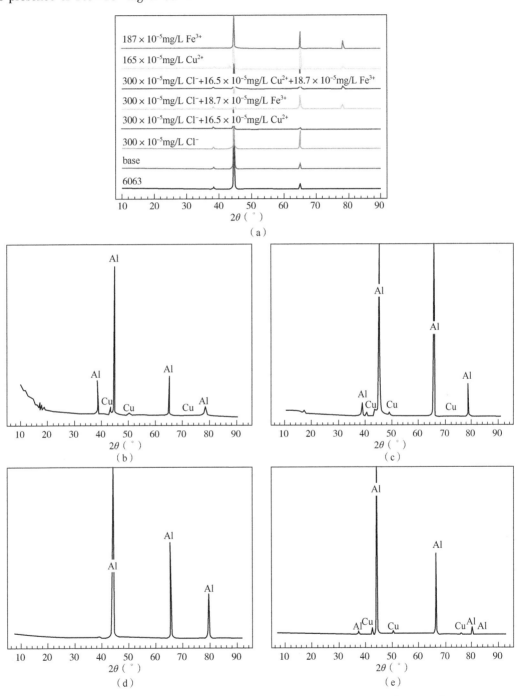

Fig. 5 XRD patterns of AA6063: (a) comprehensive; (b) 165×10^{-5} mg/L Cu^{2+}; (c) 300×10^{-5} mg/L $Cl^- + 16.5 \times 10^{-5}$ mg/L $Cu^{2+} + 18.7 \times 10^{-5}$ mg/L Fe^{3+}; (d) 187×10^{-5} mg/L Fe^{3+}; (e) 300×10^{-5} mg/L $Cl^- + 16.5 \times 10^{-5}$ mg/L Cu^{2+}

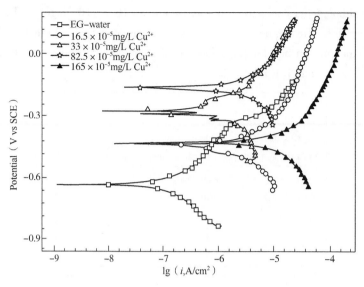

Fig. 6 Potentiodynamic polarization curves of AA6063 in basic solution containing Cu^{2+} ions

The i_{corr} and ΔE values of AA6063 in the basic solution solely containing Cu^{2+} are shown in Fig. 7. i_{corr} increased slowly when the concentration of Cu^{2+} ions was less than 82.5×10^{-5} mg/L, but it increased markedly to about $33 \mu A/cm^2$ when the concentration of Cu^{2+} ions was 165×10^{-5} mg/L, while the i_{corr} of AA6063 in the basic solution was less than $0.5 \mu A/cm^2$. ΔE decreased quickly when Cu^{2+} ions were added into the basic solution. The higher the concentration of Cu^{2+}, the more obvious the decrease of ΔE, indicating the seriously negative effect on the pitting corrosion of the alloy.

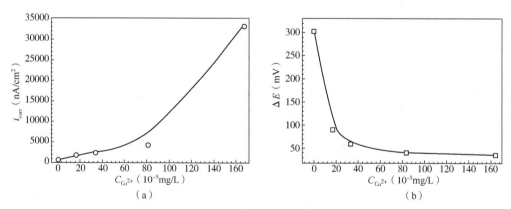

Fig. 7 Fitted results of AA6063 in basic solution containing Cu^{2+}: (a) i_{corr}; (b) ΔE values

Fig. 8 shows the corrosion morphologies of AA6063 in the basic solution solely containing Cu^{2+} varying from 10×10^{-5} to 165×10^{-5} mg/L. White spherical particles marked as A consisting of more than 60% Cu with Al and O on the alloy surface were detected by SEM and EDS. XRD pattern [Fig. 5(b)] further indicated that there was copper detected on AA6063 surface after immersion. Thus, the white spherical particles are mainly composed of copper. The appearance of copper resulted from the displacement reaction between Cu^{2+} ions and aluminium [35], and the overall reactions can simply be expressed as follows:

$$3Cu^{2+} + 2Al \longrightarrow 2Al^{3+} + 3Cu \qquad (8)$$

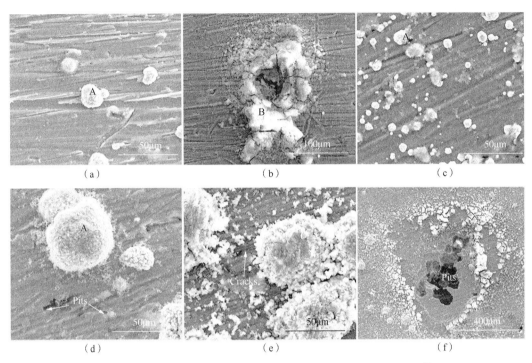

Fig. 8 SEM morphologies of AA6063 in basic solution containing Cu^{2+}:
(a, b) 16.5×10^{-5} mg/L; (c) 33×10^{-5} mg/L; (d) 82.5×10^{-5} mg/L; (e, f) 165×10^{-5} mg/L

In this case, AA6063 resembles another aluminium alloy in the deposition of copper in aqueous solutions containing cupric, i. e. these ions can precipitate on the aluminium surface and give rise to a galvanic corrosion effect [36-37], and then the corrosion of the alloy was accelerated.

Pitting occurred on AA6063 in the basic solution containing Cu^{2+} ions, similar to the experimental observation in aqueous solutions even in the complete absence of Cl^- [38]. The content of Cu around the pit marked as B in Fig. 8(b) was as high as 3.97%, while the content of Cu far away from the pit, marked as C in Fig. 8(b), was as low as 0.5%, indicating that the pits easily nucleated and developed at the sites of Cu deposited. The size and amount of the particles increased with the increase of Cu^{2+} ions concentration in the basic solution [Fig. 8(c) and (d)]. More Cu was deposited when the concentration of Cu^{2+} increased to 165×10^{-5} mg/L, and some tiny cracks formed on the surface. Some large and deep pits were observed, with plenty of broken corrosion products accumulated around the pits [Fig. 8(e) and (f)]. The rupture of the surface film may be the source of the cracks around the pits. Wang et al. [39] reviewed that the crack of the surface film is the reason of corrosion cavities of 7075-T651 alloy. Furthermore, the presence of corrosion cavities promoted the crack initiation. The chemical reaction between Cu^{2+} and aluminium generated Cu and deposits, providing the condition for forming the galvanic couple in which the copper served as a cathode and the surrounding matrix acted as an anode, and the rupture of the surface film on the substrate was accelerated. Similar galvanic corrosion between occluded micro-pit and steel substrate was reported by Li et al.[40].

3.3 Effects of Fe^{3+} ions

Fig. 9 shows the potentiodynamic polarization curves of AA6063 in the basic solution solely containing Fe^{3+} from 0 to 187×10^{-5} mg/L. The corrosion potentials of the alloy elevated to more positive regions in the presence of Fe^{3+}. The cathodic and anodic branches of the polarization curves shifted to larger current density regions, indicating that Fe^{3+} accelerated both cathodic and anodic reactions of AA6063. The detrimental acceleration effect of Fe^{3+} on the cathodic reaction of the alloy coincides with that in aqueous solutions[41-42]. AA6063 was likely to undergo active corrosion, while passive characteristic was found for the alloy in the Fe^{3+} free basic solution.

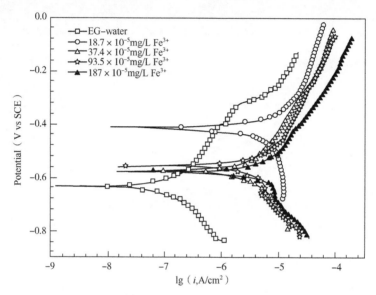

Fig. 9 Potentiodynamic polarization curves of AA6063 in basic solution containing Fe^{3+} ions

Fig. 10 shows i_{corr} and ΔE of AA6063 in the basic solution containing Fe^{3+} from 18.7 to 187×10^{-5} mg/L. i_{corr} of the alloy increased almost linearly with the increase of the concentration of Fe^{3+}. Meanwhile, the ΔE values of the alloy decreased with increasing Fe^{3+} concentration, which indicated that Fe^{3+} also increased pitting corrosion susceptibility of AA6063 in the basic solution.

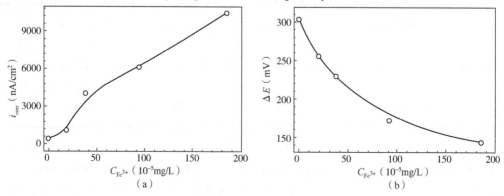

Fig. 10 Fitted results of AA6063 in basic solution containing Fe^{3+} ions:
(a) i_{corr}; (b) ΔE values

Fig. 11 shows the corrosion morphologies of AA6063 in the basic solution containing Fe^{3+} from 18.7×10^{-5} to 187×10^{-5} mg/L. Uniform corrosion occurred when the concentration of Fe^{3+} was less than 37.4×10^{-5} mg/L, and no obvious pitting corrosion was observed. EDS results indicated that the alloy surface compositions chiefly consisted of Al, O and a slight amount of Mg and Si after corrosion. Pitting corrosion occurred when the concentration of Fe^{3+} was more than 93.5×10^{-5} mg/L, and the pits were in oval shaped with relatively smooth perimeters. Some tiny cracks formed on the alloy surface around the pits. There was no Fe both inside and around the pits, and it was not detected even the concentration of Fe^{3+} was 187×10^{-5} mg/L, according to the XRD pattern [Fig. 5 (d)]. It is unlikely that the displacement reaction between Fe^{3+} ions and Al occurred to deposit metallic Fe on aluminium alloy in this environment, which was different from that in aqueous solution. In the latter case, Fe deposited owing to the displacement reaction between Al and Fe^{3+}[43]. When the concentration of Fe^{3+} ions was more than 37.4×10^{-5} mg/L, the corrosion morphologies of the alloy were similar to that of AA6061 in aqueous solutions at low pH [44].

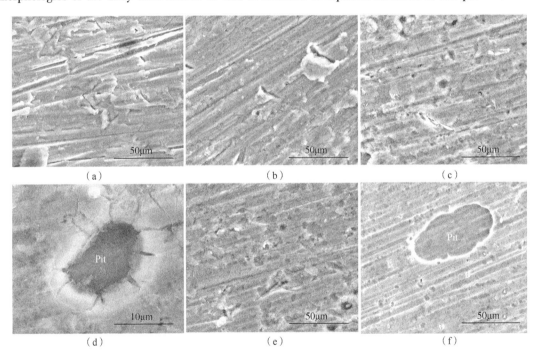

Fig. 11　SEM morphologies of AA6063 in basic solution containing Fe^{3+} ions:
(a) 18.7×10^{-5} mg/L; (b) 37.4×10^{-5} mg/L; (c, d) 93.5×10^{-5} mg/L; (e, f) 187×10^{-5} mg/L

By comparing the electrochemical results and corrosion morphologies of AA6063 in EG−water solutions solely containing Cu^{2+} and Fe^{3+}, it is concluded that the corrosion behaviours of AA6063 are different in corrosion mechanism: Fe^{3+} is inclined to play a role in accelerating uniform corrosion, while Cu^{2+} promotes pitting corrosion.

3.4　Effects of Cl^-+Cu^{2+} ions

Fig. 12 shows the potentiodynamic polarization curves of AA6063 in the basic solution containing Cl^- from 0 to 300×10^{-5} mg/L and 16.5×10^{-5} mg/L Cu^{2+}. The potentials of the alloy elevated to more positive values resembling that in the basic solution solely containing Cu^{2+}. The elevation of corrosion potential was due to the deposition of Cu resulted from the displacement reaction between Cu^{2+} and Al as indicated by the XRD pattern in Fig. 5(e). Galvanic corrosion occurred in this environment owing to the tremendous difference in electrochemical potentials between the deposited Cu and Al (about 2 V in standard electrode potential). The cathodic branches of the alloy shifted to larger current density regions, indicating an acceleration of cathodic reaction. The passive region disappeared in the case of the coexistence of Cl^- and Cu^{2+}, and the pitting corrosion potentials were located in the vicinity of the corrosion potentials, revealing a high susceptibility of AA6063 to pitting corrosion in this environment. The anodic dissolution of AA6063 was related to the concentration of Cl^-, and the higher the concentration of Cl^-, the greater the anodic current density.

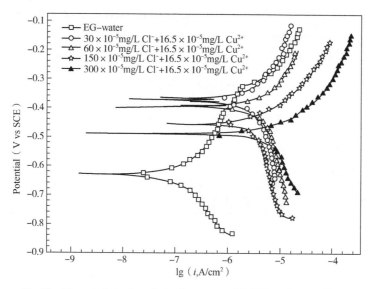

Fig. 12 Potentiodynamic polarization curves AA6063 in basic solution containing Cl^- and 16.5×10^{-5} mg/L Cu^{2+} ions

The i_{corr} and ΔE values of AA6063 in the basic solution containing Cl^- varying from 0 to 300×10^{-5} mg/L and 16.5×10^{-5} mg/L Cu^{2+} are shown in Fig. 13. i_{corr} was insignificant when the EG-water solutions solely contained 30×10^{-5} mg/L Cl^- or Cu^{2+} ions, which were about 0.5 and 0.25 $\mu A/cm^2$, respectively [Fig. 3(b), Fig. 7(a)]. i_{corr} increased when Cl^- and Cu^{2+} were simultaneously added to the basic solution. The higher the concentration of Cl^-, the larger the i_{corr}, as shown in Fig. 13 (a). The i_{corr} of AA6063 was about 5.5 $\mu A/cm^2$ when the concentration of Cl^- increased to 300×10^{-5} mg/L, and it was about 55 and 11 times larger than that in the case of solely containing 300×10^{-5} mg/L Cl^- or 16.5×10^{-5} mg/L Cu^{2+}, respectively. Therefore, the coexistence of Cl^- and Cu^{2+} accelerated the corrosion of AA6063 in the basic solution. The increase of i_{corr} with the higher Cl^- concentration was similar to that of Cu-containing aluminium alloy in aqueous solutions [45]. In a way, the effects of Cu^{2+} in EG-water solutions were similar to that of Cu in the second phase [46-47].

The ΔE values of AA6063 in the basic solution containing Cl^- and Cu^{2+} are shown in Fig. 13 (b). The ΔE value decreased rapidly from 305 mV in the basic solution to about 58 mV in the basic solution containing 30×10^{-5} mg/L Cl^- and 16.5×10^{-5} mg/L Cu^{2+}. The decrease of ΔE values slowed down when the concentration of Cl⁻ increased from 60 to 300×10^{-5} mg/L, and this phenomenon was much like that of AA6063 in the basic solution solely containing Cu^{2+} ions. The results revealed that only a small amount of Cl⁻ is needed to initiate pitting corrosion in the existence of Cu^{2+} ions.

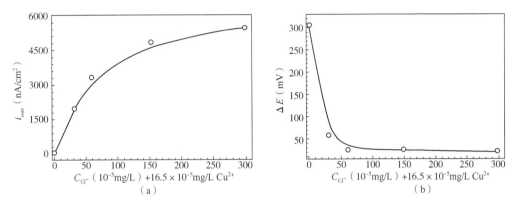

Fig. 13 Fitted results of AA6063 in basic solution containing Cl⁻ and Cu^{2+} ions: (a) i_{corr}; (b) ΔE values

Fig. 14 shows the corrosion morphologies of AA6063 in the basic solution containing Cl⁻ and Cu^{2+}. Pitting corrosion occurred in all cases. Serious pitting corrosion was observed even if the concentrations of Cl⁻ ions were as low as 30×10^{-5} mg/L when the basic solution contained 16.5×10^{-5} mg/L Cu^{2+}. The surface of AA6063 was covered by two types of corrosion products, cotton-shaped and tiny granular particles, with broken passive film dispersed around the pits [Fig. 14(a), (b)]. There was about 2.78% Cu in Fig. 14(b) marked as A at the edge of the pits. The content of Cu inside pits marked as B in Fig. 14(b) was about 0.21%. The corrosion morphologies of AA6063 after immersion in the basic solution containing 16.5×10^{-5} mg/L Cu^{2+} and 60×10^{-5} mg/L Cl⁻ were similar to that in the basic solution containing 30×10^{-5} mg/L Cl⁻ and 16.5×10^{-5} mg/L Cu^{2+} [Fig. 14 (c), (d)]. The corrosion products existed mainly in the form of tiny particles when the concentration of Cl⁻ ions was higher than 150×10^{-5} mg/L. The amount of cotton-shaped corrosion products decreased [Fig. 14(e-h)], accompanying inhomogeneously dispersed white snowflake-shaped particles containing about 4.5% copper detected by EDS. The dark grey particles were composed of Al, O and about 0.65% Cu. The deposition behaviour of Cu^{2+} ions resembled that of AA6063 in the aqueous solution containing Cl⁻ and Cu^{2+} ions [48], and a more uniform deposition of Cu was observed. Large pits formed in the case of the basic solution containing 300×10^{-5} mg/L Cl⁻ and 16.5×10^{-5} mg/L Cu^{2+} ions. The area marked as A in Fig. 14(h) was constituted by about 1.63% Cu, 0.46% Fe, 46.29% Al and 51.62% O, while the compositions of the area marked as B were solely constituted by 53.44% Al and 46.56% O respectively. It can be concluded that the pitting corrosion initiated at copper deposition sites and developed there.

Comparing Fig. 8 and Fig. 14, it is clear that the size of the deposited Cu is much smaller and dispersed in the case of coexistence of Cl⁻ and Cu^{2+} ions, while the deposited Cu in the case of solely containing Cu^{2+} ions is isolated spherical clusters. The observed phenomena resulted from the basis of the aggressive effect of Cl⁻ ions on the passive film and the displacement reaction between Al and Cu^{2+}. Generally, AA6063 is covered by a native passive film in the depth of about 5~200nm.

In the presence of only Cu^{2+} ions, they selectively reacted with Al on the active area with defects. The displacement reaction between Cu^{2+} ions and Al continued at these sites, and heavy clusters finally accumulated. On the contrary, in the presence of Cl^- ions, the passive film was inevitably ruined by these ions. As a result, more active areas were available and more fresh surfaces were exposed to the media, where the displacement reaction took place. Thus, a dispersed distribution of Cu was observed in the case of coexistence of Cl^- and Cu^{2+} ions in the base solution.

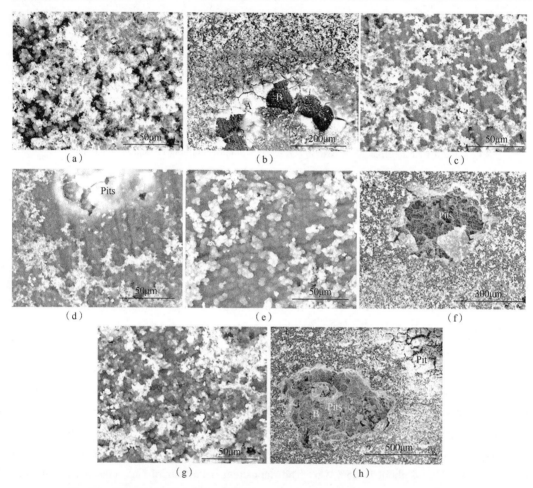

Fig. 14 SEM morphologies of AA6063 in basic solution containing Cl^- and Cu^{2+} ions:
(a, b) 30×10^{-5} mg/L Cl^- + 16.5×10^{-5} mg/L Cu^{2+}; (c, d) 60×10^{-5} mg/L Cl^- + 16.5×10^{-5} mg/L Cu^{2+};
(e, f) 150×10^{-5} mg/L Cl^- + 16.5×10^{-5} mg/L Cu^{2+}; (g, h) 300×10^{-5} mg/L Cl^- + 16.5×10^{-5} mg/L Cu^{2+}

3.5 Effects of Cl^- + Fe^{3+} ions

Fig. 15 shows the potentiodynamic polarization curves of AA6063 in the basic solution containing Cl^- and 18.7×10^{-5} mg/L Fe^{3+}. The shapes of the curves were more similar to those in the basic solution solely containing Cl^- ions, as exhibited in Fig. 1. However, the cathodic branches shifted to a larger current density region, indicating an accelerated effect on the cathodic reaction. It was clear that the cathodic reaction was greatly accelerated with the increase of Cl^- concentration. When the concentrations of Cl^- ions were less than 60×10^{-5} mg/L, the higher the concentration of Cl^- ions, the more severe the cathodic reaction. Unlike the basic solution containing both Cl^- and Cu^{2+} ions, there existed a passive region above the corrosion potential, where the anodic current density increased slowly until the pitting potential was reached. The anodic current density also increased with increasing concentration of Cl^- ions. At higher chloride concentrations, the local chemistry of pitting became aggressive and then accelerated the dissolution of the alloy [49]. The breakdown or pitting corrosion potential of AA6063 was almost identical to its corrosion potential when the concentration of Cl^- was higher than 150×10^{-5} mg/L, indicating a high pitting corrosion susceptibility of AA6063 in this environment, and the passive film on the alloy surface was prone to be broken.

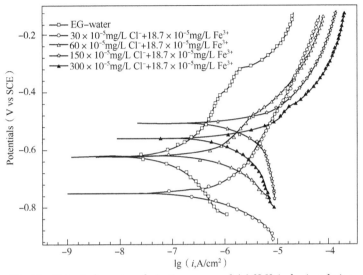

Fig. 15 Potentiodynamic polarization curves of AA6063 in basic solution containing Cl^- and 18.7×10^{-5} mg/L Fe^{3+} ions

Fig. 16 shows the i_{corr} and ΔE values of AA6063 in the basic solution containing Cl^- varying from 30×10^{-5} to 300×10^{-5} mg/L and 18.7×10^{-5} mg/L Fe^{3+}. i_{corr} was small when the concentrations of Cl^- ions were lower than 60×10^{-5} mg/L, while i_{corr} increased rapidly when the concentrations of Cl^- ions were higher than 150×10^{-5} mg/L. The i_{corr} was about $21\mu A/cm^2$ when the concentration of Cl^- was 300×10^{-5} mg/L, which is 11 times larger than that in the basic solution containing 18.7×10^{-5} mg/L Fe^{3+} ions, 380 and 277 times larger than that in the basic solution and basic solution solely containing 300×10^{-5} mg/L Cl^- ions, respectively.

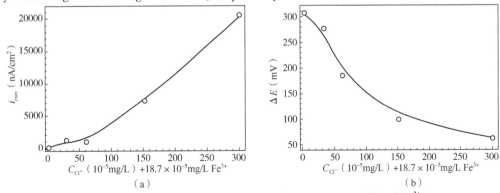

Fig. 16 Fitted results of AA6063 in the basic solution containing Cl^- and Fe^{3+} ions:
(a) i_{corr}; (b) ΔE values

Fig. 16(b) clearly indicates that the ΔE values decreased slowly with the increase of the concentration of Cl^- ions. The larger the concentration of Cl^-, the more pronounced the decrease of ΔE values, indicating an increase in susceptibility to pitting corrosion of AA6063 in these environments.

Fig. 17 shows the corrosion morphologies of AA6063 in the basic solution containing Cl^- varying from 30×10^{-5} to 300×10^{-5} mg/L and 18.7×10^{-5} mg/L Fe^{3+}. Some local passive films were broken when 30×10^{-5} mg/L Cl^- and 18.7×10^{-5} mg/L Fe^{3+} ions were present in the basic solution, but no obvious pitting corrosion occurred at the other sites [Fig. 17(a), (b)]. EDS results indicate that the compositions of the white block corrosion products [marked as A in Fig. 17(b)] contained about 0.16% Fe element, which was on the scale of the base alloy, as well as 23.51% Al and 61.74% O, indicating an adsorption of EG on these sites. Some tiny cracks were observed in the areas where the passive film was ruptured. No Fe element was detected at the sites far away from the broken passive film [marked as B in Fig. 17(b)]. It revealed that the Fe-containing second phases ($Al_9Fe_2Si_2$) are prone to be attacked during exposure owing to the thinner passive film on the alloy surface. The amount of white punctiform corrosion products increased when the concentration of Cl^- ions increased to 60×10^{-5} mg/L, and the breakdown of passive film accompanying obvious pit formation was detected [Fig. 17(c), (d)]. The composition of white punctiform corrosion products resembled that in the basic solution solely containing Cl^-. The white corrosion products around the pits marked as A in Fig. 17(d) were constituted by Fe and Si elements, about 0.49% and 0.36%, respectively, which indicated that the pits initiated at the sites of the second phases. The composition of the area apart from the pit marked as B in Fig. 17(d) was similar to that in Fig. 17(b). The corrosion behaviour of the alloy was similar to that in aqueous solutions [50]. The surface of AA6063 was covered by less amount of corrosion products after immersion in basic solution containing 150×10^{-5} mg/L Cl^- and 18.7×10^{-5} mg/L Fe^{3+} ions, and obvious pits were detected [Fig. 17(e), (f)]. Most of the alloy surfaces have almost no corrosion products. Still, pits with irregular perimeter were found after immersion in the basic solution containing 300×10^{-5} mg/L Cl^- and 18.7×10^{-5} mg/L Fe^{3+}, as shown in Fig. 17(g) and (h). In the case that higher concentration of Cl^- and 18.7×10^{-5} mg/L Fe^{3+} ions was contained, the corrosion morphology of the pits was different from that solely containing Fe^{3+} ions. It indicated that the pit growth mechanism was different. In the case of Fe^{3+} only, the pit growth might be caused by the dissolution of the Al substrate exposed to the media. In contrast, in the case of coexistence of Cl^- and Fe^{3+} ions, the aggressive Cl^- ions invaded the passive pit walls [51-53], new pit nuclei formed randomly, and the irregular perimeter was observed.

3.6 Effects of $Cl^- + Cu^{2+} + Fe^{3+}$ ions

Fig. 18 shows the potentiodynamic polarization curves of AA6063 in the basic solution containing Cl^- varying from 30×10^{-5} to 300×10^{-5} mg/L, 16.5×10^{-5} mg/L Cu^{2+} and 18.7×10^{-5} mg/L Fe^{3+} ions. The potentials of AA6063 shifted to more positive regions when the basic solution simultaneously contained Cl^-, Cu^{2+} and Fe^{3+} ions. It has been found that the uniform and pitting corrosion potentials of the alloy are influenced by the ions in the solutions [54-56]. In this study, the added Cl^- decreased the corrosion potential, while Cu^{2+} and Fe^{3+} behaved in the opposite way. In fact, the measured potentials increased to a more positive region. Thus, the Cu^{2+} and Fe^{3+} have a more significant effect on the corrosion potential than Cl^- ions. The fact is that the adsorption of Cl^- ions is promoted at high potentials [57]. Thus, the nucleation and growth of pits were accelerated in the coexistence of Cl^-, Cu^{2+} and Fe^{3+} ions. AA6063 underwent active dissolution in all cases. Both cathodic and anodic branches of curves shifted to a larger current density region, indicating an acceleration of corrosion. The pitting potential was almost identical to the corrosion potential. In the presence of Cu^{2+} and Fe^{3+} ions, the alloy had high pitting corrosion susceptibility even if the concentrations of Cl^- ions were as low as 30×10^{-5} mg/L.

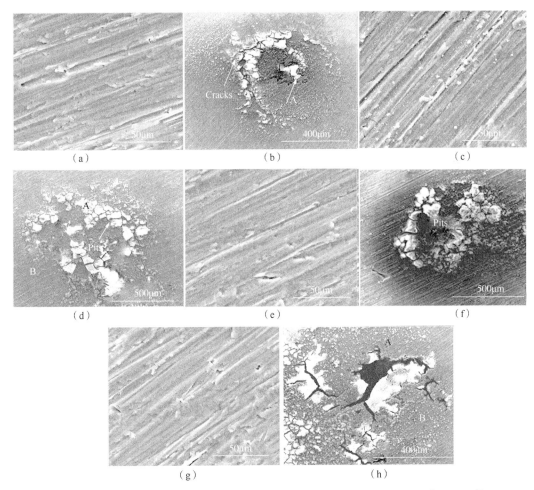

Fig. 17 SEM morphologies of AA6063 in basic solution containing Cl^- and 18.7×10^{-5} mg/L Fe^{3+} ions: (a, b) 30×10^{-5} mg/L $Cl^- + 18.7 \times 10^{-5}$ mg/L Fe^{3+}, (c, d) 60×10^{-5} mg/L $Cl^- + 18.7 \times 10^{-5}$ mg/L Fe^{3+}; (e, f) 150×10^{-5} mg/L $Cl^- + 18.7 \times 10^{-5}$ mg/L Fe^{3+}; (g, h) 300×10^{-5} mg/L $Cl^- + 18.7 \times 10^{-5}$ mg/L Fe^{3+}

Fig. 19 shows the i_{corr} and ΔE values of AA6063 in the basic solution containing Cl^-, 16.5×10^{-5} mg/L Cu^{2+} and 18.7×10^{-5} mg/L Fe^{3+} ions. In the presence of Cu^{2+} and Fe^{3+} ions, the i_{corr} of AA6063 was about 32 times larger than that in the basic solution containing 30×10^{-5} mg/L Cl^-. The i_{corr} increased slowly when Cl^- ions concentration increased to 60×10^{-5} mg/L. However, an abrupt increase in the i_{corr} was observed when the concentrations of Cl^- ions were 150×10^{-5} mg/L. The i_{corr}

was about 32 μA/cm² when the concentrations of Cl⁻ ions were 300×10⁻⁵ mg/L [Fig. 19(a)], two orders of magnitude larger than that in the basic solution and in the basic solution containing 300× 10⁻⁵ mg/L Cl⁻ ions respectively.

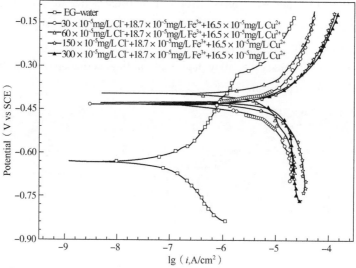

Fig. 18 Potentiodynamic polarization curves of AA6063 in basic solution containing concentrations of Cl⁻, 16.5×10⁻⁵ mg/L Cu^{2+} and 18.7×10⁻⁵ mg/L Fe^{3+} ions

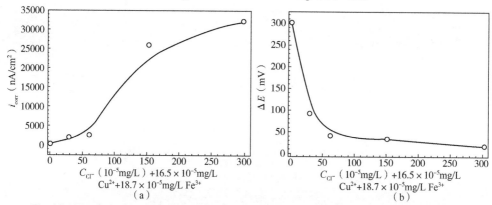

Fig. 19 Fitted results of AA6063 in the basic solution containing Cl⁻, Cu^{2+} and Fe^{3+} ions:
(a) i_{corr}; (b) ΔE values

Fig. 19(b) reveals that the ΔE values decreased suddenly in the basic solution containing Cl⁻, 16.5×10⁻⁵ mg/L Cu^{2+} and 18.7×10⁻⁵ mg/L Fe^{3+} ions. The maximum decrease took place in the case of containing 30×10⁻⁵ mg/L Cl⁻ ions were, and the decrease became extremely slow after that. Pitting corrosion susceptibility of AA6063 was increased by the coexistence of Cl⁻, Cu^{2+} and Fe^{3+} ions in the basic solutions.

Fig. 20 shows the corrosion morphologies of AA6063 in basic solution containing Cl⁻, 16.5×10⁻⁵ mg/L Cu^{2+} and 18.7×10⁻⁵ mg/L Fe^{3+} ions. It was observed that thick corrosion products accompanied with white snowflake Cu accumulated on the AA6063 surface and severe pitting corrosion occurred [Fig. 20(a), (b)]. Cl element was detected in the area marked as A in Fig. 20(b), and the amount of Cl was about 0.5%, indicating the adsorption of Cl⁻ ions, and the other elements were Al and O. The content of Cu around the pit marked as B in Fig. 20(b) was as high as 7.0%, and the region also contained Al and O. More serious corrosion occurred when the concentration of Cl⁻ ions increased to 60×10^{-5} mg/L, and the snowflake particles in Fig. 20(c) were Cu. The deep grey areas were composed of Al and O. About 1.3% Cu was detected around the pit [marked as A in Fig. 20(d)], and no Fe element was detected. There was no Cu or Fe element detected inside the pit [marked as B in Fig. 20(d)], which was mainly composed of Al, O and Mg, and their content was 95.7%, 3.8% and 0.5%, respectively. The corrosion morphologies of AA6063 after immersion in the basic solution containing 150×10^{-5} mg/L Cl⁻ resembled that in the basic solution containing 60×10^{-5} mg/L Cl⁻ [Fig. 20(e), (f)]. Thicker corrosion products covered the AA6063 surface and pitting corrosion occurred when the concentration of Cl⁻ was 300×10^{-5} mg/L as exhibited in Fig. 20(g) and (h). The composition in the region marked as A also contained 5.2% Cu, while no Cu and Fe elements were detected inside the pits marked as B.

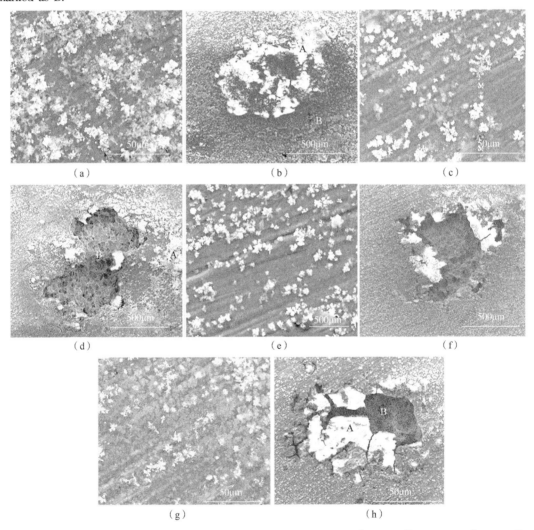

Fig. 20 SEM morphologies of AA6063 in basic solution containing Cl⁻, 16.5×10^{-5} mg/L Cu²⁺ and 18.7×10^{-5} mg/L Fe³⁺ ions: (a, b) 30×10^{-5} mg/L Cl⁻ + 16.5×10^{-5} mg/L Cu²⁺ + 18.7×10^{-5} mg/L Fe³⁺; (c, d) 60×10^{-5} mg/L Cl⁻ + 16.5×10^{-5} mg/L Cu²⁺ + 18.7×10^{-5} mg/L Fe³⁺; (e, f) 150×10^{-5} mg/L Cl⁻ + 16.5×10^{-5} mg/L Cu²⁺ + 18.7×10^{-5} mg/L Fe³⁺; (g, h) 300×10^{-5} mg/L Cl⁻ + 16.5×10^{-5} mg/L Cu²⁺ + 18.7×10^{-5} mg/L Fe³⁺

3.7 Comprehensive analysis

Fig. 21 shows i_{corr} and ΔE values of AA6063 in the basic solution containing Cl^-, Cu^{2+} and Fe^{3+} Evidently, the i_{corr} increased in the order of $Cl^- < Cl^- + Fe^{3+} < Cl^- + Cu^{2+} \approx Cl^- + Cu^{2+} + Fe^{3+}$ [Fig. 21 (a)] when the concentrations of Cl^- ions were less than 60×10^{-5} mg/L. In these environments, the ΔE values decreased in the order of $Cl^- \approx Cl^- + Fe^{3+} \gg Cl^- + Cu^{2+} \approx Cl^- + Cu^{2+} + Fe^{3+}$. The i_{corr} of the alloy increased in the order of $Cl^- < Cl^- + Cu^{2+} < Cl^- + Fe^{3+} < Cl^- + Cu^{2+} + Fe^{3+}$ when the concentrations of Cl^- were more than 60×10^{-5} mg/L. ΔE decreased slowly with the increase of the concentration of Cl^- ions for the alloy in the case of containing Fe^{3+}, but ΔE quickly decreased to almost 0 mV in the cases of containing Cu^{2+} and $Cu^{2+} + Fe^{3+}$ ions, and almost maintained that level until 300×10^{-5} mg/L, and ΔE decreased in the order of $Cl^- > Cl^- + Fe^{3+} \gg Cl^- + Cu^{2+} \approx Cl^- + Cu^{2+} + Fe^{3+}$.

Accordingly, Cl^- had less influence on the uniform corrosion, but affected the pitting corrosion of AA6063 significantly. The coexistence of $Cl^- + Cu^{2+}$, $Cl^- + Fe^{3+}$, especially the combination of $Cl^- + Cu^{2+} + Fe^{3+}$, markedly increased the corrosion of AA6063. And the pitting corrosion of the alloy was extremely accelerated. In the cases that Cu^{2+} and Fe^{3+} ions were contained, the higher the concentration of Cl^- ions, the more serious the corrosion. Moreover, Cu^{2+} ions accelerated pitting corrosion of AA6063 more than Fe^{3+} both in EG-water solutions with and without Cl^- ions.

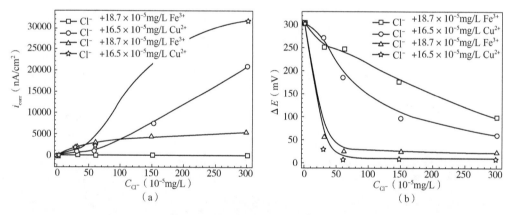

Fig. 21 Fitted results of AA6063:
(a) i_{corr}; (b) ΔE values

3.8 Corrosion rates of AA6063

Fig. 22 shows the corrosion rate of AA6063 after immersed for 385 h in the basic solutions containing Cl^-, Cu^{2+} and Fe^{3+} ions at 50 ℃. The mass of AA6063 was obtained after immersion in the basic solution solely containing Cl^- ions varying from 30×10^{-5} to 300×10^{-5} mg/L [Fig. 22(a)], and the mass gain was in a range of 0.06~0.012 mg/cm². The mass gain of the alloy in the Cl^- ions free basic solution resulted from the adsorption of ethylene glycol to the alloy surface and the adhesion of the corrosion products. Bazeleva et al.[58] studied the adsorption of EG on the metal surface in detail. It was found that EG adsorbed on the surface of AMg_3 alloy, in such a way, retards the hydration of aluminium oxide and leads to the formation of surface complexes of aluminium with EG and the products of its oxidation. A decreased mass gain was detected when Cl^- ions were present in the basic solution. Two aspects should be taken into account to explain the role of Cl^- ions on the corrosion of AA6063. On the one hand, the corrosivity of the solutions increased with the addition of Cl^- ions. In fact, some tiny white cotton silk-shaped corrosion products were observed floating in the solutions containing Cl^- ions during immersion, which resulted in mass loss of the alloy. On the other hand, Cl^- ions were adsorbed easily on the alloy surface [59-60], resulting in less adsorption of ethylene glycol. Its molecular weight is larger than that of Cl^- ions. Accordingly, the adsorption of EG is the main reason for the mass gain in the immersion procedure. Cl^- ions have a slight effect on uniform corrosion but play an important role in the pitting corrosion of AA6063, which is indicated by the mass change of the alloy.

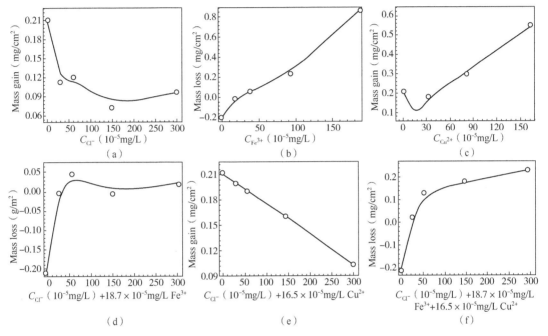

Fig. 22 Corrosion rates of AA6063:
(a) Cl^-; (b) Fe^{3+}; (c) Cu^{2+}; (d) $Cl^- + 18.7 \times 10^{-5}$ mg/L Fe^{3+};
(e) $Cl^- + 16.5 \times 10^{-5}$ mg/L Cu^{2+}; (f) $Cl^- + 18.7 \times 10^{-5}$ mg/L $Fe^{3+} + 16.5 \times 10^{-5}$ mg/L Cu^{2+}

The mass of AA6063 in the basic solution containing Fe^{3+} ions decreased after immersion for 385 h in the basic solution containing Fe^{3+} [Fig. 22(b)]. The mass loss of the alloy increased linearly with the increase of Fe^{3+} ions, which revealed that Fe^{3+} had an accelerating effect on the corrosion of the alloy.

In the basic solution, $Fe_2(SO_4)_3$ hydrolyses as follows:

$$Fe_2(SO_4)_3 + 6H_2O \rightleftharpoons 2Fe(OH)_3 + 3H_2SO_4 \quad (9)$$

According to Eq. (9), the generated H_2SO_4 leads to an decrease in pH value of the solution. In fact, the pH value of the solution was 4 when the concentration of Fe^{3+} ions was up to 187×10^{-5} mg/L. Some reports have revealed that active dissolution of aluminium alloys occurred in acidic solutions [61-62]. With the concentration increase of $Fe_2(SO_4)_3$, more H_2SO_4 generated according to Eq. (9), and the dissolution rate of AA6063 increased. This phenomenon coincided with the results of aluminium corrosion in an aqueous citric acid solution [63]. Moreover, the displacement reaction between Fe^{3+} and aluminium was not likely to occur in these circumstances. Thus, AA6063 was more prone to undergo uniform corrosion in the basic solution only containing $Fe_2(SO_4)_3$ due to the decrease of pH.

The mass of AA6063 increased markedly after immersion for 385h [Fig. 22(c)] in the basic solution solely containing Cu^{2+}, which was different from that in the basic solution containing only Fe^{3+}. The mass gain of the alloy increased almost linearly with the increase of the Cu^{2+} concentration. In the basic solution containing $CuSO_4$, the pH value was 4 when the concentrations of Cu^{2+} ions were up to 165×10^{-5} mg/L. However, the displacement reaction between Cu^{2+} and aluminium as described in Eq. (8) was the main reaction in this case, which was proved by XRD, SEM and EDS, and the hydrolysis of Cu^{2+} ions could be neglected. The deposited Cu resulted in a mass gain. More Al^{3+} ions dissolved in the solution owing to the displacement reaction, and their hydrolysis brought about an environment with a low pH value. The consequence was the dissolution of the passive film on the alloy surface, which resulted in more fresh Al was exposed to the solution. The displacement reaction between Cu^{2+} ions and Al was sped up. According to Eq. (8), the deposited Cu is heavier than the dissolved Al. Thus, AA6063 exhibited a large mass gain.

The mass of the alloy was obtained after immersion for 385 h in the basic solution containing Cl^- and 18.7×10^{-5} mg/L Fe^{3+} ions, as shown in Fig. 22(d). A slight mass gain of the alloy was observed, even in the case of containing 187×10^{-5} mg/L Fe^{3+} ions, and the mass gain of the alloy was 1~2 order of magnitude smaller than that in the basic solution solely containing Fe^{3+}. Thus, the alloy behaved as well in the coexistence of Cl^- and Fe^{3+} ions as it did in the basic solution solely containing Cl^-. It is unlikely for the alloy to corrode uniformly in the presence of Cl^- and Fe^{3+} ions. The addition of Fe^{3+} ions primarily accelerated the pitting corrosion of AA6063 in the presence of Cl^- ions, as indicated by SEM results.

The mass of AA6063 increased after immersion for 385 h in the basic solution containing Cl^- and 16.5×10^{-5} mg/L Cu^{2+} ions [Fig. 22(e)]. The mass gain of the alloy decreased almost linearly with the increase of the Cl^- ions concentration. In these cases, the displacement reaction between cupric ions and Al also occurred, but the Cl^- ions easily attacked the fresh aluminium surface, and a large number of cotton-shaped corrosion products floating in the solution were observed during immersion, which resulted in the mass loss of the alloy. There is a possibility that the deposited Cu is taken away from the alloy surface by the corrosion products and dissolves in solutions.

The mass of AA6063 alloy decreased after immersion for 385 h in the basic solution containing Cl^-, 16.5×10^{-5} mg/L Cu^{2+} and 18.7×10^{-5} mg/L Fe^{3+} ions [Fig. 22(f)]. Mass loss of AA6063 increased with the increase of the Cl^- ions concentration. It should be noticed that the mass increased when the alloy was in the basic solution solely containing Cl^-, and the corrosion process may be the same as the case of coexistence of Cl^-, Cu^{2+} and Fe^{3+} ions. Cl^- initiated the pit nucleation. At the same time, the displacement reaction between Cu^{2+} and Al resulted in the deposition of Cu, leading to the galvanic corrosion of aluminium alloy owing to the electrochemical difference of Cu and Al. The chemical reaction, not displacement reaction, may also occur between Fe^{3+} and Al as described in Eq. (10).

$$3Fe^{3+} + Al = Al^{3+} + 3Fe^{2+} \tag{10}$$

The forming Fe^{2+} remained in the solutions and was further oxidized by oxygen dissolved in solutions to form Fe^{3+} ions again. Another effect of Fe^{3+} in the solution is that its hydrolysis resulted in a decrease in the pH value of the solution, leading to the dissolution of aluminium. In these cases, more white cotton-shaped corrosion products were observed during immersion. Therefore, Cl^- is the main cause of pitting corrosion, and Cu^{2+} and Fe^{3+} accelerate the pitting corrosion of the alloy. The mass loss of the alloy is more attributable to the increased ionic strength of the solutions and to the more H^+ from the hydrolysis of Fe^{3+} and dissolved Al^{3+}.

4 Conclusion

The effects of Cl^-, Cu^{2+} and Fe^{3+} on the corrosion behaviour of AA6063 in EG and water solutions at 50℃ were investigated. The results revealed that the corrosion behaviour of AA6063 in EG-water solutions is similar to that in aqueous solutions, and Cl^- plays an important role in pitting corrosion of the alloy. In EG-water solutions solely containing Fe^{3+} or Cu^{2+} ions, pitting corrosion will occur when the concentration of Fe^{3+} is more than 93.5×10^{-5} mg/L, while pits initiate readily even at a concentration of Cu^{2+} as low as 16.5×10^{-5} mg/L. Fe^{3+} is likely to promote uniform corrosion by its hydrolysis. In contrast, Cu^{2+} is more prone to accelerate pitting corrosion of AA6063 owing to the deposition of copper. Serious pitting corrosion occurs on the alloy in the presence of Cl^-+Cu^{2+} and Cl^-+Fe^{3+}, especially in the case of coexistence of $Cl^-+Cu^{2+}+Fe^{3+}$.

Acknowledgements

This work was financially supported by the National Natural Science Foundation of China (No. 51531007), the Basic Research and Strategic Reserve Technology Research Fund (project of CNPC) (2020D-5008(2020Z-07)) and the Major Science and Technology Project of CNPC (2019D-2311).

References

[1] SAMIENTO-BUSTOS E, GONZÁLEZ-RODRIGUEZ J, G, URUCHURTU J, et al. Corrosion behavior of iron-based alloys in the LiBr + ethylene glycol + H_2O mixture [J]. Corrosion Science, 2009, 51(5): 1107-1114.

[2] SONG G L, STJOHN D. Corrosion behaviour of magnesium in ethylene glycol [J]. Corrosion Science, 2004, 46(6): 1381-1399.

[3] ANDREATTA F, DRUART M E, LANZUTTI A, et al. Localized corrosion inhibition by cerium species on clad AA2024 aluminium alloy investigated by means of electrochemical micro-cell [J]. Corrosion Science, 2012, 65: 376-386.

[4] TAN Y J, AUNG N N, LIU T. Evaluating localised corrosion intensity using the wire beam electrode [J]. Corrosion Science, 2012, 63: 379-386.

[5] ARIBO S, FAKOREDE A, IGE Q, et al. Erosion-corrosion behavior of aluminum alloy 6063 hybrid composite [J]. Wear, 2017, 376: 608-614.

[6] XU W F, LIU J H, ZHU H Q. Pitting corrosion of friction stir welded aluminum alloy thick plate in alkaline chloride solution [J]. Electrochimica, Acta, 2010, 55(8): 2918-2923.

[7] MEHDIZADE M, EIVANI A R, SOLTANIEH M. Effects of reduced surface grain structure and improved particle distribution on pitting corrosion of AA6063 aluminum alloy [J]. Journal of Alloys and Compounds,

2020, 838: 155464.

[8] LIU Y, MENG G Z, CHENG Y F. Electronic structure and pitting behavior of 3003 aluminum alloy passivated under various conditions[J]. Electrochimica, Acta, 2009, 54: 4155-4163.

[9] MACDONALD D D. The point defect model for the passive state [J]. Journal of the Electrochemical Society, 1992, 139(12): 3434-3449.

[10] KHARITONOV D S, ÖRNEK C, CLAESSON P M, et al. Corrosion inhibition of aluminum alloy AA6063-T5 by vanadates: microstructure characterization and corrosion analysis [J]. Journal of the Electrochemical Society, 2018, 165(3): C116-C126.

[11] WONG D, SWETTE L, COCKS F H, Aluminum corrosion in uninhibited ethylene glycol-water solutions [J]. Journal of the Electrochemical Society, 1979, 126(1): 11-15.

[12] ABIOLA O K, OTAIGBE J O E. Effect of common water contaminants on the corrosion of aluminium alloys in ethylene glycol-water solution [J]. Corrosion Science, 2008, 50(1): 242-247.

[13] ZHOU W, AUNG N N, CHOUDHARY A, et al. Heat-transfer corrosion behaviour of cast Al alloy [J]. Corrosion Science, 2008, 50(12): 3308-3313.

[14] MAZHAR A A, ARAB S T, NOOR E A. Electrochemical behaviour of Al-Si alloys in acid and alkaline media [J]. Bulletion of Electrochemistry, 2001, 17(10): 449-458.

[15] ESENIN V N, DENISOVICH L I. Contact corrosion of metals in aqueous and organic-aqueous environments. II. concentrated glycol-water solutions [J]. Protection of Metals and Physical Chemistry of Surfaces, 2009, 45(1): 95-99.

[16] KHOMAMI M N, DANAEE I, ATTAR A A, et al. Effects of NO_2^- and NO_3^- ions on corrosion of AISI 4130 steel in ethylene glycol + water electrolyte [J]. Transcations of the Indian Institute of Metals, 2012, 65(3): 303-311.

[17] NGUYEN T H, FOLEY R T. The chemical nature of aluminum corrosion: II. the initial dissolution step [J]. Journal of the Electrochemical Society, 1982, 129(1): 27-32.

[18] MCCAFFERTY E. Sequence of steps in the pitting of aluminum by chloride ions [J]. Corrosion Science, 2003, 45(7): 1421-1438.

[19] WANG Y B, LI H F, CHENG Y, et al. Corrosion performances of a nickel-free Fe-based bulk metallic glass in simulated body fluids [J]. Electrochemistry Communications, 2009, 11(11): 2187-2190.

[20] DEROSE J A, SUTER T, BAŁKOWIEC A, et al. Localised corrosion initiation and microstructural characterisation of an Al 2024 alloy with a higher Cu to Mg ratio [J]. Corrosion Science, 2012, 55: 313-325.

[21] GHEEM E V, VEREECKEN J, PEN C L. Influence of different anions on the behaviour of aluminium in aqueous solutions [J]. Journal of Applied Electrochemistry, 2002, 32(11): 1193-1200.

[22] EL-ETRE A Y. Inhibition of aluminum corrosion using opuntia extract [J]. Corrosion Science. 2003, 45(11): 2485-2495.

[23] ZHANGA G A, XUA L Y, CHENG Y F. Mechanistic aspects of electrochemical corrosion of aluminum alloy in ethylene glycol-water solution [J]. Electrochimica Acta 2008(53): 8245-8252.

[24] SHERIFA E M, PARK S M. Effects of 1, 5-naphthalenediol on aluminum corrosion as a corrosion inhibitor in 0.50 M NaCl [J]. Journal of the Electrochemical Society, 2005, 152(6): B205-B211.

[25] BADAWY W A, AL-KHARAFI F M, EL-AZAB A S. Electrochemical behaviour and corrosion inhibition of Al, Al-6061 and Al-Cu in neutral aqueous solutions [J]. Corrosion Science. 1999, 41(4): 709-727.

[26] LIU M R, LU X, YIN Q, et al. Effect of acidified Aerosois on initial corrosion behavior of Q235 carbon steel [J]. Acta Metallurgica Sinica(English Letters), 2019, 32(8): 995-1006.

[27] SZKLARSKA-SMIALOWSKA Z. Pitting corrosion of aluminum [J]. Corrosion Science. 1999, 41(9):

1743-1767.

[28] WANG B, ZHANG L, SU Y, et al. Corrosion behavior of 5A05 aluminum alloy in NaCl solution [J]. Acta Metallurgica Sinica (English Letters), 2013, 26(5): 581-587.

[29] BIRBILIS N, BUCHHEIT R G. Electrochemical characteristics of intermetallic phases in aluminum alloys: an experimental survey and discussion [J]. Journal of the Electrochemical Society, 2005, 152(4): B140-B151.

[30] BUCHHEIT R G, GRANT R P, HIAVA P F, et al. Local dissolution phenomena associated with S phase (Al_2CuMg) particles in aluminum alloy 2024-T3 [J]. Journal of the Electrochemical Society, 1997, 144(8): 2621-2628.

[31] LIU Y, CHENG Y F. Role of second phase particles in pitting corrosion of 3003 Al alloy in NaCl solution [J]. Materials and Corrosion, 2010, 61(3): 211-217.

[32] BOAG A, TAYLOR R J, MUSTER T H, et al. Stable pit formation on AA2024-T3 in a NaCl environment [J]. Corrosion Science, 2010, 52: 90-103.

[33] KHEDR M G A, LASHIEN A M S. The role of metal cations in the corrosion and corrosion inhibition of aluminium in aqueous solutions [J]. Corrosion Science, 1992, 33(1): 137-151.

[34] SZABÓ S, BAKOS I. Cathodic protection with sacrificial anodes [J]. Corrosion Reviews, 2006, 24(3-4): 231-280.

[35] BAKOS I, SZABÓ S. Corrosion behaviour of aluminium in copper containing environment [J]. Corrosion Science, 2008, 50(1): 200-205.

[36] OBISPO H M, MURR L E, ARROWOO R M, et al. Copper deposition during the corrosion of aluminum alloy 2024 in sodium chloride solutions [J]. Journal of Materials Science, 2000, 35(14): 3479-3495.

[37] MURTHY K S N, AMBAT R, DWARAKADASA E S. The role of metal cations on the corrosion behavior of 8090-T851 alloy in a pH 2.0 solution [J]. Corrosion Science, 1994, 36(10): 1765-1775.

[38] BLACKWOOD D J, CHONG A S L. Pitting corrosion on aluminium in absence of chloride [J]. British Corrosion Journal, 1998, 33(3): 219-224.

[39] WANG L W, LIANG J M, UI H, et al. Quantitative study of the corrosion evolution and stresscorrosion cracking of high strength aluminum alloys in solution and thin electrolyte layercontaining Cl^- [J]. Corrosion Science, 2021(178).

[40] LIA G X, WANG L W, WU H L, et al. Dissolution kinetics of the sulfide-oxide complex inclusion and resulting localized corrosion mechanism of X70 steel in deaerated acidic environment [J]. Corrosion Science, 2020(174).

[41] NISANCIOGLU K. Electrochemical behavior of sluminum-base intermetallics containing iron [J]. Journal of the Electrochemical Society; 1990, B7(1): 69-77.

[42] AMBAT R, DAVENPORT A J, SCAMANS G M, et al. Effect of iron-containing intermetallic particles on the corrosion behaviour of aluminium [J]. Corrosion Science, 2006, 48(1): 3455-3471.

[43] BÖHNI H, UHLIG H H. Environmental factors affecting the critical pitting potential of aluminum [J]. Journal of the Electrochemical Society, 1969, 116(7): 906-910.

[44] ZAID B, SAIDI D, BENZAID A, et al. Effects of pH and chloride concentration on pitting corrosion of AA6061 aluminum alloy [J]. Corrosion Science, 2008, 50(7): 1841-1847.

[45] AMBAT R, DWARAKADASA E S. Studies on the influence of chloride ion and pH on the electrochemical behaviour of aluminium alloys 8090 and 2014 [J]. Journal of Applied Electrochemistry, 1994, 24(9): 911-916.

[46] VUKMIROVIC M B, DIMITROV N, SIERADZKI K, Dealloying and corrosion of Al alloy 2024-T3 [J]. Journal of the Electrochemical Society, 2002, 149(9): B428-B439.

[47] BUCHHEIT R G, MARTINEZ M A, MONTES L P. Evidence for Cu ion formation by dissolution and

dealloying the Al$_2$CuMg intermetallic compound in rotating ring-disk collection experiments [J]. Journal of the Electrochemical Society, 2000, 147(1): 119-124.

[48] BLACKWOOD D J, CHONG A S L. Influence of chloride on deposition of copper on 6063 aluminium alloy [J]. British Corrosion Journal, 1998, 33(3): 225-229.

[50] CECCHETTO L, AMBAT R, DAVENPORT A J, et al. Emeraldine base as corrosion protective layer on aluminium alloy AA5182, effect of the surface microstructure [J]. Corrosion Science, 2007, 49(2): 818-829.

[51] BALÁZS L. Corrosion front roughening in two-dimensional pitting of aluminum thin layers [J]. Physical Review E, 1996, 54(2): 1183-1189.

[52] FRANKEL G S. The growth of 2-D pits in thin film aluminum [J]. Corrosion Science, 1990, 30(12): 1203-1218.

[53] FRANKEL G S, DUKOVIC J O, BRUSIC V, et al. Pit growth in NiFe thin films [J]. Journal of the Electrochemical Society, 1992, 139(8): 2196-2201.

[54] CHEN L, MYUNG N, SUMODJO P T A, et al. A comparative electrodissolution and localized corrosion study of 2024Al in halide media [J]. Electrochemica Acta, 1999, 14(16): 2751-2764.

[55] FRANKELT G S. Pitting corrosion of metals-a review of the critical factors [J]. Journal of the Electrochemical Society, 1998, 145(6): 2186-2198.

[56] SERI O. The role of NaCl concentration on the corrosion behavior of aluminum containing iron [J]. Corrosion Science, 1994, 36(10): 1789-1803.

[57] KOLICS A, POLKINGHORNE J C, WIECKOWSKI A. Adsorption of sulfate and chloride ions on aluminum [J]. Electrochemica Acta, 1998, 43(18): 2605-2618.

[58] BAZELEVA N A, HERASYMENKO Y S. Corrosion-electrochemical behavior of aluminum alloys in aqueous ethylene glycol media [J]. Materials Science, 2007, 43(6): 851-860.

[59] NATISHAN P M, O'GRADY W E, MARTIN F J, et al. The effect of chloride on passive oxide film breakdown on stainless steels and aluminum [J]. ECS Transactions, 2012, 41(25): 49-57.

[60] MI C W, LAKHERA N, KOURIS D A, et al. Repassivation behaviour of stressed aluminium electrodes in aqueous chloride solutions [J]. Corrosion Science, 2012, 54: 10-16.

[61] MERCIER D, HERINX M, BARTHÉS-LABROUSSE M G. Influence of 1, 2-diaminoethane on the mechanism of aluminium corrosion in sulphuric acid solutions [J]. Corrosion Science, 2010, 52(10): 3405-3412.

[62] ROSLIZA R, WAN NIK W B, SENIN H B. The effect of inhibitor on the corrosion of aluminum alloys in acidic solutions [J]. Materials Chemistry and Physics, 2008, 107(2-3): 281-288.

[63] ŠERUGA M, HASENAY D. Electrochemical and surface properties of aluminium in citric acid solutons [J]. Journal of Applied Electrochemistry, 2001, 21(9): 961-967.

本论文原发表于《Acta Metallurgica Sinica(English Letters)》2022年第35卷第2期。

Failure Analysis about Girth Weld of Gathering Pipelines

Li Fagen[1] Feng Quan[2] Fu Anqing[1] Cai Rui[1]

(1. CNPC Tubular Goods Research Institute, State Key Laboratory of Performance and Structural Safety for Petroleum Tubular Goods and Equipment Materials; 2. Petro China Tarim Oilfield Company)

Abstract: Through failure generalization, fracture feature analysis and material performance test, comprehensively analysis was made on the fracture failure analysis about girth weld of gathering pipelines containing H_2S gas. The results showed that the fracture failure might be mainly due to sulfide stress cracking in the girth weld. The crack originated from the fusion line on the inner surface of girth weld and extended along the girth weld to outside close to bends. Sulfide stress cracking of girth weld was caused by the intersection of multiple factors. The service condition was located in SSC 3 zone and the SSC risk of girth weld was high. The girth weld itself was not been stress-relieved, and its ability to resist SSC was poor. Due to low wall thickness, welding defects, welding stress and additional load, the actual stress of weld was higher.

Keywords: Gas gathering pipelines; Girth weld; Fracture failure analysis; Transition cuts; Stress concentration; Sulfide stress cracking

1 Introduction

Gathering pipelines containing wet H_2S gas should not only protect from electrochemical corrosion, but also resist sulfide stress corrosion cracking(SSC or SCC)[1]. Relatively, more attention was paid to cracking failure for oilfield engineers due to short occurrence time and great damage[2-7]. In this paper, aiming at cracking failure of girth weld about the gathering pipeline for sour service, a series of analysis were made on the cracked area, including macro- and micro-analysis, chemical, hardness, Charpy impact microstructure tests and SSC test. Failure process review, fracture characteristics analysis and material performance test were carried out firstly, and then a comprehensive analysis was carried out to find out the failure causes.

2 Failure process review

The gathering pipeline about 16.6km long, whose service condition was shown in Table 1. After 1.5 service years, one girth weld was located in about 3m outside of the station. The failure

---

Corresponding author: Li Fagen, lifg@ cnpc. com. cn.

girth weld was the butt joint of ϕ406.4mm × 8.8mm L360MCS induction bend and L360QCS seamless pipe.

Table 1 Service condition of gathering pipeline

H_2S content(mg/m^3)	Operating pressure(MPa)	Gas flow rate(m^3/d)	T(℃)
2184	5.32	$25×10^4$	23

The buried depth of the pipeline shownin Fig.1(a) was about 0.8m. A crack with a length of 105mm could be found on the outside of the weld in Fig.1(b), and RT results showed that the crack length is 175mm inside the weld. In addition, it was obvious that transition cuts had been carried out on the induction bend in Fig.2, whose thickness (6.1mm) was less than the nominal thickness of pipeline (8.8 mm).

Fig. 1 Macro-morphologies of the failure girth weld

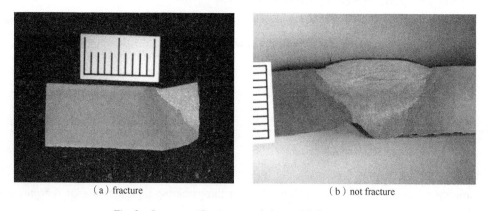

Fig. 2 Low magnification morphology of failure fracture

3 Fracture characteristics inspection

The fracture morphology and crack propagation trend was shown in Fig. 3. Most areas on both sides of the fracture were dark brown, and there were serious corrosion marks in the middle. There was no obvious plastic deformation on the failure fracture, and the side near the inner wall was flat while the middle part was rough. A series of characteristics showed that brittle fracture occurred in the weld. Further, the crack opening area of the inner wall was wider than that of the outer wall, and the crack might originate from the inner surface of girth weld and be driven by typical multi-source.

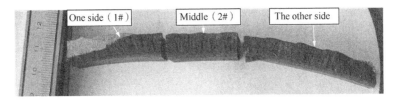

Fig. 3 Morphology of failure fracture

Radial patterns marks on the inner side of the source zone was found by SEM and EDS investigations of the failures, and the radial patterns converged to the inner surface of the girth weld (Fig. 4). In addition, weld defects was found in the source area from Fig. 4. The corrosion traces and intergranular cracking marks were obvious in the crack propagation zone (Fig. 5). EDS results showed that more elemental sulfur were found from corrosion products in the source region and the propagation region, and the atomic proportion was up to 25.24% (Fig. 6). It was inferred that there were a lot of FeS_x corrosion products on the fracture surface.

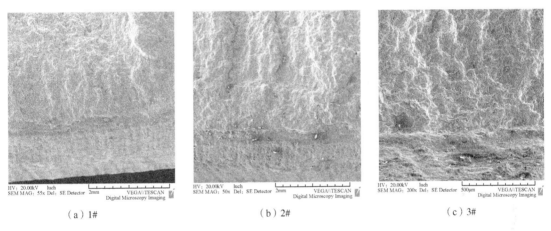

(a) 1# (b) 2# (c) 3#

Fig. 4 SEM morphology of crack source zone

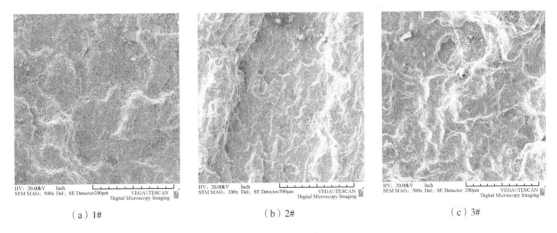

(a) 1# (b) 2# (c) 3#

Fig. 5 SEM morphology of crack propagation zone

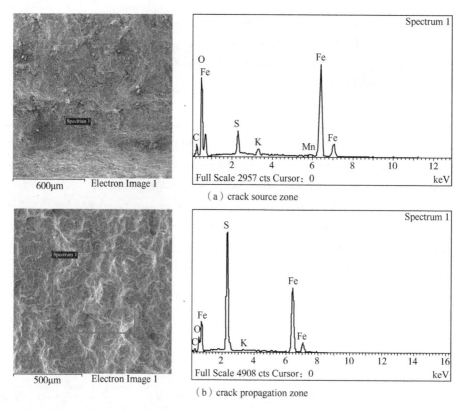

Fig. 6 EDS results of corrosion products

4 Materials performance test

4.1 Physical and chemical properties test

Chemical, hardness, Charpy impact properties of girth weldfar away from the fracture were tested. The icrostructure of failure fracture and welding is shown in Fig. 7. The test results in Table 2 to Table 4 could meet the requirements of GB/T 9711—2017[8], SY/T 5257—2012[9] and GB/T 31032—2014[10]. The metallographic specimens were taken from the fracture and far away from the fracture. Compared with microstructure test far away from the fracture, no abnormal metallographic structure was found at the fracture, and no coarse structure or martensitic structure was found. It could be concluded that no microstructure change occurred during the welding process and no obvious performance degradation occurred during the service period.

Table 2 Test results of chemical composition (wt. %)

	C	Si	Mn	P	S	V	Nb	Ti
L360MCS induction bend	0.065	0.21	1.05	0.014	<0.002	<0.005	0.032	0.013
SY/T 5257—2012	≤0.10	≤0.45	≤1.45	≤0.020	≤0.002	≤0.05	≤0.05	≤0.04
L360QCS seamless pipe	0.13	0.26	1.06	0.0076	0.0029	0.038	<0.005	0.0017
GB/T 9711	≤0.16	≤0.45	≤1.65	≤0.020	≤0.003	≤0.07	≤0.05	≤0.04

Table 3 Test results of CVN (specimen sizes: 5mm×10mm×55mm, test temperature: 20℃)

Specimen No.	KV$_2$(J)			FA(%)		
Weld	66	49	43	100	80	80
HAZ(induction bend side)	110	105	109	100	100	100
HAZ(seamless pipe side)	97	98	95	100	100	100
GB/T 31032	average value ≥20; single value ≥15			/		

Table 4 Test results of HV10

Test location	1	2	3	4	5	6	7	8	9	10	11	12	13	14
Specimen A	194	171	166	210	218	162	162	191	191	161	190	187	169	205
Specimen B	184	162	159	212	209	174	168	183	193	164	175	178	150	177
Specimen C	190	168	159	181	183	170	166	187	176	161	173	179	170	197
GB/T 31032	≤250HV10													

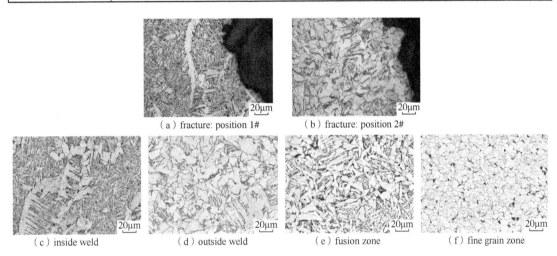

(a) fracture: position 1#   (b) fracture: position 2#
(c) inside weld   (d) outside weld   (e) fusion zone   (f) fine grain zone

Fig. 7 Microstructure of failure fracture and welding

4.2 SSC test

The maximum value of 16 test results is 346MPa in the stress test report about other girth weld of the failure pipeline[11]. Considering the influence of transition cuts, the actual stress load value was 460MPa in the SSC test. Two specimens were fractured after SSC test in solution A according to NACE TM 0177—2016 and ISO 7539-2: 1989. The macro-morphology of specimens after SSC test is shown in Fig. 8. The girth with weld transition cuts could not meet the SSC resistance requirements[12-13,6].

Fig. 8 Macro-morphology of specimens after SSC test

5 Discussion and analysis

The test results of physical and chemical properties indicated that no microstructure change occurred during the welding process and no obvious performance degradation occurred during the service period. However, the fracture characteristics showed that there were intergranular crack originated from the fusion line position of the induction bend and propagated along the weld to the outside of the weld, and sulfide stress cracking occurred in the girth weld.

The service condition shown in Table 1 (H_2S content: 2184mg/m^3, T: 23℃) was located in SSC 3 zone specified in SY/T 0599—2018, and the risk of sulfide stress cracking was very high. In addition, the current service condition exceeded the design requirements and the design condition was located in the SSC 1 zone. Weldments located in SSC 3 zone should be stress-relieved at a minimum temperature of 620℃ after welding according ISO 15156-2: 2020[6]. The failure weldment could not be stress-relieved, and the girth weld itself was sulfide stress cracking sensitive materials[14].

The internal pressure and elastic laying of gathering pipeline would both bring axial stress to the pipeline. Meanwhile, the welding residual stress was not eliminated, and the pipe beard large axial load. In addition, the buried depth of the pipeline(0.8m) could not meet the requirements of the buried depth at the crossing of soil road and gravel road (1.8m). Moreover, the pipe was bearing additional load repeatedly rolled by heavy vehicles before failure.

The local thickness of weldment was less than the nominal thickness of pipeline as transition cuts, which could not meet the requirements of the welding process design. In fact, the wall thickness on both sides of the girth weld was close to each other, so transition cuts treatment should not be carried out. In addition, welding defects were found in the failure fracture. Combined with lower wall thickness and welding defects, the stress concentration at the fusion line of the bend side was caused, and the actual stress was further enlarged and may significantly exceed the maximum detection stress level (346MPa).

Based on the above analysis, the girth weld failure was mainly caused by three factors: (1) The service condition was located in SSC 3 zone and the SSC risk of girth weld was high. (2) The girth weld itself was not been stress-relieved, and its ability to resist SSC was poor. (3) Due to the low wall thickness, welding defects, welding stress and additional load, the actual stress of weld was higher. The intersection of various factors contributed to the initiation and propagation of sulfide stress cracking, and the SSC test results under high stress conditions also verified that the girth weld could not meet the requirements of acid environment cracking resistance.

6 Conclusion

(1) Sulfide stress cracking occurred in the girth weld of gas gathering pipeline, which originated from the inner surface of fusion line and propagated to the outside of the weld along the girth weld.

(2) Sulfide stress cracking of girth weld was caused by the intersection of multiple factors. The service condition was located in SSC 3 zone and the SSC risk of girth weld was high. The girth weld

itself was not been stress-relieved, and its ability to resist SSC was poor. Due to low wall thickness, welding defects, welding stress and additional load, the actual stress of weld was higher.

Acknowledgements

The authors acknowledge the financial support of Key Research and Development Program of Shaanxi Province (No. 2018ZDXM-GY-171, 2019KJXX-091).

References

[1] HU Y B, GU T. Corrosion features and corrrosion crontrol technologies in high-sulfur gas fields[J]. Natureal Gas Industry, 2012, 32(12): 92-96.

[2] SHAN G B, LIU X H, QI J, et al. Failure Analysis of Natural Gas Pipeline[J]. Pressure Vessel Technology, 2013, 30(11): 47-51.

[3] SUN J W, ZHAN X L, YAO F, et al. Cause Analysis for Cracking of Hydrogen Gas Pipeline[J]. Surface Technology, 2016, 45(2): 50-56.

[4] QIAO X Y. Failure Analysis about Joint Pipes for Natural Gas End Station[J]. Pressure Vessel Technology, 2012, 29(3): 59-64.

[5] SY/T 0611. Requirements of controlling internal corrosion in gathering pipeline system for highly hydrogen sulphide gas field[S]. Beijing: Petroleum Industry Press, 2018.

[6] ISO 15156-2. Petroleum and natural gas industries-Materials for use in H_2S-containing environments in oil and gas production-Part 2: Cracking-resistant carbon and low alloy steels, and the use of irons[S]. ISO Technical Committee, 2020.

[7] EFC Publication No. 16, Guidelines on materials requirements for cabon and low alloy steels for H_2S-containing environments in oil and gas production.

[8] GB/T 9711. Petroleum and natural gas industries-Steel pipe for pipeline transportation systems[S]. Beijing: Standards Press of China, 2017.

[9] SY/T 5257. Steel bends for oil and gas transmission[S]. Beijing: Petroleum Industry Press, 2012.

[10] GB/T 31032. Welding and acceptance standard for steel pipings and pipelines[S]. Beijing: Standards Press of China, 2014.

[11] FU A Q, LI F G, YIN C X, et al. The detection project of gas gathering[R]. Xi'an: CNPC Tubular Goods Research Institute, 2016.

[12] NACE TM 0177. Laboratory Testing of Metals for Resistance to Sulfide Stress Cracking and Stress Corrosion Cracking in H_2S Environments[S]. American Society of Corrosion Engineers, 2016.

[13] ISO 7539-2. Corrosion of metals and alloys-Stress corrosion testing-Part 2: Prepartation and use of bent-beam specimens[S]. ISO Technical Committee, 1989.

[14] SY/T 0599. Metallic material requirements on resistance to sulfide stress crackingand stress corrosion cracking for natural gas surface equipment[S]. Beijing: Petroleum Industry Press, 2018.

本论文原发表于《Materials Science Forum》2021年第1035卷。

Failure Analysis of Cracked and Corroded Tubings in Sudong Block of Changqing Oilfield

Zhao Xuehui[1, 2]　Li Mingxing[3]　Liu Junlin[4]　Liu Man[5]

(1. CNPC Tubular Goods Research Institute, State Key Laboratory for Performance and Structure Safety of Petroleum Tubular Goods and Equipment Materials;
2. Xi'an Jiaotong University, State Key Laboratory for Mechanical Behavior of Materials;
3. Research Institute of oil and gas technology of Changqing Oilfield Company;
4. Drilling and production technology research institute of Qinghai Oilfield;
5. The 11th oil extraction plant of Changqing Oilfield)

Abstract: In the process of layer inspection and hole mending, it was found in Changqing gas production plant that the tubing of a well was seriously corroded, and there were perforation and fracture of the tube body. It was found that some of the tubes were cracked and the pitted corrosion pits on the outer wall surface were serious. The fracture morphology and corrosion products were analyzed by means of macro analysis, metallographic microscope, SEM and EDS. The results show that the mechanical damage on the outer wall of the tubing was the primary condition for corrosion acceleration, and the serious corrosion and thinning of the inner and outer walls of the tubing under the corrosive service medium were the main reasons for the failure of the tubing string, and oil pipe corrosion perforation was mainly caused by internal corrosion.

Keywords: Failure analysis; Tubing; Damage; Pitting

1 Introduction

In the process of oil and gas exploration, development and production, the oil casing string is subjected to complex loads such as tension/compression, internal pressure/external pressure, bending, etc. Meanwhile, it is subjected to downhole media and temperature effects such as oil/gas/water, $H_2S/CO_2/Cl^-$, etc[1]. Therefore, the common and interactive effects of various factors in the string service environment aggravate the corrosion of oil casing[2-3]. However, severe corrosion will lead to tube wall thinning, which directly affects the load bearing capacity of the string[4]. Especially with the increasing of well water content in the middle and later stages, the synergistic effect of erosive gas and Cl^- environment leads to intensified corrosion, and the failure of string becomes more and more prominent[5-6].

Corresponding author: Zhao Xuehui, zhaoxuehui@cnpc.com.cn.

The failed string originated from the pre-exploration well, located in Bayinwendu Gacha, Dabukecha Town of Wushen Qi, belonging to 41~33 block of East Su. The completion depth of the well is 3620.0m, and the artificial bottom hole is 3591.80m. Production tubing size was ϕ73.02mm× 5.51mm, and tubing material was N80 EUE, the total shaft depth was 3369.45m. A maximum CO_2 content of 1.33% was detected, the formation water type was $CaCl_2$ and pH value was 6. The salinity of the formation water was 118.32g/L, as shown in Table 1. The string service period from completion to July 2018 is about 7 years. From July 12 to 23, 2018, 225 oil pipelines were discharged, and the 226 of them were found to have broken off (the depth of the well was 2.135ms, corresponding to the formation temperature of 72℃). The location of the fault was about 0.4m from the parent buckle end, and the 226 of them cracked at the fracture point, with a length of about 0.5m.

Table 1 Statistical table of water quality analysis

PH	Colour	Ion content(mg/L)						Mineralization (g/L)	Type
		$K^+ + Na^+$	Ca^{2+}	Mg^{2+}	Cl^-	SO_4^{2-}	HCO_3^-		
6.0	brown	24133	15090	3052	63752	12056	238	118.32	$CaCl_2$

2 Test methods and results

2.1 Macro analysis

The submitted samples were the 226 cracked tubing (A#) and the short-connected sample (B#) with local perforation extracted from the site. The macroscopic morphology is shown in Figs 1~2. It can be seen that the cracked pipe body of A# is obviously thinner and the pipe body in the cracked area has torsion. Both the inner wall and the outer wall are reddish-brown corrosion products, and the outer wall is partially ulcerated corrosion pits. The sample B# was taken from the same section of the pipe body corroded and perforated, which showed the characteristics of small inside and large outside corrosion pits. The basic physical and chemical properties of the pipe samples were tested, and the corrosion morphology of the samples at the perforation and cracking sites was observed and analyzed.

(a) macroscopic morphology of cracked pipe

(b) corrosion morphology of outer wall　　　(c) corrosion morphology of inner wall

Fig. 1 Fracture morphology of the 226 tubing (A#)

Fig. 2　Macro-morphology of perforated tubing (B#)

2.2　Physical and chemical properties

The physical and chemical properties of the samples were all taken from the area without local corrosion of B# samples. The mechanical properties of the samples were tested according to the standard GB/T 228.1—2021, and the testing equipment was UTM 5305 material testing machine. The test results are shown in Table 2. The tensile strength results show that the tubing has a 3% loss after about 7 years of service. The chemical composition was detected by a straightness spectrometer, and the test results were shown in Table 3.

Table 2　Mechanical properties of test samples

Project	R_m(MPa)	$R_{t0.5}$(MPa)	A(%)
Results	668	582	27.3
API 5CT	≥689	552-758	—

Table 3　Chemical composition detection results of the samples　(wt.%)

Element	C	Si	Mn	P	S	Cr	Mo	Ni	V	Ti	Cu
Result	0.26	0.25	0.57	0.0084	0.0014	0.67	0.091	0.040	0.0036	0.023	0.035
API 5CT	—	—	—	≤0.030	≤0.030	—	—	—	—	—	—

2.3　Metallographic structure

Metallographic observation was carried out on the failure samples, and abnormal fracture tissues were also observed. The detection equipment is OLS 4100 laser confocal microscope. From the observation of failure morphology, scratches similar to mechanical scratches with an interval of about 32mm were found on the surface of the failed tube body, and the corrosion at the scratch was grooved, some scratches were shallow and short in longitudinal direction, and some scratches were seriously corroded through the length of the failed sample, and the cracking of the tube body occurred at this serious corrosion place. Therefore, we chose to take samples from the two scratches for transverse tissue observation. The sampling locations of the metallographic samples, numbered 1#, 2# and 3#, are shown in Fig 3. The results of metallographic detection are shown in Table 4. The metallographic structures on the outer surface are tempered Soxhlet and ferrite.

Fig. 3　Schematic diagram of metallographic sample

Table 4 Metallographic structure and inclusion

Samples	Nonmetallic inclusion	Organization
1#	A0.5, B0.5, D0.5	outer surface: Tempered Sorbite +Ferrite, rest: Tempered Sorbite
2#	A0.5, B0.5, D0.5	outer surface: Tempered Sorbite +Ferrite, rest: Tempered Sorbite
3#	A0.5, B0.5, D0.5	outer surface: Tempered Sorbite +Ferrite, rest: Tempered Sorbite

Fig. 4—Fig. 13 shows the tissue analysis diagram of three samples. The results show that the internal and external surface structure of sample 1# is basically normal, the external surface structure is tempered soxhlet and ferrite, the internal surface structure is tempered soxhlet, and the tissue has no obvious deformation. Corrosion pits were obviously visible on the inside and outside surfaces, indicating that corrosion damage occurred inside and outside the tube under service environment. There were gray nonmetallic substances in the corrosion pits, and the deepest corrosion pits were 1.02mm.

Fig. 4   Outer surface microstructure of 1#

Fig. 5   Innter surface microstructure of 1#

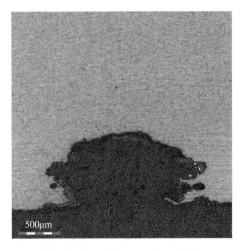

Fig. 6   Pit of inner surface of 1#

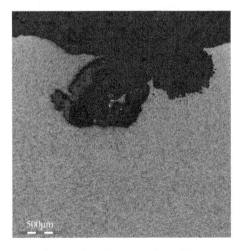

Fig. 7   Pit of outer surface of 1#

The outer surface microstructure of 2# sample is tempered soxhlet and ferrite, and the inner surface is tempered soxhlet. There is obvious tissue deformation at the cracking fracture, indicating that the cracking area is subjected to tensile force. Obvious corrosion pits can be seen on the inner and outer surfaces, and there are gray nonmetallic substances in the pits, again indicating that corrosion thinning occurs simultaneously inside and outside. The outer surface microstructure of 3# sample has tempered soxhlet and ferrite, the inner surface has tempered soxhlet, and the cracking end has slight structural deformation. Corrosion pits are visible on the inner and outer surfaces. There are gray nonmetallic substances in the pits.

Fig. 8 Outer surface microstructure of 2#

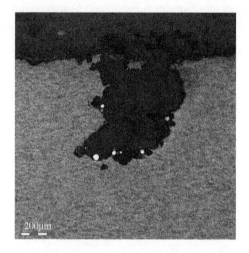

Fig. 9 Corrosion pits on the outer surface of 2#

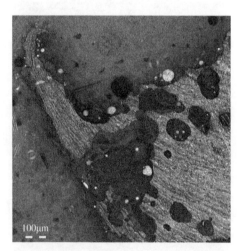

Fig. 10 Edge deformation structure of 2# sample

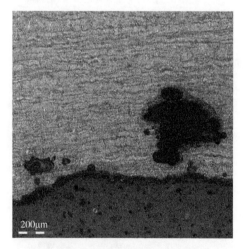

Fig. 11 Corrosion pits on the inner surface of 3# sample

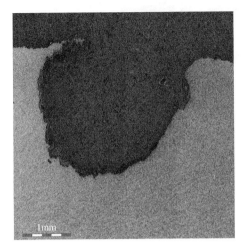
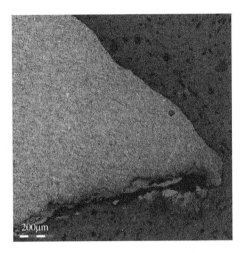

Fig. 12 Corrosion pits on the outer surface of 3#

Fig. 13 Deformed structure of the outer surface of 3#

2.4 Corrosion morphology observation

The corrosion morphology of the inner and outer walls of the failed sample was observed. Fig. 14 shows the corrosion morphology of the outer wall of the sample A#, the surface is seen to be honeycombed and continuously corroded pits [Fig. 14(a)]. The wall thickness at the corrosion pits was severely thinned, and the remaining wall thickness was about 1.5mm. The most severe thinning occurred where the pipe cracked [Fig. 14(b)]. Fig. 15 shows the macroscopic morphology of corrosion thinning at the cross section of the corrosion sample, in which Fig. 15(a) shows the depth morphology of the corrosion pit on the external surface, and the residual wall thickness at the pitting pit is 1.64mm (original wall thickness is 5.51mm). Fig. 15(b) morphology shows that the pipe is corroded and thinned from the inside wall to the outside wall until it penetrates. The results show that there is corrosion thinning in both inner and outer walls of the pipe during service.

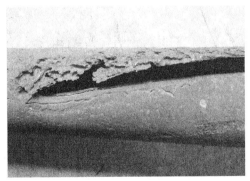

(a) honeycomb pits

(b) crack morphology

Fig. 14 Corrosion morphology of outer wall of A# tube

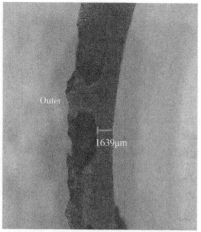
(a) morphology of thickness reduction of outer wall

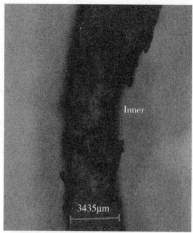
(b) simultaneous thinning of inner and outer wall thickness

Fig. 15　corrosion thinning morphology of the cross section of sample A#

The micro-corrosion morphology and corrosion products of the corrosion samples were analyzed, as shown in Fig. 16. The corrosion products on the non-penetrated outer surface were relatively thick, and the product film was mainly composed of C, O and Fe by energy spectrum analysis [Fig. 16 (a)]. The analysis of the corrosion products at the edge of the corrosion perforation [Fig. 16 (b)] was also mainly composed of C, O and Fe, indicating that the corrosion products of the inner and outer walls were mainly iron oxides or carbonate compounds.

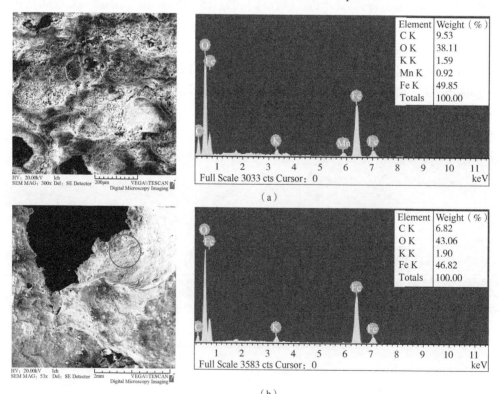

Fig. 16　Microstructure of outer wall of A# sample

Fig. 17 shows the inner wall corrosion appearance of the cracked pipe body of A# sample after being cut open. The surface corrosion products are obvious and pitted, and corrosion pits and perforations are found locally. There is a hole of 18mm in length and 10mm in width near the outside thickening of the threaded coupling. The inner surface morphology of the hole after cutting is shown in Fig. 18 (b). The observation of the inner surface morphology shows the characteristics of corrosion perforation from the inside out with large inside and small outside. The corrosion products at the bottom of the corrosion pits were observed under scanning electron microscope and their composition was analyzed. As shown in Fig. 19, element S was present in the corrosion products.

(a) inner wall   (b) corrosion pits and perforations

Fig. 17 Morrosion morphology of inner wall of A# tube

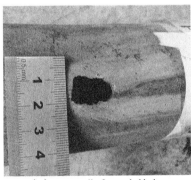 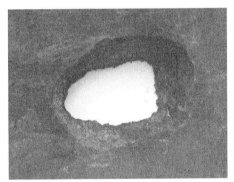

(a) outer wall of corroded hole   (b) inner wall of corroded hole

Fig. 18 Macro-morphology of B# sample

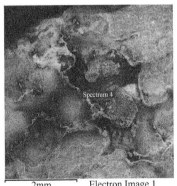 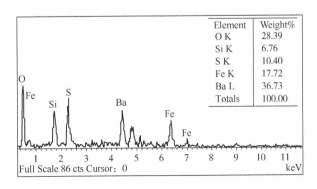

(a) morphology   (b) energy spectrum

Fig. 19 Morphology and energy spectrum of inner wall corrosion pits of B# sample

2.5 Simulation verification experiment

In order to determine whether the string has corrosion damage under acidizing environment after reservoir transformation, the corrosion performance of the string under the conditions of gelled acid and drag reduction acid was simulated in the laboratory. Combined with the field conditions, the experimental temperature was 72℃, the test period was 168h, and the total test pressure was 1.5MPa. Since the content of CO_2 and H_2S was detected in the field, CO_2 and H_2S were not added in the simulation test. The gelled acid and drag-reducing acid were added to the autoclavator according to the process and composition. Then heat up and pressure to the operating environment.

Fig. 20 shows the morphology of the sample when it was taken out from the acidizing solution after the experiment [Fig. 20(a)]. The acid solution presented foam viscosities and adhered to the surface of the sample. After cleaning, the surface of the sample is basically metallic luster, and there is no obvious local corrosion characteristic [Fig. 20(b)]. The weight loss method was used to calculate the average corrosion rate of the sample V_{corr} = 0.31mm/a, which was judged as extremely severe corrosion according to NACE 0775—2013 standard, indicating that extremely severe corrosion damage had occurred to the material in the solution of gelled acid + retarder acid without CO_2 and H_2S corrosive gas. According to the composition of acidizing liquid containing corrosion inhibitor, but the sample corrosion is still serious, it is necessary to consider the corrosion inhibition effect of corrosion inhibitor.

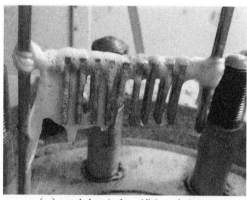
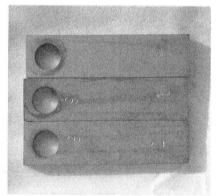

(a) morphology in the acidizing solution　　(b) morphology after cleaning acid solution

Fig. 20　Macro morphology of the sample after simulated test

3　Failure cause analysis

The chemical composition of the failed pipe meets the requirements of API 5CT, the yield strength of the material meets the requirements of API 5CT, and the tensile strength is relatively slightly lower than the standard requirements. As it is a failed sample to be tested, the mechanical performance is not evaluated. The metallographic structure of 2# and 3# samples were taken from the area with cracking edge and mechanical damage on outer wall. The results showed that obvious corrosion pits were visible on the inner and outer surfaces of 2# samples, and there were obvious structural deformation at the cracked fracture end, indicating that the pipe wall was deformed by external force. From the appearance of the outer wall, it can be seen that the wall is in the area with severe mechanical damage and corrosion, and the wall thickness is severely thinned. Obvious corrosion pits can be seen on the inside and outside surface of 3# sample, and the cracked fracture end has slight structural deformation, indicating that the pipe is deformed by external force. Corrosion pits on the inner and outer walls indicate that corrosion exists on the inside and outside of the string under service conditions.

From the observation of the failure morphology of the pipe string, scratches similar to mechanical scratches were found on the surface of A# tubing at an interval of about 32mm, and the corrosion at the scratches was grooved. Some scratches were shallow and the longitudinal direction was short. Some scratches were seriously corroded and ran through the length of the failure sample, and the cracking of the pipe body occurred at this serious corrosion point (Fig. 21). From the crack section, it can be observed that the wall thickness is severely thinned and the corrosion pit penetrates the wall. Meanwhile, there were honeycomb corrosion pits on the outer surface of the tube, with wall thickness remaining about 1mm in many places. String has served seven years, and the pipe wall scratches are relatively obvious, the transformation of reservoir after acidification process, and combined the process of annulus of acidizing fluid is emptying has certain hysteresis, the annulus (the outer wall of the tubing environment) longer than tube stranded[7], thus infer that the outer wall of tubing with damage in the acidification environment corrosion gradually. On the other hand, with the extension of service time and the medium containing CO_2 and H_2S corrosion, when thinned to a certain thickness of the tube body and the residual strength can not meet the needs of the service conditions of strength, the cracking in the weakest place to fracture[8,9]. When the pipe is cracked, the fracture surface is stretched and torn, and the deformation characteristics of microstructure can be observed. Element S was detected in the inner wall corrosion products, indicating that H_2S also played a certain role in corrosion.

The results of laboratory simulated field environmental experiments showed that the material was severely corroded during acidification without CO_2 and H_2S, and there was no obvious local corrosion phenomenon. Therefore, the corrosion thinning of acidizing fluid on the string during acidification was also the main reason for the string failure.

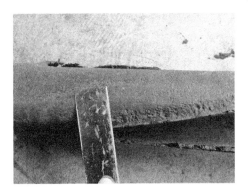 

Fig. 21 Morphology of surface damage and cracking

4 Conclusions

(1) Mechanical damage to the outer wall of the tube is the main reason for the corrosion

thinning of the tube string in service medium.

(2) When the pipe wall thins and the residual strength of the pipe string is insufficient to withstand the service strength demand, the pipe body will crack in the mechanical damage area.

(3) The partial perforation of the pipe body is caused by the simultaneous erosion and thinning of the pipe wall inside and outside.

(4) The tube body perforation of B# corroded from the inside out, and H_2S promoted the corrosion perforation.

(5) It is suggested that the corrosion inhibition effect evaluation of corrosion inhibitors should be carried out for the field acidification fluid, and the application effect of corrosion inhibitors should be further improved.

Acknowledgments

The authors greatly acknowledge the financial support from the National Key R&D Program of China and the CNPC Science and Technology Project, and also acknowledge the state key laboratory for its support of high-temperature and high-pressure experiments.

References

[1] FENG Y R, FU A Q, WANG J D, et al. Research progress and prospect of failure control and integrity technology of oil casing string in complex working conditions [J]. The Natural Gas Industry, 2020, 40(2): 106-114.

[2] LV S L, LI H L, TENG X Q. et al. Summarizing of Failure Analysis on Tubing and Casing Galling and Leakage [J]. Oil Field Equipment, 2011, 40(4): 21-25.

[3] WANG S L, FEI J Y, LIN X H, et al. Research progress of high performance corrosion resistant pipe and super 13Cr[J]. Corrosion Science and Protection Technology, 2013, 25(4): 322-326.

[4] YUAN Y, ZHANG J J, GUO Y J, et al. Analysis on Finite element of residual strength of the oil pipe with uniform corrosion defects[J]. New Technology and New Process, 2020, 4: 60-66.

[5] LIN N M, XIE F Q, WU X Q, et al. Research status and prospect of oil casing surface protection technology [J]. Corrosion and Protection, 2009, 30(11): 801-805.

[6] HAN Y, JI R, LV Y H, et al. Failure analysis of corrosion perforation of a N80S tubing [J]. Corrosion and Protection, 2019, 40(12): 933-937.

[7] LIU Z Z, WANG Y, WANG J. Research on Corrosion Behavior of Release Sewage of Acidizing in the Exploration of LD10-1 Oilfield[J]. Total Corrosion Control, 2013, 27(6): 35-39.

[8] LI W F, XIA W N. Numerical Simulation Analysis on Corroded Casing Residual Strength[J]. Natural Gas and Oil, 2013, 31(6): 70-75.

[9] SHI H Y. Hole-type Corrosion Damage Pipelines Residual Strength Evaluation[J]. Pipeline Technique and Equipment, 2017, 1: 47-50.

本论文原发表于《Materials Science Forum》2021年第1035卷。

Failure Analysis on Fiber Reinforced Thermoplastic Pipe

Song Chengli[1, 2]　Liu Xinbao[2]　Bai Zhenquan[1]
Kuang Xianren[1]　Wang Shuai[1]

(1. Tubular Goods Research Institute, China National Petroleum Corporation & State Key Laboratory for Performance and Structure Safety of Petroleum Tubular Goods and Equipment Materials;
2. School of Chemical Engineering, Northwest University)

Abstract: In recent years, fiber reinforced thermoplastic pipe (FRTP) has been widely used in oil fields of China due to its excellent corrosion resistance and coiling performance. With the further expansion of application scope, some failure accidents have occurred. Based on the inspection of the wall thickness, chemical composition and physical properties of a failed FRTP in the western oilfield of China, the causes are systematically analyzed. The results show that the main cause of the sample fracture is the decrease of the strength of FRTP due to the oil and gas media constantly penetrate into the lining and reinforced layer during the long-term service. Moreover, it is found that the outer sheath is worn and penetrated to the reinforced layer which further reduces the compressive performance of FRTP and finally causes the fracture failure.

Keywords: FRTP; Cracking; Lining; Reinforced layer; Outer sheath; Failure analysis

1 Introduction

As oil and gas exploration deepens, corrosion and scaling of steel pipe have been common problems which lead to a series of adverse effects, such as shutdown or decrease yield, loss and waste of oil and gas resources, serious pollution of the ecological environment, rising costs of maintenance and pollution control, and life threatening caused by H_2S leakage, etc[1-3]. Therefore, FRTP has become one of the important solutions to solve the corrosion problem of gathering pipe due to its high pressure resistance, corrosion resistance, light weight, good flexibility and coiling transportation performance[4-5]. It is a three-layer structure of composite pipe in which the fiber is tightly wound around the outer wall of the plastic liner, and the plastic outer sheath is mounted on the outer wall of the fiber layer. The lining pipe is used for anti-corrosion, the reinforced layer is used for bearing the internal pressure, and the outer sheath is used to protect the fiber from

---

Corresponding author: Liu Xinbao, xbliu2011@163.com.

wearing. Moreover, FRTP has become the fastest increase in the consumption of non-metallic pipe under the advocacy of "vigorously enhance the exploration and development and benefit construction requirements in the next few years" in china[6-7].

In the end of 2018, the leakage of a buried FRTP occurred in western oilfield of china after 8 years and 2 months service. Fig. 1 shows the on-site photo of failed FRTP after excavation and the basic information as shown in Table1. The leakage point is located at 9 o'clock of the pipe body, about 400mm away from the end of the steel casing which used to protect FRTP during crossing the highway.

Fig. 1 On-site photo of the failed FRTP

Table 1 Information of FRTP

Length (km)	Specifications	Design temperature(℃)	Operating temperature(℃)	Design pressure(MPa)	Operating pressure(MPa)	Medium	Production standard
0.8	DN80	60	60	2.5	1.5	oil and gas	SY/T 6662.2—2020[8]

2 Experimental analysis

2.1 Macroscopic fracture detection

Fig. 2 presents the macroscopic morphology photo of the perforation of FRTP sample which is a typical three-layer structure flexible composite pipe, namely the outer sheath, reinforced layer and lining. Outward opening cracking of three-layer structure indicates obvious plastic deformation.

Fig. 3 presents the macroscopic morphology photo of the outer sheath of FRTP sample. The outer sheath has undergone significant deformation and fracture [Fig. 3(a)], and the fracture direction is mainly along the fiber winding direction [Fig. 3(b)]. On the yellow circle as shown in Fig. 3(a) exists clearly traces of wear and concave deformation [red box in Fig. 3(c)] where the wall thickness becomes thinning, and the fracture edge is flat which is different from other fracture edges with obvious plastic deformation. Meanwhile, there is a small hole, a length of about 8mm triangular hole and a "extrusion edge" which should be formed after repeated extrusion [yellow circle in Fig. 3(c)]. And no abnormality is observed on the dorsal surface of the fracture.

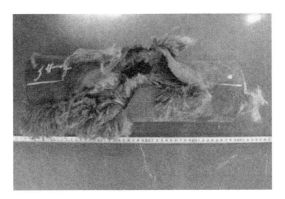

Fig. 2 Macroscopic morphology photo of failed FRTP sample

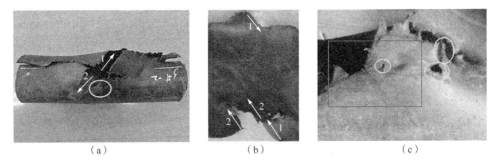

Fig. 3 The macroscopic morphology photo of the outer sheath

Fig. 4 is the macroscopic morphology photo of the reinforced layer which is composed of two layers of winding. It is observed that the fiber has become dark brown and hard tactile feeling.

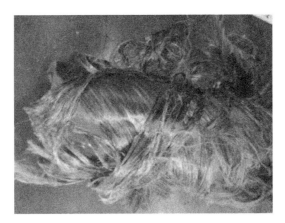

Fig. 4 The macroscopic morphology photo of the reinforced layer

Fig. 5 shows the macroscopic morphology photos of the lining which the outer wall is smooth and the direction of the fracture is at a certain angle with the longitudinal direction of the pipe body. The length of the fracture is about 250mm, and the maximum opening width is about 50mm. There is a defect at the edge of the fracture [yellow box in Fig. 5(b)] which is also located under the triangular hole in the outer sheath [Fig. 5(c)]. After the lining was cut longitudinally along the fracture [Fig. 6(d)], it can be seen that the inner wall is smooth, but the color is black and brown which is different from the color of the outer wall.

· 407 ·

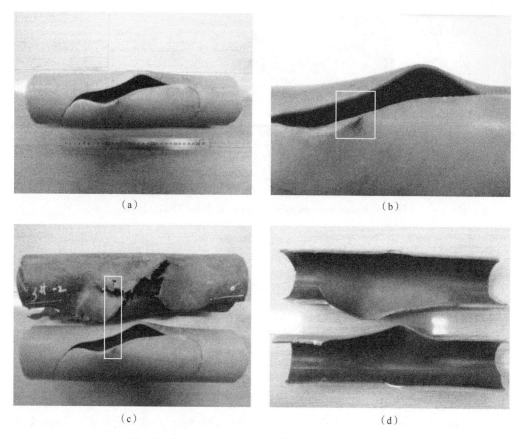

Fig. 5 The macroscopic morphology photo of the lining

2.2 Wall thickness measurement

Fig. 6 shows the location map of wall thickness measurement at the edge of the fracture with a vernier caliper (accuracy 0.02mm). Table 2 shows the obtained results, which the "g" and "f" point is the smallest (2.70mm and 2.72mm). It is observed that the wall thickness gradually increases along this two points to both ends.

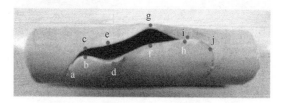

Fig. 6 The location map of wall thickness measurement (mm)

Table 2 shows the thickness measurement results in four directions along the annulus of the two ends of the sample which are 6.58 ~ 6.80mm, indicating that the wall thickness of pipe body is relatively uniform.

Table 2 The thickness measurement results at the fracture (mm)

Location	a	b	c	d	e	f	g	h	i	j
Thickness	6.48	5.52	5.68	5.00	5.30	2.72	2.70	5.98	6.12	6.50

Table 3 The thickness measurement results of the two ends of the sample   (mm)

Location	3 o'clock	6 o'clock	9 o'clock	12 o'clock
Left end	6.68	6.70	6.58	6.76
Right end	6.80	6.76	6.78	6.68

2.3 Infrared spectroscopic analysis

A sample of 10mm×10mm×t (t represents wall thickness) is manually sawn down the lining layer and a small bundle of reinforcement layer fibers is selected which are analyzed by Fourier transform infrared spectrometer. Fig. 7 illustrates the obtained results, which the similarity between the infrared spectrum of the tested lining and the standard spectrum of polyethylene (PE) is 97.8%, the similarity between the infrared spectrum of the tested reinforced layer and the standard spectrum of polyester is 95.32%. It is indicated that the lining is PE and the reinforced layer is polyester.

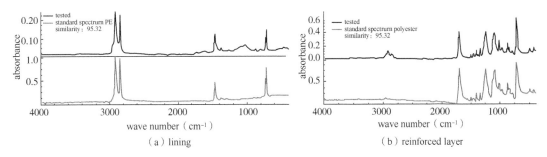

Fig. 7 Infrared spectroscopic analysis results

2.4 Vicat softening temperature testing

Three samples of 20mm×20mm×t are manually sawn down the lining layer to test vicat softening temperature by RV-300FW vicat softening temperature testing machine according to the standard of GB/T 1633—2000 of B50 method (50N, 50℃/h)[9]. Table 4 shows the obtained results, which the vicat softening temperature of lining is 62.23℃, greater than the operating temperature.

Table 4 Vicat softening temperature test results

Sample	Initial temperature (℃)	Vicat softening temperature (℃)	Average of vicat softening temperature (℃)
1#	indoor temperature	61.77	
2#	indoor temperature	62.74	62.23
3#	indoor temperature	62.18	

2.5 Hardness testing

Six samples (20mm×20mm×t) are taken from the lining layer by hand sawing, and hardness tests were carried out on the inner wall and the outer wall respectively by TIME 5410 Shore durometer according to GB/T 2411—2008[10]. Table 5 shows the obtained results. It is observed that the hardness of the outer wall is greater than that of the inner wall.

Table 5 Hardness test results

Sample		Hardness (HD)	Average of hardness (HD)
Inner wall	1#	44.0	46.7
	2#	46.8	
	3#	49.4	
Outer wall	4#	52.2	54.2
	5#	53.8	
	6#	56.7	

2.6 Density testing

A 1~2g flake sampleis taken from the lining layer for density detection by ET-120SL electronic densitometer according to GB/T 1033.1—2008 of liquid pyknometer method[11]. Table 6 shows the obtained results, which are consistent with the requirements of SY/T 6662.2—2020.

Table 6 Density test results

Sample	Density (g/cm³)	Average of density (g/cm³)
1#	0.942	0.945
2#	0.948	
3#	0.945	
SY/T 6662.2—2020	≥0.930	

3 Discussion

In order to analyze and find the cause of FRTP failure, the following analysis will be made from three aspects: lining, reinforced layer and outer sheath.

3.1 Lining analysis

Under the long-term operation of FRTP (8 years and 2 months), the oil and gas medium keeps spreading to the lining matrix. The macro inspection also found that the color from the outer wall to the inner wall is getting darker and darker (Fig. 8), which will lead to significant swelling of the lining and decrease its pressure bearing capacity and heat resistance [12]. Further hardness test results show that the hardness of the inner wall is less than that of the outer wall, which verifies that the strength of the lining has decreased. Once the fiber layer is loosened or fractured locally, it will crack at this point of the lining.

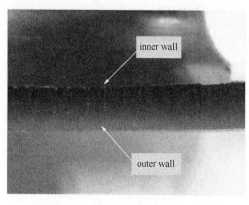

Fig. 8 Color distribution along wall thickness of lining

3.2 Reinforced layer analysis

The reinforced layer has turned dark brown in color and has a hard touch, indicating that oil and gas media has penetrated into the reinforcement layer. Generally, the polyester fiber is usually white and soft [Fig. 9 (a)], it is composed of linear macromolecules of partial crystallization of polymer. With the infiltration of oil and gas media, the interaction between the macromolecules of the fibers is weakened, so that the distance between molecules increases, the pore increases, and the fiber tensile strength will decline[13-14]. In addition, when the reinforced layer fibers at the fracture are bent by hand, partial fracture occurs [Fig. 9(b)], further indicating that the bearing pressure performance of the reinforced layer fibers decreases. As fiber reinforced layer is the most important pressure bearing part of FRTP, once its strength decreases, the pipe will burst when the internal pressure cannot be carried.

 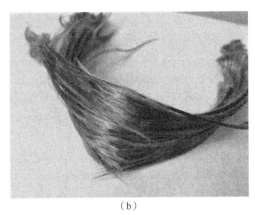

(a) (b)

Fig. 9 Polyester fiber morphology

3.3 Outer sheath analysis

Because of inserting into the steel casing pipe, outer sheath has two holes along with the severe wear marks and compression marks in that it is easy to cause scratches and wear in the process of construction installation and operation vibration. Since the fracture at the wear place is flat and the other parts of the fracture show typical plastic deformation characteristics of PE material, it can be judged that the fracture at the wear place is the most original. Furthermore, when the outer sheath is worn out and the reinforced layer is broken, the bearing capacity of FRPT will be significantly reduced.

4 Conclusions and recommendations

(1) The lining of FRTP is PE and the reinforced layer is polyester fiber. However, the hardness of the inner and outer wall of the lining is significantly different, showing uneven distribution of hardness. In addition, obvious wear marks and defects appeared on the surface of the outer sheath.

(2) In the long-term operation of FRTP, the oil and gas medium constantly infiltrates into the lining and reinforced layer, resulting in the decrease of the strength the reinforced layer. In addition, the outer sheath wears and penetrates into the reinforcing layer, further reducing the pressure performance of the pipe and finally causing fracture failure.

(3) It is recommended to test the permeability of lining and reinforcing layer to oil, gas and water media during the procurement of FRTP, especially to evaluate whether the strength of lining and reinforcing layer decreases after infiltration for a period of time. Meanwhile, the oil and gas

permeability resistance of lining and reinforcing layer can also be included in the relevant product standards.

References

[1] XIAO W W, SONG C L, BAI Z Q, et al. Risk analysis on surface gathering pipeline corrosion in oil fields [J]. Oil-gas field surface engineering, 2017, 36(4): 81-85.

[2] SONG F M A. Comprehensive model for predicting CO_2 corrosion rate in oil and gas production and transportation systems[J]. Electrochimica Acta, 2010, 55(3): 689-700.

[3] LIU Q Y, MAO L J, ZHOU S W. Effect of chloride content on CO_2 corrosion of carbon steel in simulated oil and gas well environments [J]. Corrosion Science, 2014, 84: 165-171.

[4] LI H B, YAN M L, QI D T, et al. Failure analysis of steel wire reinforced thermoplastics composite pipe [J]. Engineering Failure Analysis, 2012, 20(8): 88-96.

[5] KUANG Y, MOROZOVE V, ASHRAFM A, et al. Analysis of flexural behavior of reinforced thermoplastic pipes considering material nonlinearity [J]. Composite Structures, 2015, 119.

[6] LI H B, LI H L, QI D T, et al. Analysis on non-metallic pipes used for oil and gas gathering and transportation [J]. Petroleum instruments, 2014, 28(6): 4-8.

[7] ZHANG G J, QI G Q, QI D T. Present conditions and prospect for application of non-metallic and composite materials for oil pipeline [J]. Petroleum Science and Technology Forum, 2017, 36(02): 26-31.

[8] SY/T 6662.2—2020. Non-metallic composite pipe for petroleum and natural gas industries. Part 2: Flexible composite pipe for high pressure transmission [S]. Beijing: Petroleum Industry Press, 2020.

[9] GB/T 1633—2000. Plastics-thermoplastic materials-determination of Vicat softening temperature [S]. Beijing: Standards Press of China, 2000.

[10] GB/T 2411—2008. Plastics and ebonite: determination of indentation hardness by means of a duronmeter (shore hardness) [S]. Beijing: Standards Press of China, 2000.

[11] GB/T 1033.1—2008. Plastics: methods for determining the density of non-cellular plastics—Part1: Immersion method, liquid pyknometer method and titration method [S]. Beijing: Standards Press of China, 2000.

[12] QI G Q, CUI X H, QI X B. Cracking failure of reinforced thermoplastic pipe used in ground gathering and transportation [J]. Petroleum tubular goods and instruments, 2020, 6(2): 79-82.

[13] KOTOMIN S V, AVDEEV N N. Compaction and consolidation of aramid and composite fibers [J]. Mechanics of composite meterials, 2002, 38(5): 461-470.

[14] QI G Q, QI D T, LI X J, et al. High temperature performance test of PET fiber-reinforced flexible composite pipe [J]. Oil & gas storage and transportation, 2015, 34(6): 616-620.

本论文原收录于 IOP Conference Series 2021 年第 706 卷第 1 期。

H_2S Dissociation on Defective or Strained Fe(110) and Subsequent Formation of Iron Sulfides: A Density Functional Theory Study

Li Fagen[1] Zhou Zhaohui[2] He Chaozheng[3]
Li Yufei[4] Zhang Lin[4] Zhu Dajiang[4]

(1. CNPC Tubular Goods Research Institute, State Key Laboratory of Performance and Structural Safety for Petroleum Tubular Goods and Equipment Materials; 2. Chemical Engineering and Technology, School of Water and Environment, Key Laboratory of Subsurface Hydrology and Ecological Effects in Arid Region, Ministry of Education, Chang'an University; 3. Institute of Environmental and Energy Catalysis, Shaanxi Key Laboratory of Optoelectronic Functional Materials and Devices, School of Materials Science and Chemical Engineering, Xi'an Technological University; 4. Engineering Technology Research Institute of PetroChina Southwest Oil & Gas Field Company, Guanghan)

Abstract: In this paper, spin-polarized periodic density functional theory-based simulations were performed to investigate H_2S adsorption and dissociation on the defective and strained Fe(110) surfaces which exist in real oil exploitation and transport. It was found that the defective surface facilitates H_2S adsorption and dissociation with respect to the perfect surface, while the homogeneous external stresses, giving rise to uniform lattice expansion, show negligible effects. More interestingly, the simulations predicted a correct order of phase transition at the Fe(110) surface by using the chemical potential based thermodynamic model, that is, from the elementary Fe crystal to the FeS crystal to the FeS_2 crystal to the elementary S crystal with the S coverage on the Fe(110) surface increasing.

Keywords: Fe(110); Defect; Strain; H_2S adsorption; H_2S dissociation; Density functional theory

1 Introduction

H_2S gas is a common companion in oil and gas exploitation, which is highly toxic and strongly corrodes drilling equipment and transport pipelines. Stress corrosion is one of the most destructive corrosion types in acidic oil and gas fields. It has been shown that H_2S is one type of gases

Corresponding author: Zhou Zhaohui, zzhlax@chd.end.cn; He Chaozheng, hecz2019@xatu.edu.cn.

responsible for the most serious acidic corrosion, where electrochemical corrosion and sulfide stress corrosion could occur[1]. The cracking accident due to sulfide stress corrosion can suddenly happen without showing any foreseeable signs, thus resulting in extremely serious consequences[1].

Literature reports that steel oil and gas pipe corrosion by H_2S begins with the adsorption and decomposition of H_2S wet gas on the surface of the pipe, followed by adsorption and deposition of sulfur and formation of the Fe-S compounds on the steel surface[2]. The initial corrosion product is the mackinawite FeS which is further transformed into the intermediate pyrrhotite $Fe_{1-x}S$, and finally the stable product of pyrite FeS_2.

Density functional theory (DFT)-based simulations of the interaction between H_2S molecules and Fe surfaces have been reported in the last two decades. Jiang and Cater studied the adsorption, diffusion and decomposition of H_2S molecules and HS free radicals on the surfaces of Fe (100) and (110)[3-4]. They predicted that H_2S molecules and HS free radicals had low activation energies to dissociate on the two Fe surfaces. In addition, they explored the effects of alloying methods on the steel surface corrosion by CO and H_2S gases and uncovered the fundamental reaction paths and energy barriers for CO and H_2S gas on the surface of Fe-Al and Fe_3Si alloys[5]. They found that the Fe-Al alloy prevented the H_2S decomposition but accelerated the CO decomposition but the Fe_3Si alloy favored the decomposition of H_2S over CO gas. H_2S adsorption and decomposition on Fe(100), (111) and (310) were also reported by several investigations[6-9]. Mahyuddin et al. examined the effects of alkali metal adsorption on CO and H_2S adsorption on Fe(100), a process commonly found in Fischer-Tropsch synthesis[10]. They uncovered how Na and K affected CO and H_2S adsorption on the Fe(100) surface, and pointed out that Na and K adsorption promoted CO adsorption but prevented H_2S adsorption. Recently, the effect of Fe vacancy on H_2S adsorption and dissociation on Fe(100) was reported, which indicated that the Fe vacancy lowered the dissociation energy barrier of H_2S[11]. Ren et al. revealed the adsorption properties of H_2S on intact Fe (100) while compared to NH_3 and HCN[12]. Akande et al. calculated the reaction of H_2S with Fe(110) and determined the rate determining step to be the first deprotonation step of H_2S dissociation[13].

Besides the static calculations above, the first-principles molecular dynamics are also used to study the interaction of H_2S with Fe surfaces and the dissociation products. Spencer and Yarovsky reported temperature-dependent adsorption and decomposition behavior of H_2S on Fe (110) and (100)[14-15]. They observed that H_2S spontaneously decomposed on the Fe surface at room temperature (298K), H spread into the Fe substrate at higher temperatures (800K and 1000K), and the dissolved H recombined to form H_2 molecules and only S was left to adsorb on the Fe surface at Fe melting point temperature (1600K). Meanwhile, Todorova, Spencer and Yarovsky reported the dynamic behavior of S on the Fe (110) and (100) surfaces and the migration of S on the Fe surface was observed[16].

Although dozens of theoretical investigations have been conducted about the H_2S adsorption and dissociation on Fe surfaces, the effects of stress and vacancy defect in Fe (110) have not been reported to date. Furthermore, it has been reported that the initial stage of the H_2S corrosion is the adsorption and dissociation of H_2S molecules and the final product forms Fe-S compounds which have many stoichiometries and polymorphs[17-18]. It is still a challenge to accurately predict these Fe-S compounds by simulations[19-21]. According to general observation in experiments, the initial product of H_2S corrosion is the mackinawite FeS which is eventually transformed into the final product of pyrite FeS_2[2].

In this paper, we performed density functional theory-based calculations to investigate H_2S adsorption and dissociation on the defective and strained Fe(110) surface. It is found that the Fe vacancy facilitates H_2S adsorption and dissociation, while the effect of homogeneous external stresses, giving rise to the lattice expansion, is negligible. More interestingly, the correct order for phase transition at the Fe(110) surface was predicted by using the thermodynamic chemical potential model, that is, from the elementary Fe crystal to the FeS crystal to the FeS_2 crystal to the elementary S crystal with the S coverage increasing.

2 Computational details

All the computations were performed by using the VASP code[22] within the framework of spin-polarized density functional theory[23]. The generalized gradient approximation in the formalism of Perdew-Burke-Ernzerhof[24] was employed to treat the exchange-correlation interaction. The kinetic energy cutoff of 400 eV was set to truncate the plane wave basis set. The electron-core interaction was described by the projector augmented wave potentials[25]. The valence configurations for Fe, S and H atoms were chosen to be $3d^74s^1$, $3s^23p^4$, and $1s^1$, respectively. The initial magnetic moment on each Fe atom was set to be 3 μ_B which guarantees that the correct magnetic moment on Fe atoms in both bulk and surface calculations can be achieved after electronic optimization.

The periodic slab model was used to simulate adsorption and dissociation of an H_2S molecule on the Fe(110) surface. Each slab model consists of 7 Fe atomic layers with 2×2 periodicity in the surface plane and a vacuum layer of 16Å. Two H_2S molecules, as well as the dissociated products of HS and S, were symmetrically adsorbed onto the top and bottom surfaces of the (110) slab. The slab of the perfect Fe(110) surface contains 56 Fe atoms in total, with the dimension of 5.662× 8.007Å^2 along the surface. For the defective Fe(110) surface, one Fe atom was removed from each surface of the slab. For the strained surface, four different lattice constants in the surface plane, namely 0%, 0.2%, 0.4%, and 5% with respect to the equilibrated lattice constant, were taken into consideration to mimic the lattice expansion caused by external stresses on petroleum pipelines in real exploitation and transport conditions. For the (110) surface model with 2×2 periodicity, the k-point grid mesh was set to be 7×5×1 which well converges the single point energy.

The adsorption energy was defined as

$$E_a = [E(2H_2S^*) - 2E(H_2S) - E(^*)]/2 \quad (1)$$

Here, $E(H_2S)$, $E(^*)$, and $E(2H_2S^*)$ represent the energies of an H_2S molecule and the Fe(110) slabs without and with adsorption of an H_2S molecule on each side, respectively.

The dissociation energy was defined by using the following equation

$$E_d = [E(H^* + HS^* \text{ or } H^* + H^* + S^*) - E(H_2S^*)]/2 \qquad (2)$$

$E(H^* + HS^* \text{ or } H^* + H^* + S^*)$ is the energies of the Fe(110) slabs with adsorption of an H^* and an HS^* or adsorption of two H^* and an S^*.

To explore the S adsorption on the Fe(110) surface, five different coverages were considered, namely, 0.125, 0.5, 1.0, 1.5, and 2.0 monolayer (ML). For the 0.125 ML, the 2×2 slab model was used, while for the other coverages, the 1×1 slab model with the dimension of 2.831× 4.003Å² was employed. Similarly, the k-point grid mesh for the 1×1 slab model was chosen to be 15×11×1.

Besides, the geometric structure of the unit cell for elementary Fe, ferromagnetic α-Fe in body centered cubic symmetry, was fully optimized with a k-point grid mesh of 15×15×15, giving rise to the lattice constant of 2.83Å and the magnetic moment on Fe atoms of 2.19μ_B. Our results agree well with the previously computed values of 2.83Å and 2.20 μ_B and the experimental values of 2.86Å and 2.22 μ_B[4]. The geometric structure of the relatively large unit cell for elementary S in monoclinic symmetry was partially relaxed with the constant lattices (a = 10.38Å, b = 12.76Å, and c = 24.41Å) fixed, and a k-point grid mesh of 2×2×1 was used. The H_2S and S_2 molecules were placed in the boxes of 10×10×10Å³ and 12×12×12Å³, respectively, for geometric optimization, and only the Γ-point was used. The computed S-H bond length and H-S-H angle of H_2S is 1.349Å and 91.6°, equal to the values reported by a theoretical computation and close to the experimental values cited therein of 1.328Å and 92.2°[3]. The computed S-S bond length of S_2 is 1.90Å which is very close to the measured bond length of 1.89Å which was retrieved from Computational Chemistry Comparison and Benchmark Database, NIST on Oct. 13, 2019 (https://cccbdb.nist.gov/alldata1.asp).

The crystal structures of mackinawite FeS and pyrite FeS_2 were optimized with the k-point grid meshes of 9×9×7 and 7×7×7, respectively. The experimental lattice constants for FeS ($a = b$ = 3.67Å, c = 5.03Å)[19,26] were fixed during the structural optimization, while the lattice constants for FeS_2 were optimized to be 5.40Å, very close to the experimental value of 5.42Å[19]. The lattice parameters and atomic coordinates for α-Fe, elementary S, mackinawite FeS, pyrite FeS_2 crystals, and H_2S and S_2 molecules were listed in Part 1 to Part 6 of Supplementary Materials.

3 Results and discussion

3.1 Adsorption and dissociation of H_2S on defective Fe(110)

Adsorption and dissociation of H_2S on Fe(110) is assumed to proceed by following the procedure: $H_2S + ^* \longrightarrow H_2S^*$ (adsorption), $H_2S^* \longrightarrow H^* + HS^*$ (first step of dissociation), and $H^* + HS^* \longrightarrow H^* + H^* + S^*$ (second step of dissociation). Fig. 1 shows the geometries for H_2S adsorption and dissociation on the perfect and defective Fe(110). Both Jiang et al[4]. and Akande et al[13]. have reported first-principles studies of H_2S adsorption and dissociation on the perfect Fe(110) surface, where they indicated that the high symmetry adsorption site of long bridge was favored for all the H_2S, HS, and S adsorption. As a consequence, only the long bridge adsorption site was considered here. Upon an Fe vacancy created nearby the long bridge site, however, the geometries of H_2S adsorption and dissociation change. The H_2S molecule moves away from the Fe vacancy and stabilizes at an on-top adsorption site due to lack of the attraction by the missing Fe atom, while the HS and S sit on top of the Fe vacancy. Meanwhile, the two H atoms dissociated from H_2S adsorb at two threefold sites on both the perfect and defective Fe(110), consistent with the simulation result reported by Jiang et al[4].

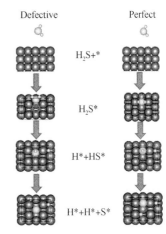

Fig. 1 Adsorption and Dissociation of an H_2S molecule on the Fe(110) surface with (left panel, defective) and without (right panel, perfect) an Fe vacancy (red circle). On the perfect surface, the high symmetry adsorption site of long bridge was used. Fe, S, and H atoms are depicted in brown, yellow, and white, respectively. (Color online)

Accordingly, the H_2S adsorption and dissociation energies were listed in Table 1. The H_2S adsorption energy on perfect Fe(110) is −0.62eV which is very close to −0.60eV given by Akande et al[13], but larger than −0.49eV reported by Jiang et al[4]. The discrepancy might stem from the insufficient slab thickness therein. The Fe vacancy enhances the H_2S adsorption on Fe(110) by increasing the adsorption energy to −1.39eV. The H_2S dissociation on both the perfect and defective Fe(110) surface are found to be easy to proceed with the large energy gains for both the first and second step of the dissociation reaction. The energies for H_2S adsorption (H_2S^*) and dissociation (H^*+HS^*, $H^*+H^*+S^*$) with respect to isolated H_2S molecule and Fe(110) (i.e., H_2S+^*) are depicted in Fig. 2. All the three resulting species are much more stable on the defective surface with the energy gain exceeding 0.7eV, indicating that the surface Fe vacancy significantly promotes the H_2S adsorption and dissociation. In contrary, the Fe vacancy weakens the H_2S adsorption and dissociation on Fe(100) by reducing the adsorption energy from −0.53 to −0.40eV and the dissociation energy from −2.25 to −2.10eV, as reported by Wen et al[11], implying facet-dependent defect chemistry for H_2S adsorption and dissociation on steel.

Table 1 Adsorption energy (E_a) and dissociation energy (E_d) for H_2S on the perfect and defective Fe(110) surface

	Fe(110)	E_a or E_d(eV)
H_2S^*	perfect	−0.62
	defective	−1.39
H^*+HS^*	perfect	−1.60
	defective	−1.83
$H^*+H^*+S^*$	perfect	−2.96
	defective	−3.03

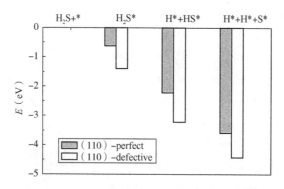

Fig. 2 Energies for H_2S adsorption (H_2S^*) and dissociation (H^*+HS^*, $H^*+H^*+S^*$) with respect to isolated H_2S molecule and Fe(110), i.e., H_2S+*. (Color online)

3.2 Adsorption and dissociation of H_2S on strained Fe(110)

In real oil exploitation and transport, there exist various external stresses, for example the gravitational force of oil pipes and fluid pressure which can stretch the Fe lattice. The effect of external stresses on H_2S adsorption and dissociation on Fe(110) was investigated by adjusting the Fe lattice constants in the surface plane. Four sets of lattice constants, derived by expanding the lattice by 0, 0.2%, 0.4%, and 5%, respectively, were considered. H_2S adsorption and dissociation follow the same procedure as on the perfect Fe(110). The computed adsorption and dissociation energies were summarized in Table 2. Generally, lattice expansion facilitates H_2S adsorption and dissociation on Fe(110). However, in our oil exploitation experiments with the oil well pipe of 5km in length, a maximum lattice expansion caused by the gravitational force was estimated to be only 0.2%, as shown in Part 7 of Supplementary Materials. Such a small strain shows negligible influence on H_2S adsorption and dissociation. Therefore, the effect of homogeneous external stresses can be safely neglected, but in the case of stress concentration, the effect may not be excluded.

Table 2 Adsorption energy (E_a) and dissociation energy (E_d) for H_2S on Fe(110) surface under different extents of lattice expansion

	Lattice expansion percentage (%)	E_a or E_d (eV)
H_2S^*	0	−0.62
	0.2	−0.62
	0.4	−0.63
	5	−0.63
H^*+HS^*	0	−1.60
	0.2	−1.60
	0.4	−1.61
	5	−1.76
$H^*+H^*+S^*$	0	−2.96
	0.2	−2.97
	0.4	−2.97
	5	−3.03

3.3 Coverage dependence of S adsorption on Fe(110)

There coverages of S atom on Fe(110) was considered, namely, 0.125ML for both the defective and perfect surfaces, 0.5ML and 1.0ML for the perfect surface. The most stable S adsorption geometries at the three coverages were shown in Fig. 3. As mentioned above, the S atom sits on the Fe vacancy of the defective surface at the coverage of 0.125ML, while it adsorbs at the long bridge site of the perfect surface at the same coverage. For the coverage of 0.5ML, the S atom adsorbs at the long bridge site stronger than at the on-top site. For the coverage of 1.0ML, a mixed adsorption pattern made up of one S atom at the long bridge site and the other one at the on-top site is favored with respect to the patterns of both the two S atoms at the long bridge sites or at the on-top sites. The adsorption geometries with the S coverage greater than 1.0ML, for example 1.5ML and 2.0ML, failed to be stabilized, as shown in Part 8 of Supplementary Materials. The S adsorption geometries with the coverage of 1.5ML and 2.0ML degenerate to the geometries with the coverage of 0.5ML and 1.0ML, respectively, and one S_2 molecule far away from the Fe(100) surface. As a result, the S coverages of 1.5ML and 2.0ML were not considered further in the present investigation.

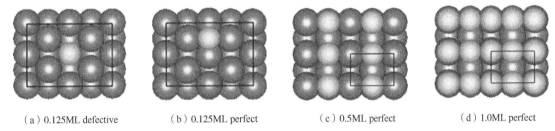

(a) 0.125ML defective   (b) 0.125ML perfect   (c) 0.5ML perfect   (d) 1.0ML perfect

Fig. 3 The most stable adsorption geometries for S atom (a) on the defective Fe(110) with the coverage of 0.125ML, on the perfect Fe(110) with the coverages of (b) 0.125ML, (c) 0.5ML, and (d) 1.0ML. (Color online)

The S chemical potential is approximated by the average adsorption energy of S on Fe(110),

$$\Delta\mu_S \approx [E(S^*) - n^*E(S) - E(^*)]/n \qquad (3)$$

where n stands for the number of S atoms absorbed on the surface, and $E(S)$ is the energy per atom for elementary S crystal. The computed S chemical potentials were listed in Table 3. On the defective Fe(110), the S chemical potential is minimum, while the S chemical potential in a S_2 molecule is maximum. On the perfect Fe(110) surface, the S chemical potential increases with the coverage. The negative value of the S chemical potential calculated here indicates that it is likely to form the corresponding adsorption geometry at the conditions of thermodynamic equilibrium. However, experiments reported that the saturation coverage was 1/3 ML for S on Fe(110)[4], which is lower than the predicted coverage. The reason is unclear to date, but might be related to the simulation method, i.e., overestimation of the S-Fe bond on Fe(110) or negligence of the thermal effect.

Table 3   S chemical potentials at various coverages of S atoms adsorbed at defective and perfect Fe(110) surface, as well as the S_2 molecule

	Defective	Perfect			S_2
Coverage (ML)	0.125	0.125	0.5	1.0	—
$\Delta\mu_S$(eV)	-3.33	-2.71	-2.09	-0.36	0.52

3.4 Formation of FeS and FeS_2 phases

The formation energies for FeS and FeS_2 crystals were calculated by using the following two equations,

$$\Delta H_f(FeS) = E(FeS) - E(Fe) - E(S) \tag{4}$$

$$\Delta H_f(FeS_2) = E(FeS_2) - E(Fe) - 2E(S) \tag{5}$$

$E(FeS)$ and $E(FeS_2)$ are the energies per formula for the FeS and FeS_2 crystals, respectively, and $E(Fe)$ and $E(S)$ are the energies per atom for elementary Fe and S crystals, respectively. The computed $\Delta H_f(FeS)$ is -1.02eV and $\Delta H_f(FeS_2)$ is -1.57eV, comparable to the experimental values of -1.06 and -1.74eV, respectively.[27] The larger formation energy for FeS_2 suggests better stability.

Under the conditions of thermodynamic equilibrium, the Fe and S chemical potentials are required to meet some constraints for formation of stable FeS and FeS_2 phases. For FeS, the Fe and S chemical potentials with respect to their respective elementary crystal follow

$$\Delta\mu_S + \Delta\mu_{Fe} = \Delta\mu_{FeS} = \Delta H_f(FeS) = -1.02\text{eV} \tag{6}$$

To prevent formation of bulk Fe and S crystals, the Fe and S chemical potentials are limited by

$$\Delta\mu_{Fe} \leq 0\text{eV} \tag{7}$$

$$\Delta\mu_S \leq 0\text{eV} \tag{8}$$

For this reason, the S chemical potential can only be taken in the range

$$-1.02\text{eV} \leq \Delta\mu_S \leq 0\text{eV} \tag{9}$$

For FeS_2, similarly, the relative Fe and S chemical potentials follow

$$2\Delta\mu_S + \Delta\mu_{Fe} = \Delta\mu_{FeS_2} = \Delta H_f(FeS_2) = -1.57\text{eV} \tag{10}$$

In the same way, the Fe and S chemical potentials should be less than 0eV, i.e.,

$$\Delta\mu_{Fe} \leq 0\text{eV} \tag{11}$$

$$\Delta\mu_S \leq 0\text{eV} \tag{12}$$

Finally, the S chemical potential is limited in the range

$$-0.79\text{eV} \leq \Delta\mu_S \leq 0\text{eV} \tag{13}$$

Fig. 4 shows the allowed ranges of the S chemical potential to stabilize the elementary Fe and S crystals and the FeS and FeS_2 crystals. As the coverage of the S adsorbate on Fe(110) increases, the stable phase is predicted to come out in the order from the elementary Fe, compound FeS, compound FeS_2, to elementary S crystal. This prediction agrees well with the observation in the experiment.[2] The FeS and FeS_2 phases were predicted to occur at the S coverage from 0.5 to 1.0 ML which exceeds the saturation coverage.

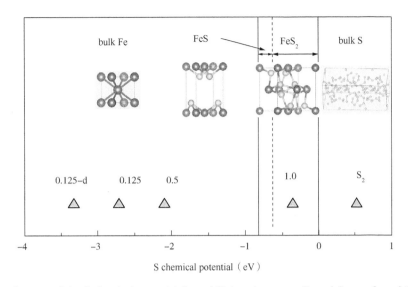

Fig. 4 Allowed ranges of the S chemical potential for stabilizing elementary Fe and S crystals and FeS and FeS$_2$ crystals, as well as the S chemical potentials for S adsorption on the defective and perfect Fe(110) surfaces at the coverages of 0.125, 0.5 and 1.0ML, and the S chemical potentials of an S$_2$ molecule. (Color online)

4 Conclusions

In summary, we performed spin-polarized periodic density functional theory-based calculations to investigate H$_2$S adsorption and dissociation on the defective and strained Fe (110) surfaces with respect to the perfect surface, as well as adsorption of S atoms and formation of Fe-S compounds. Fe vacancy defect in the Fe (110) facilitates H$_2$S adsorption and dissociation by increasing both the H$_2$S adsorption and dissociation energy. On the contrary, the homogeneous stresses acting on the oil well pipe, leading to uniform lattice expansion, show negligible effects on the H$_2$S adsorption and dissociation in the conditions under investigation. However, the strain effect may not be excluded in case of stress concentration. The adsorption of S atoms on the Fe (110) is stronger on the defective Fe (110) than on the perfect one and becomes weak with the S coverage increasing. Finally, by using the chemical potential of S which was defined as the adsorption energy of S on the Fe (110), the simulations predicted a correct order of phase transition at the Fe (110) surface, that is, from the elementary Fe crystal to the FeS crystal to the FeS$_2$ crystal to the elementary S crystal with the S coverage on the Fe (110) surface increasing.

Acknowledgements

This work is supported by the Key Research and Development Program of Shaanxi Province (Grant No. 2018ZDXM - GY - 171), the Natural Science Basic Research Program of Shaanxi Province (2019JQ - 440), the National Natural Science Foundation of China (Grant No. 21603109), and the Henan Joint Fund of the National Natural Science Foundation of China (Grant No. U1404216). The work was carried out at LvLiang Cloud Computing Center of China, and the simulations were performed on TianHe-2.

References

[1] ZHANG F C. First principles study on the cracking behaviour of duplex stainless steel induced by H_2S [D]. Chengdu: Southwest Petroleum University, 2016.

[2] WEN X, LIANG Y, BAI P, et al. First-principles calculations of the structural, elastic and thermodynamic properties of mackinawite (FeS) and pyrite (FeS_2) [J]. Physica B: Condensed Matter, 2017, 525: 119-126.

[3] JIANG D E, CARTER E A. Adsorption, Diffusion, and Dissociation of H_2S on Fe(100) from First Principles [J]. The Journal of Physical Chemistry B, 2004, 108(50): 19140-19145.

[4] JIANG D E, CARTER E A. First principles study of H_2S adsorption and dissociation on Fe(110) [J]. Surface Science, 2005, 583(1): 60-68.

[5] JIANG D E, CARTER E A. Effects of Alloying on the Chemistry of CO and H_2S on Fe Surfaces [J]. The Journal of Physical Chemistry B, 2005, 109(43): 20469-20478.

[6] LUO Q, TANG B, ZHANG Z, et al. First principles calculation of adsorption for H_2S on Fe(100) surface [J]. Acta Physica Sinica, 2013, 62(7).

[7] ZHANG Q, HU P, LUO Q, et al. First principles study on adsorption for different concentration of H_2S on Fe(100) [J]. Advanced Materials Research, 2014, 1015: 521-525.

[8] ZHANG F C, LI C F, ZHANG C L, et al. Surface absorptions of H_2S, HS and S on Fe(111) investigated by density functional theory [J]. Acta Physica Sinica, 2014, 63: 127101-127130.

[9] CARONE FABIANI F, FRATESI G, BRIVIO G P. Adsorption of H_2S, HS, S, and H on a stepped Fe(310) surface [J]. The European Physical Journal B, 2010, 78(4): 455-460.

[10] MAHYUDDIN M H, BELOSLUDOV R V, KHAZAEI M, et al. Effects of Alkali Adatoms on CO and H_2S Adsorptions on the Fe(100) Surface: A Density Functional Theory Study [J]. The Journal of Physical Chemistry C, 2011, 115(48): 23893-23901.

[11] WEN X, BAI P, HAN Z, et al. Effect of vacancy on adsorption/dissociation and diffusion of H_2S on Fe(100) surfaces: A density functional theory study [J]. Applied Surface Science, 2019, 465: 833-845.

[12] REN L, CHENG L, SHAO R, et al. DFT studies of adsorption properties and bond strengths of H_2S, HCN and NH_3 on Fe(100) [J]. Applied Surface Science, 2020, 500: 144232.

[13] AKANDE S O, BENTRIA E T, BOUHALI O, et al. Searching for the rate determining step of the H_2S reaction on Fe(110) surface [J]. Applied Surface Science, 2020, 532: 147470.

[14] SPENCER M J S, YAROVSKY I. Ab Initio Molecular Dynamics Study of H_2S Dissociation on the Fe(110) Surface [J]. The Journal of Physical Chemistry C, 2007, 111(44): 16372-16378.

[15] SPENCER M J S, TODOROVA N, YAROVSKY I. H_2S dissociation on the Fe(100) surface: An ab initio molecular dynamics study [J]. Surface Science, 2008, 602(8): 1547-1553.

[16] TODOROVA N, SPENCER M J S, YAROVSKY I. Ab initio study of S dynamics on iron surfaces [J]. Surface Science, 2007, 601(3): 665-671.

[17] LENNIE A R, REDFERN S A T, SCHOFIELD P F, et al. Synthesis and Rietveld crystal structure refinement of mackinawite, tetragonal FeS [J]. Mineralogical Magazine, 1995, 59(397): 677-683.

[18] BERNER R A, Tetragonal Iron Sulfide [J]. Science, 1962, 137(3531): 669-669.

[19] LIU J, XU A, MENG Y, et al. From predicting to correlating the bonding properties of iron sulfide phases [J]. Computational Materials Science, 2019, 164: 99-107.

[20] ZHANG M Y, CUI Z H, JIANG H. Relative stability of FeS_2 polymorphs with the random phase approximation approach [J]. Journal of Materials Chemistry A, 2018, 6(15): 6606-6616.

[21] LI Y, CHEN J, CHEN Y, et al. DFT+U study on the electronic structures and optical properties of pyrite and marcasite [J]. Computational Materials Science, 2018, 150: 346-352.

[22] KRESSE G, FURTHMULLER J. Efficient iterative schemes for ab initio total-energy calculations using a plane-wave basis set[J]. Physical Review B, 1996, 54(16): 11169-11186.

[23] KOHN W, SHAM L J. Self-Consistent Equations Including Exchange and Correlation Effects[J]. Physical Review, 1965, 140(4A): 1133-1138.

[24] PERDEW J P, BURKE K, ERNZERHOF M. Generalized gradient approximation made simple[J]. Physical Review Letters, 1996, 77(18): 3865-3868.

[25] KRESSE G, JOUBERT D. From ultrasoft pseudopotentials to the projector augmented-wave method[J]. Physical Review B, 1999, 59(3): 1758-1775.

[26] WITTEKINDT C, MARX D. Water confined between sheets of mackinawite FeS minerals[J]. The Journal of Chemical Physics, 2012, 137(5): 054710.

[27] CHASE M W. NIST-JANAF Themochemical Tables[M]. Washington, DC: American Chemical Society, 1998.

本论文原发表于《Surface Science》2021年第709卷。

Hydrogen Sulfide Stress Cracking in a Q345R Welded Joint

Han Yan[1] Luo Jingbin[2] Fu Anqing[1] Yin Chengxian[1]

(1. CNPC Tubular Goods Research Institute, State Key Laboratory for Performance and Structure Safety of Petroleum Tubular Goods and Equipment Materials;
2. Drilling and production technology research institute of Qinghai Oilfield)

Abstract: In this paper, the failure reason of Q345R Welded Joint was studied through macroscopic observation, chemical properties, metallurgical analysis, scanning electron microscope (SEM) and EDS test method. The results showed that there were a large number of microcracks in the fracture surface. The reason of cracking is severe banded structure in base metal microstructure, which provided opportunity for hydrogen atoms to enter into the inside of the steel. The existence of tensile stress promotes the entry of hydrogen atoms and the propagation of cracks. The welding products of this procedure are not suitable use under sour conditions.

Keywords: Q345R welded joint; Hydrogen sulfide stress cracking; Banded structure

1 Introduction

With the rapid development of economy, oil and gas field development has been towards to high temperature, high pressure, high sulfur and other harsh reservoirs, and the demand for oil and gas storage steel has increased significantly. At the same time, the service conditions of pressure vessel steel have become increasingly harsh[1]. Hydrogen sulfide is one of the most corrosive and harmful media in petroleum and natural gas. Its existence will lead to corrosion and cracking of metal pipes and even lead to catastrophic accidents[2-6]. Similar hazards may occur in containers and tanks that store sulfur-containing media, as in pipelines. In addition, the high residual stress caused by welding may promote the occurrence of sulfide stress cracking (SSC) even in the weld with low hardness[7]. In this paper, the fracture of the welded joint of a Q345R acid resisting vessel occurred in SSC test was analyzed systematically, in order to provide relevant reference for the solution of this problem.

2 Experimental methods

To test the resistance of SSC of Q345R welded joint, the evaluation test was conducted

---

Corresponding author: Han Yan, hanyan003@cnpc.com.cn.

according to NACE TM0177—2016[8] method A under constant tensile load. The test solution was H₂S saturated aqueous brine solution consist of 5% NaCl+0.5% glacial acetic acid dissolved in deionized water. The loading stress was 247MPa, the test temperature was 24℃. Three samples were fetched from the weld joint, the diameter of sample was 6.35mm.

To find the reasons of the fracture, the following tests were conducted: (1) visual examination of fracture surface, (2) chemical composition analysis, (3) metallographic structural characterization, (4) scanning electron microscopy (SEM) observation and energy – dispersive spectrometry (EDS).

3 Results and discussion

3.1 SSC evaluation

All three samples were fractured at 46, 56 and 168h respectively during the SSC test, shown in Fig. 1. The welded joint cannot meet the no fracture requirements during 720h' test in NACE TM0177 standard.

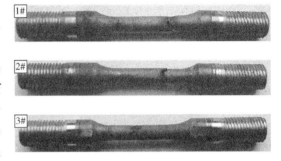

From thelow magnification of the fracture morphology, the fracture has the characteristics of stratification, and the typical fracture morphology is shown in Fig. 2. There was no obvious plastic deformation on the samples. The fracture position was located in the working section of the samples

Fig. 1 SSC test results of Q345R welded joint

and presented a certain tilt angle with the direction of tensile stress. The surface of the fracture was rough with the black color, and there were many secondary cracks on the fracture surface.

Fig. 2 Low magnification of fracture surface

3.2 Chemical composition analysis

The chemical composition of the Q345R is shown in Table 1. The elements content accorded with the requirements in GB 713—2014 standard[9], the chemical composition is qualified.

Table 1 Chemical composition of the Q345R (Wt. %)

Elements	C	Si	Mn	P	S	Cr	Mo	Ni	Nb	V	Ti	Cu
Q345R	0.18	0.24	1.25	0.0060	<0.002	0.13	<0.005	0.11	0.015	<0.005	0.0019	0.0054
GB 713—2014	≤0.20	≤0.55	1.20–1.70	≤0.025	≤0.010	≤0.30	≤0.08	≤0.30	<0.050	<0.050	≤0.30	≤0.30

3.3 Metallographic structural characterization

The metallographic structural of the failed sample was shown in Fig. 3. The fracture location was in the fine grained area near the weld.

Fig. 3 Metallographic structural of the failed sample

The microstructure of base metal was tempered sorbate and bainite, the microstructure of weld zone was PF + IAF + B + P, the microstructure of fusion zone was B + PF + P + WF, and the microstructure of fine grained zone was PF+P, as shown in Fig. 4. The inclusion grade of sample was A 0.5, B 0.5, D 0.5.

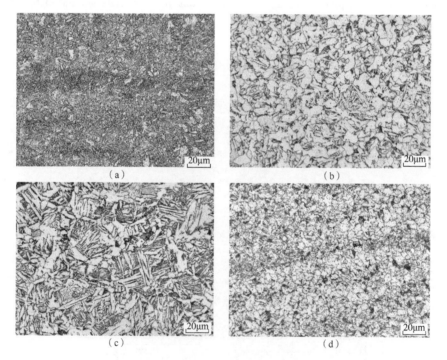

Fig. 4 Microstructure of welding joint
(a—base metal, b—weld zone, c—fusion zone, d—fine grained zone)

A 377μm long and 89μm wide crack was found in the fine crystal region. The crack distribution was along the banded segregation trend, as shown in Fig. 5. The metallographic structure near the fracture is the same as the fine grained area, and there is obvious banded segregation, as shown in Fig. 6.

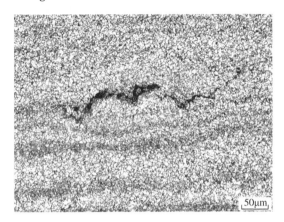
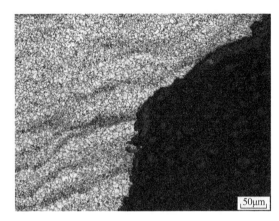

Fig. 5 Crack morphology in fine grained region   Fig. 6 Microstructure near the fracture

Tukon 2100B microhardness tester was used to conduct hardness test. The test results were shown in Table 2. The hardness of sample was less than 22HRC, it was qualified to the requirement specified in ISO 15156-2: 2020[10].

Table 2 Microvickers hardness test results

Sample	$HV_{0.5}$		
	Base metal near the weld zone	Fine grained zone	Fine grained zone near the fracture
Q345R Welded joint	190, 189, 195	162, 156, 155, 161, 163	165, 166, 150, 156, 158

3.4 SEM observation and EDS analysis

The surface of the fracture was uneven, and there were many microcracks and sectorial shapes on the fracture, as shown in Fig. 7. The cracks originated from the inside of the sample and distributed in a layered manner. Many multiple cracks were connected together then it leading to the final fracture. The cracks have typical hydrogen-induced cracking (HIC) crack morphology, which is in step shape and distributed along the wall thickness direction. Each sectorial shape was a small crack propagation surface, the radial edge shrinkage direction of the sectorial propagation surface is the source of crack, most of the sectorial propagation surface originates from the internal crack.

The EDS results show that the corrosion products mainly contain Fe, O and S elements, as shown in Fig. 8.

According to the test results, the chemical composition of the sample meets the requirements of GB 713—2014 standard, and there was obvious banded segregation in the metallographic structure of the sample base metal and fine grained zone. Banded segregation is one of the internal defects of hot rolled low carbon structural steel. Ferrite and pearlite grains are arranged in parallel, stratified distribution and in the same strip along rolling direction. The existence of banded structure makes the structure of steel uneven, and affects the properties of steel, forming anisotropy and reducing the

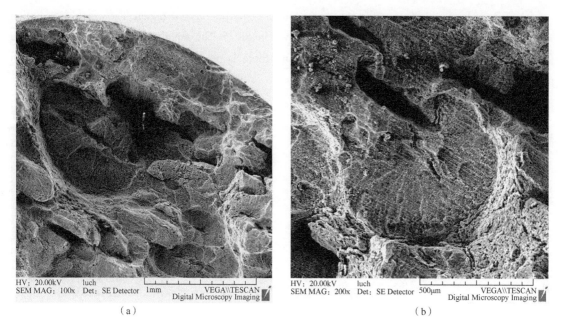

(a)   (b)

Fig. 7　Micro-morphology of fracture surface

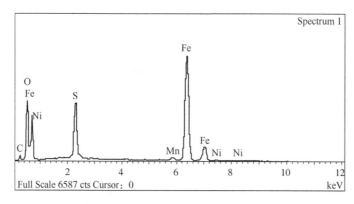

Fig. 8　Energy spectrum analysis of corrosion products on the fracture surface

plasticity, impact toughness and section shrinkage of steel. The banded structure also had a significant effect on the hydrogen – induced cracking resistance of the material, and the HIC sensitivity increased with the increase of banded structure[11]. At the same time, bainite also exists in the sample tissue. As the dislocation density in the bainite sturcture is higher than that of ferrite, HIC cracks are more likely to form around bainite[12].

When metal materials contact with wet H_2S environment, a large number of ion in solution state of hydrogen through electrochemical reaction to the interior of the material, supersaturated hydrogen in steel defects such as dislocations, inclusion and other interface gathered together, and form hydrogen molecules, resulting in a huge internal pressure, combined with tensile stress makes hydrogen induced crack nucleation and propagation[13]. In the study of the fracture morphology, a large number of layered distribution of crack appeared in conformity with the banded structure distribution. And the metallographic analysis also indicated that the cracking was along the banded structure. The morphology and microscopic analysis reavealed that sector is on the surface of the micro-cracks extension stage. This also explained that the micro-cracks formation has experienced a certain amount of time, in accordance with the principle of lagging hydrogen induced cracking. The formation of internal micro-cracks, combined with external tensile load, resulted in premature failure of the material under external stress far below its yield strength.

4 Conclusions

The main reason for the fracture of the Q345R sample is the severe banded segregation in thebase material, which provides conditions for hydrogen atoms to accumulate inside the steel. Under the action of wet hydrogen sulfide environment and external tensile stress, internal micro-cracks generates, and finally fracture occurs.

Acknowledgments

This work was supported by the National Key R&D Program of China (2016YFC0802101) and Major science and technology project of CNPC (2016YFC0803201).

References

[1] HUANG Z C, YANG J, ZHOU K. Practice to F120t BOF-LF-RH-260 mm×2070mm CC melting technology of Q345R acid resistant steel[J]. Special steel, 2020, 41(4): 30.

[2] HAN Y, ZHAO X H, BAI Z Q, et al. Review of Sour Gas Field on Corrosion Factors and Development Progress[C]. Advanced Materials Research, 2011, 201: 438-447.

[3] LIU Z Y, LI H, JIA Z J, et al. Failure analysis of P110 steel tubing in low-temperature annular environment of CO_2 flooding wells[J]. Engineering Failure Analysis, 2016, 60: 296-306.

[4] LONG Y, WU G, FU A Q, et al. Failure analysis of the 13Cr valve cage of tubing pump used in an oilfield [J]. Engineering Failure Analysis, 2018, 93: 330-339.

[5] HAN Y, ZHAO X H, ZHANG J T, et al. Failure Analysis on Fracture of a 3½ in P110SS Tubing[J]. Advanced Materials Research, 2012, 524: 1412-1417.

[6] LIU W Y, SHI T H, LU Q, et al. Failure analysis on fracture of S13Cr-110 tubing[J]. Engineering Failure Analysis, 2018, 90: 215-230.

[7] QIAN Z J, GUO P, BRUCE D. Craig: SPE Monograph Volume 15 Sour-Gas Design Considerations, 2003, 133.

[8] ANSI/NACE TM0177—2016, Laboratory testing of metals for resistance to sulfide stress cracking and stress corrosion cracking in H_2S environments [S]. Houston: NACE International, 2016.

[9] GB 713—2014, Steel plates for boilers and pressure vessels[S]. Beijing: China Standards Press, 2014.

[10] ISO 15156-2-2020, Petroleum and natural gas industries- Materials for use in H_2S containing environments in oil and gas production- Part 2: Cracking-resistant carbon and low-alloy steels, and the use of cast irons [S]. Switzerland: ISO, 2020.

[11] ZHOU Q, JI G S, YANG R C, et al. Banded structure and hydrogen-induced cracking in pipe-lines steel [J]. Journal of Gansu University of Technology, 2002, 28(2): 30-33.

[12] TONG K, HAN X L, SONG H, et al. Research on Mechanical Property and HIC, SSC Test Analysis of L245NS Anti-sulfur Bending Pipe [J]. HAN GUAN, 2012, 35(12): 45-49.

[13] CHU W Y. New progress in hydrogen-induced cracking and stress corrosion mechanism [J]. Progress in natural science, 1991, 1(5): 393-399.

本论文原发表于《Materials Science Forum》2021年第1035卷。

Investigations of Polyethylene of Raised Temperature Resistance Service Performance Using Autoclave Test under Sour Medium Conditions

Qi Guoquan[1, 2]　Yan Hongxia[1]　Qi Dongtao[2]
Li Houbu[2]　Kong Lushi[2]　Ding Han[1, 2]

(1. School of Chemistry and Chemical Engineering, Northwestern Polytechnical University;
2. State Key Laboratory of Performance and Structural Safety for Petroleum Tubular Goods and Equipment Materials, CNPC Tubular Goods Research Institute)

Abstract: The performance evaluation of the polyethylene of raised temperature resistance (PE-RT) and polyethylene (PE) using autoclave test under sour oil and gas medium conditions. The analyses of performance changes, showed that PE-RT has good media resistance at 60℃. As the temperature increases, its mechanical properties decrease, accompanied by an increase in weight. Comparative analyses showed that no matter what temperature conditions are, PE-RT media resistance is better than PE80. The better media resistance of PE-RT depends on its higher degree of branching. Short branches are distributed between the crystals to form a connection between the crystals, thereby improving its heat resistance and stress under high temperature conditions. PE-RT forms an excellent three-dimensional network structure through copolymerization, ensuring that it has better media resistance than PE80. However, the mechanical performance will be attenuated due to the high service temperature.

Keywords: Plastic; Stress; Chain; Degradation; Swelling

1 Introduction

Due to the excellent corrosion resistance, non-metallic and composite pipes have become one of the most important solutions to solve the corrosion problem of oil and gas fields[1-3]. As the amount addition increases, the trend of using non-metallic and composite pipes to replace carbon steel pipes is becoming more and more obvious. Thermoplastics can be used directly as pipes or as the inner lining of composite pipes, which has an important position[4-5]. Since the thermoplastic is in direct contact with the conveying medium, the media resistance of the non-metallic composite pipe

---

Corresponding author: Yan Hongxia, hongxiayan@ nwpu. edu. cn.

depends on the thermoplastics. Among them, PE has become the most used thermoplastic due to its high-cost performance[6-8]. In the microstructure, the crystalline region is formed by the folding of the chain segments and the amorphous region coexist. The tie molecules connecting the crystalline and amorphous regions of the polymer play a key role in the long-term high temperature creep properties of the PE pipe[9-11].

With the expansion of oil and gas fields, the higher medium transportation temperature puts forward higher requirements on the temperature resistance of thermoplastic, PE-RT came into being. In the service process of various types of pipes, the inner wall of the pipe is chemically corroded by the contacting oil-gas coupling medium under high temperature, high pressure, and other multi-physical field conditions[12]. Especially in the oil and gas medium with high acidity (H_2S, CO_2, etc.), it will cause performance degradation of plastic pipes such as swelling, bulging, and softening[13]. The combination of medium and internal hydraulic load will cause pipe cracking failure behavior[14]. Although thermoplastic plastic pipes have excellent corrosion resistance, they will experience aging under long-term service conditions and are subjected to the multiple effects of temperature, pressure, and media, simultaneously[15-16]. There is a risk of performance degradation or even failure under long-term service conditions. Once pipe failure occurs, it may cause serious casualties and environmental damage. The evaluation is an important link in the development and application of new products. For the new products developed, it is necessary to carry out long-term service performance evaluation under simulated working conditions and verify the relevant performance indicators before application. Only in this way can the product be safe and reliable during the service period. However, currently there are few reports on the performance evaluation of thermoplastics under sour oil and gas medium conditions[6,9,11,13,17].

In order to understand the interacting mechanisms involved in this study, it will be necessary to consider all the underlying processes, such as the effects of material structure, the effects of temperature and stress, and degradation in specific environments. In this study, the applicability and aging mechanism of PE-RT for the transportation of sour oil and gas media are also studied.

2 Experimental

2.1 Sample preparation

The PE-RT pipe with the specification of DN100mm×6mm produced by SABIC Co. is used for research. The sample is prepared according to type 1 of ISO 6259-3: 2015 *Thermoplastics pipes-Determination of tensile properties-Part 3: Polyolefin pipes*. Four specimens were prepared under each condition, three of which were used for tensile tests and one was used for other performance tests. For comparison, under the same conditions, PE80 pipes were also tested.

2.2 Experimental process

Before conducting the exposure test, the specimens of PE-RT and PE80 should be tested. Test items include sample size and weight, tensile properties, thermal stability, composition, and microstructure.

The exposure test is completed with the autoclave system, and the medium selection in the autoclave refers to ISO 23936-1: 2009 *Petroleum, petrochemical and natural gas industries Non-*

metallic materials in contact with media related to oil and gas production Part 1: Thermoplastics. The requirements for the test medium are as follows: 30% gas phase (10% CO_2, 10% H_2S, and 90% CH_4), and 70% liquid phase (70% heptane, 20% cyclohexane, and 10% toluene); In this exposure test, the temperature is 90℃, the pressure is 8MPa, and the test period is 0, 1, 3, 5, and 7 weeks.

The sample is taken out and tested after theexposure test is completed. The test samples are the same as those before the test. Among them, visual inspection of external damage includes but is not limited to the following forms: cracks, delamination, swelling, blistering, etc. Meanwhile, cutting is performed to check the internal structure of the sample.

2.3 Performance test and characterization

2.3.1 Weight change

The weight is tested by electronic analytical balance (CPA225D, Sartorius, Germany), with an accuracy of 10^{-4} g. Five groups of samples with different exposure time (0 ~ 7 weeks) were subjected to weight test, and the weight change of different exposure time was compared and analyzed.

2.3.2 Macro mechanical performance

The tensile properties are tested by the universal test machine (UH – F500KNI, Shanghai Xinsansi, China) with the accuracy of 1N and the tensile rate of 50mm/min according to ASTM D638. The test data included strain and stress.

2.3.3 Vicat softening temperature analysis

The Vicat softening temperature is one of the indexes to evaluate the heat resistance of materials and reflect the physical and mechanical properties of samples under heated conditions. The Vicat softening temperature is tested by the Vicat softening point tester (Chengde Precision, China) under the force of 50N and the heating rate of 50℃/h refers to ISO 306. The specimen is a square with a side length of 10mm. The temperature ranges from 25℃ to 200℃.

2.3.4 Composition analysis

The composition of functional group is analyzed by the fourier transform infrared spectroscopy (FT – IR, Thermo Nicolet Avator 360, American). FT – IR analysis is performed by means of attenuated total reflection, with the spectral range of 400 to 4000cm^{-1} and the spectral resolution of 4cm^{-1}.

2.3.5 Microstructure characterization

The microstructure is characterized by Scanning electron microscope (SEM, Hitachi S4800, Japan). In order to avoid the damage to the sample in the process of preparing the sample, the surface of the inner wall of the pipe after cleaning is observed in this test.

2.3.6 Crystallinity analysis

The crystallinity is the ratio of the mass or volume of the crystalline part of the polymer to the total mass or volume, which is one of the important physical quantities. The degree of crystallinity is directly related to the mechanical and heat resistance properties of polymer materials. In this study, the differential scanning calorimetry (DSC) is used to test the crystallinity by measuring the enthalpy of fusion. For recording the crystalline melting curve of PE-RT and PE80, the DSC study is carried out. In the DSC analysis, two heating and cooling cycles were performed at a rate of 10℃/min under N_2 environment protection, and the temperature ranges from 25℃ to 240℃.

3 Results and discussions

3.1 Weight change

The test result of weight change rate is shown in Fig. 1. A comparative analysis of the weight change rate of PE-RT and PE80 shows that the quality change rate of PE-RT is significantly smaller than that of PE80, and this trend becomes more obvious as the exposure time increases. In addition, the analysis of the weight change of PE-RT under different temperature conditions demonstrates that in the middle period of exposure time (from 1 to 5 weeks), the weight change rate at 90℃ is slightly greater than that at 60℃. As the exposure time further increases, that is, when it reaches 7 weeks, the weight change rate of the two samples is relatively close. Through the above analysis, it can be seen that the exposure time and temperature will have a significant effect on the weight change rate, and both will promote the increase of the sample quality. However, the temperature has different effects on the different materials. Due to the better temperature resistance of PE-RT, the effect of temperature on PE-RT is not obvious. During the exposure test, owing to the penetration and swelling of the oil and gas medium, part of the medium enters the sample, which increases the weight of the sample[18].

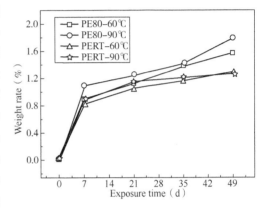

Fig. 1 The weight change rate of PE-RT and PE80 after the exposure test

3.2 Macro mechanical performance

The stress-strain curves of PE-RT and PE80 are shown in Fig. 2. It can be clearly seen that the yield strengths of all samples decrease with the increase of exposure time. Comparing with the yield strength of the original sample, it can be seen that PE-RT's yield strength is larger than PE80's, and its strength values are 22.6MPa and 21.8MPa, respectively. And the tensile performance of the former is better than that of the latter. Comparing with the tensile strengths of the two samples at 60℃, the yield strength of PE-RT after exposure for 7 weeks is reduced to 19.6MPa, which is still greater than that of PE80 after 7 weeks of immersion (18.8MPa). At 90℃, after the 7-week of the exposure test, the yield strength of PE-RT decreases to 18.3MPa, which was still higher than that of PE80 (16.9MPa). Through the above analyses, whether it is the original sample or the sample exposed to oil and gas media, the mechanical properties of PE-RT are better than those of PE80. However, after the exposure to oil and gas media, the two types of the samples' mechanical properties still showed a trend of decreasing to varying degrees. This is because that during the exposure process, the medium penetrates the sample and extracts the small uncured resin molecules and decomposed molecules. Due to extraction, the various defects such as concentrated holes and micro-cracks are formed inside the pipe. The stress on the molecular chain near the cavities far exceeds the average stress of the actual material, causing the mechanical properties of the place to decrease, thereby affecting the overall mechanical properties of the pipe[19-21].

As the temperature rises, the movement unit of the chain segment gradually increases, and the total free volume increases. In this state, the restrictive effect of polymer chain on the molecules' movement is weakened, which makes the medium penetration or extraction easier.

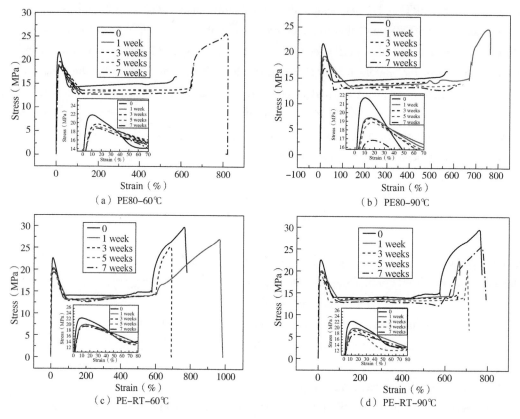

Fig. 2 The stress-strain curves

3.3 Vicat softening temperature analysis

The Vicat softening temperature test result is shown in Fig. 3. The analysis shows that the Vicat softening temperature of all samples at different immersion temperatures showed a downward trend. As the exposure time increases, the downward trend becomes flatter. As mentioned earlier, during the exposure test, due to the swelling and extraction of the medium, the temperature resistance of the sample is reduced. As the exposure time increases, this change gradually reaches a steady state, so the Vicat softening temperature change trend is relatively flat. Comparing the Vicat softening temperature at different temperatures (60℃ and 90℃), it can be seen that the temperature increase in the test can significantly reduce the Vicat softening temperature of the sample. This is because that the temperature accelerates the swelling and extraction. Regardless of the test temperature, the Vicat softening temperature of PE-RT is higher than that of PE80, indicating that the temperature resistance of the former is better than that of the latter under contact with the medium. This is also the result of the unique molecular chain structure of PE-RT. The reason for this change is the unique molecular chain structure of PE-RT.

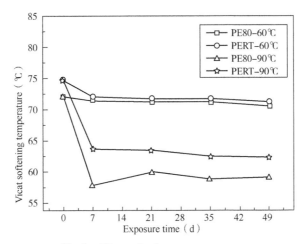

Fig. 3 Vicat softening temperature test

3.4 Composition analysis

The infrared spectrum is shown in Fig. 4. The comparison with the samples under different test conditions and the initial samples shows that the composition between them is basically unchanged, and the characteristic peaks are shown as typical polyethylene types. Therefore, during the exposure test, there was no chemical reaction leading to changes in functional groups and physical phenomena such as swelling and extraction due to the medium entering the inside of the pipe.

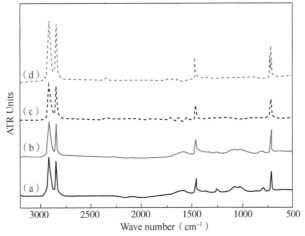

Fig. 4 Infrared spectrum of PE-RT and PE80
(a) PE-RT-60℃; (b) PE-RT-90℃; (c) PE80-60℃; (d) PE80-90℃

3.5 Microstructure characterization

The microstructure of the two types of thermoplastic during the exposure test is shown in Fig. 5. The surface of the original sample of the two samples is relatively smooth, without obvious erosion. With the increase of exposure time, micro-holes appeared on the surfaces of both, and this phenomenon became more obvious with the increase of exposure time and temperature. The appearance of this feature also shows that the medium extracts small molecules in the amorphous region during the exposure test. Under 60℃, with the increase of exposure time, the microscopic morphology of PE–RT is basically unchanged, showing good resistance to the media. When the temperature of the exposure test was increased to 90℃, island-like micro-cracks appeared in PE–RT as that did in PE80, but the number of micro-cracks was less than that of the latter, indicating that its performance under this condition is better than PE80.

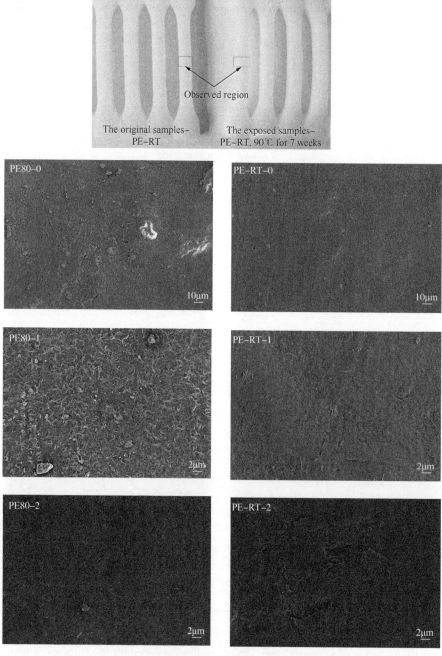

Fig. 5 SEM micrographs of PE80 and PE–RT
(−0): the original samples; (−1): 60℃ for 3 weeks; (−2): 60℃ for 7 weeks;
(−3): 90℃ for 3 weeks; (−4): 90℃ for 7 weeks

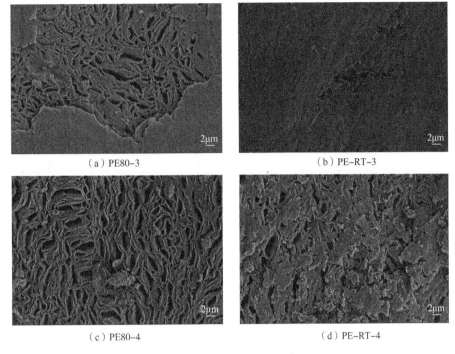

Fig. 5 SEM micrographs of PE80 and PE-RT (continued)
(-0): the original samples; (-1): 60℃ for 3 weeks; (-2): 60℃ for 7 weeks;
(-3): 90℃ for 3 weeks; (-4): 90℃ for 7 weeks

3.6 DSC analysis

The thermal scan curves of the two initial samples and the samples afterexposuring to the test conditions at 90℃ for 7 weeks are shown in Fig. 6. Comparing the enthalpy of fusion before and after the test, it can be seen that the enthalpy of fusion after the exposure test has an increasing trend. Since the enthalpy of heat fusion is proportional to the degree of crystallinity, it can be seen that the degree of crystallinity is also increasing, which also proves that the physical phenomenon of the medium extracting substances in the amorphous region does exist during the exposure test.

3.7 Comprehensive analysis

The attenuation of pipe performance is mainly affected by three factors, the conveying medium, temperature, and pressure. The exposure test simulates the service conditions of PE-RT and PE80 pipes under actual working conditions. The reasons of PE-RT service performance changes under this working condition are studied through the comparison of physical properties, composition, and micro-morphology analysis before and after the exposure test. During long-term operation, the medium will enter the inside of the pipe due to penetration. Under microscopic view, the medium molecules entering the inside of the pipe are located at the crystal boundary, causing swelling, and appearing as weight gain. In addition, the medium extracts the small resin molecules which is uncured or decomposed inside the pipe, showing weight loss. Since the swelling effect is significantly stronger than the extraction, the final manifestation is that the mass increases with the prolonged exposure time. The degradation on PE or PE-RT, such as other semicrystalline polymers, has almost exclusively shown that degradation is concentrated in the amorphous regions[22-24].

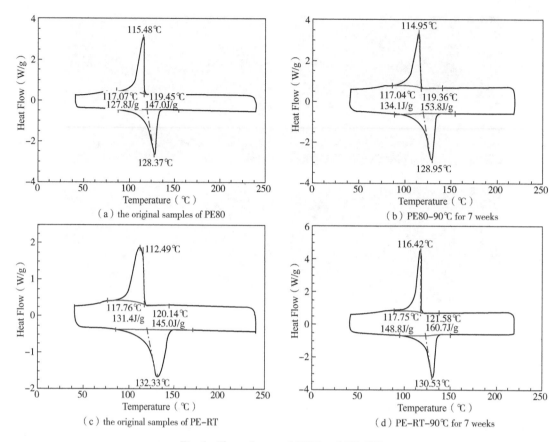

Fig. 6　Thermal scan of PE80 and PE-RT

Due to the penetration and swelling of the medium, various defects, such as cavities and micro-cracks are formed inside the pipe, can cause the stress concentration of the molecular chain at this place and are higher than the stress that the molecular chains at other positions. Fracture failure occurs firstly at this stress concentration, which eventually leads to a decrease in overall mechanical properties. Since the swelling process is only a physical process and no chemical reaction occurs, which can also be proved by infrared analysis. Through microscopic morphology analysis, the existence of micro-crack defectsis confirmed [25]. However, because the defects are micro-sized morphology, they have little effect on reducing the macro-mechanical properties.

The mechanical and temperature resistance of PE-RT are better than those of PE80, the reason is that its molecular chain is different in the structure. The PE-RT is copolymerized by ethylene monomer and 1-octene monomer to form a longer branched olefin monomer containing 6 carbon atoms in the branch, which has a higher degree of branching. PE-RT in this study is a high-density PE produced by a special molecular design and synthesis process. It adopts a copolymerization method of ethylene and 1-octene monomer and obtains a unique molecular structure by controlling the number and distribution of side chains to improve the heat resistance of PE. Due to the existence of short octene chains, the macromolecules of PE-RT cannot be crystallized in a flaky crystal but penetrate through several crystals to form a connection between the crystals. The short branches (the tie molecules) run through several crystals to form the connection between the crystals

(Fig. 7), thus forming a "three-dimensional network structure". The formation of this structure retains the good flexibility of PE and high thermal conductivity and inertness, while making it more pressure-resistant, thereby the PE-RT pipe obtain excellent thermal stability and mechanical properties[26].

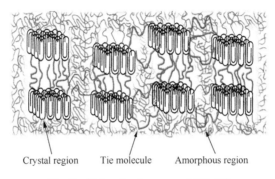

Fig. 7 Molecular Structure of PE-RT

4 Conclusion

The results of the high-temperature autoclave test show that PE-RT exhibits good media resistance under the test conditions of 60℃. However, when the test temperature is increased to 90℃, its media resistance slightly decreases. Comparative analysis shows that no matter what temperature conditions are, media resistance of PE-RT is better than that of PE80. The excellent resistance to media of PE-RT depends on its higher degree of branching. Short branches as the tie molecules are distributed between the crystals to form a connection between the crystals, thereby improving its heat resistance and stress under high temperature conditions.

Acknowledgements

This work was supported by the State Key Laboratory for Performance and Structure Safety of Petroleum Tubular Goods and Equipment Materials, China.

References

[1] MIAO J, YUAN J T, HAN Y, et al. Corrosion Behavior of P110 Tubing Steel in the CO_2-saturated Simulated Oilfield Formation Water with Element Sulfur Addition[J]. Rare Metal Materials and Engineering, 2018, 47(7): 1965-1972. doi: CNKI: SUN: COSE. 0. 2018-07-003.

[2] QI D T, LI H B, CAI X H, et al. Application and Qualification of Reinforced Thermoplastic Pipes in Chinese Oilfields[C]. ICPTT, Beijing, 2011. doi: 10. 1061/41202(423)31.

[3] MANO J F, SOUSA R A, REIS R L, et al. Viscoelastic behaviour and time-temperature correspondence of HDPE with varying levels of process-induced orientation[J]. Polymer, 2001, 42(14): 6187-6198. doi: 10. 1016/S0032-3861(01)00090-8.

[4] QI G Q, WU Y, QI D T, et al. Experimental study on the thermostable property of aramid fiber reinforced PE-RT pipes[J]. Natural Gas Industry, 2015, 2(5): 461-466. doi: 10. 1016/j. ngib. 2015. 09. 023.

[5] JOURNAL G. Polyamide Pipe Selected to Carry High-Volume Gas to Chemical Plant[J]. Pipeline & gas journal, 2015, 242(8): 72-73. doi: 10. 1533/9781845694609. 1. 171.

[6] HAFIZ U K, MOKHTAR C I, NORLIN N. Permeation Damage of Polymer Liner in Oil and Gas Pipelines: A Review. Polymers, 2020, 12(10): 2307. doi: 10.3390/polym12102307.

[7] MICHAEL K G, GERNOT M W, KLEMENS G, et al. Aging behavior and lifetime assessment of polyolefin liner materials for seasonal heat storage using micro-specimen[J]. Sola Energy, 2018, 170: 988-990. doi: 10.1016/j.solener.2018.06.046.

[8] ALBERTO V, NAZDANEH Y, IGNACY J. Determination of the long-term performance of district heating pipes through accelerated ageing [J]. Polymer degradation and stability, 2018, 153: 15-22. doi: 10.1016/j.polymdegradstab.2018.04.003.

[9] KALLIO K J, NAGEYE A S, HEDENQVIST M S. Ageing properties of car fuel-lines: accelerated testing in "close-to-real" service conditions [J]. Polymer Testing, 2010, 29(1): 41-48. doi: 10.1016/j.polymertesting.2009.09.003.

[10] QI G Q, QI D T, YAN H X, et al. Advance in Environmental Stress Cracking Study of Thermoplastic Pipes Used in Oil and Gas Field[J]. Polymeric Materials Science and Engineering, 2018, 34(6): 179-183. doi: 10.16865/j.cnki.1000-7555.2018.06.031.

[11] RITUMS J E, MATTOZZI A, GEDDE U W, et al. Mechanical properties of high-density polyethylene and crosslinked high-density polyethylene in crude oil and its components[J]. Journal of Polymer Science Part B Polymer Physics, 2010, 44(4): 641-648. doi: 10.1002/polb.20729.

[12] QI G Q, QI D T, YAN H X, et al. Advance in Environmental Stress Cracking Study of Thermoplastic Pipes Used in Oil and Gas Field[J]. Polymeric Materials Science and Engineering, 2018, 34(6): 179-183. doi: 10.16865/j.cnki.1000-7555.2018.06.031.

[13] QI G Q, YAN H X, QI D T, et al. Analysis of cracks in polyvinylidene fluoride lined reinforced thermoplastic pipe used in acidic gas fields [J]. Engineering Failure Analysis, 2019, 99: 26-33. doi: 10.1016/j.engfailanal.2019.01.079.

[14] SAHARUDIN M S, ATIF R, SHYHA I, et al. The degradation of mechanical properties in polymer nano-composites exposed to liquid media - a review [J]. Rsc Advances, 2015, 6(2): 1076-1089. doi: 10.1039/C5RA22620A.

[15] GRABMAYER K, WALLNER G M, BEIβMANN S, et al. Accelerated aging of polyethylene materials at hig oxygen pressure characterized by photoluminescence spectroscopy and established aging characterization methods [J]. Polymer degradation and stability, 2014, 109: 40-49. doi: 10.1016/j.polymdegradstab.2014.06.021.

[16] TORRES A H U, D'ALMEIDA J R M, HABAS J P. Aging of HDPE pipes exposed to diesel lubricant [J]. Polymer-Plastics Technology and Engineering, 2011, 50(15): 1594-1599. doi: 10.1080/03602559.2011.578297.

[17] WEI X F, KALLIO K J, BRUDER S, et al. Diffusion-limited oxidation of polyamide: Three stages of fracture behavior [J]. Polymer Degradation and Stability, 2018, 154: 73-83. doi: 10.1016/j.polymdegradstab.2018.05.024.

[18] VOULTZATIS I S, PAPASPYRIDES C D, TSENOGLOU C J, et al. Diffusion of Model Contaminants in High-Density Polyethylene[J]. Macromolecular Materials and Engineering, 2007, 292(3): 272-284. doi: 10.1002/mame.200600420.

[19] GACOUGNOLLE J L, CASTAGNET S, WERTH M. Post-mortem analysis of failure in polyvinylidene fluoride pipes tested under constant pressure in the slow crack growth regime[J]. Engineering Failure Analysis, 2006, 13(1): 96-109. doi: 10.1016/j.engfailanal.2004.10.007.

[20] SCHOEFFL P F, BRADLER P R, LANG R W. Yielding and crack growth testing of polymers under severe liquid media conditions[J]. Polymer Testing, 2014, 40: 225-233. doi: 10.1016/j.polymertesting.2014.09.005.

[21] KIASS N, KHELIF R, BOULANOUAR L, et al. Experimental approach to mechanical property variability through a high-density polyethylene gas pipe wall[J]. Journal of Applied Polymer Science, 2005, 97(1): 272-281. doi: 10.1002/app.21713.

[22] YARYSHEVA A Y, BAGROV DV, RUKHLYa E G, et al. First direct microscopic study of the crazed polymer structure stabilized by a liquid medium[J]. Doklady Physical Chemistry, 2011, 440(2). doi: 10.1134/S0012501611080045.

[23] BAGROV D V, YARYSHEVA A Y, RUKHLYA E G, et al. Atomic force microscopic study of the structure of high-density polyethylene deformed in liquid medium by crazing mechanism[J]. Journal of Microscopy, 2014, 253(2): 151-160. doi: 10.1111/jmi.12104.

[24] ZHANG Y, BEN J P Y, NGUYEN K C T, et al. Characterization of ductile damage in polyethylene plate using ultrasonic testing[J]. Polymer Testing, 2017, 62: 51-60. doi: 10.1016/j.polymertesting.2017.06.010.

[25] BREDÁCS M, FRANK A, BASTERO A, et al. Accelerated aging of polyethylene pipe grades in aqueous chlorine dioxide at constant concentration[J]. Polymer Degradation and Stability, 2018, 157: 80-89. doi: 10.1016/j.polymdegradstab.2018.09.019.

[26] CHENG J J, POLAK M A, PENLIDIS A. Influence of micromolecular structure on environmental stress cracking resistance of high density polyethylene[J]. Tunnelling and Underground Space Technology, 2011, 26: 582-593. doi: 10.1016/j.tust.2011.02.003.

本论文原发表于《e-ploymers》2021年第21卷第1期。

Performance Evaluation of Polyamide-12 Pipe Serviced in Acid Oil and Gas Environment

Qi Guoquan[1,2]　Yan Hongxia[1]　Qi Dongtao[2]　Li Houbu[2]
Kong Lushi[2]　Ding Han[1,2]

(1. School of Chemistry and Chemical Engineering, Northwestern Polytechnical University;
2. State Key Laboratory of Performance and Structural Safety for Petroleum Tubular Goods and Equipment Materials, CNPC Tubular Goods Research Institute)

Abstract: To study the compatibility of PA12 with oil and gas medium, the exposure test was carried out by the HTHP autoclave under the simulated medium environment of oil and gas fields. At different exposure periods (test period is 0, 1 week, 3 weeks, 5 weeks, and 7 weeks) in different temperatures (60℃ and 90℃), the physical properties of the samples before and after the test, such as weight and tensile strength were investigated. And then the applicability of the pipes in the simulated medium environment was evaluated comprehensively by analyzing the changes of the micro-morphology and composition of the samples. Compared with the original samples, the weight of samples after test did not change significantly with the average rate of 1.1% to 2.4% at 60℃, while 1.3% to 3.0% at 90℃. The yield strength showed a downward trend with the increase of exposure time. In the early stage of exposure (0~3 weeks), the temperature had little effect on the tensile properties, but with the increase of exposure time (3~7 weeks), the effect became more obvious. When the exposure time reaches 7 weeks, the corresponding yield strength at 60℃ is 27.65MPa, while 27.29MPa at 90℃, which is 27.01% and 27.96% lower than the yield strength of the original sample (37.88MPa), respectively. According to the microscopic morphology analysis, there are cracks on the surface of the specimen under the simulated working conditions. The infrared analysis shows no change in the composition of the sample after immersion compared with the original one, and no damage to the composition and structure of the sample after the exposure test. The above analyses show that the service performance of PA12 pipes is reduced due to penetration and swelling.

Keywords: PA12; Exposure test; Simulated medium; Yield stress; Macro mechanical performance; Thermal performance

With the rapid development of the oil and gas industry, the construction of oil and gas

---

Corresponding author: Qi Guoquan, qgqstar@163.com.

pipelines has advanced rapidly. At present, pipeline transportation has become the main method of oil and gas transportation onshore and offshore. The metal pipelines have always been the main varieties of high and medium pressure pipes in oil fields due to its high strength and good pressure resistance, but it also has many shortcomings such as heavy weight, easy scaling, and easy corrosion[1-2]. The occurrence of metal pipeline failure accidents has caused a large number of casualties, environmental pollution, and capital losses. To solve the problem of corrosion in oil and gas fields, thermoplastic pipes and their composite pipes have become one of the important choices[3-4]. The material of polylaurolactam (PA12), as a type of thermoplastic, has excellent resistance to mechanical stress, stress cracking, and chemical corrosion[5-7], thus its applications expand from the civil industry to ground, offshore oil and gas pipelines.

Due to environmental factors (temperature, pressure, and media), the thermoplastics will be aging or failure when they serve in an oil and gas environment. There are three forms of aging: The first is the appearance of changes, such as stains, cracks and color change. The second is changes in physical properties, such as solubility, swelling, and changes in cold or heat resistance, and water or gas permeability. The third is changes in mechanical properties, such as changes in tensile strength, bending strength, shear strength, impact strength, and elongation. This type of aging or failure is caused by changes in the chemical structure and aggregate structure of the material. Due to the change of the intermolecular force, the break of the molecular chain or the shedding of certain groups will eventually destroy the aggregated structure of the material. Because of the structure change, the physical properties of the material have changed, which ultimately leads to the macroscopic phenomenon of pipe failure. The aging behaviour of PA12 involves changes in crystallinity, extraction of plasticizers and low-molar mass PA12-related species. Reactions with oxygen and fuel components may cause chain scission or crosslinking[8].

Since 2006, since PA12 entered the oil and gas industry as a non-metallic pipeline solution, thousands of kilometers of composite pipes using PA12 have been put into operation. With the increasingly harsh conditions of oil and gas fields, it is doubtful whether PA12 composite pipes can be used in such conditions. Before the pipe is used in special working conditions, effective material selection and evaluation will be of great significance for improving the application effect of the pipe and reducing the risk of failure[8-9].

In this study, the high temperature autoclave equipment is used to evaluate the performance of PA12 materials in the oil and gas environment. According to simulated actual working conditions, the service performance of PA12 pipes is analyzed through the comparison of various performance parameters before and after the exposure time. The autoclave system is mainly composed of an autoclave, control unit, transmission system, heating unit, and monitoring unit. The autoclave system can meet the functions of a simulated environment. According to the field conditions and expected parameters of material service, the following environment variables can be controlled: total system pressure, H_2S partial pressure, CO_2 partial pressure, temperature, test solution composition, test solution composition pH value, and relative movement speed of the sample. Due to the rapid and accurate evaluation of material properties, the exposure test has become a research hotspot in recent years.

1　Experimental

1.1　Sample preparation

The PA12 pipe with the specification of DN100mm×7mm and grade of NRG 3001 nc produced by Evonik is used for research. The sample is prepared according to type 1 of ISO 6259-3: 2015 *Thermoplastics pipes - Determination of tensile properties - Part 3: Polyolefin pipes*. Four specimens were prepared under each condition, three of which were used for tensile tests and one was used for other performance tests.

1.2　Test process

Before conducting the exposure test, the specimen should be tested. Test items include sample size and weight, tensile properties, thermal stability, composition, and microstructure.

The exposure test is completed with the autoclave system, and the medium selection in the autoclave refers to ISO 23936-1: 2009 *Petroleum, petrochemical and natural gas industries - Non-metallic materials in contact with media related to oil and gas production - Part 1: Thermoplastics*. The requirements for the test medium are as follows: 30% gas phase (10% CO_2, 10% H_2S, and 80% CH_4), and 70% liquid phase (70% heptane, 20% cyclohexane, and 10% toluene); In this exposure test, the pressure is 8MPa, the temperatures are 60℃ and 90℃, and the test period is 0, 1 week, 3 weeks, 5 weeks and 7 weeks, respectively.

The sample is taken out and tested after theexposure test is completed. The test items are the same as those before the test. Among them, visual inspection of external damage includes but is not limited to the following forms: cracks, delamination, swelling, blistering, etc. Meanwhile, cutting is performed to check the internal structure of the sample.

1.3　Performance test and characterization

1.3.1　Weight change

The weight is tested by electronic analytical balance (CPA225D, Sartorius, Germany), with an accuracy of 10^{-4} g. Five groups of samples with different exposure time (0~7 weeks) were subjected to weight test, and the weight change of different exposure time was compared and analyzed.

1.3.2　Macro mechanical performance

The tensile properties are tested by the universal test machine (UH-F500KNI, Shanghai Xinsansi, China) with an accuracy of 1N according to ASTM 638. The same batch of the mechanical tensile test was performed under the tensile rate of 50 mm/min. The test data included strain and stress.

1.3.3　Thermal performance analysis

The thermal stability is tested by the Vicat softening point tester (Chengde Precision, China) and the differential scanning calorimeter (DSC, TA Q200, USA). The Vicat softening temperature test was conducted using a force of 50N and a heating rate of 50℃/h refers to ISO 306. The specimen is a square with a side length of 10mm. The temperature range is from 25℃ to 200℃. When the depth of the pressure needle piercing the sample exceeds the initial position by

1mm ± 0.01mm, the temperature of the oil bath measured by the sensor is the Vicat softening temperature of the sample.

To check the possible alteration of the crystallinity due to the exposure test, the DSC has been performed to check the crystallinity of the samples in comparison. In the DSC analysis, two heating and cooling cycles were performed at a rate of 10℃/min under N_2 protection, and the temperature range is from 25℃ to 240℃.

1.3.4 Composition analysis

The composition isanalyzed by the Fourier transform infrared spectroscopy (FT-IR, Thermo Nicolet Avator 360, American). FT-IR analysis is performed using attenuated total reflection, with a spectral range of 400 to 4000cm^{-1} and a spectral resolution of 4cm^{-1}.

1.3.5 Microstructure characterization

The microstructure is characterized by a scanning electron microscope (SEM, Zeiss Supra 55/55VP, Germany). The observation target is the inner wall of the pipe after spraying gold treatment.

2 Results and discussion

2.1 Weight change

The test result of the weight change rate is shown in Fig. 1. As the exposure time increases, the weight change rate at different temperatures (60℃ and 90℃) both shows an increasing trend. When the exposure time is 7 weeks, the weight change rates under 60℃ and 90℃ conditions reach the maximum values with the average values of 2.41% and 3.03% respectively. The result of the weight change rate meets the requirements of the standard ISO 23936-1 regarding the quality change rate (±5%). Comparing the weight changes at 60℃ and 90℃ respectively, the rate of weight change at 90℃ is higher than that at 60℃ under different exposure time. It can be seen that the temperature has a promoting effect on the weight increase. As the exposure time continues to increase, the degree of increase in the rate of weight change tends to decrease, as the medium in the sample tends to saturate due to sufficient penetration[10]. Thermoplastics have limitations in heat resistance, and their performance varies significantly with the increase of temperature. When the temperature increases, it promotes the activity of molecular movement and accelerates the physical process of penetration and swelling[11]. This is because that the temperature has a very significant effect on the movement of the thermoplastic segment. At the low temperature, the molecular segment of the polymer material in a "freezing" state. As the temperature rises, the movement unit of the chain segment gradually increases, and the total free volume increases. In this state, the restrictive effect of the polymer chain on the movement of molecules is weakened, which makes the medium penetration easier.

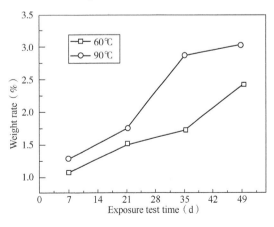

Fig. 1 The weight change rate at different temperature

2.2 Mechanical performance

The stress-strain curves at 60℃ and 90℃ are shown in Fig. 2 and Fig. 3. All stress-strain curves show typical changes in the tensile properties of polymer materials, which is shown in three regions: elastic deformation, forced rubber-like deformation, and viscous flow. In region 1, the bond length and angle move, but recoverable; in region 2, the chain segments are oriented in the direction of external force and can be recovered by heating above T_g; in region 3, the entire chain slips or breaks with each other and are not recoverable. Regardless of whether the temperature is 60℃ or 90℃, the yield strength of the exposed specimen is lower than that of the initial specimen, and which decreases with the increase of the exposure time. The average value of the yield strength is shown in Fig. 4. The yield strength shows a clear decreasing trend with the increase of the exposure time. As the exposure time further increases, the decreasing trend is slower than the initial stage. In addition, the temperature during the exposure test also affects the tensile properties of the specimen, especially at longer exposure times, such as 5 weeks and 7 weeks. The reason for the decrease in yield strength is that the concentration effect is caused by various defects such as holes and micro-cracks formed inside the pipe due to the penetration and swelling of the medium during the exposure test[12]. When the test temperature increases, the penetration and swelling process is accelerated slightly.

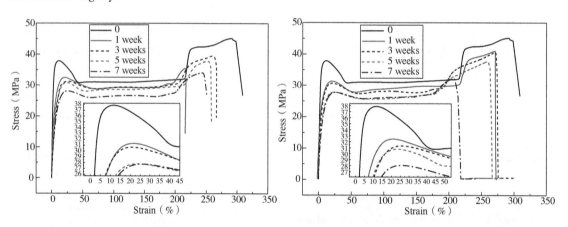

Fig. 2 The stress-strain curves at 60℃   Fig. 3 The stress-strain curves at 90℃

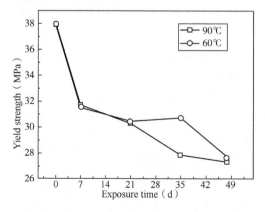

Fig. 4 The average value of the yield strength

2.3 Thermal performance analysis

2.3.1 Vicat softening temperature

The Vicat softening temperature test results are shown in Fig. 5. At different temperatures (60℃ and 90℃), the Vicat softening temperature continued to decrease with increasing exposure time. In the initial stage (0 to 1 week), the downward trend is more obvious. As the exposure time increases (1 week to 7 weeks), the decline becomes slow. As the test temperature increases, the Vicat softening temperature corresponding to different exposure time tends to decrease. This is because, during the exposure test, the increase in the test temperature accelerates the penetration and swelling of the medium, resulting in poor dimensional stability of the material under heating conditions[13].

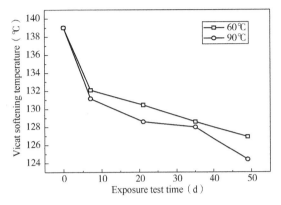

Fig. 5 The Vicat softening temperature test results

2.3.2 Melting and crystallization curve

The DSC curves of the three test conditions (the original sample, exposing at 60℃ for 7 weeks and 90℃ for 7 weeks) are shown in Fig. 6. The characteristic parameters of the thermal spectrum include melting peaks, crystallization peak, and enthalpies of fusion and crystallization. By comparing the characteristic parameters, the three samples are basically the same. It shows that the exposure test has little effect on the order of the internal structure of the polymer.

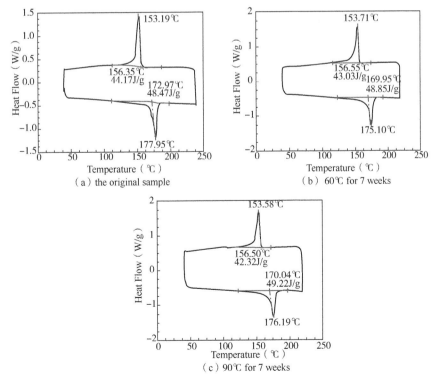

Fig. 6 The DSC curves of the three test conditions

2.4 Composition analysis

The results show that the peaks at around 3299cm^{-1} are assigned to —NH$_4$ stretching vibration, at 2934cm^{-1} is —CH$_2$-asymmetrical stretching vibration, at 2858 cm^{-1} is —CH$_2$- symmetrical stretching vibration, respectively. The peaks at 1633cm^{-1}, 1540cm^{-1}, and 1276cm^{-1} are the bending vibration peak of the —CO—NH—, while the peaks at 1473cm^{-1} and 720cm^{-1} are the bending vibration peaks of the —CH$_2$-. Compared with the results of three samples, the peaks of samples with an exposure time of 60℃ and 90℃ is similar to the other two. It can be seen that the functional group of the PA12 did not change significantly, that is, the material was not modified by environmental medium corrosion. It shows that in the oil field simulation environment, the internal structure is still stable and has excellent corrosion resistance with the increase of exposure time at 60℃ and 90℃, respectively.

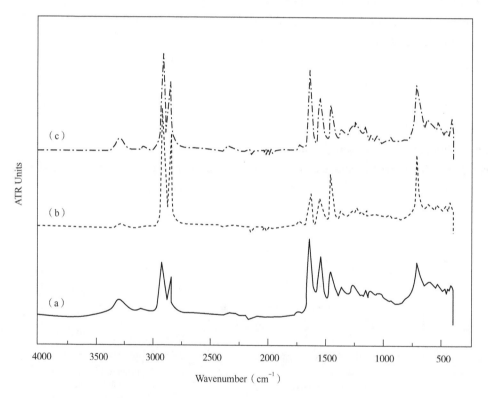

Fig. 7 Infrared spectrum of PA12 after exposure test
(a) the original sample; (b) 60℃ for 7 weeks; (c) 90℃ for 7 weeks

2.5 Microstructure characterization

Fig. 8 shows the SEM observations of samples before and after the exposure test. We can find that under the condition of 60℃, the morphology has not changed obviously after the exposure test. However, under the condition of 90℃, the island-like pitting appeared on the surface of the sample after the 3 weeks of exposing. With the extension of the exposure time, the pitting phenomenon becomes more and more obvious. The island-like pitting is caused by the dissolution of small molecule additives (such as plasticizer) and amorphous materials in the pipe by the medium. During the exposure test, the medium will penetrate the inside of the pipe body, and then will occupy its original position after dissolving additives or amorphous materials, and then form a swelling phenomenon[14-16]. The temperature increase will promote the migration of molecules. Compared with the test condition of 60℃, the swelling effect at 90℃ is obviously stronger than the former. Due to the occurrence of penetration and swelling, the microscopic morphology of the pipe is affected[17-18], which in turn affects its mechanical properties, which is consistent with the previous description. However, since the crystal structure of the material has not been damaged, the degree of decrease in mechanical properties is not serious.

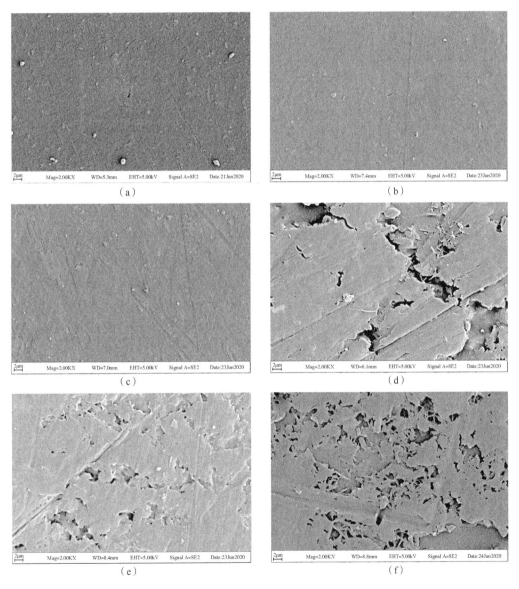

Fig. 8 The SEM micrographs of PA12: (a) the original sample; (b) 60℃ for 7 weeks;
(c) to (f) 90℃ for 1, 3, 5, 7 weeks, respectively

2.6 Comprehensive analysis

In the service course of the pipe, its performance will be attenuated by external factors, which are medium, temperature, and pressure. Under the synergy of temperature and pressure, the medium enters the inside of the pipe due to penetration. To study the influence of external factors on the performance of service pipes, theexposure test was used to simulate the service state of PA12 pipes under actual working conditions. Through the comparison of physical properties before and after the exposure test, composition, and micro-morphology analysis, the declining trend of PA12 performance with the increase of exposure time under this working condition was studied. Through the study, it was found that the phenomenon of weight increase and mechanical properties decreased at different test temperatures (60℃ and 90℃), and this trend became more obvious as the temperature increased. However, after analyzing its composition, the composition of the sample after the exposure test did not change, indicating that the pipe properties only changed physically during the exposure test.

From a microscopic point of view, the medium molecules entering the pipe are located at the crystal boundary, causing swelling phenomenon[19], which presents as weight increment. In addition, the medium extracts the uncured resin small molecules and the decomposed small molecules inside the resin, presenting as weight loss. Since the swelling effect is significantly stronger than extraction, the final performance is an increase in mass with exposure time increase.

Because of the effect of medium penetration and swelling, various defects such as micro-pits and holes are formed inside the pipe to cause a concentration effect[20]. The stress of the molecular chain is concentrated at the position of the defects, where the endurance is lower than in other positions. Fracture failure occurs first in the stress concentration area, which ultimately leads to a decrease in the overall mechanical properties. As the temperature increases, the molecular chains of plastic materials are more likely to move, and the media molecules' mobility increases. The increase in temperature accelerates the process of penetration and swelling, making the attenuation of mechanical properties more obvious eventually. The swelling process is only a physical process and no chemical reaction occurs, it can also be proved from the infrared analysis of the structure. The analysis of micro-morphology confirmed the existence of micro-pitting defects. However, the pits are micro-sized, which have a minor effect on reducing the macro-mechanical properties.

3 Conclusion

Through the analysis of physical properties, structural composition, and micro-morphology after theexposure test, the PA12 pipe formed the island-like micro-pitting defect due to the penetration and swelling of the medium. This defect caused a downward trend in the mechanical properties of the pipe, and this trend became more obvious when the exposure time and the temperature increased. The reason for the decline in mechanical properties is only caused by physical changes, not chemical changes. It can be seen that in the service process of the pipe, under the condition of a specific medium, as the service time increases, the mechanical properties of the pipe will decline. Therefore, the performance of the pipe should be evaluated to determine its applicability under specific working conditions before and during service.

Acknowledgments

This work was supported by the State Key Laboratory for Performance and Structure Safety of Petroleum Tubular Goods and Equipment Materials, China.

References

[1] MIAO J, WANG Q. Corrosion rate of API 5L Gr. X60 multipurpose steel pipeline under combined effect of water and crude oil[J]. Metals & Materials International, 2016, 22(5): 797-809.

[2] MIAO J, YUAN J T, HAN Y, et al. Corrosion Behavior of P110 low alloy steel in the CO_2 – saturated Simulated Oilfield Formation Water with Element Sulfur Addition[J]. Rare Metal Materials and Engineering, 2018, 47(7): 1965-1972.

[3] QI G, QI D, YAN H, et al. Advance in Environmental Stress Cracking Study of Thermoplastic Pipes Used in Oil and Gas Field[J]. Polymeric Materials Science and Engineering, 2018, 34(6): 179-183.

[4] QI D T, LI H B, CAI X H, et al. Application and Qualification of Reinforced Thermoplastic Pipes in Chinese Oilfields[C]. ICPTT, Beijing, 2011.

[5] MESSIHA M, FRANK A, BERGER I, et al. Determination of the Slow Crack Growth Resistance of PA12 Pipe Grades[C]. Plastic Pipes XIX Conference, 2018.

[6] JOURNAL G. Polyamide Pipe Selected to Carry High – Volume Gas to Chemical Plant[J]. Pipeline & gas journal, 2015, 242(8): 72-73.

[7] KALLIO K J, NAGEYE A S, HEDENQVIST M S. Ageing properties of car fuel-lines: accelerated testing in "close-to-real" service conditions[J]. Polymer Testing, 2010, 29(1): 41-48.

[8] TEYSSANDIER F, CASSAGNAU P, GÉRARD J F, et al. Morphology and mechanical properties of PA12/ plasticized starch blends prepared by high-shear extrusion[J]. Materials Chemistry & Physics, 2012, 133(2-3): 913-923.

[9] WEI X F, KALLIO K J, BRUDER S, et al. Diffusion-limited oxidation of polyamide: Three stages of fracture behavior[J]. Polymer Degradation and Stability, 2018, 154: 73-83.

[10] KALLIO K J, HEDENVIST M. Degradation of Polyamide-12 pipes aged in fuel[J]. Polymer Degradation & Stability, 2010.

[11] KALLIO K J, HEDENQVIST M S. Effects of ethanol content and temperature on the permeation of fuel through polyamide-12-based pipes[J]. Polymer Testing, 2010, 29(5): 603-608.

[12] BAI L, HONG Z, WANG D, et al. Deformation-induced phase transitions of polyamide 12 in its elastomer segmented copolymers[J]. Polymer, 2010, 51(23): 5604-5611.

[13] ZHANG J, ADAMS A. Understanding thermal aging of non-stabilized and stabilized polyamide 12 using 1H solid-state NMR[J]. Polymer Degradation & Stability, 2016, 134: 169-178.

[14] PLUMMER C J G, ZANETTO J E, BOURBAN P E., et al. The crystallization kinetics of polyamide-12 [J]. Colloid and Polymer Science, 2001, 279(4): 312-322.

[15] WANG D, SHAO C, ZHAO B, et al. Deformation-Induced Phase Transitions of Polyamide 12 at Different Temperatures: An in Situ Wide – Angle X – ray Scattering Study [J]. Macromolecules, 2010, 43 (5): 2406-2412.

[16] CHEN L, CHEN W, ZHOU W, et al. In Situmicroscopic infrared imaging study on deformation-induced spatial orientation and phase transition distributions of PA12[J]. Journal of Applied Polymer Science, 2014, 131(17).

[17] KALLIO K J, HEDENQVIST M S. Ageing properties of polyamide-12 pipes exposed to fuels with and without ethanol[J]. Polymer Degradation & Stability, 2008, 93(10): 1846-1854.

[18] PAOLUCCI F, VAN MOOK M J H, GOVAERT L E, et al. Influence of post-condensation on the crystallization kinetics of PA12: From virgin to reused powder[J]. Polymer, 2019, 175: 161-170.

[19] WEI X F, DE V L, LARROCHE P, et al. Ageing properties and polymer/fuel interactions of polyamide 12 exposed to (bio)diesel at high temperature[J]. Npj Materials Degradation, 2019, 3(1): 1-11.

[20] NASATO D S, PSCHEL T. Influence of particle shape in additive manufacturing: Discrete element simulations of polyamide 11 and polyamide 12[J]. Additive Manufacturing, 2020, 36: 101421.

本论文原发表于《Journal of Failure Analysis and Prevention》2021年第21卷第6期。

Simulation Analysis of Limit Operating Specifications for Onshore Spoolable Reinforced Thermoplastic Pipes

Li Houbu[1]　Zhang Xuemin[2]　Huang Haohan[2]
Zhou Teng[2]　Qi Guoquan[1]　Ding Han[1]

(1. State key Laboratory for Performance and Structure Safety of Petroleum Tubular Goods and Equipment Materials, CNPC Tubular Goods Research Institute; 2. School of Materials Science and Engineering, Chang'an University)

Abstract: Spoolable reinforced plastic line pipes (RTPs), exhibiting a series of advantages such as good flexibility, few joints, long single length, light weight, easy installation, etc., have been widely used in the onshore oil and gas industry such as oil and gas gathering and transportation, high pressure alcohol injection, water injection, sewage treatment, and other fields. However, due to the lack of clear standard specification of the limit operating properties for RTPs, three typical failure modes, i. e., tensile, flexure, and torsion, frequently occur in terrain changes, construction operation, and subsequent application, which seriously affects the promotion and use of RTPs. In this paper, the stress distribution of a non-bonded polyester fiber reinforced high-density polyethylene (HDPE) pipe (DN150 PN2.5MPa) was systematically studied by the finite element method (FEM), and then the limit operating values under the axial tensile, coiled bending, and torsion load were determined. The corresponding experiments were conducted to validate the reliability and accuracy of the FEM model. The FEM results showed that the critical strain for axial tensile was 3%, the minimum respooling bend radius was 1016.286mm, and the limit torsion angle of this RTP was 58.77°, which are very close to the experimental results. These limit values will be useful to establish normative guidelines for field construction and failure prevention of onshore RTP.

Keywords: Reinforced thermoplastic pipe; Onshore; Limit operating specification; Finite element method

1 Introduction

Carbon steels and low alloy steels are the most widely used in ground gathering and

---

Corresponding author: Li Houbu, lihoubu@ cnpc. com. cn, +86 29 8188 7680.

transportation system of onshore oilfields. However, in recent years, with the increasingly harsh environment of oilfields, the water content and temperature have gradually increased. On the other hand, the content of corrosive media such as Cl^- and CO_2 has increased, and H_2S has even appeared in oil and gas wells in some areas, which has brought great risks to the application of steel pipe[1]. Non-metallic and composite pipes have a series of advantages such as corrosion resistance (without internal and external anti-corrosion), smooth inner walls, anti-scaling and waxing, low fluid resistance, anti-wear and anti-erosion, electrical insulation, light weight, easy transportation and construction, and low maintenance costs, etc., which have become one of the most important solutions for anti-corrosion of the onshore gathering pipelines[2-7].

Currently, the non-metallic and composite pipes used in China's onshore oilfield systems mainly include glass fiber reinforced thermosettingpipes (GRP), anticorrosion plastic alloy composite pipes, steel skeleton reinforced polyethylene composite pipes, and spoolable reinforced plastic pipes (RTPs)[8]. Among them, RTPs have been widely used in oil and gas fields due to their good flexibility, excellent impact resistance, few joints, light weight, low transportation cost, quick and easy installation, etc. Recently, it has become the fastest growing non-metallic composite pipe in China. RTPs are mainly used for oil and gas gathering and transportation, high-pressure alcohol injection, oilfield water injection, and sewage treatment, etc.[9]. Since 2011, RTPs have been tested in gas transportation (mixed oil and gas transportation), downhole water injection, and other fields[10-15]. Therefore, application of the RTP products in China is developing towards serialization and diversification.

RTPs used in onshore oil fields mainly consist of a multi-layer structure: internal thermoplastic liner, reinforcement layer, and outer thermoplastic layer. The typical layer structure of an RTP is shown in Fig. 1. For liners, various grades of high-density polyethylene (HDPE) are being used for product temperatures up to 65℃. If required, RTPs can be designed for higher operating temperatures by using other appropriate thermoplastic materials such as cross-linked polyethylene (PEX), polyamides (PA), and polyvinylidene fluoride (PVDF). The reinforcement layer is usually made of continuous fiber (or tape) and steel wire (or tape); the fiber materials can be polyester fiber, glass fiber, carbon fiber, and aramid fiber. The outer protective layer is usually high-density polyethylene (HDPE).

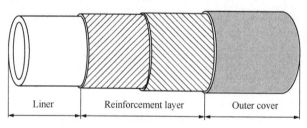

Fig. 1 Typical structure of RTP

RTPs can be constructed as bonded or non-bonded structures based on the manufacturing process. In bonded construction, the reinforcement fiber tape is fused to the linerand the outer cover to form a solid wall. RTPs withbonded construction offer greater resistance to linercollapse on depressurization or cover blow-off due todiffused gases. However, in non-bonded construction, different layers are mutually independent and relativemovement is allowed. Un-bonded RTPs have the advantages of a simple manufacturing process, high manufacture efficiency, and good flexibility. Therefore, based on consideration of manufacturing efficiency, product cost, and application conditions, the most used RTP in onshore oil fields is the non-bonded polyester fiber reinforced HDPE composite pipe, especially in China.

As mentioned above, the non-bonded polyester fiber reinforced HDPE composite pipe has been popularized and applied in various fields of onshore oil fieldsin China. The total use exceeds 40,000 km, and the annual growth rate is higher than 10%. However, with the increasing quantity, three typical failure modes are gradually exposed, as shown in Fig. 2, which seriously affect the product promotion and the user's confidence. One is that the end fitting or pipe body of the RTP fails frequently due to the tensile overload during the drag construction or subsequent ground subsidence [Fig. 2(a)]. The other is buckling failure of the RTP caused by bending overload under the condition of coiling transportation, laying or service with a small bending radius such as sand dunes and hillsides, as shown in Fig. 2(b). The third is that the pipe body fails due to torsional overload when the RTP rolls along in the process of desert or mountain movement[Fig. 2(c)].

 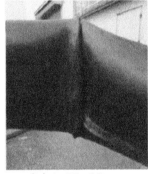 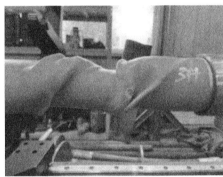

(a) tensile failure　　　　(b) buckling failure　　　　(c) torsional failure

Fig. 2　Three typical failure modes of RTP

Existing standards and manufacturershave not provided clear specific requirements for the limit operating performance of RTPs such as tensile, bending, and torsion, so there is no useful guideline for oilfield users to effectively control the above failure risks and prevent them. The effective method to determine the limit operating performance of RTPs is to conduct a large number of full-scale laboratory tests. However, due to a variety of the RTP specifications (diameter range of 50~150mm, normal pressure range of 1.6~32MPa), various material types (various thermoplastics and reinforcement materials and types), and different manufacturing processes (number of winding layers, angles, tension, etc.), this method of testing is time-consuming and expensive, and it is difficult to give specific data quickly and accurately.

In orderto establish normative guidelines for field construction operation and failure prevention of RTPs, a fast and accurate method to obtain the limit tensile displacement, the minimum respooling bend radius, and the limit torsional radian of the RTP was explored in this paper. Polyester fiber reinforced HDPE pipe, which is the most widely used RTP in onshore oil fields in China . A pipe body model of the RTP (DN150 PN2.5MPa) was established by the finite element method, and the reliability of this model was verified by a burst pressure test. On this basis, the stress distributions of the RTP under axial tension, coiled bending, and torsion load were systematically studied by the finite element method. Furthermore, based on the yield strength of the RTP materials such as the thermoplastic liner and reinforced fiber, the limit operating performance of the RTP was calculated and analyzed and further verified by the corresponding experiments.

2 Numerical simulations

A finite element (FE) model was established by using ABAQUS software (ABAQUS software Ctd., USA), as shown in Fig. 3. The geometrical parameters of the FE model were in accordance with the ones listed in Table 1. To eliminate the end effect of the pipeline subject to a fixed constraint or connection, a pipe length of 2000mm was adopted, which is 10 to 20 times of the outside diameter of the RTP (DN150 PN2.5MPa) used in this study. The properties of the HDPE and polyester fiber are shown in Table 2.

Fig. 3 FEM model of RTP

Table 1 The geometrical parameters of RTP

Parameter	Value	Parameter	Value
Inner radius (mm)	148	Thickness of outer PE pipe(mm)	4
Outer radius (mm)	180	Winding angle of polyester fiber (°)	+55/−55
Thickness of inner PEpipe(mm)	10	Reinforced layer Number	4

Table 2 Parameters of materials used in the RTP

Layer	Parameter	Value
Inner HDPE	Elastic modulus (MPa)	403.1
	Yield strength (MPa)	17.33
Outer HDPE	Elastic modulus (MPa)	300.8
	Yield strength (MPa)	8.66
Polyester fiber	Linear density (Dtex)	20, 370
	Breaking force (N)	1650
	Breaking strength (MPa)	525.2
	Elongation at break (%)	15.2

2.1 The reinforced layers model and element

Considering the anisotropy of reinforced polyesterfiber, the composite material homogenization Halpin−Tsai model[16] was used to simulate the reinforced layer. The whole reinforced layer was regarded as a kind of orthogonal anisotropic homogeneous material that was interlaced, and four layers of polyester fibers were radially stacked from inside to outside at a winding angle of ±55°, as shown in Fig. 4. The solid element C3D8R was chosen to model each layer and to complete the mesh division; 14364 elements, 19298 nodes, and 6 degrees of freedom for each node were assigned for the HDFE model. One layer of solid elements across a layer of material was used because the increase inlayers and finite element number had little effect on the numerical result[17].

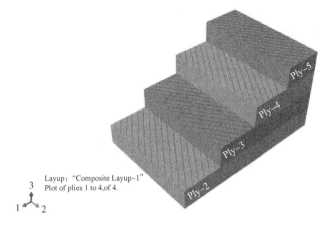

Fig. 4　Polyester fiber layup model (4 layers)

2.2　Interaction

Surface-to-surface contact was used to simulate the interactions between each layer. The surface of the HDPE pipe was specified as the slave surface due to its softer properties compared with those of the polyester fiber for the reinforced layer. The normal mechanical behavior was defined as "Hard Contact" with ''Allow separation after contact", and the Penalty was taken as the friction formulation for tangential behavior. The friction coefficients were selected as 0.2 for the fiber-HDPE surface contact.

2.3　Boundary conditions and load

The right end of the pipe was fixed, while the other was coupled to a reference point located at the center of the cross section. All of the six degrees of freedom were coupled to this point to ensure the deformation of the sectionwas consistent and rigid. The load of inner pressure, bending, torsion, and tension was applied on the RTP, as shown in Fig. 5.

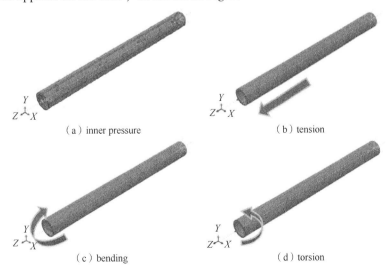

Fig. 5　Schematic diagram of the loading of the RTP

2.4 Verification of FEM simulation

Based on the RTP model established above, the mechanical behavior of the DN150 polyester fiber-reinforced HDPE composite pipe under internal pressure was simulated by the finite element method. The stress distribution of each layer of the RTP under 3.75MPa (1.5 times thenominal pressure of 2.5MPa) was obtained, as shown in Fig. 6. It can be seen that the cross-section remained circular after being subjected to the internal pressure, and the simulated outer diameter of the RTP increased by 3.17mm. After the actual hydrostatic test (holding pressure of 3.75MPa for 4h) according to standard SY/T 6662.2—2020[18], the measured outside diameter of the RTP increased by 3.24mm. The consistency of the simulation and experimental results preliminarily showed that the established RTP model is reliable.

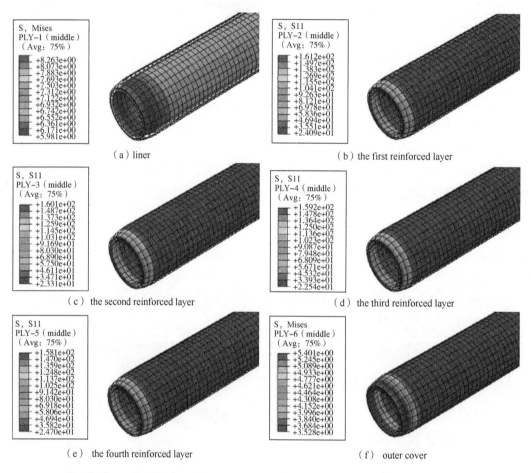

Fig. 6 Stress distributions of each layer of the RTP at the inner pressure of 3.75MPa

The stress distributions of each layer of the RTP are presented in Fig. 6. All layers exhibited a similar homogeneous state of stress, and the stresses of reinforced layers were much higher than those of the inner and outer HDPE layer. It can be considered that the fiber reinforced layer is the main load bearing part of the RTP when subjected to the internal pressure. Moreover, the stress of the innermost fiber layer [the first reinforced layer in Fig. 6(b)] was the largest one among the four reinforced layers. This means that the first reinforced fiber layer next to the inner HDPE pipe is the most liable part to fail. As a result, the stress of the innermost fiber layer was taken as the criterion to estimate the failure of the pipe.

In order to determine the burst pressure of the RTP, the maximum stress criterion was used as the failure criterion of the innermost polyester fiber. That is, when any of the stress components in the main direction of polyester fiber reaches its ultimate strength, the fiber would be broken, and the corresponding internal pressure is the burst pressure of the pipe. As shown in Fig. 7, the stress of the polyester fiber increased with the internal pressure. When the stress value of polyester fiber reached its breaking strength of 525.2MPa,

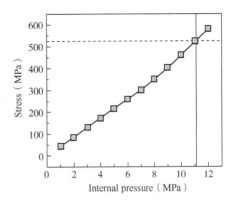

Figure 7 Stress-internal pressure curve for the innermost polyester fiber layer of the RTP

the corresponding internal pressure (i.e., burst pressure) of the RTP was 11.105MPa. The simulated burst pressure of 11.11MPa is basically consistent with the experimental measured average value of 10.75MPa, and the relative error was only 3.3%, which further confirms the availability and accuracy of the established model of the RTP.

3 Limit operating performance analysis of RTPs

3.1 Tensile analysis of the RTP

The axial displacement tension was used to study the tensile behavior of the RTP in which the incremental step of the axial displacement was 10mm. Fig. 8 shows the stress distributions of each layer of the RTP at the axial tensile displacement with 100mm. It can be seen that the cross-section of each layer tended to shrink under the tension of the RTP. The maximum stresses of the inner liner and the outer cover were 17.69MPa and 10.15MPa, respectively. There were both higher than the yield strength value 17.33MPa and 8.66MPa of their corresponding materials. However, the stresses of the polyester fiber in the innermost and outermost layer were 33.42MPa and 34.52MPa, respectively, which were much lower than their breaking strength of 525.2MPa. Therefore, the stress of the HDPE used in the inner liner and the outer cover was taken as the criterion to estimate the tension failure of the pipe.

In order to determine the critical strain fortension, the stresses of the inner liner and the outer cover with tensile displacement were calculated and are shown in Fig. 9. It can be seen that both of the stresses for the two layers increased with the displacement. Because the tensile strength of the HDPE used for outer cover wasl ower than that of the inner liner, the plastic deformation first occurred at the outer cover. Accordingly, the critical axial displacement for the RTP with a length of 2000mm was determined by the outer cover to be 60mm[Fig. 9(b)], and its corresponding critical strain was 3%.

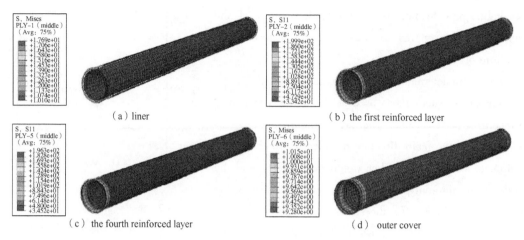

Fig. 8　Stress distributions of each layer of the RTP at the axial tensile displacement of 100mm

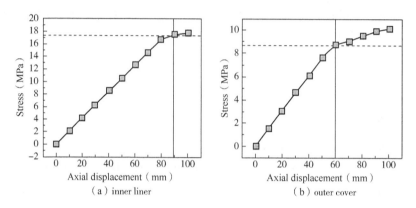

Fig. 9　Curves of stress with axial tensile displacement

3.2　Flexure analysis of RTP

The RTP being reeled onto a spool experiences large deformations. The geometrically non-linear approach was employed to simulate the mechanical behavior of the RTP under coiled bending. The Nlgeom option was chosen in the general static solution procedure of the ABAQUS/Standard to solve the non-linear problem. Riks analysis step was used to simulate the bending process. It uses the load magnitude as an additional unknown and solves for loads and displacements simultaneously. The Riks method follows an eigen value buckling analysis to provide complete information about a structure's failure.

At the eigen value calculated by Riks, the stress distributions of each layer of the RTP are shown in Fig. 10. An obvious stress concentration and buckling were observed in the middle of each layer due to bending action. The maximum stress value of the reinforced fiber layer was 209.4MPa [Fig. 10(a)], which was much lower than its breaking strength of 525.2MPa. Therefore, it is difficult for the fiber layer to break even if it is dented during the bending process. However, the maximum stress values of the inner liner and the outer cover HDPE layers were 20.2MPa and 13.47MPa respectively, both of which were higher than the yield strength of their corresponding material. Hence, the plastic deformation occurred at the inner liner and the outer cover layers, forming an unrecoverable depression, which led to the instability failure of the RTP. In a word, the failure criterion of bending was whether the pipe had buckling or not.

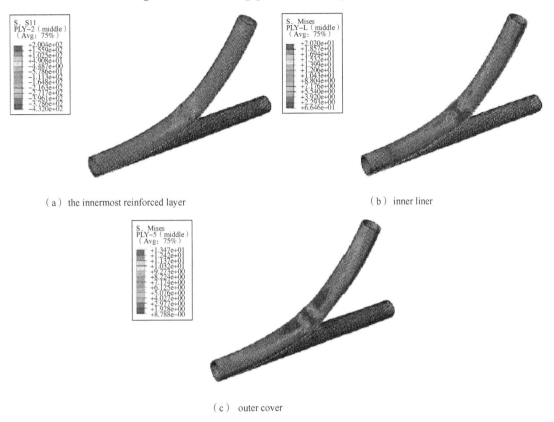

(a) the innermost reinforced layer

(b) inner liner

(c) outer cover

Fig. 10 Stress distributions of each layer of the RTP at the bending load

The minimum respooling bend radius R is one of the key parameters of the RTP in transportation and storage process, which can be calculated by Eq. (1):

$$R = L/\theta_c \tag{1}$$

where L is the length of the RTP (mm), θ_c is the critical bending radian (rad), as shown in Fig. 11.

The critical bending radian θ_c of the RTP can be obtained by FEM simulation according to above-mentioned failure criterion. In this study, the critical bending radian of the RTP (DN150 PN2.5MPa) was calculated to be 1.97rad, and the minimum respooling bend radius was calculated to be 1016.29mm accordingly.

3.3 Torsion analysis of the RTP

The static analysis step was used to simulate and analyze the torsion process of the RTP in which the torsional radian can be increased uniformly from zero to the critical torsional load. As shown in Fig. 5(d), the left end of the RTP was rotated counterclockwise with the z-axis, and the torsional load was loaded in

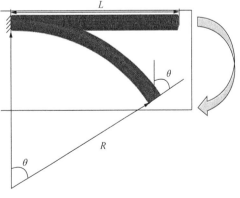

Fig. 11 Schematic diagram of the bending radius of the RTP

increments of 0.1rad.

The stress distributions of the RTP under a torsion load of 0.2rad are presented in Fig. 12. The stress value of the outermost fiber was +102.4MPa, which was slightly higher than that of the innermost fiber of −101.3MPa (the positive and negative signs of the fiber stress value corresponding to the wound angle of ±55°). At the same time, the stress of the inner liner and the outer cover was much smaller than that of the yield strength of the corresponding HDPE material. Therefore, the polyester fiber layer was still the main bearing part of the pipe during the torsion process, and the stress of the outermost fiber layer was taken as the criterion to estimate the failure of the pipe. As the torsional radian increased, the stress of the outermost fiber increased linearly, as shown in Fig. 13. When the stress of the outermost fiber reached its breaking strength of 525.2MPa, the corresponding critical torsional radian of the RTP was 1.03rad, and the corresponding torsional angle was 58.77°.

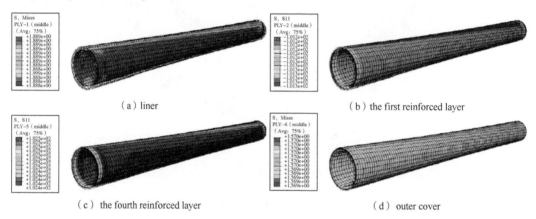

(a) liner (b) the first reinforced layer

(c) the fourth reinforced layer (d) outer cover

Fig. 12 Stress distributions of each layer of the RTP at the torsion load of 0.2rad

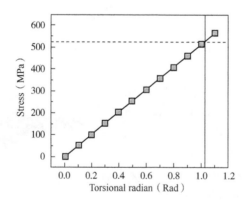

Fig. 13 Curve of fiber stress with increasing torsional radians

4 Experimental verification

4.1 Tensile tests

The tensile property of the RTP was determined according to ASTM D2105-01(2019)[19] by using a hydraulic universal testing machine(WAW-500D-JG, SHENZHEN SUNS TECHNOLOGY STOCK CO., LTD., Shenzhen, China) with a tensile velocity of 10mm/min, as shown in

Fig. 14. The length of the sample with two end fittings was 2000mm. The comparison between the load-strain curves from the experiment and FEM is given in Fig. 15. It can be seen that the curves shared the same trend in general. The maximum error of loads at the same strain between the simulation results and the experimental results was basically within 10%. Considering the simplification of the FEM model, the simulation results can be considered to be within an acceptable range.

Fig. 14 Tensile test of the RTP sample

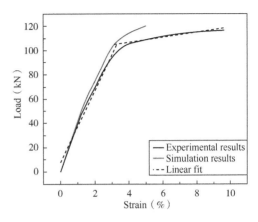
Fig. 15 Load-strain curves obtained from the experiment and FEM

In practical engineering, when the tensile strain of an RTP is more than a certain value, it is considered to have failed[12]. Bai et al. reported that the strain of the HDPE layers should not exceed 7.7% for MSFP (metallic strip flexible pipes)[20]. Generally, once the stress of any layer of the RTP is higher than its yield stress, it might carry a safety risk as the pipe would lose some proportion of its stiffness during the usage. It can be noticed from Fig. 15 that the tensile process of the RTP (DN150 PN2.5MPa) consisted of a linear elastic stage and a plastic elongation stage. The critical strain is the turning point of these two stages. The determined strain values of simulation and experimental results were 3% and 3.28%, respectively. These two critical strains are very close, further implying the accuracy and reliability of the FEM model used in this study.

4.2 Flexure tests

The flexure test was conducted on the Bending Tester (CHENGDE PRECISION TESTING MACHINE CO. LTD., Chengde, China). The bending properties of the RTP were tested starting with a bending radius of 2000mm [Fig. 16(a)] and then tested at a decreasing interval of 100mm. Appearance of the RTP was observed by visual inspection after conditioning for four hours according to ISO 291: 2008[21] at each interval. There was an obvious morphology change [as shown in Fig. 16(b)] when the bending radius of the RTP decreased to 1100mm. Continued bending of the RTP to a bending radius of 1000mm yielded an obvious buckling phenomenon that can be observed as shown in Fig. 16(c). It is very similar to the simulation results shown in Fig. 10. Based on the experimental results, it can be concluded that the irreversible buckling failure occurs when the bending radius is less than 1100mm. Therefore, the minimum respooling bend radius of the RTP (DN150 PN2.5MPa) is 1100mm, which is basically the same as the simulated value of 1016.29mm.

(a) bending radius of 2000mm　　(b) bending radius of 1100mm　　(c) bending radius of 1000mm

Fig. 16　Bending test of the RTP sample

4.3　Torsion tests

The torsion property of the RTP was determined by using a Torsion Tester (Jinan Bangwei Mechanical & Electrical Equipment Co., Ltd., Jinan, China) with a velocity of 10°/min, as shown in Fig. 17. The torsional behavior of the RTP is shown by the torque-torsion angle curves given in Fig. 18. The torque increased linearly as the torsion angle increased. When a significant torque decline was observed on the test curve, its corresponding torsion angle was the limit torsion angle. As shown in Fig. 18, the simulation curve exhibited similar characteristics to the experimental result. The simulated limit torsion angle of the RTP (DN150 PN2.5MPa) was 56.17°, which is close to the experimental result of 58.77°.

Fig. 17　Torsion test of the RTP sample

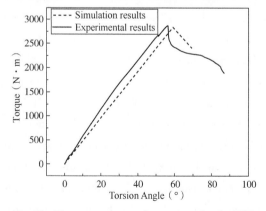

Fig. 18　Torque-torsion angle response for the RTP

5 Conclusions

In this paper, the finite element model of a non-bonded RTP (polyester fiber reinforced HDPE pipe, DN150 PN2.5 MPa) subjected to inner pressure, axial tensile, coiled bending, and torsion load was established and verified by experiments. FEM simulation determined the limit operating performances of the RTP under these different loadsquickly and accurately.

The burst pressure of the RTP obtained by simulation was 11.11MPa, which is basically consistent with the experimentaltest of 10.75MPa in the laboratory. The limit axial tensile displacement of the RTP (DN150 PN2.5MPa) with a length of 2000mm was determined to be 60mm by simulation, and the corresponding critical strain was 3%. The critical strain obtained by the tensile test was 3.28%, indicating the accuracy and reliability of the FEM model used in this study. The minimum respooling bend radius of the RTP (DN150 PN2.5MPa) was considered to be 1100mm from the flexure test, which is basically the same as the simulated value of 1016.29mm. The critical torsion angle of the RTP was 58.77°, similar to the experimental result of 56.17°.

For the operating performances analyzed in this study, it is concluded that FEM is a reliable method toestablish specifications of the RTP forfield construction operation and failure prevention. However, in practical engineering, the RTP is often subjected to complicated loads such as inner pressure-tensile, inner pressure-bending, tensile-torsion, and bending-torsion, etc. The limit operating performances of the RTP will be changed and should be systemically studied further.

Acknowledgments

The authors acknowledge the Northwestern Polytechnical University High Performance Computing Center for the allocation of computing time on their machines.

References

[1] LI H B, LI H L, QI D T, et al. Analysis on non-metallic pipes used for oil and gas gathering and transportation[J]. Petroleum Instruments, 2014, 28: 4-8.

[2] CONLEY J, WELLER B, SAKR A. Recent Innovations in Reinforced Thermoplastic Pipe[M]. Berlin, Germany: Springer International Publishing, 2014.

[3] YU K, MOROZOV E V, ASHRAF M A, et al. Analysis of flexural behavior of reinforced thermoplastic pipes considering material nonlinearity[J]. Composite Structures, 2015, 119: 385-393.

[4] BAI Y, LIU S, HAN P, et al. Behaviour of steel wire-reinforced thermoplastic pipe under combined bending and internal pressure[J]. Ships Offshore Structures, 2018, 13(7): 696-704.

[5] ZhANG X, LI H, QI D, et al. Failure analysis of anticorrosion plastic alloy composite pipe used for oilfield gathering and transportation[J]. Engineering Failure Analysis, 2013, 32: 35-43.

[6] TOH W, BIN T L, TSE K M, et al. Material characterization of filament-wound composite pipes[J]. Composite Structures, 2018, 206: 474-483.

[7] LIU W, WANG S. An elastic stability-based method to predictthe homogenized hoop elastic moduli of reinforced thermoplastic pipes(RTPs)[J]. Composite Structures, 2019, 230: 111560.

[8] LI H, YAN M, QI D, et al. Failure analysis of steel wirer einforced thermoplastics composite pipe[J].

Engineering Failure Analysis, 2012, 20: 88-96.
[9] QI D, YAN M L, DING N, et al. Application of polymer composite pipes inoilfield in China[C]. In Proceedings of the 7thInternational MERL Oilfield Engineering with Polymers Conference, London, UK, 21-22 September, 2010.
[10] YUE Q, LU Q, YAN J, et al. Tension behavior prediction of flexible pipelines in shallow water[J]. Ocean Engineering, 2013, 58: 201-207.
[11] XIA M, KEMMOCHI K, TAKAYANAGI H. Analysis of filament-wound fiber-reinforced sandwich pipe under comb inedinternal pressure and thermomechanical loading[J]. Composite Structures, 2001, 51: 273-283.
[12] LOU M, WANG Y, TONG B, et al. Effect of temperature on tensile properties of reinforced thermoplastic pipes[J]. Composite Structures, 2020, 241: 112119.
[13] BAI Y, CHEN W, XIONG H, et al. Analysis of steel strip reinforced thermoplastic pipe under internal pressure[J]. Ships Offshore Structures, 2016, 11: 766-773.
[14] BAI Y, LIU T, CHENG P, et al. Buckling stability of steel strip reinforced thermoplastic pipe subjected to external pressure[J]. Composite Structures, 2016, 152: 528-537.
[15] ASHRAF M A, MOROZOV E V, SHANKAR K K. Flexure analysis of spoolable reinforced thermoplastic pipes for off shore oil and gas applications[J]. Journal of Reinforced Plastic and Composites, 2014, 33(6): 533-542.
[16] MENG X J, WANG S Q, YAO L. Research on equivalent simplification model of glass fiber-reinforced flexible pipes[J]. Ocean, Engineering, 2017, 35: 71-83
[17] GAO L, LIU T, SHAO Q, et al. Burst pressure of steel reinforced flexible pipe[J]. Marine Structures, 2020, 71: 102704.
[18] SY/T 6662.2—2020. Non-Metallic Composite Pipe for Petroleum and Natural Gas Industries—Part 2: Flexible Composite Pipe for High Pressure Transmission[S]. National Energy Board: Beijing, China, 2020.
[19] ASTM D2105-01. Standard Test Method for Longitudinal Tensile Properties of "Fiberglass" (Glass-Fiber-Reinforced Thermosetting-Resin) Pipe and Tube[S]. ASTM: West Conshohocken, PA, USA, 2019.
[20] BAI Y, LIU T, RUAN W, et al. Mechanical behavior of metallic strip flexible pipe subjected to tension[J]. Composite Structures, 2017, 170: 1-10.
[21] ISO 291: 2008. Plastics-Standard Atmospheres for Conditioning and Testing. International Standard[S]. The International Organization for Standardization: Geneva, Switzerland, 2008.

本论文原发表于《Ploymers》2021年第13卷第20期。

Study on the Hydrogen Embrittlement Susceptibility of AISI 321 Stainless Steel

Xu Xiuqing[1,2] **An Junwei**[1] **Wen Chen**[1] **Niu Jing**[3]

(1. Nanchang Institute of Technology, Nanchang; 2. State Key Laboratory of Performance and Structural Safety for Petroleum Tubular Goods and Equipment Materials, Xi'an 710077, China; R&D Center of TGRI, CNPC; 3. Xi'an Jiaotong University)

Abstract: AISI 321 stainless steel is widely used in hydrogenation refining pipes and hydrogen storage vessel and so on owing to its excellent performance of creep stress resistance and high-temperature resistance. In this study, slow strain rate tensile tests (SSRT) were conducted under the condition of electrolytic hydrogen charging (EHC). The hydrogen embrittlement mechanism of AISI 321 stainless steel was analyzed in detail by means of scanning electron microscopy (SEM), X-ray diffraction spectrometer (XRD) and transmission electron microscope (TEM). The results show that hydrogen can change the fracture mode of tensile specimens from ductile fracture to brittle fracture mode. The main reason is that dislocations slide carrying hydrogen continuously in the material under the action of slow strain rate stress, resulting hydrogen cracks are preferentially produced at austenite grain boundaries, inclusions and the interface between δ ferrite and austenite. Additionally, it is easy to induce the transformation and growth of α' martensite in the process of hydrogen charging and reduce the plasticity of the material.

Keywords: AISI 321 stainless steel; Electrolytic hydrogen charging; Crack; Hydrogen embrittlement mechanism

1 Introduction

AISI 321 austenitic stainless steel is usually used as refining hydrogenation pipes, hydrogen storage vessel and lining of hydrogenation reactor bearing high temperature, high pressure and hydrogen environment due to its excellent high-temperature mechanical properties. Under mechanical loading in hydrogen environment, metastable austenitic stainless steel is prone to hydrogen embrittlement due to the formation of α' martensite induced by strain[1-4]. It is well known that hydrogen embrittlement leads to failure at lower loads and in a shorter time. The potential result of this phenomenon is catastrophic failure and economic loss[5].

At present, hydrogen damage has become one of the most active research fields in the world[6-8]. As we known, the essence of hydrogen induced cracking is the nucleation and

Corresponding author: Xu Xiuqing, 86-029-81887905, xuxiuqing00@126.com, xuxiuqing@cnpc.com.cn.

propagation of cracks caused by hydrogen atoms entering the interior of materials. Now scholars at home and abroad generally think that hydrogen embrittlement occurs only in BCC metals and the austenitic stainless steel of face centered cubic structure has low hydrogen sensitivity[9-11]. However, hydrogen embrittlement still occurs after long-term service in hydrogen containing environment, such as hydrogen induced cracking of austenitic stainless steel structure in hydrogenation reactor[12-14]. Rozenak[15] studied hydrogen loss during surface stress and martensitic α'-bcc and ε-hcp by electrochemical hydrogen charging method. He considered that the maximum stress lead to the formation of α' phase and ε-martensite, both of which participated in the plastic deformation process and promoted the surface crack growth. Pu et al.[16] studied the hydrogen desorption behavior of austenitic steel before and after deformation, and directly observed the hydrogen release caused by deformation for the first time, which is also the key factor leading to embrittlement of austenitic steel. Li et al.[17] studied the effect of pre-strain on hydrogen embrittlement of 304L welded joint. He believed that in the high pre-strain joint, severe strain induced α' martensitic transformation occurs in some regions of the base metal, and cracks appear at the phase boundary, which may provide more places for the aggregation of hydrogen atoms and dislocations.

In summary, the interaction between hydrogen and metastable austenitic stainless steel is a special case. Many studies have identified martensite phase on the surface of samples produced by hydrogen embrittlement, but these studies have not explicitly addressed the relationship between hydrogen induced stress and hydrogen related martensitic transformation. Therefore, the study of hydrogen induced martensitic transformation and the effect of hydrogen on austenitic stainless steel are of great significance for the application of austenitic 321 stainless steel. In this paper, the hydrodrogen embrittlement mechanism of AISI 321 stainless steel was analyzed in detail under the condition of SSRT and EHC.

2 Materials and methods

The experimental material was AISI 321 stainless steel and its chemical composition and mechanical properties are shown in Table 1.

Table 1 Chemical composition of AISI 321 stainless steel (Wt. %)

C	Cr	Ni	Mn	Mo	Ti	Si	S	P	Fe
0.028	17.52	8.99	1.41	0.60	0.19	0.50	0.004	0.031	Bal.

Because hydrogen is only distributed near the specimen surface layer after the static EHC of tensile specimens, dynamic EHC and slow strain rate tests (using INSTRON 1195) were used to study the hydrogen embrittlement sensitivity of AISI 321 stainless steel. The specimens were processed into tensile specimens according to the ASTM A370 standard, and the dimensions of specimens are shown in Fig. 1. The electrolyte for EHC test was a mixture of 0.5mg/L H_2SO_4, 1.85mg/L $Na_4P_2O_7$ and deionized water. SSRTs were conducted under the optimized test condition of cathodic current density (30 mA/cm^2) and a constant strain rate of 0.1 mm/min.

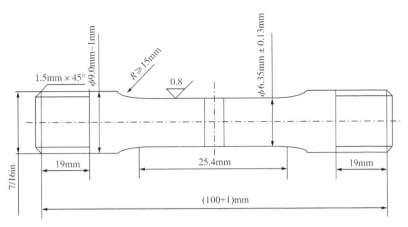

Fig. 1　Shape and dimensions of specimens for standard SSRT

The tensile specimens were wet ground to a 1200-grit finish on SiC emery papers, followed by polishing and ultrasonically cleaned with alcohol. Then sample was immersed in the test environment in a specially designed chamber mounted in tensile testing machine grips. Hydrogen was introduced into the samples by the cathodic current method under a galvanostatic condition at room temperature. SSRT tests were carried out at the same time of hydrogen charging until the specimens failed. Three specimens were measured in each group in the slow strain rate test to guarantee the reliability of test data.

The macroscopic morphology of slow strain tensile specimens before and after dynamic EHC were observed using a Nikon ECLIPSE MA200 optical microscope. Characterizations of the fracture surface morphology and the elemental composition of samples were analyzed by scanning electron microscopy (SEM, VEGA II XMUINCA) and energy-dispersive X-ray spectrometer (EDS), respectively. The crystal structure and microstructures of samples were determined by X-ray diffraction spectrometer (XRD) and transmission electron microscope (TEM, JEM-2000CX), respectively.

3　Experimental results and discussion

3.1　Effect of hydrogen on tensile performance of AISI 321 stainless steel

Fig. 2 shows the effect of hydrogen on the tensile strength of AISI 321 stainless steel. As seen, the tensile strength of AISI 321 stainless steel decreased significantly from 730MPa to 660MPa due to the existence of hydrogen. Fig. 3 shows the fracture morphologies of tensile specimens before and after EHC. The percentage reduction area of AISI 321 stainless steel decreased obviously after EHC. A clear necking phenomenon was observed and no clear crack on the specimen surface which shows the common fracture characteristics of smooth specimens. Contrary to fracture morphologies of

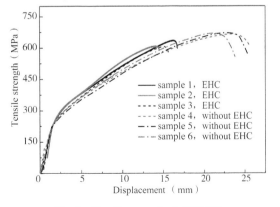

Fig. 2　Tensile strength of AISI 321 stainless steel before and after EHC

specimens before EHC, the tensile specimens after EHC showed many cracks vertical to the tensile direction and no clear necking phenomenon was observed, as seen in Fig. 3 (b). These characteristics indicates that the final fracture of specimens is directly related to these cracks.

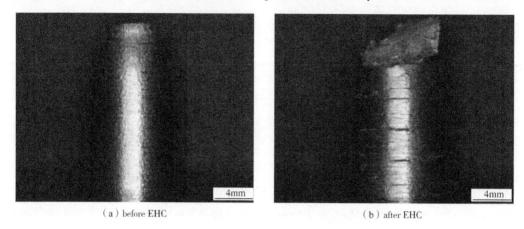

(a) before EHC (b) after EHC

Fig. 3 Macroscopic fracture morphology of AISI 321 stainless steel

3.2 Fracture morphology analysis

Fig. 4 shows the low-power fracture morphology for AISI 321 stainless steel under the condition of EHC. As shown, the fracture mode after EHC changes from ductile fracture to brittle fracture mode. The fracture initiates from the specimen surface or near-surface under the EHC condition and gradually propagate to the interior with a continuous increase in load, finally leading to an unstable fracture of specimens. The fracture morphologies of crack initiation and propagation region near the surface mainly exhibit intergranular and transgranular fracture morphologies which is typical hydrogen brittleness fracture morphology, as seen the inset in Fig. 4(b).

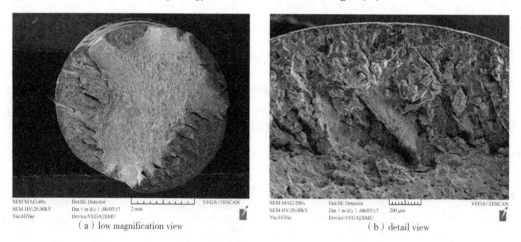

(a) low magnification view (b) detail view

Fig. 4 Fracture surface morphologies of AISI 321 stainless steel under the EHC condition

3.3 Analysis of crack initiation and propagation characteristics

Fig. 5 shows the metallographic images of cracks in AISI 321 stainless steel after EHC. Typical hydrogen crack which is perpendicular to the stretching direction appeared near the surface of stainless steel. Additionally, some crack appeared in the matrix and the crack direction is irregular neither along the tensile direction nor along the rolling direction, as seen in Fig. 5(b). In order to analyze the crack initiation clearly, the metallographic samples were observed by SEM.

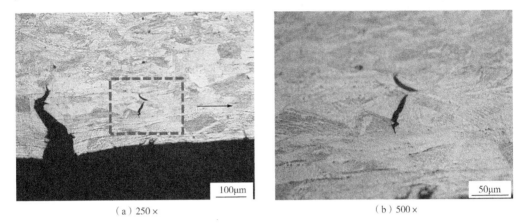

(a) 250×   (b) 500×

Fig. 5 Crack morphologies of AISI 321 stainless steel under the condition of EHC

Fig. 6 displays the SEM images and EDS analysis for different regions of AISI 321 stainless steel under the condition of EHC. During the dynamic hydrogen charging process, the hydrogen sensitivity of the material increased and cracks were found on the surface and inside of samples, as seen in Fig. 6. EDS analysis around the crack in Fig. 6(a) shows that the content of Cr is 24.5% which is much higher than the normal of 18% and the content of Ni is 4.8% which is much lower than the normal of 9%. As we known, Cr is ferritizing elements whereas Ni is austenitizing elements. Thus, it can be confirmed that the bone-shaped structures [area a in Fig. 6(a)] is δ ferrite. The content of elements in area b is the normal level of 321 stainless steel, which should be austenite. The area c is mainly composed of Ti and N elements, which is titanium nitride inclusions. The crack appeared at the inclusions [Fig. 6(a)] where exist stress concentration and has high hydrogen sensitivity, which is prone to hydrogen induced cracking. Fig. 6(b) shows the crack initiation along austenite grain boundary. EDS results perpendicular to the crack exhibit that the main elements Fe, Cr and Ni on both sides of the crack have little change.

In this paper, there are three ways for hydrogen to enter the material as follows: i) diffusion along austenite grain boundary, ii) diffusion along δ ferrite and iii) dislocations motion carrying hydrogen. The internal slip mechanism of austenitic stainless steel begins with the increasing slow strain tensile stress. Dislocations carrying hydrogen atoms slips directionally in austenite grains, which leads to a large amount of dislocations piled up around the grain boundary. And then the stress concentration and hydrogen concentration increased. When it increases to a certain extent, the crack will begin and expand along the grain boundary, as shown in Fig. 5(b) and Fig. 6(b). When a large number of dislocations pile up at grain boundary to a certain extent, dislocations will diffuse through the grain boundary to the next grain and also carry hydrogen atoms through the grain boundary, resulting in hydrogen expansion in depth. In this way, the dislocations carry hydrogen continuously into the austenite stainless steel. Fig. 7 displays the schematic diagram of hydrogen diffusion path in AISI 321 stainless steel under the condition of SSRT and EHC.

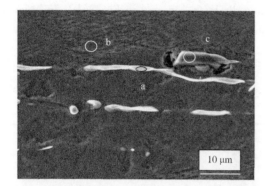

Elements (Wt.%)	Area a	Area b	Area c
Fe	69.7	70.9	3.9
Cr	24.7	17.7	3.6
Ni	4.8	8.6	
C			
Si	0.8	1.8	
Mn		1.2	
Ti			68.5
N			24.0

(a) crack initiation along inclusion

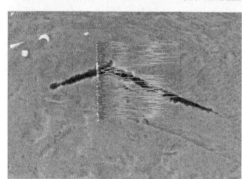
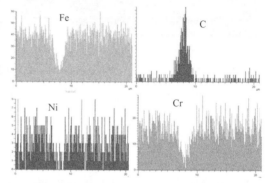

(b) crack initiation along austenite grain boundary

Fig. 6 SEM images and EDS analysis for of AISI 321 stainless steel under the condition of EHC

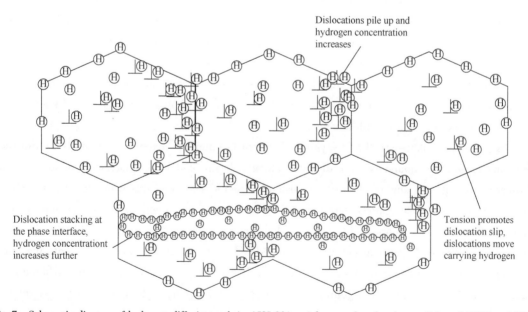

Fig. 7 Schematic diagram of hydrogen diffusion path in AISI 321 stainless steel under the condition of SSRT and EHC

3.4 Transformation of crystal structure of AISI 321 stainless steel

Fig. 8 shows the XRD spectra for different states of AISI 321 stainless steel. As seen, they were identified as austenite and δ ferrite for base material [Fig. 8(a)], which is further proved the banded structure in SEM image [Fig. 6(a)] is δ ferrite. Under the condition of SSRT without EHC, the peaks of α′ martensite were also observed except austenite and δ ferrite in Fig. 8(b). After EHC and SSRT, crystal structure of AISI 321 stainless steel were still composed of austenite, δ ferrite and α′ martensite. However, the diffraction peaks intensities of δ ferrite and α′ martensite increased and that of austenite decreased obviously, as shown in Fig. 8(c). Combined with the results of tensile test, we deduced that hydrogen may promote the formation and growth of α′ martensite.

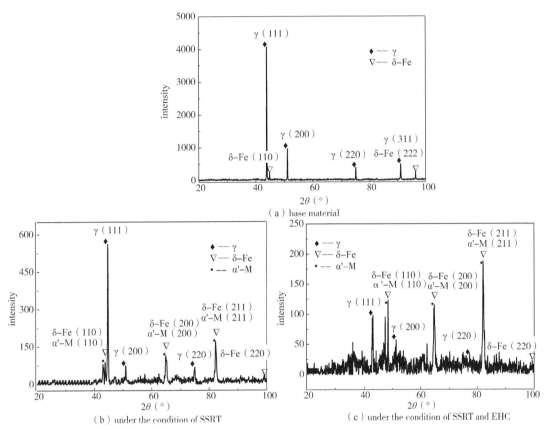

Fig. 8 XRD spectra for different states of AISI 321 stainless steel

In order to confirm this conclusion, TEM analyses for different states of AISI 321 stainless steel were carried out. The results are shown in Fig. 9. The main characteristic of microstructure for base material was massive structure and twin crystal, which was identified as austenite [Fig. 9(a), the crystal band axis is [011]]. Under the condition of SSRT without EHC, much lath structure with the width of 150nm was observed around twin or grain boundary in Fig. 9(b). The TEM analysis and electron diffraction pattern [inset in Fig. 9(b)] revealed the lath structure is α′ martensite (the crystal band axis is [11-1]). Because a large number of dislocations are accumulated near the twin or grain boundary resulting in the stress concentration. It is easier to induce the transformation of α′ martensite and promote the further growth of α′ martensite. After EHC and SSRT, α′ martensite becomes denser and its width is up to 200nm, as seen in Fig. 9(c) (the crystal band axis is [-112]). This result shows that hydrogen greatly promotes the transformation and growth of α′ martensite.

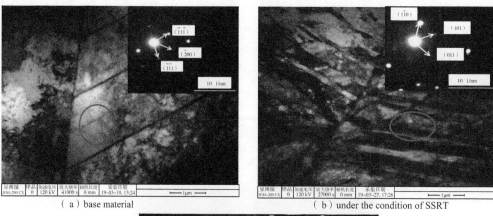

Fig. 9 TEM images for different states of AISI 321 stainless steel

3.5 Fracture mechanism analysis of AISI 321 stainless steel in hydrogen environment

Based on the above analysis, a large amount of hydrogen accumulates in the process of hydrogen charging. Under the action of slow strain tensile stress, dislocations slide carrying hydrogen continuously in the material, resulting in a large number of dislocations piling up near the grain boundary and sub grain boundary. So the hydrogen cracks are preferentially produced at austenite grain boundaries, inclusions and the interface between δ ferrite and austenite. Once the crack initiated and defects formed, more hydrogen will accumulate and further accelerate the crack growth. Additionally, it is easy to induce the transformation of α′ martensite in the process of hydrogen charging. As we known, α′ martensite is a brittle phase and its appearance and growth can reduce the plasticity of the material, which leads to the further hydrogen damage of AISI 321 stainless steel. Finally, the brittle fracture of the material happens under the condition of SSRT and EHC.

4 Conclusions

The fracture behavior of AISI 321 stainless steel in hydrogen environment was analyzed through dynamic EHC and SSRT, and the hydrogen embrittlement mechanism was discussed by means of the microstructure analysis, SEM and EDS analysis, XRD and TEM analysis. The main conclusions are as follows:

(1) After EHC, the fracture mode of specimen under a slow strainrate test changed fundamentally, from ductile fracture to brittle fracture mode.

(2) The hydrogen cracks are preferentially produced at austenite grain boundaries, inclusions and the interface between δ ferrite and austenite. Once the crack initiated and defects formed, more hydrogen will accumulate and further accelerate the crack growth.

(3) The transformation and growth of α′ martensite in the process of hydrogen charging happens which reduces the plasticity of the material and leads to the further hydrogen damage of AISI 321 stainless steel.

Acknowledgements

The authors acknowledge the financial support ofthe National Key Research and Development Program of China (2018YFC0808800) and Key Project of PetroChina (2018E-1809) and CNPC Project of Science Research and Technology Development No. 2017D-2307.

References

[1] MICHALSKA J, CHMIELA B J, et al. Hydrogen Damage in Superaustenitic 904L Stainless [J/OL]. springerlink.com, 2014, 23(12): 2760-2765.

[2] PU S D, TURK A, LENKA S, et al. Study of hydrogen release resulting from the transformation of austenite into martensite[J]. Materials Science and Engineering: A, 2019, 754: 628-635.

[3] PERNG T P, JOHNSON M, ALTSTETTER C J. Influence of plastic deformation on hydrogen diffusion and permeation in satinless steels[J]. Acta Meallurgica, 1989, 37(12): 3393-3397.

[4] HARDIE D, XU J, CHARLES E A, et al, Hydrogen Embrittlement of Stainless Steel Overlay Materials for Hydrogenators[J]. Corrosion Science, 2004, 46(12): 3089-3100.

[5] GIBBS P J, SAN M C, NIBUR K A, et al. Comparison of Internal and External Hydrogen on Fatigue-Life of Austenitic Stainless Steels [C]. ASME 2016 Pressure Vessels and Piping Conference. American Society of Mechanical Engineers, 2016: V06BT06A033-V06BT06A033.

[6] WANG Z, LUO Z C, HUANG M X. Revealing hydrogen-induced delayed fracture in ferrite-containing quenching and partitioning steels[J]. Materialia, 2018, 4: 260-267.

[7] YAN Y, HE Y, et al. Hydrogen-induced cracking mechanism of precipitation strengthened austenitic stainless steel weldment[J]. International journal of hydrogen energy, 2015, 40(5): 2404-2414.

[8] MINE Y, KOGA K, KRAFT O, et al. Mechanical characterisation of hydrogen-induced quasi-cleavage in a metastable austenitic steel using micro-tensile testing[J]. Scripta Materialia, 2016, 113: 176-179.

[9] FARRELL K, QUARRELL A G. Hydrogen embrittlement of ultra-high-tensile steel [J]. Journal of the Iron and Steel Institute, 1964, 202(12): 1002-1011.

[10] HERMS E, OLIVE J M, PUIGGALI M. Hydrogen embrittlement of 316L type stainless steel[J]. Materials Science and Engineering, 1999, 272(2): 279-283.

[11] ELIEZER D, CHAKRAPANI D G, ALTSTETTER C J, et al. The influence of austenite stability on the hydrogen embrittlement and stree-corrosion cracking of stainless steel [J]. Metallurgical and Materials Transactions A, 1979, 10(7): 935-941.

[12] TROIANO A R. The role of hydrogen and other interstitials in the mechanical behavior of metals [J]. trans. ASME, 1960, 51(2): 54-80.

[13] PECKNER D, BERNSTEIN, M. Handbook Stainless Steel[M]. New York Press, 1977: chao 6. 1322.
[14] WEST A J, HOLBROOK J H. Hydrogen effects in metals[M]. Warrendale: AIME, 1981.
[15] ROZENAK P. Stress Induce Martensitic Transformations in Hydrogen Embrittlement of Austenitic Stainless Steels[J]. Metallurgical and Materials Transcations A, 2014, 45(1): 162-178
[16] PU S D, TURK A, LENKA S, et al. Study of hydrogen release resulting from the transformation of austenite into martensite[J]. Materials Science and Engineering, 2019, 86(1): 21-29.
[17] LI X G, GONG B, DENG C, et al. Effect of pre-strain on microstructure and hydrogen embrittlement of K-TIG welded austenitic stainless steel[J]. Corrosion Science, 2019, 149: 1-17.

本论文原发表于《Engineering Failure Analysis》2021 年第 122 卷。

CH_4 在 PVDF 中的渗透特性及机理

李厚补[1]　张学敏[2]　马相阳[3]　王品[2]　杨晓辉[3]　张冬娜[1]

(1. 中国石油集团石油管工程技术研究院石油管材及装备材料服役行为与结构安全国家重点实验室；2. 长安大学材料科学与工程学院；3. 长庆油田分公司第一采油厂采油工艺研究所)

摘　要：气体组分在热塑性塑料中的渗透导致起泡失效，给油气集输用热塑性塑料内衬复合管的安全运行带来极大隐患。为了了解气体介质在热塑性塑料中的渗透特性并明确其渗透机理，从根本上提出控制气体渗透的有效措施，研究了典型气体 CH_4 在聚偏氟乙烯(PVDF)中的渗透特性，检测分析了渗透样品结构成分和耐热性能，并基于蒙特卡洛法(GCMC)和分子动力学模拟了 30℃、0.1MPa 下 CH_4 在 PVDF 中的吸附扩散行为，探讨了 CH_4 在 PVDF 中的吸附及扩散机理。结果表明：CH_4 在 PVDF 中的渗透系数随温度升高而提高，渗透样品的结构成分和耐热性能未发生明显变化，气体渗透主要为物理渗透过程，不存在化学侵蚀破坏。模拟计算得到的渗透系数与实验结果吻合，30℃、0.1MPa 下 CH_4 有选择性地聚集吸附在 PVDF 晶胞中，扩散过程中 CH_4 分子均在 PVDF 空穴内振动，并未出现空穴间的跃迁。

关键词：聚偏氟乙烯；渗透；吸附；扩散；集输

含 H_2S 天然气已成为我国天然气资源的重要组成部分[1]。对于集输用管线，普通碳钢管在这类酸性环境中(含 H_2S/CH_4)腐蚀非常严重。近年来，施工方便且具有优异耐腐蚀、柔韧性、抗疲劳性能等特点的增强热塑性塑料复合管正成为含硫油气输送的一个重要选择[2-3]。该产品是以热塑性塑料管为内衬层，以金属(钢丝、钢带等)或非金属(芳纶纤维、聚酯纤维等)材料作为增强层，外敷热塑性塑料保护层复合而成。但鉴于热塑性塑料的本质特性，气体分子的自由运动不可避免地会在材料中发生渗透现象。气体渗透不仅会导致输送气体的浪费，还可能造成热塑性塑料起泡、坍塌等失效现象[4-7]。而吸附在内衬表面的气体介质会沿内衬厚度方向发生扩散，进而增加了增强层(尤其是金属增强层)的腐蚀失效风险，造成复合管承压能力下降及使用寿命缩短[8]。

聚偏氟乙烯(PVDF)具有良好的耐温、耐化学腐蚀、耐候性以及优良的机械性能，因此成为增强热塑性塑料复合管在高温集输领域替代传统高密度聚乙烯(HDPE)内衬层的最佳材料[9]。为了预防气体渗透损伤破坏 PVDF 内衬，降低增强热塑性塑料复合管在油气输送过程中的失效风险，必须了解气体介质在 PVDF 中的渗透特性并明确其渗透机理，以便从根本上

基金项目：中央高校基本科研业务费专项基金(300102310201)，中石油超前储备研究基金(2017D5008)。

作者简介：李厚补(1981—)，男，博士，从事油气田用非金属管材研究。E-mail: lihoubu@cnpc.com.cn。

提出控制气体渗透的有效措施，改善 PVDF 的气体阻隔性能。在此基础上，采用压差法气体渗透仪，研究了不同温度下典型气体 CH_4 在 PVDF 中的渗透特性，结合渗透样品结构成分、耐热性能分析以及分子模拟，探讨了 CH_4 在 PVDF 中的吸附和扩散机理，为油气介质在热塑性塑料中的渗透性评价及控制提供支撑。

1 实验部分

1.1 试验过程

试验用 PVDF 样品为挤塑制备的 ϕ60mm 圆状薄膜，厚度为 300μm。参考 GB/T 1038—2000[10]标准，采用 VAC-V2 型压差法气体渗透仪测试 CH_4 气体在 PVDF 中的渗透系数。试验温度为 30℃、40℃、60℃和 80℃，每种条件下测试 3 个样品。

1.2 测试与表征

采用 VERTEX 70 型傅里叶红外光谱仪测试 PVDF 结构成分变化；利用 DTG204F1 热分析系统测试 PVDF 的 TG-DSC 曲线，升温速率 10℃/min，流动 N_2 保护；采用 Material Studio 6.0(MS)分析 CH_4 在 PVDF 中的渗透机理。

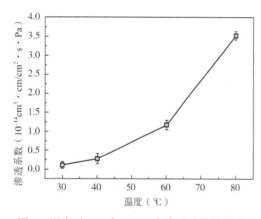

图 1 温度对 CH_4 在 PVDF 中渗透系数的影响

2 结果与讨论

2.1 CH_4 在 PVDF 中的渗透系数

在不同温度条件下，CH_4 在 PVDF 中的渗透系数如图 1 所示。随着温度的升高，CH_4 气体在 PVDF 中的渗透系数不断增加。在恒定压力下，温度对气体在聚合物中扩散系数的影响服从阿累尼乌斯(Arrhenius)方程[11-12]。根据自由体积模型，低温时(如靠近或低于聚合物材料的玻璃化转变温度时)，聚合物材料自由体积减小，激活能升高。随温度升高，聚合物材料总的自由体积增加，渗透气体分子可在材料中更加自由地运动，此时介质分子遵循"似液体"渗透机理，激活能较低，气体渗透系数随之增加[11]。

2.2 渗透样品性能分析

2.2.1 结构成分分析

不同条件下，CH_4 气体在 PVDF 样品中渗透前后的 FTIR 图谱见图 2。PVDF 分子式为：—[CH_2—CF_2]$_n$—。由图 2 可以看出，PVDF 原始样品的红外光谱图主要特征吸收峰分别在 1402 cm^{-1}、1180 cm^{-1}、976 cm^{-1}、880 cm^{-1}、840 cm^{-1}、796 cm^{-1}、763 cm^{-1}、613 cm^{-1} 等处。其中 1402 cm^{-1} 是 PVDF 中与 CF_2 相连的 CH_2 的变形振动吸收峰；1180 cm^{-1} 是 PVDF 中 CF_2 的伸缩振动吸收峰；976 cm^{-1}、796 cm^{-1}、763 cm^{-1}、613 cm^{-1} 处的尖锐吸收峰是结晶相的振动吸收峰，当样品熔融和溶解时，这些吸收峰会明显消失。880 cm^{-1}、840 cm^{-1} 是无定形相的特征吸收峰[13]。与 PVDF 原始样品的 FTIR 图谱相比，不同温度下 CH_4 气体渗透后样品 FTIR 图谱中的各处特征吸收峰位置和强度均未发生明显变化(图 2)，表明不同温度下(30℃、40℃、60℃、80℃)，CH_4 气体的渗透并未造成 PVDF 样品结构成分破坏。可以推

断，气体在 PVDF 中的渗透过程为单纯的物理渗透过程，并未明显侵蚀破坏其分子结构和元素构成。

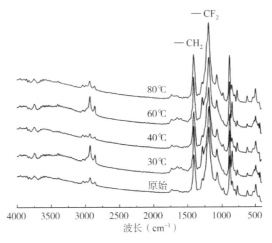

图 2　不同温度下 CH_4 气体渗透对 PVDF 样品 FTIR 图谱的影响

2.2.2　耐热性能分析

不同条件下，CH_4 气体在 PVDF 样品中渗透前后的 TG-DSC 图谱见图 3。由 TG 曲线可以看出，PVDF 原始样品起始失重温度为 445.1℃；拐点（最大失重速率点）为 466.4℃，失重终止点为 481.4℃；质量损失 55.6%（图 3）。PVDF 原始样品的 DSC 曲线在 165.4℃ 左右出现第 1 个吸热峰，该峰是 PVDF 的熔融峰，表明当温度达到 165.4℃ 时 PVDF 开始熔化（图 3）。随后，聚合物树脂侧链小分子受热断裂逸出导致缓慢放热，使曲线上升。在 462.5℃ 时（失重拐点附近）出现一个较为明显放热台阶峰，这是由于此时 PVDF 发生了脱 HF 反应，失重速率迅速增长[9]。与 PVDF 原始样品的 TG-DSC 曲线相比，不同温度下 CH_4 气体在 PVDF 样品渗透后的 TG-DSC 曲线变化趋势类似（图 3），吸热峰和放热峰的位置及强度基本保持一致，并未出现其他吸热峰或放热峰，表明不同温度下 CH_4 气体的渗透并未影响 PVDF 样品的耐热性能和热解行为。

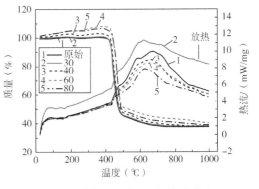

图 3　不同温度下 CH_4 气体渗透对 PVDF 样品 TG-DSC 的影响

2.3　CH_4 在 PVDF 中渗透的分子模拟

气体在聚合物中的渗透过程通常可以用溶解-扩散机理来描述[11]，即高压气体吸附（溶解）进入材料的高压侧表面，渗透的气体分子在压差作用下向材料内部进行扩散，最后从低压侧表面脱附。基于分子模拟方法在原子尺度上研究 30℃、0.1MPa 时 CH_4 气体分子在 PVDF 中的吸附扩散行为，以获得相应的溶解度及扩散系数，并与实验结果对比，验证分子模拟手段的可行性，并分析其渗透机理。

2.3.1　模拟理论

气体在聚合物中的渗透系数，可由以下公式计算获得：

$$P = S \times D \tag{1}$$

其中：S 为溶解系数，反映了渗透分子与聚合物相互作用的热力学特性；D 为扩散系数，反映了渗透分子与聚合物体系的动力学特性；P 为渗透系数，代表了气体通过聚合物的难易程度。

求解气体分子在聚合物中的溶解系数，一般采用吸附等温线法。通过在一定的压力范围内进行一系列的巨正则系综蒙特卡洛（GCMC）模拟，得到渗透分子的浓度 C 随压力 p 变化的曲线，求出压力为 0 时的极限斜率，从而得到溶解系数，即：

$$S = \lim_{p \to 0} \frac{C}{p} = K_D + C_H b \tag{2}$$

其中：K_D 为亨利常数，C_H 为 Langmuir 吸附容量，b 为 Langmuir 常数。

扩散系数 D 的计算基于分子动力学模拟分子的均方位移（MSD）随时间 t 变化的 Einstein 方程[14]求得，即：

$$D = \frac{1}{6N} \lim_{t \to \infty} \frac{d}{dt} \left\{ \sum_{i=1}^{N} [r_i(t) - r_i(0)] \right\}^2 \tag{3}$$

其中：N 为体系中所有的扩散粒子数目，t 为模拟时间，$r_i(t)$、$r_i(0)$ 分别为粒子 i 在 t 时刻和初始时刻的位置向量。

求解出 MSD-t（时间）曲线对时间的微分即曲线的斜率 a，便可简化式（3）为式（4），进而获得扩散系数 D：

$$D = a/6 \tag{4}$$

2.3.2 模型构建

采用 Materials Studio（MS）软件首先构建偏氟乙烯单体和 CH_4 分子模型，并采用 Charge groups 方法进行非键参数的确定，随后对偏氟乙烯单体和 CH_4 分子进行几何优化及能量最小化处理。将优化后的偏氟乙烯单体作为 PVDF 的重复单元，构建聚合度为 10 的 PVDF 分子链并进行能量最小化处理。为减少分子模拟过程中的链端效应，选用 4 条 PVDF 分子链利用 Amorphous Cell 模块在 30℃ 和 0.1MPa 条件下构建远程无序、近程有序，带有三维周期边界条件的 PVDF 晶胞，随后对该晶胞进行能量优化，得到最终稳定构型。

吸附体系中选取稳定构型的 PVDF 晶胞作为吸附剂，优化后的 CH_4 模型为吸附质分子，采用巨正则系综蒙特卡洛（GCMC）分子模拟方法，利用 Sorption 模块中固定逸度的方法研究在特定的压强下 CH_4 分子在 PVDF 晶胞中的吸附量。

在扩散体系中，利用 Amorphous Cell 模块构建包含 4 个 CH_4 分子和 4 个聚合度为 10 的 PVDF 分子链的无定形晶胞（CH_4 和 PVDF 分子链均为上述的稳定构型），设定最终构型的目标密度为 1.75g/cm³。为了矫正无规则晶胞生成时分子随机分布而造成的真空区，采用能量最小化来优化晶胞，使其势能降低。能量最小化后，通过正则系综（NVT）+微正则系综（NVE）动力学模拟来平衡体系，即对结构进行弛豫。动力学模拟在 Discover 模块中选定 Dynamics 执行。

整个模拟过程采用 COMPASS 力场，选用 Atom-Based 法计算范德华相互作用力，采用 Ewald 法计算静电相互作用力。在吸附模拟过程中，前 100000 步用于使体系达到平衡，后 1000000 步用于统计所需的热力学性质；在 NVE 动力学模拟时，模拟步长为 1fs，模拟总时间为 200ps，并保存体系所有的运动轨迹。

2.3.3 模拟结果

基于以上理论及模拟分析，获得 30℃、0.1MPa 下 CH_4 在 PVDF 中的吸附热为 26.56kJ/mol，小于40kJ/mol，说明该过程属于物理吸附[14]，与试验结果相一致。模拟获得的吸附等温线如图4所示。从图4中可以看出，其与Langmuir吸附等温线趋势一致，对应于Langmuir单层可逆吸附。利用Langmuir式(2)对其进行非线性最小二乘法拟合，得到溶解系数S为0.298×10^{-6} cm^3(STP)/(cm$^3\cdot$Pa)。

CH_4分子在PVDF晶胞体系中扩散的均方位移(MSD)曲线如图5所示。CH_4分子均方位移曲线的线性关系良好，对其进行最小二乘法线性拟合，得到CH_4在PVDF中的扩散系数D为1.1796×10^{-8}cm^2/s。

基于式(1)，获得30℃、0.1MPa条件下CH_4在PVDF中的渗透系数为0.35×10^{-14} cm^3(STP)·cm^2/cm$^3\cdot$s·Pa，与该条件下实验测试值0.105×10^{-14}cm^3·cm/cm^2·s·Pa基本吻合，表明所建立的模型是可靠的。

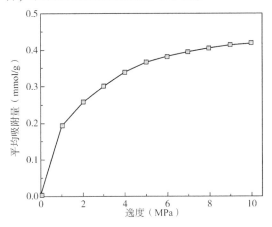 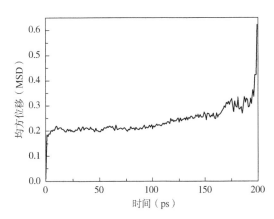

图4 30℃、0.1MPa下CH_4在PVDF中的吸附等温线　　图5 CH_4分子在PVDF中扩散的均方位移图

2.4 渗透机理分析

为了研究CH_4在PVDF中的吸附机理，基于Sorption模块得到了吸附质分子的密度场分布图和等密度面显示图，如图6所示。其中，图6(a)代表的是CH_4在PVDF模拟单元中的密度分布集中区域，也就是溶解的分子集中吸附的位置，可以看到CH_4的吸附位置比较少，区域比较集中。图6(b)为CH_4溶解在PVDF晶胞中的等密度面显示图，颜色越深(数值越大)代表该区域的分子间作用能越高，相应的在该区域气体越不容易吸附；反之气体越容易吸附，这与密度场分布图反映的情况是一致的。因此，从密度场分布图和能量场分布图都可以看出，CH_4分子在PVDF中的吸附不是均匀吸附，而是有选择性地聚集吸附。

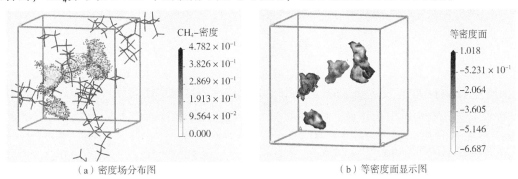

(a)密度场分布图　　　　　　　　　　(b)等密度面显示图

图6 CH_4在PVDF中的密度场分布图和等密度面显示图

气体分子在聚合物内扩散时,其自身不断振动,并能在振动位置附近扩散或在不同空穴间跃迁[15]。为了更直观地观察 CH_4 分子的扩散过程,编写 Perl 脚本提取出原子坐标,经处理后绘制出 CH_4 分子在聚合物盒子内的三维运动轨迹图和扩散位移时间关系图,如图7所示。从图7(a)中可以看出,在30℃、0.1MPa下 CH_4 分子基本都在某一区域内来回运动,这一区域就是聚合物基体的空穴,表明该条件下 CH_4 大部分时间的运动都在空穴内进行,并未出现空穴间的跃迁[图7(a)]。从位移时间图[图7(b)]中也可以看出 CH_4 分子的运动基本都是局限在一个较小的范围内来回振动,运动位移始终保持在0.1nm内,再次证明在30℃、0.1MPa下 CH_4 分子的扩散属于空穴内运动。

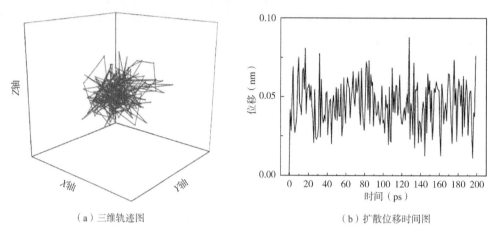

(a)三维轨迹图　　　　　　　　　　(b)扩散位移时间图

图7　CH_4 分子在 PVDF 中的三维轨迹图和扩散位移时间图

3　结论

(1) CH_4 气体在 PVDF 中的渗透系数随温度升高而增加。渗透样品的结构成分和耐热性能未发生明显变化,表明 CH_4 在 PVDF 中渗透的主要模式为物理渗透过程,不存在化学侵蚀破坏。

(2) 30℃、0.1MPa下利用分子动力学模拟得到的 CH_4 分子在 PVDF 中的渗透系数为 $0.35 \times 10^{-14} \, cm^3(STP) \cdot cm^2/cm^3 \cdot s \cdot Pa$,与实验结果基本吻合,表明所建立的模型具有较好的可靠性。

(3) 分子动力学模拟结果表明,30℃、0.1MPa下 CH_4 分子在 PVDF 中的吸附不是均匀吸附,而是有选择性地聚集吸附;CH_4 扩散时大部分时间的运动都在空穴内进行,并未出现空穴间的跃迁,属于空穴内运动。

参 考 文 献

[1] 朱光有,张水昌,梁英波,等. 四川盆地高含 H_2S 天然气的分布与 TSR 成因证据[J]. 地质学报,2006,80(8):1208-1218.

[2] QI D T, YAN M L, DING N, et al. Application of Polymer Composite pipes in Oilfields in China [C]. The 7th Internatioanl MERL Oilfield Engineering with Polymers Conference, 20 – 22 September 2010, London, UK. New-York: SPE, 2010.

[3] 韩方勇,丁建宇,孙铁民,等. 油气田应用非金属管道技术研究[J]. 石油规划设计,2012,23(6):5-9.

[4] SCHEICHL R, KLOPFFER M, BENJELLOUN-DABAGHI Z, et al. Permeation of gases in polymers: parameter identification and nonlinear regression analysis [J]. Journal of Membrane Science, 2005, 254: 275-293.

[5] RUEDA F, TORRES J P, MACHADO M, et al. External pressure induced buckling collapse of high density polyethylene (HDPE) liners: FEM modeling and predictions [J]. Thin-Walled Structures, 2015, 96: 56-63.

[6] YERSAK T A, BAKER D R, YANAGISAWA Y, et al. Predictive model for depressurization-induced blistering of type IV tank liners for hydrogen storage [J]. International Journal of Hydrogen Energy, 2017, 42: 28910-28917.

[7] PEPIN J, LAINE E, GRANDIDIER J C, et al. Determination of key parameters responsible for polymeric liner collapse in hyperbaric type IV hydrogen storage vessels [J]. International Journal of Hydrogen Energy, 2018, 43: 16386-16399.

[8] MAKINO Y, OKAMOTO T, GOTO Y, et al. The problem of gas permeation in flexible pipe [C]. Houston: Offshore Technology Conference, 1988.

[9] 张学敏, 李厚补, 戚东涛, 等. 油气田模拟环境下聚偏氟乙烯的适用性研究[J]. 天然气与石油, 2016, 34(2): 68-71.

[10] 国家质量技术监督局. GB/T 1038—2000 塑料薄膜和薄片气体透过性试验方法压差法 [S]. 北京: 中国标准出版社, 2000.

[11] MOZAFFARI F, ESLAMI H, MOGHADASI J. Molecular dynamics simulation of diffusion and permeation of gases in polystyrene [J]. Polymer, 2010, 51: 300-307.

[12] SONEY C G, THOMAS S. Transport phenomena through polymeric systems [J]. Progress in Polymer Science, 2001, 26: 985-1017.

[13] ROSMA N, ISMAIL A F, CHEER N B. Polyvinylidene fluoride and polyetherimide hollow fiber membranes for CO_2 stripping in membrane contactor [J]. Chemical Engineering Research and Design, 2014, 92(7): 1391-1398.

[14] 张廷山, 何映颉, 杨洋, 等. 有机质纳米孔隙吸附页岩气的分子模拟[J]. 天然气地球科学, 2017, 28(1): 146-155.

[15] 钟颖. 分子模拟研究气体在高渗透性膜中扩散溶解行为[D]. 重庆: 西南大学, 2012.

本论文原发表于《塑料》2021年第50卷第2期。

饱和 CO_2 溶液中 Cl^- 浓度对马氏体不锈钢应力腐蚀敏感性的影响

赵雪会[1,2] 刘君林[3] 曾瑞华[4] 昝聪敏[5] 徐秀清[1]

(1. 中国石油集团石油管工程技术研究院 石油管材及装备材料服役行为与结构安全国家重点实验室；2. 西安交通大学金属材料强度国家重点实验室；3. 青海油田钻采工艺研究院；4. 南方石油勘探开发有限责任公司工程技术处；5. 西安石油大学材料学院)

摘 要：本文利用电化学动电位测试技术对比分析了 Cl^- 浓度对超级 13Cr 腐蚀电位的影响关系；采用慢应变速率拉伸（SSRT）应力腐蚀开裂（SCC）实验方法和应力-应变曲线（σ-ε）、扫描电镜（SEM）等分析手段，研究了饱和 CO_2 环境下在一定慢应变速率条件下 Cl^- 浓度的变化对超级 13Cr 马氏体不锈钢的抗拉强度、延伸率、应力腐蚀敏感指数（I_{SSRT}）的影响，并结合断口形貌分析材料的断裂特征。结果显示：在 Cl^- 浓度 ≤60g/L 的饱和 CO_2 溶液中，超级 13Cr 相对于空气中强韧性变化不大，属于韧性断裂模式；随着 Cl^- 浓度的增大，材料的断裂寿命和伸长率均逐渐降低；材料断裂模式由韧性断裂逐渐转变为韧性—脆性混合断裂模式；当 Cl^- 浓度增大到 90g/L 时，断口侧面开始出现二次裂纹；当 Cl^- 浓度 ≥120g/L 时，材料点蚀电位明显降低，点蚀敏感性增大，同时断口侧面出现点蚀现象，进一步促进应力腐蚀开裂。

关键词：超级 13Cr 不锈钢；SCC；应力-应变；断裂；点蚀

随着油气能源需求的不断加大和油气田勘探开发的不断深入，原油的开采环境也变得越来越恶劣，地层压力、开采温度都大幅度升高，尤其地层水的矿化度、油气中具有较强腐蚀性的伴生气 CO_2、H_2S 等也越来越复杂，并且由于 CO_2/H_2S 气体引起的油套管腐蚀失效也逐渐成为制约油田发展的重要因素[1-2]。从材料防腐方面油套管多年来逐渐从碳钢升级到不锈钢，更苛刻的甚至考虑镍基合金管材[3-5]。目前针对 CO_2 环境腐蚀问题 13Cr 马氏体不锈钢体现出优良的耐蚀性，在油气环境中得到越来越广泛的应用，然而日益苛刻的工况条件对超级 13Cr 管材的服役安全性带来挑战[6-7]。随着原油开采逐渐进入中后期阶段，原油含水率日益增大，地层采出液更趋复杂，油套管接触的不仅是伴生气，同时含有 Cl^-、HCO_3^- 等大量侵蚀性离子地层水对油套管存在严重的腐蚀损伤，因此在复杂环境下材料的抗应力腐蚀开裂问题逐渐凸现出来[8-10]。吕祥宏等[11]、Zhao[12] 等分别研究发现高压 CO_2 环境下及随着 Cl^- 浓度的增加显著降低 13Cr 马氏体不锈钢的点蚀电位，明显增加 13Cr 马氏体不锈钢的 SSC 敏感性，雷冰[13] 等发现在高氯（Cl^->150g/L）环境下，注入饱和 CO_2、相对单一的高氯环境加

基金项目：本研究工作获得国家自然科学基金青年科学基金项目(51904331)和中国石油天然气集团公司重大专项科技项目(2014E-3603)资助。

作者简介：赵雪会(1973—)，硕士，主要从事油气田复杂工况环境石油管材腐蚀与防控技术研究及油田综合防治技术服务工作，电话：15991166636，E-mail：zhaoxuehui@cnpc.com.cn。

大了材料的腐蚀速率,并使得13Cr的腐蚀速率在高氯环境下从点蚀转变为晶间腐蚀和均匀腐蚀[13]。多种单一腐蚀性因素对材料腐蚀行为的影响规律已有成熟的认识[14-16],而多种腐蚀因素对材料的协同作用近几年才逐渐得到关注[17-18]。本文主要是模拟油田环境在典型地层水高Cl^-环境且饱和CO_2环境下,采用控制变量法,结合管柱在苛刻服役环境下承受应力作用并受到腐蚀介质的侵蚀情况,研究超级13Cr油管在慢应变速率及含CO_2溶液介质协同影响下的SCC敏感性,掌握材料在不同Cl^-浓度变化下腐蚀敏感性的变化规律,研究结果对油田管材的优化筛选、科学使用以及工业安全生产提供技术支撑和理论依据。

1 试验方法及材料

1.1 材料及试样制备

试验材料为超级13Cr马氏体不锈钢,采用的是无缝钢管TN公司生产的商业用油管材料,图1为13Cr不锈钢材料的金相组织,主要为回火马氏体,材料的化学成分见表1所示。试样按照慢应变拉伸试验机的要求制作,试样的形状和尺寸如图2所示。试验开始前,试验工作区(标距区)表面经过200~1000号金相砂纸打磨,并经过抛光处理防止试样表面存在预制划痕影响试验,并用丙酮脱脂,无水乙醇清洗、除水干燥处理后待用。

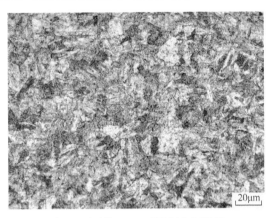

图1 马氏体13Cr不锈钢金相组织

表1 马氏体13Cr不锈钢化学成分组成

元素	C	Si	Mn	P	S	Cr	Mo	Ni	Ti	Cu
$w(\%)$(质量分数)	0.029	0.18	0.18	0.013	0.0011	12.40	2.03	5.65	<0.001	0.031

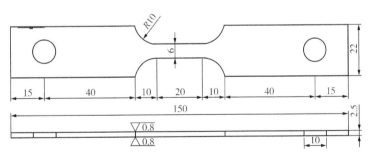

图2 慢应变速率试验中试样尺寸图(单位:mm)

1.2 试验设备及方法:

试验所用设备为某公司生产的PLO-1000慢应变速率应力腐蚀试验机。试验介质为模拟塔里木油田某井地层水,溶液主要离子为Cl^-,实验室用去离子水和分析纯NaCl配制,均在饱和CO_2气氛中进行;通过控制变量法来探究超级13Cr的抗SCC性能,试验参数为:试验温度25℃,应变速率为1×10^{-6} mm/s,Cl^-浓度分别为60g/L(条件1)、90g/L(条件2)、120g/L(条件3)、150g/L(条件4);通过试验参数的变化,分析超级13Cr马氏体不锈钢材料的应力腐蚀敏感性。通过比对试验前后试样,根据应力腐蚀前后的强度变化和断裂延伸率的变化来评定应力腐蚀的程度,如式(1)所示。

$$k\sigma = \frac{\sigma b - \sigma b'}{\sigma b}$$

$$k\varepsilon = \frac{\varepsilon f - \varepsilon f'}{\varepsilon f} \tag{1}$$

式中:σb 和 $\sigma b'$ 分别是试样在基准条件和试验条件下的抗拉强度;εf 和 $\varepsilon f'$ 分别是试样在基准条件和试验条件的应变量。一般情况下,试样的应变敏感性指数越大,该材料-介质体系的应力腐蚀敏感性越强;$k\sigma$ 为以抗拉强度损失率表示的应力腐蚀开裂敏感性指数,$k\varepsilon$ 为以断裂延伸率损失率表示的应力腐蚀开裂敏感性指数,应力腐蚀开裂敏感性指数越大,表明材料的应力腐蚀开裂的倾向性越大,应力腐蚀越严重。

电化学动电位极化曲线采用三电极体系测试,试样为工作电极;参比电极为氯化银电极(Ag-AgCl);对石墨棒为辅助电极,动电极扫描速率为 0.5mV/s。极化曲线是在四种不同 Cl^- 浓度溶液下进行测试,试验温度 60℃,试验时通入 CO_2 气体至饱和,待体系稳定 30min 后进行测试。

1.3 试验后形貌观察

试验后立即取出断裂的试样,防止断口污染或二次损伤,先用去离子水冲洗表面附着的腐蚀产物,然后在超声波清洗仪中使用丙酮溶液清洗断口,以除去腐蚀产物,吹干后放入干燥器中密封保存,在 JSM-6390A 型扫描电子显微镜(SEM)下进行断口形貌和断口侧面形貌观察,分析油管柱的 SCC 敏感性。

2 结果与讨论

2.1 动电位极化曲线

图3为材料在不同 Cl^- 浓度且饱和 CO_2 条件下的极化曲线变化特征,可见在不同 Cl^- 浓度下材料的自腐蚀电位区别不大,变化范围为 -440~ -410mV。阳极曲线均出现稳定的钝化区,钝化电流密度随着 Cl^- 浓度的增大向右移动增大趋势。材料的点蚀电位 E_{pit} 随着 Cl^- 浓度的增大发生明显变化,Cl^- 浓度 ≤90g/L 时,点蚀电位基本相似,$E_{pit} \approx -125mV$;当 Cl^- 浓度 ≥120g/L 时,E_{pit} 明显降低为约 -220mV。因此结果表明在相同试验条件下随着 Cl^- 浓度的逐渐增大,到达 120g/L 时材料的点蚀敏感性增大,阳极极化在相对较低的电位时 Cl^- 即可穿破钝化膜,发生点蚀。

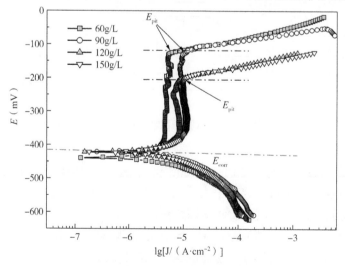

图3 不同 Cl^- 浓度下材料的极化曲线

2.2 SSRT 应力—应变曲线

SSRT 试验是在室温下进行，慢应变速率为 $1×10^{-6}$ mm/s。图 4 是超级 13Cr 不锈钢材料的 SSRT 应力-应变曲线。由图 4 可见不同试验条件下的 SSRT 试验结果，其应力-应变曲线形状基本相同，不同条件下的抗拉强度稍有区别。表 2 为不同试验条件下具体的应力腐蚀断裂参数，可以看出，与空气中的应力腐蚀断裂参数相比，材料在 Cl^- 浓度相对低的 CO_2 饱和的实验溶液条件下的断裂寿命 T_F 和断裂强度 σ_f 与空气中的差值不大。而随着试验条件下 Cl^- 浓度的增大，材料的断裂寿命、断裂强度以及延伸率总体呈现明显降低趋势，表明在试验条件下，由于在 CO_2 气氛中生成的弱酸环境对 Cl^- 的活性具有促进作用，材料在应力以及溶液介质中 Cl^- 的共同作用下表面活性发生变化，影响材料的断裂寿命及强度。当 Cl^- 浓度增大到一定范围时，使得试样裂纹顺利扩展，从而降低了材料的塑性。

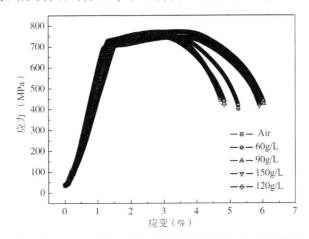

图 4 超级 13Cr 在空气及不同离子浓度的应力—应变曲线

表 2 超级 13Cr 在不同试验条件下应力腐蚀断裂参数

介质	断裂寿命 T_F(h)	断裂强度 σ_f(MPa)	延伸率 δ(%)	强度损失系数 I_σ(%)	延伸率损失系数 I_δ(%)	应力腐蚀指数 I_{SSRT}
空气	66	756	24.50	—	—	—
条件 1	71.15	751	20.92	0.53	14.61	0.04
条件 2	66.72	768	20.87	-1.59	14.82	0.01
条件 3	63.71	748	16.92	1.06	30.94	0.07
条件 4	60.71	728	18.60	3.70	28.16	0.06

将空气条件下和低 Cl^- 浓度的试验结果比较，空气中为潮湿的且含氧的环境，相比低 Cl^- 浓度环境下含氧环境的腐蚀性相对对材料的性能影响较大，导致空气中材料的较早断裂失效。超级 13Cr 在不同 Cl^- 浓度中的延伸率均低于其在空气中的，延伸率损失系数 I_δ 最大为 30.94%，I_δ 的大小排列为条件 1<条件 2<条件 4<条件 3。I_{SSRT} 是应力腐蚀指数，是将 SSRT 试验所获得的各项力学性能指标进行数学处理的结果，相比单项力学性能能更好地反映腐蚀断裂敏感性，计算式为[19]：

$$I_{SSRT} = 1-[(\sigma_{fw}(1+\delta_{fw}))/[\sigma_{fA}(1+\delta_{fA})] \tag{2}$$

式中：σ_{fw} 为在环境介质中的断裂强度，MPa；σ_{fA} 为在空气中的断裂强度，MPa；δ_{fw} 为

在环境介质中的断裂伸长率,%;δ_{fA} 为在空气中的断裂伸长率,%;I_{SSRT} 从 0 到 1 表示应力腐蚀断裂敏感性增加。I_{SSRT} 计算结果见表 2,I_{SSRT} 的值很小,最大为 0.07,I_{SSRT} 的大小顺序排列为条件 2<条件 1<条件 4<条件 3,表明材料的应力腐蚀敏感性逐渐增大。由上述试验结果可知。13Cr 马氏体不锈钢材料在试验条件下 Cl⁻浓度范围内应力腐蚀敏感性较低,但随着 Cl⁻浓度的增大,材料应力腐蚀敏感性逐渐增大。

2.3 断口形貌观察

断口形貌是最能体现材料断裂模式的典型特征。图 5 为材料在不同试验条件下的拉伸断口形貌。从整个宏观形貌观察,不同条件下的断口都有不同程度的颈缩。微观断口基本都以韧窝形貌为主,这代表不同条件下材料的断裂仍然包括韧性断裂在内,但随着试验条件的变化材料也表现出一些有别于典型韧性断裂的特征。

图 5(a)是材料在空气中的 SSRT 断口的宏观形貌,断口发生了明显的颈缩现象,整个微观断口形貌主要以韧窝为主[图 5(b)],同时韧窝间存在着微孔,局部韧窝壁上有明显的蛇形滑移特征,呈现为韧窝—微孔型的韧性断裂,属于典型的韧性断裂特征。图 6(a)是材料在 Cl⁻浓度为 60g/L 的溶液介质中的 SSRT 断口的宏观形貌,试样断口颈缩现象比较明显,整个微观断口形貌[图 6(b)]主要以小而浅的韧窝为主,同时韧窝间存在着大的微孔,大的韧窝壁上有明显的蛇形滑移特征,为韧窝—微孔型的韧性断裂,属于典型的韧性断裂特征。

(a) 宏观断口

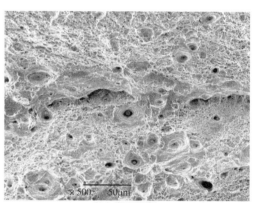
(b) 微观形貌

图 5　材料在空气环境下的断口形貌

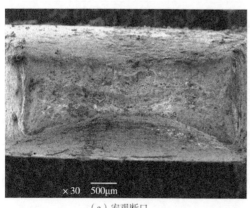
(a) 宏观断口

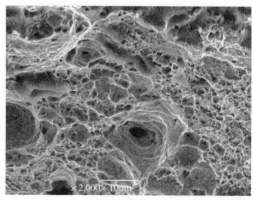
(b) 微观形貌

图 6　材料在 Cl⁻浓度为 60g/L 条件下的断口形貌

图 7(a)是材料在 Cl⁻浓度为 90g/L 的溶液介质中的 SSRT 断口的宏观形貌,试样断口颈缩现象比较明显。整个微观断口形貌[图 7(b)]主要以韧窝为主,左右两边区域的同时韧窝间存在着大的微孔,大的韧窝壁上有明显的蛇形滑移特征,为韧窝—微孔型的韧性断裂,属于典型的韧性断裂特征。图 8(a)是材料在 Cl⁻浓度为 120g/L 的溶液介质中的 SSRT 断口的宏观形貌,试样断口颈缩现象较明显。断口左、右两侧的微观形貌[图 8(b)(c)]主要以浅而大的韧窝为主,同时韧窝间存在着大量微孔,大的韧窝壁上有明显的蛇形滑移特征,属于典型的韧性断裂特征,而断口中间区域[图 8(d)]由小而浅的韧窝和大的孔洞以及裂纹组成,孔洞周围表现为准解理的脆性断裂特征,此现象表明脆性断裂所占比重逐渐增大,以上特征可知材料的断口属于韧—脆混合断口。

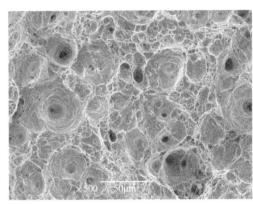

(a) 宏观断口　　　　　　　　　　　　(b) 微观形貌

图 7　材料在 Cl⁻浓度为 90g/L 条件下的断口形貌

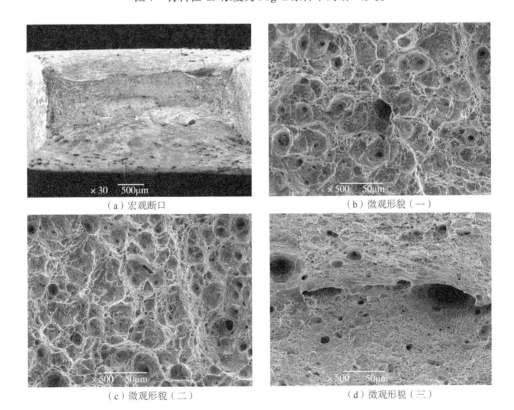

图 8　材料在 Cl⁻浓度为 120g/L 条件下的断口形貌

图9(a)是材料在Cl⁻浓度为150g/L的溶液介质中的SSRT断口的宏观形貌，可见试样断口颈缩比例相对较小。微观观察，断口左侧[图9(b)]和中间区域[图9(c)]的微观形貌由大量韧窝和孔洞组成，大的韧窝壁上有明显的蛇形滑移特征，孔洞周围为准解理的特征，断口右侧区域[图9(d)]表现为河流状的准解理断裂特征，其上分布着许多孔洞，有些孔洞已成为开裂的裂纹源，综合以上特征可知材料的断口属于韧—脆混合断口。因此随着溶液Cl⁻浓度的增大，材料脆性断裂特征明显。

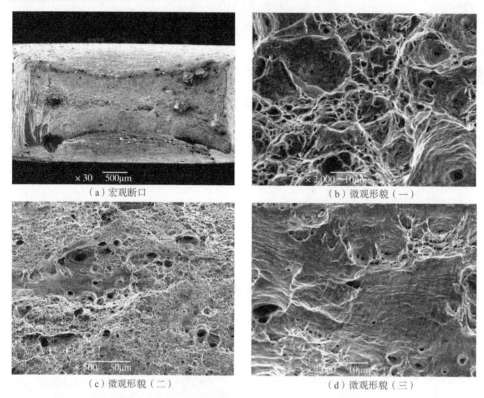

(a)宏观断口　　　　　　　　(b)微观形貌（一）

(c)微观形貌（二）　　　　　　(d)微观形貌（三）

图9　材料在Cl⁻浓度为150g/L条件下的断口形貌

2.4　断口侧面形貌观察

应力腐蚀的一个主要特征就是在主裂纹之外，会有二次裂纹的存在。一般认为，如果在腐蚀介质中拉伸时试样断口侧面有二次裂纹存在，则表明材料SCC是敏感的[20]，因此断口侧面二次裂纹也是评价应力腐蚀敏感性的重要标准[21]。

图10是超级13Cr在空气中的SSRT断口侧面的宏观与微观SEM形貌，可见空拉时材料断口侧面为典型的塑性变形，没有二次裂纹产生，不具有SCC敏感性。分析表明超级13Cr在空气环境下的SSRT试验伴有塑性形变，当应力大于材料的屈服强度后，材料开始发生塑性形变，在材料内部夹杂物、析出相、晶界、亚晶界等部位发生位错塞积，形成应力集中，进而形成微孔洞，且随着形变增加，显微孔洞相互吞并并变大，最后发生颈缩和断裂。

图11是超级13Cr在Cl⁻浓度为60g/L溶液中的SSRT断口侧面的宏观与微观SEM形貌，材料断口侧面可看见二次微裂纹，但比较短而小，说明SCC敏感性相对很低。图12是材料在Cl⁻浓度为90g/L溶液中的SSRT断口侧面的宏观与微观SEM形貌，试样断口侧面可看见二次微裂纹，且这些裂纹均与拉伸方向垂直，大部分的二次微裂纹比较短小，但其中有一条裂纹很长，表明试样的开裂属于穿晶应力腐蚀开裂。综上所述超级13Cr的SCC敏感性有相

对增大的趋势，随着 Cl⁻ 浓度的升高，材料的 SCC 敏感性略有升高。

（a）宏观　　　　　　　　　　　　　　　（b）微观

图 10　超级 13Cr 在空气中的断口侧面形貌

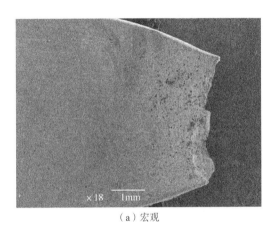 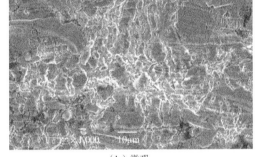

（a）宏观　　　　　　　　　　　　　　　（b）微观

图 11　超级 13Cr 在 60g/L Cl⁻ 溶液中的断口侧面形貌

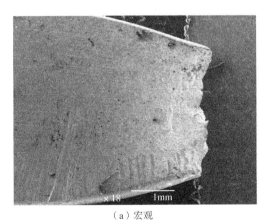 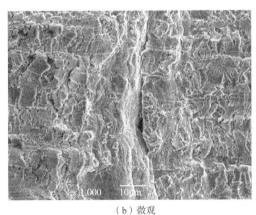

（a）宏观　　　　　　　　　　　　　　　（b）微观

图 12　超级 13Cr 在 90g/L Cl⁻ 溶液中的断口侧面形貌

图 13 是 Cl⁻ 浓度增大为 120g/L 时材料的 SSRT 断口侧面的宏观与微观 SEM 形貌，试样断口侧面可看见大量的二次微裂纹和腐蚀坑，表明在该试验条件下材料点蚀敏感性相对增大，此现象与电化学测试的结果一致（当 Cl⁻ 浓度≥120g/L 时点蚀电位降低）。同时腐蚀坑成为这些二次裂纹的裂纹源，且二次裂纹均与拉伸方向垂直，表明试样的开裂属于穿晶应力腐

蚀开裂，并且随着 Cl⁻ 浓度的进一步升高，超级 13Cr 的 SCC 敏感性进一步增加。材料在 Cl⁻ 浓度为 150g/L 溶液中的 SSRT 断口侧面的宏观与微观 SEM 形貌如图 14 所示，同样可见断口侧面有二次微裂纹和腐蚀坑，数量较多且长短不一，腐蚀坑成为这些二次裂纹的裂纹源，且这些二次裂纹扩展方向基本上均与外加应力轴方向垂直，而且微观断口形貌无完整晶粒暴露，因此可以判断试样的开裂模式属于穿晶 SCC 断裂，并且随着 Cl⁻ 浓度的进一步升高，超级 13Cr 的 SCC 敏感性进一步逐渐增大。

(a) 宏观

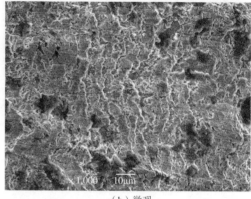
(b) 微观

图 13　超级 13Cr 在 120g/L 的 Cl⁻ 溶液中的断口侧面形貌

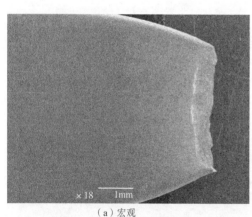
(a) 宏观

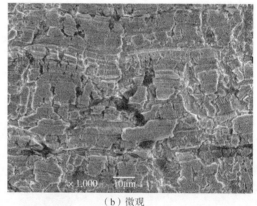
(b) 微观

图 14　超级 13Cr 在 150g/L 的 Cl⁻ 溶液中的断口侧面形貌

2.5　讨论

不锈钢在氯离子含量相对高的环境下，点蚀、缝隙腐蚀、应力腐蚀敏感性都会不同程度增大，因此材料发生晶间应力腐蚀和点蚀的概率增大，进而导致发生应力腐蚀开裂。由于氯离子半径小，容易穿透氧化膜内极小的空隙，到达金属表面，并与金属相互作用形成可溶性化合物，使钝化膜的结构发生变化[22]。因此具有自钝化特性的金属在含有氯离子的介质中经常发生孔蚀，超级 13Cr 不锈钢由于合金元素 Cr、Ni 的易钝性特点在材料表面形成致密的钝化膜，提高了抗腐蚀能力，而 Cr、Ni 元素在钝化膜中是以氧化物形式存在[23]。因此当氯离子优先被金属吸附并从金属表面把氧排掉，可以取代吸附中的钝化离子与金属形成氯化物，而氯化物与金属表面的吸附并不稳定，形成了可溶性物质，结果在基底金属上生成小蚀坑，导致了腐蚀的加速。因此材料的点蚀敏感性随着 Cl⁻ 浓度不同而发生变化。在本试验中，当 Cl⁻ 浓度≤90g/L 时，Cl⁻ 对材料阳极钝化膜的击穿电位相对较高一些，而当 Cl⁻ 浓度增大到 120g/L 时，材料钝化膜的击穿电位相对降低，点蚀敏感性增大。

另一方面由于氯离子在金属表面的吸附性以及提高材料点蚀敏感性，同时材料在慢应变拉伸应力作用下，使得材料表面钝化膜破裂，点蚀成为裂纹的起裂源，因此材料在 CO_2 溶液中，在拉伸应力、氯离子环境下的协同作用下，导致材料发生应力腐蚀开裂。因此从断口形貌看，随着 Cl⁻ 浓度的增大断裂模式转为脆性特征。

3 结论

（1）Cl⁻ 浓度在 60~150g/L 变化时，超级 13Cr 自腐蚀电位变化不明显，当 Cl⁻ 浓度≥120g/L 时，材料点蚀电位明显降低，点蚀敏感性增大。

（2）在 Cl⁻ 浓度≤60g/L 的饱和 CO_2 溶液中，超级 13Cr 在慢应变速率为 1×10^{-6}mm/s 的作用下相对于空气中强韧性变化不大，属于韧性断裂模式；随着 Cl⁻ 浓度的增大，材料的断裂寿命和伸长率均逐渐降低。

（3）饱和 CO_2 溶液中且在慢应变速率为 1×10^{-6}mm/s 条件下，随着 Cl⁻ 浓度的逐渐增大，材料断裂模式由韧性断裂逐渐转变为韧性—脆性混合断裂模式；当 Cl⁻ 浓度≥90g/L 时，断口侧面开始出现二次裂纹，当 Cl⁻ 浓度≥120g/L 时断口侧面二次裂纹及点蚀现象较明显，促进应力腐蚀开裂。

参 考 文 献

[1]《油气田腐蚀与防护技术手册》编委会. 油气田腐蚀与防护技术手册（下册）[M]. 北京：石油工业出版社，1999：471-492.

[2] 张亚明，臧晗宇，董爱华，等. 13Cr 钢油管腐蚀原因分析[J]. 腐蚀科学与防护技术，2009，21(5)：499-501.

[3] 吕祥鸿，路民旭. N80 钢动态和静态 CO_2 腐蚀行为对比研究[J]. 腐蚀科学与防护技术，2003，15(1)：5-8.

[4] NINGSHEN S, SAKAIRI M, SUZUKI K, et al. The corrosion resistance and passive film compositions of 12% Cr and 15% Cr oxide dispersion strengthened steels in nitric acid media[J]. Corrosion Science, 2014(78): 322-334.

[5] ZHAO X H, HAN Y, BAI Z Q, et al. The experiment research of corrosion behaviour about Ni-based alloys in simulant solution containing H_2S/CO_2[J]. Electrochimica Acta. (SCI), 2011, 56(22): 7725-7731.

[6] LIU W Y, SHI T H, LU Q, et al. Failure analysis on fracture of S13Cr-110 tubing[J]. Engineering Failure Analysis, 2018(90): 215-230.

[7] LEI X W, FENG Y R, FU A Q, et al. Investigation of stress corrosion cracking behavior of super 13Cr tubing by full-scale tubular goods corrosion test system[J]. Engineering Failure Analysis, 2015(50): 62-70.

[8] 唐玮，朱华，王勇. 应力腐蚀断口的分形行为[J]. 钢铁研究学报，2007，19(8)：56-58.

[9] QU Y P, WANG R K, WANG C, et al. Experimental study on the stress corrosion cracking behavior of AISI347 in acid chloride ion solution[J]. Results in Physics, 2016(6): 690-697.

[10] SALAZAR M, ESPINOSA-MEDINA M A, HERNA´NDEZ P, et al. Evaluation of SCC susceptibility of super-martensitic stainless steel using slow strain rate tests[J]. Corrosion Engineering, Science and Technology, 2011, 46(4): 464-470.

[11] 吕祥鸿，赵国仙，樊治海，等. 高温高压下 Cl⁻、CO_2 分压对 13Cr 不锈钢点蚀的影响[J]. 材料保护，2004，37(6)：34-37.

[12] ZHAO J J, LIU X B, HU S, et al. Effect of Cl⁻ Concentration on the SCC Behavior of 13Cr Stainless Steel in High-Pressure CO_2 Environment[J]. Acta Metallurgica Sinica, (English Letters), 2019(32): 1459-1469.

[13] 雷冰, 马元泰, 李瑛. 高氯环境中饱和 CO_2 对13Cr油管钢腐蚀行为的影响[J]. 腐蚀科学与防护技术[J], 2013, 25(2): 95-99.

[14] 吴领, 谢发勤, 姚小飞, 等. 13Cr-N80油管钢在不同浓度NaCl溶液中的电偶腐蚀行为[J]. 材料导报, 2013, 27(12): 117-120.

[15] BANAS J. Effect of CO_2 and H_2S on the composition and stability of passive film on iron alloys in geothermal water [J]. Electrochim Acta, 2007(52): 5704-5710.

[16] KRISHNAN, VARSHNEY A, PARAMESWARAN V, et al. Effect of Dynamic Change in Strain Rate on Mechanicaland Stress Corrosion Cracking Behavior of a Mild Steel[J]. Journal of Materials Engineering and Performance, 2017(26): 2619-2631.

[17] VELOZ M A, GONZ'ALEZ I. Electrochemical study of carbon steel corrosion in buffered acetic acid solution with chlorides and H_2S [J]. Electrochim Acta, 2002(48): 135-141.

[18] DING J H, ZHANG L, LU M X, et al. The electrochemical behaviour of 316L austenitic stainless steel In Cl⁻ containing environment under different H_2S partial pressures[J]. Applied Surface Science, 2014(289): 33-41.

[19] 路浩东, 袁鸽成, 冷文兵, 等. 极化电位对5083铝合金型材应力腐蚀行为的影响.[J]. 广东工业大学学报, 2010, 27(2): 43-46.

[20] 褚武扬, 乔利杰, 陈奇志. 断裂与环境断裂[M]. 北京: 科学出版社, 2000: 25-31.

[21] 刘明, 张瑜. X80钢在近中性pH溶液中的应力腐蚀开裂行为[J]. 管道技术与设备, 2016(3): 4-6.

[22] 张鸣伦, 王丹, 王兴发, 等. 海水环境中Cl⁻浓度对316不锈钢腐蚀行为的影响[J]. 材料保护, 2019, 52(1): 34-39.

[23] FAN Z, WANG Z Y, LIU J Y, et al. Influence of Elemental Sulfur on the Corrosion Behavior of Alloy G3 in H_2S+CO_2 Saturated Chloride Solution[J]. Rare Metal Materials and Engineering, 2019, 48(10): 3167-3174.

本论文原发表于《材料保护》2021年第54卷第1期。

基于矫顽力的早期蠕变损伤智能诊断方法

李文升[1,2]　吕运容[3]　尹成先[1,2]　李伟明[3]

(1. 石油管材及装备材料服役行为与结构安全国家重点实验室；
2. 石油管工程技术研究院；3. 广东石油化工学院广东省石化装备故障诊断重点实验室)

摘　要：乙烯是国民经济的重要物资，其生产的核心单元之一就是乙烯裂解炉，而裂解炉炉管则是引发裂解炉安全问题的重要单元，蠕变则是引发乙烯裂解炉管失效的最重要机理之一。按照渗碳体形态特征和蠕变孔洞的状态，裂解炉管的高温损伤可分为7个阶段，通过电子扫描显微镜及能谱分析发现，蠕变早期损伤在第4阶段才发生，而该阶段的矫顽力特征曲线显示，在损伤进入该阶段时，矫顽力曲线会出现一个U形拐点，因此可以利用这一特征来进行蠕变早期阶段的损伤诊断。利用PAA和ODSS算法对实测矫顽力曲线进行了降噪和光滑处理，从而实现了蠕变早期阶段的智能诊断。

关键词：蠕变；早期蠕变阶段；矫顽力；饱和磁化强度；损伤诊断

乙烯作为重要的化工原料，目前为止已经是世界上用量最大的化学品之一[1]。目前，我国已成为世界上仅次于美国的第二大乙烯生产国[2]，并且仍处于一个迅猛发展的时期[3]。目前的乙烯的生产工艺仍以裂解石油烃为主[4]，且绝大部分采用使用管式炉裂解技术[5]。作为乙烯裂解炉最核心部件的裂解炉管，其花费占到乙烯装置总投资的10%，并且若在运行过程中发生炉管失效，则会带来巨大的经济损失，甚至人员伤亡[6]。虽然裂解炉管的设计使用寿命一般在10^4 h以上，但是其寿命期内的失效事故时有发生，其中主要原因之一就是蠕变损伤[7]。目前，国内外研究主要集中在损伤机理定性研究[8]和检测技术研究[9]上，其中涡流检测主要用于表面或近表面缺陷检测[10-12]，并可结合电磁场数学模型进行渗碳层厚度等参数的软测量[13-17]。而矫顽力作为一种电磁参数，由于其大小与金属含碳量存在明显密切关系[18]，因此被用来进行渗碳损伤机理的检测技术方法开发[19-21]。另外，巴克豪森噪声法[22]和超声检测技术[23-25]也被用于开展高温损伤检测研究，但目前还无法实际应用。

1　乙烯裂解炉管各阶段高温损伤特征研究

乙烯裂解炉管高温损伤过程中，蠕变与渗碳往往为伴生关系，蠕变孔洞的位置与渗碳晶界有关，因此本文以渗碳体形态特征和蠕变孔洞的状态为划分标准，根据电子扫描显微镜和能谱分析结果，将炉管从原始状态到失效之间分为7个阶段，即基本完好阶段、晶内二次渗碳体析出阶段、骨架状晶界开始粗化溶解阶段、蠕变早期萌生阶段、骨架状晶界全面溶解阶段、失效临界状态阶段和失效状态阶段，并以晶界形态、晶内二次渗碳体、晶内Cr/Ni元素

基金项目：国家重点研发计划资助项目(2017YFF0210406)。
作者简介：李文升(1986—)，男，山东诸城人，博士，工程师，研究领域为多相冲刷腐蚀相关研究。

分布、蠕变孔洞形态、晶内 Cr/Ni 含量 5 个特征描述各阶段的基本特征，如表 1 所示。

表 1　各高温损伤状态的材料微观特征描述

损伤状态	基本特征
基本完好阶段(A)	晶界形态：骨架和块状形态，且纤细、完整 晶内二次渗碳体：几乎无任何析出的二次渗碳体 晶内 Cr/Ni 元素分布：Cr 含量符合标准要求，且分布基本均匀，仅在晶界有稍微集中的分布 蠕变孔洞：无 晶内 Cr/Ni 含量：Cr 含量(wt%)较扫描范围内的平均值低 2%~2.5%，Ni 含量(wt%)扫描范围内的平均值高 1.5% 以下
晶内二次渗碳体析出阶段(B)	晶界形态：骨架和块状形态，且较之前有所粗化，但仍然完整 晶内二次渗碳体：少量、点状的细小二次渗碳体析出 晶内 Cr/Ni 元素分布：Cr 含量无明显变化，且分布基本均匀，仅在晶界有稍微集中的分布 蠕变孔洞：无 晶内 Cr/Ni 含量：Cr 含量(wt%)较扫描范围内的平均值低 2.5%~4%，Ni 含量(wt%)扫描范围内的平均值高 1.5%~2.0%
骨架状晶界开始粗化溶解阶段(C)	晶界形态：骨架状晶界开始溶解且开始明显变粗，出现断续的不连续晶界 晶内二次渗碳体：弥散大量、点状的细小二次渗碳体，且数量和密度逐渐增加 晶内 Cr/Ni 元素分布：Cr 含量较最初状态微量减少，差别不大，但晶界有明显的集中分布，同时晶界上开始出现相对孤立的贫 Ni"黑洞" 蠕变孔洞：无 晶内 Cr/Ni 含量：Cr 含量(wt%)较扫描范围内的平均值低 6%~8%，Ni 含量(wt%)扫描范围内的平均值高 3%~4%。
蠕变早期萌生阶段(D)	晶界形态：骨架状晶界大部分已发生溶解，且粗化明显，整体呈现为由块状渗碳体组成的链状晶界 晶内二次渗碳体：弥散大量碳化物，但是细小碳化物明显出现合并和粗化 晶内 Cr/Ni 元素分布：Cr 含量减少百分量保持在个位数以内，且晶界 Cr 元素集中程度进一步增加，同时在晶界上出现连续的贫 Ni"黑洞" 蠕变孔洞：部分晶界边缘可以看到独立的蠕变孔洞，但孔洞并未连接成链状 晶内 Cr/Ni 含量：Cr 含量(wt%)较扫描范围内的平均值低 7.5%~8.5%，Ni 含量(wt%)扫描范围内的平均值高 4.0%~5.0%
骨架状晶界全面溶解阶段(E)	晶界形态：大量溶解并粗化为断续的条块状晶界 晶内二次渗碳体：二次渗碳体相互融合，粗化并团聚为大块的团状粗化二次渗碳体 晶内 Cr/Ni 元素分布：Cr 含量减少量达到双位数，但其含量仍为两位数，同时晶界和晶内出现大量且连续的贫 Ni"黑洞" 蠕变孔洞：大量独立的蠕变孔洞，开始出现少量连续蠕变孔洞 晶内 Cr/Ni 含量：Cr 含量(wt%)较扫描范围内的平均值低 8.0%~15.0%，Ni 含量(wt%)扫描范围内的平均值高 5.0%~13.0%
失效临界状态阶段(F)	晶界形态：残留粗化晶界和晶内二次渗碳体重新连接形成网状晶界 晶内二次渗碳体：二次渗碳体进一步相互溶解，团聚，且数量大量减少 晶内 Cr/Ni 元素分布：Cr 含量明显减少至个位数，导致晶内出现明显的贫 Cr"黑洞"，同时晶界上出现呈条块状、网格状的贫 Ni"黑洞" 蠕变孔洞：大量连续蠕变孔洞出现在晶界上(孔径小于 $10\mu m$) 晶内 Cr/Ni 含量：晶内 Cr 含量(wt%)较扫描范围内的平均值 15.0%~22.0%，Ni 含量(wt%)扫描范围内的平均值高 12.0%~18.5%

续表

损伤状态	基本特征
失效状态阶段(G)	晶界形态：网状晶界进一步粗化，但较之前的状态变化不明显 晶内二次渗碳体：二次渗碳体进一步团聚且粗化加重，且数量持续减少 晶内 Cr/Ni 元素分布：Cr 含量继续减少，导致晶内出现块状的贫 Cr "黑洞"，同时晶界上出现网格化的贫 Ni "黑洞" 蠕变孔洞：大量连续蠕变孔洞几乎布满晶界，晶内也出现连续孔洞，(孔径小于 10μm)，甚至出现微裂纹 晶内 Cr/Ni 含量：Cr 含量(wt%)较扫描范围内的平均值低 20.0%~26.5%，Ni 含量(wt%)扫描范围内的平均值高 17.0%~21.0%

从表 1 可以看出，晶内 Cr/Ni 含量是唯一随着损伤程度增加连续变化的特征参量，即高温损伤程度越深，晶内的 Cr 含量越低，而 Ni 含量越高，因此对于一个高温损伤过程而言，可以用晶内 Cr 含量或 Ni 含量与平均 Cr 含量或 Ni 含量之差表征损伤程度。

蠕变早期萌生的 SEM 图像如图 1 所示，蠕变孔洞仅在 D 阶段开始出现。从图 2 所示的蠕变早期特征可知，独立蠕变孔洞是蠕变进入早期阶段的主要微观特征之一，这与表 1 所示的蠕变早期特征一致，故可认为 D 阶段为蠕变早期萌生阶段。

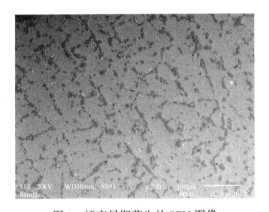
图 1 蠕变早期萌生的 SEM 图像

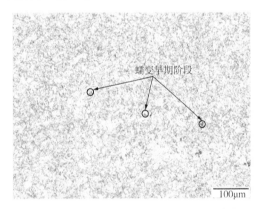
图 2 蠕变早期阶段的独立孔洞特征图[25]

2 乙烯裂解炉管各损伤状态的磁特征信号分析

从乙烯裂解炉管的磁特征来说，材料中的 Ni 元素属于铁磁性材料，但是 Cr 元素和奥氏体相铁，即 γ-Fe 相，均为顺磁性，因此在各元素均匀分布状态下，含量占优的 Cr 元素和 γ-Fe 使得材料整体表现出顺磁性。但是，随着高温损伤过程的不断深入，晶内脱碳 C 使晶内的 γ-Fe 转变为铁磁性的 α-Fe，脱 Cr 则导致晶内 Ni 元素的铁磁性无法得到抑制，最终因晶内铁磁性的 Ni 和 α-Fe 含量增加而使材料的磁特性由顺磁性转化为铁磁性。在极端情况下，晶内的 Cr 含量接近零的"黑洞"处可以形成接近强磁性坡莫合金的 Ni/Fe 比例。

因此，晶内 Ni/Cr 比即可以表征材料高温损伤严重程度，也可以表示材料的磁特性变化程度，从而建立材料高温损伤严重程度与材料磁特征信号之间的定量关联。本文中采用晶内 Ni/Cr 比与材料 Ni/Cr 比均值(含晶内和晶界在内)之差作为高温损伤过程微观特征变量，研究高温损伤程度与材料磁特征信号之间的定量关系和变化规律。

将矫顽力做为磁特征信号，针对 7 个损伤阶段的多个试样进行测量，并以晶内 Ni/Cr 比与材料 Ni/Cr 比均值(含晶内和晶界在内)之差作为微观损伤变量，建立起高温损伤全寿命周期内矫顽力的变化规律曲线，如图 3 所示，其中 A~F 分别表示表 1 中从基本完好阶段到失效临界状态阶段的矫顽力测量结果。

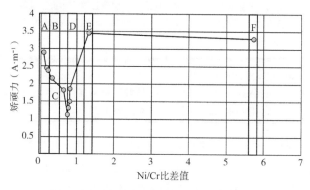

图 3　矫顽力随高温损伤程度的变化规律

从图 3 中可以发现，蠕变早期阶段，矫顽力开始由下降趋势转变为上升趋势，出现一个 U 形曲线，其原因在于 D 阶段晶内大量形成密集的二次渗碳体，由于这些二次渗碳体的直径均较小，会产生钉扎效应，大量的钉扎效应作用导致该阶段矫顽力快速上升。因此，可以将这个 U 形拐点作为识别乙烯裂解炉管早期蠕变损伤的特征信号。

3　乙烯裂解炉管蠕变早期智能诊断算法研究

虽然矫顽力规律曲线中的 U 形拐点可以作为识别蠕变早期阶段的特征信号，但是如图 4 所示，真实的测量曲线中仍然存在很多干扰拐点信号。为在实际应用中有效识别蠕变早期特征信号，需引入数学优化算法，有效剔除干扰信号的影响，从而实现蠕变早期阶段的智能诊断。

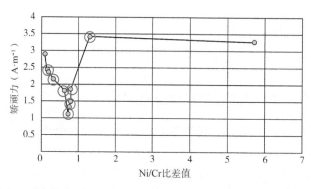

图 4　高温损伤程度矫顽力变化规律中的特征拐点和干扰拐点

为排除干扰信号的影响，首先必须分析干扰信号与特征信号的数据特征，采用线性数值差分法识别各拐点，并对拐点的变化差分率进行定量数值分析，其分析算法如式(1)至式(3)所示。

$$r_k = \arctan[(y_2-y_1)/(x_2-x_1)] \quad k=0,1,\cdots,n \quad (1)$$

$$r_k = r_0 + kh \quad k=0,1,\cdots,n \quad (2)$$

$$\Delta f(r_k) = f(x_{k+1}) - f(x_k) \quad (3)$$

式中：h 为差分步长。

根据式(1)至式(3)，并将图4中所示拐点从左到右分别标号为0~8号，则各拐点的线性差分结果如表2所示。从表2中可以看出，矫顽力测量结果的数据特征为干扰信号的差分结果远明显比特征信号要低一个数量级（编号4和8），利用这一数据特征，只要通过合理弱化差分数据特征的方法，即可消除差分特征明显微弱的干扰信号，又保留特征信号的现有特征。本文采用分段聚集近似算法(Piecewise Aggregate Approximation，PAA)来消除噪声拐点信号。

表2 各拐点差分数值特征

拐点编号	$\Delta f(r_k)$
0	0.4298
1	0.0857
2	0.2428
3	0.5855
4	2.8859
5	0.0891
6	0.1550
7	0.2761
8	1.2931

假设在线检测周期为 N，每个单元(如炉管)每次测量 m 个点位，每个点位获得 k 个数据，同时，令被检测单元第 k 次测量的第 i 个点位的第 j 个测量结果为 d_{kij}，$k=1, 2, 3\cdots, n$，其中 n 为测量的总次数，$i=1, 2, 3\cdots, u$，其中 u 为每次测量各单元所选测点的总数，$j=1, 2, 3\cdots, v$，其中 v 为每个测点的数据总数。则在 d_{kij} 和 $d_{(k-1)ij}$ 之间用于弱化差分的填充数据可由式(4)加以计算。

$$d_{kij-l} = \frac{d_{kij} - d_{(k-1)ij}}{Q}(1-Q) + d_{kij} \tag{4}$$

式中：d_{kij-l} 为第 k 组填充数据的第 l 个填充数据；Q 为 d_{kij} 和 $d_{(k-1)ij}$ 之间所需填充数据的个数。

在数据处理过程中，则利用 d_{kij} 和 d_{kij-l} 并以 M 为数据窗口数量进行原始数据处理，得到用于数据分析的中间数据 $\overline{d}_k = \{\overline{d}_1, \overline{d}_2, \overline{d}_1\cdots, \overline{d}_M\}$，其中 $M<n$，如式(5)所示。

$$\overline{d}_k = \frac{M}{n}\sum_{k=n/M(k-1)+1}^{(n/M)[n+(n-1)l]}\sum_{j=1}^{v}\sum_{i=1}^{u}d_{kij} \tag{5}$$

经过降噪处理的数据点及对应的折线如图5所示，从图中可以看出，经过数据处理，除了特征拐点之外，干扰拐点基本都被剔除。

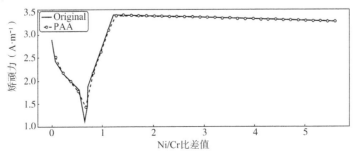

图5 数据处理前后矫顽力测量特征曲线的对比

为了更为智能的提取蠕变早期的特征拐点，针对式(5)处理之后的中间数据，采用一维光滑样条算法(One-dimensional Smoothing Spline，ODSS)拟合出 d_{kij} 与测量时间跨度 t 之间的规律曲线，其拟合公式如式(6)和式(7)所示：

$$RSS(f, \lambda) = \sum_{s=1}^{M} [\bar{d}_s - f(t_s)]^2 + \lambda \int [f''(x)]^2 dx \tag{6}$$

$$f(t) = \sum_{s=1}^{M} t_s \theta_s \tag{7}$$

式中：θ_s 为拟合公式 $f(x)$ 的参数集合；λ 为平滑拟合参数。

确定参数的方程如式(8)所示：

$$\frac{\partial RSS}{\partial \theta_s} = 0 \tag{8}$$

经过式(4)至式(8)处理得到的光滑数据曲线，如图6所示。

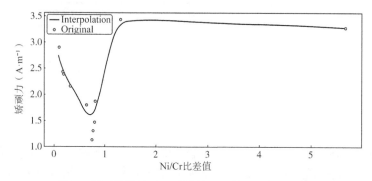

图6　经过数据处理和平滑处理后的测量特征曲线

对于图6所示的光滑曲线，蠕变早期阶段智能诊断算法可以通过如式(9)和式(10)所示的方程判断矫顽力测量曲线是否存在 U 形曲线的极值点，如果存在极值点，则判断材料的高温损伤已进入蠕变早期阶段，根据表1所示的内容，材料内部可能已经开始出现少量的独立蠕变孔洞，材料大量二次渗碳体开始强化材料，材料韧性下降，从而实现蠕变早期阶段的智能诊断。

$$\frac{df}{dt} = 0 \tag{9}$$

$$\frac{d^2 f}{dt^2} > 0 \tag{10}$$

4　结论

乙烯裂解炉管，按照渗碳体形态特征和蠕变孔洞的状态，从原始状态到失效之间大致可分为7个状态，即基本完好阶段、晶内二次渗碳体析出阶段、骨架状晶界开始粗化溶解阶段、蠕变早期萌生阶段、骨架状晶界全面溶解阶段、失效临界状态阶段和失效状态阶段，各状态之间可按照晶界形态、晶内二次渗碳体、晶内 Cr/Ni 元素分布、蠕变孔洞形态、晶内 Cr/Ni 含量5个特征参量加以区分。其中蠕变早期萌生阶段是蠕变早期阶段，进入该阶段之前，矫顽力规律曲线会出现一个 U 形拐点，该特征可以作为识别蠕变早期阶段的物理特征信号，但是实际的矫顽力测量曲线，除了该 U 形特征信号之外，还存在大量干扰拐点信号，这些干扰信号较特征信号之间存在明显的数量差距，根据这个特点本文采用 PAA 和 ODSS

算法处理，从而消除所有干扰信号，并实现 U 形特征信号的智能识别，进而实现蠕变早期损伤的智能诊断。

参 考 文 献

[1] 刘方涛．我国乙烯工业现状及发展前景[J]．化学工业，2010，28(1)：1-4．
[2] 钱伯章．中国乙烯工业市场和原料分析[J]．中外能源，2011，16(6)：62-73．
[3] 王红秋，郑轶丹．我国乙烯工业强劲增势未改[J]．中国石化，2019(1)：27-30．
[4] 王可，张洪林．当代乙烯技术进展[J]．当代化工，2006(2)：117-120，125．
[5] 王国清，曾清泉．裂解技术进展[J]．化工进展，2002(2)：92-96．
[6] 曹菊勇．多种工况下乙烯裂解管蠕变损伤模拟[D]．天津：天津大学，2012．
[7] 郑显伟．我国乙烯裂解炉辐射炉管的使用状况[J]．压力容器，2013，30(5)：45-52．
[8] 黄雷．裂解炉炉管长期高温组织损伤研究[D]．大庆：东北石油大学，2011．
[9] 沈功田．中国无损检测与评价技术的进展[J]．无损检测，2008(11)：787-793．
[10] 程晓敏．材料的渗碳层特性及其涡流测试系统的研究[D]．武汉：武汉理工大学，2003．
[11] 吴步宁，陈志祥．钢材硬度涡流无损检测技术的研究[J]．无损检测，2000(6)：243-245，268．
[12] 李强，刘学文．铁磁材料表面硬度无损测量方法的研究[J]．无损检测，2001(3)：93-95．
[13] 靳亚鹏，萨殊利，温伟刚．涡流无损检测淬火钢轨踏面硬度的数值分析[J]．北方交通大学学报，2001(1)：80-83．
[14] 程晓敏，方华斌，游风荷，等．涡流检测渗碳层深度中磁感应强度的计算[J]．武汉理工大学学报(信息与管理工程版)，2003(4)：174-177．
[15] 方华斌．涡流检测渗碳层深度的有限元分析[D]．武汉：武汉理工大学，2003．
[16] 程晓敏，刘凤娟，方华斌．渗碳材料涡流测试系统的研究[J]．无损检测，2004(8)：396-398．
[17] 陈祯．小波分析在渗碳层深度涡流检测信号处理中的应用[D]．武汉：武汉理工大学，2003．
[18] 史新民，李耀明．基于矫顽力的电磁无损检测仪的设计[J]．南通职业大学学报，2007(4)：61-64．
[19] 贾健明．钢铁件渗碳层深度的电磁无损检测[J]．机械工人(热加工)，2003(12)：24-25．
[20] 刘德宇，韩利哲，湛晓林，等．磁滞无损评估技术在裂解炉管渗碳检测中的应用[C]．远东无损检测新技术论坛——基于大数据的无损检测，2015．
[21] 谢国山，韩志远，付芳芳，等．乙烯裂解炉管渗碳损伤磁滞检测方法的影响因素研究[J]．化工机械，2018，45(1)：35-39．
[22] 屈辰鸣．18CrNiMo7-6 钢渗碳层深度的巴克豪森噪声无损检测方法研究[D]．郑州：郑州大学，2018．
[23] 杨那．高温炉管数字化超声检测系统研究[D]．大连：大连理工大学，2012．
[24] 曲明盛．高温炉管无损检测系统的研制与开发[D]．大连：大连理工大学，2013．
[25] KUMAR H, LAMBERT J W, NAGESWARAN C, et al. Towards a Viable Field Deployable Ultrasonic Technique For Detection of Type IV Creep Damage In Csef Steels at an Early Stage[C]. Proceedings of the ASME 2019 Pressure Vessels & Piping Conference, San Antonio Texas in USA, 2019.

本论文原发表于《机电工程技术》2021 年第 50 卷第 1 期。

井下用非金属复合材料连续管研究进展

李厚补[1]　张学敏[2]　马相阳[3]　丁楠[1]　丁晗[1]　戚东涛[1]

(1. 中国石油集团石油管工程技术研究院石油管材及装备材料服役行为与结构安全国家重点实验室；2. 长安大学材料科学与工程学院；3. 长庆油田分公司第一采油厂采油工艺研究所)

摘　要： 非金属复合材料连续管具有优异的耐腐蚀性能，足够的承压强度和抗拉伸、耐弯曲性能，又可连续成型，能够大幅降低施工人员劳动强度，提高起下井作业效率，因此成为传统钢制油管的良好替代品。基于良好的可设计性及连续成型工艺特征，非金属复合材料连续管为动力缆、信号缆、辅热缆等缆线的复合敷设提供了有利条件，制备的智能化油管在国内油田也得到成功应用。本文总结了国内外井下注水、采油用非金属复合材料连续管产品研发现状，分析了复杂油井工况下遇到的主要技术问题，从结构设计及选材、全寿命周期检验评价、配套施工工艺及装备、智能化制造及控制和标准化建设等方面提出了规范化研究建议，力求助推国内井下用非金属复合材料连续管产品和技术的快速进步。

关键词： 非金属复合材料；连续管；油井；智能管

油管是油井中用于采油、采气、注水和酸化压裂的管子，它是地下油气层到地表的通道，在油气田勘探开发中发挥着极其重要的作用。当前，国内外油田使用的油管大多是钢材制成。工作时钢制油管不但要承受拉伸力、压缩力和内外压力，还要经受原油、气体、油田污水、土壤等各种介质的腐蚀作用[1]。随着油气田的深入开发，地层液体中的H_2S、CO_2、Cl^-、水及微生物等伴生物质含量逐渐升高，导致井下钢制油管腐蚀加剧，穿孔失效现象频现，管材使用寿命大大降低，成为制约油气田安全、高效生产的巨大瓶颈[2]。

非金属复合材料连续管在具有优异耐腐蚀性能的同时，通过结构设计，可获得足够的抗内外压强度和抗拉伸、耐弯曲性能，同时又因其连续成型，单根可达数百米、接头少，安装快速简单等一系列优点，使其成为传统钢制油管的良好替代品[3]。在制造技术和应用技术逐渐成熟的基础上，随着油井注水、采油和注CO_2等对耐腐蚀管材需求的不断增加，非金属复合材料连续管的应用迎来广阔的发展前景。本文综述了国内外非金属复合材料连续管在井下注水、采油系统中的研发及应用现状，汇总提出该类产品的规范化研究方向及建议，为井下管柱的选择和应用提供借鉴。

基金项目：中国石油天然气集团公司基础研究和战略储备技术研究基金项目"非金属复合材料智能连续油管研究"(2017D5008)，长安大学中央高校基本科研业务费专项基金"油气输送用热塑性塑料的气体渗透机理及失效控制研究"(300102310201)。

作者简介：李厚补(1981—)，博士，高级工程师，2009年毕业于西北工业大学材料学专业。主要从事油气田用非金属与复合材料管材研究。电话：029-81887680，E-mail：lihoubu@cnpc.com.cn。

1 注水用非金属复合材料连续管

高品质非金属复合材料连续管比金属连续油管更轻、更耐疲劳和抗腐蚀,比钛合金连续油管对 H_2S、CO_2 环境的适应能力更强。井下非金属复合材料连续管通常由内衬层、中间层和外保护层等构成[4]。内衬层通常采用热塑性塑料管,如高密度聚乙烯(HDPE)、交联聚乙烯(PEX)、尼龙(PA)或聚偏氟乙烯(PVDF)等;基于可设计性强的特点,中间层可采用金属(如钢丝)或非金属(如纤维)材料或其复合材料作为增强结构层;外保护层通常采用热塑性塑料,起到防止外损伤的作用。按用途分,非金属复合材料连续管可用于钻井、增注、完井、重新完井、打捞等井下作业管,也可用作衬管、注入管、地面管线管等油田永久管线。

1.1 国外研究情况

国外对于复合材料连续油管的研究开始于 20 世纪 80 年代末[5]。Conoco 公司率先研制出一种耐高压、长距离、无腐蚀的复合材料连续油管,取代原有近海注水钢制管线,并开展了静液压、拉伸、弯曲、疲劳等各类性能测试[6]。与美国应用材料公司(AMAT)合作开发的数百英尺长小口径高性能复合材料连续油管盘卷在直径为 2.7m 的滚筒上,用于挪威 Sandefjord 油田注水和注化学剂领域。近年来,美国 Fiberspar、挪威 NAT Compipe 和美国 Hydril 成为 3 个最重要的非金属复合材料连续管制造商。

如图 1 所示,美国 Fiberspar 公司生产的非金属复合材料连续管内衬层采用 HDPE、PEX 或 PVDF;结构层(同时作为外保护层)为纤维(高强度玻璃纤维、碳纤维或芳纶纤维等)浸渍环氧树脂缠绕成型[7]。该公司产品主要有三类:管线管、修井用连续油管和油田固定装置管,内径范围为 50.8~142mm,耐压等级为 5.1~17.2 MPa,工作温度为 −30~150℃。Fiberspar 的首批非金属复合材料连续管于 1999 年投放市场,主要用于铺设天然气举升管线、油气修复管以及地面临时管线等。

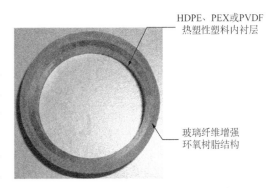

图 1 Fiberspar 非金属复合材料连续管

NAT Compipe 是挪威主要的非金属复合材料连续管供应商,已生产出内径 76.2mm、全长 15km 世界最长的非金属复合材料连续管。该产品采用 PEX 作为内衬管,表面缠绕玻璃纤维和环氧树脂复合层,额定工作压力为 38MPa,设计寿命 20 年[8]。

美国 Hydril 生产的非金属复合材料连续管已成系列,为进一步提高管材的耐化学介质性能和耐温性,该产品采用壳牌化学公司生产的 Carilon 系列酯族聚酮聚合物作为衬管,价格在 13Cr 和 22Cr 管之间,具备 Hastelloy G3 或 Inconel825 合金的高性能[5,8]。

1.2 国内研究情况

非金属复合材料连续管在我国油气田地面集输系统得到规模化推广应用。2011 年,基于地面用管材的研发及应用经验,河北恒安泰油管有限公司研制的非金属复合材料连续管产品首次在长庆油田混注井中开展了现场试验,最大井深 2743m,最大井斜 45.3°[9]。截至 2018 年 3 月,在长庆油田区域内使用的混注型非金属复合材料连续管共计 150 余口井,管线累计长度近 300 km。试验应用的混注型非金属复合材料连续管产品主要由聚合物内衬层、增强层、抗拉层、外护套构成,其中增强层和抗拉层采用涤纶纤维。不同服役时间起井后的性能评价结果表明，非金属复合材料连续管各项指标仍符合油田使用要求，显示出良好的适用性[10]。

2017年，河北恒安泰油管有限公司在混注型非金属复合材料连续管产品基础上，通过增设抗压层及功能层开发出适用于分层注水的分注型非金属复合材料连续管[11]和水源井用采水管道，并在长庆油田第一采油厂王394-51井、王394-52井和长庆油田五里湾二区24号井成功进行了带套管保护封隔器下井试验和电潜泵采水试验，下入深度分别为1478m、1493m和450m，当前仍在正常运行中。

近些年，威海鸿通管材股份有限公司开发的非金属复合材料连续管也在井下注水领域得到成功应用。该产品结构分为6层，由内到外分别为内衬层、环向增强层、骨架层、右旋纵向拉伸层、左旋纵向拉伸层和外保护层[12-13]。其中，环向增强层、骨架层和纵向拉伸层采用了玻璃纤维增强热固性树脂（玻璃钢）复合材料带材。产品口径范围为DN40mm～DN65mm，压力等级范围为1.6~32 MPa，使用温度范围为-30~110℃，先后在胜利油田临盘采油厂的临71-斜12井、临22-3井、盘12-斜48井成功完成了混注试验，在长庆油田采油九厂的旗93-80井成功进行了分层注水试验。

2 采油用复合材料智能连续管

目前，我国陆地采油技术以有杆抽油为主，随着斜井、水平井等的增多，有杆抽油更加困难，普遍存在偏磨、断杆、空抽、结蜡、结垢等问题。解决这些问题较为理想的方法是采用无杆采油方式。近年来，潜油电泵采油在国内各油田得到广泛应用，成为最为常用的无杆采油方式。一般情况下，潜油电泵采油是以金属管一节一节续接，然后在金属管外侧捆绑铠装电缆提供动力。这种工作方式只能解决管材的偏磨现象，而结垢、结蜡、腐蚀情况依然存在；同时在起下油管过程中，发生的碰撞及井液的腐蚀都会对外置电缆造成损害，从而成为检泵的重要原因[14]。

另外，如果能实现对潜油电泵或管材系统的实时监测，可随时掌握油井生产动态，帮助作业者改善电潜泵性能、优化管材运行工况，进而可以降低作业成本并提高油井采油效率。因此，在井下油管捆绑动力电缆的同时，仍需添加温度、压力等监测装置，并通过信号缆将测试信号传输至地面二次仪表，进行实时记录和读数，从而实现智能化的采油和控制。

基于此，研究开发出既能解决传统碳钢油管耐腐蚀性、抗结垢结蜡性和耐磨性能差的问题，又能消除捆绑线缆的烦琐工艺，避免线缆的碰撞损伤和井液侵蚀损坏，同时还兼备电源动力正常供给和监测信号高效传输功能的复合材料智能连续管，成为当前无杆采油工艺发展的新思路。

2.1 国外研究情况

20世纪90年代初，美国Smart Pipe Company. Inc公司生产的SmartPipe®产品可实现对腐蚀破坏的金属旧管线进行非开挖穿插修复[15]。SmartPipe®的管体是多层柔性结构，由于分别采用了轴向和环向增强材料，因此可以采用更高的拖拽牵引力，而且可以独自承载内部流体的输送压力，降低了对外部钢管的剩余强度要求。与此同时，SmartPipe®管体内部植入了光纤，可对管道的运行状态实施监测，及时发现并定位渗漏或外界破坏位置，是采油用复合材料智能连续管的雏形（图2）。

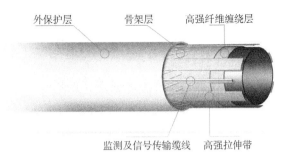

图 2 SmartPipe®结构图

1998年，Fiberspar公司为连续油管钻井系统设计并建造了带动力供给和数据传输且连续长度超过6400m的非金属复合材料智能连续油管（图3）。该产品直径为73mm（2⅞in），设计压力50MPa。产品制造时，在管体内植入6个20#AWG导线（美国线规）[7]，并通过在热塑性塑料内衬管上添加玻璃纤维耐磨层，提升了管体抗外压能力。管体最外部使用的黏合聚丙烯材料具有优良的耐磨性能，在保护导线的同时，解决了复合管外部碰撞及腐蚀问题。

在此基础上，Fiberspar公司于2002年设计了一种新型非金属复合材料智能连续油管并通过现

图 3 Fiberspar智能管管体结构

场试验。新的非金属复合材料智能连续油管在保留大部分设计性能基础上，通过使用玻璃纤维作为增强材料和使用HDPE作为内衬材料使其制造成本更低，管材的可盘绕性大大提升。

基于现有技术成果，Fiberspar公司还在持续发展适合不同钻井环境的非金属复合连续油管技术[7]：在实现数据和电力传导的同时，持续研发大口径产品；开发适用于大斜度井和深井中应用的新型管材；解决非金属复合材料智能连续油管的外部磨损问题等。目标产品的成本将控制在同等级普通非金属复合材料智能连续油管成本的1.5倍，但其使用寿命要达到普通连续油管的3倍。

2.2 国内研究情况

随着井下注水用非金属复合材料连续管产品质量和应用技术的不断进步，国内企业也开启了采油用复合材料智能连续管产品的研究及应用。2010年，河北恒安泰油管有限公司设计开发出口径DN40mm~DN55mm，公称压力2.5~16MPa的井下用柔性复合连续油管产品（图4）。该产品为非黏结结构，电伴热带、信号传输电缆及动力供给电缆可分别实现电加热、实时监测井下管柱运行信息和提供电潜泵动力等功能。

2014年以来，威海鸿通管材股份有限公司在注水用非金属复合材料连续管产品的基础上，研制出非金属复合材料智能连续油管，其基本结构如图5所示。该产品同样为非黏结结构，主要由内衬层、环向层、骨架层、内外拉伸层和外护套组成，内嵌动力电缆、信号电缆、辅热缆或光纤。环向层、骨架层和拉伸层同样采用玻璃纤维增强热固性树脂（玻璃钢）复合材料带材缠绕而成。该产品配套潜油往复泵，先后在大庆油田采油八厂的芳葡84-斜026井、芳葡86-斜026井、芳葡86-斜024井、太东136-116井等完成了稀油高含蜡井况

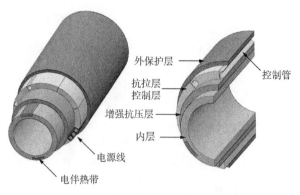

图 4 柔性复合连续油管结构图

试验，配套潜油往复泵及潜油螺杆泵在新疆油田准东采油厂吉祥作业区的稠油区块应用 100 余口油井，最大下井深度 2000m[16]。

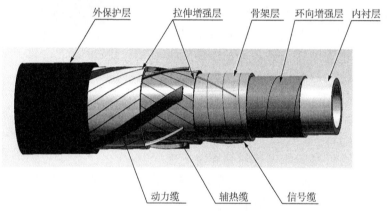

图 5 非黏结型非金属复合材料智能连续油管结构

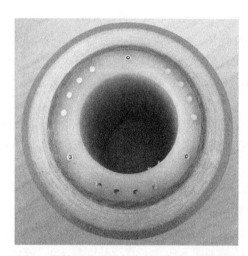

图 6 黏结型非金属复合材料智能连续油管

上海飞舟博源石油装备技术有限公司开发出黏结型非金属复合材料智能连续油管，典型管层结构如图 6 所示。该产品主要由智能内衬层、增强层、抗拉层、外护套组成。智能内衬层内嵌动力电缆、信号电缆、辅热电缆或光纤，增强层和抗拉层主要采用连续纤维增强热塑性塑料带缠绕熔结而成，承受管材的各种力学性能。2016 年至今，该产品及其采油系统在长庆油田、大庆油田、华北油田进行了 20 口井的现场试验，当前运行情况良好[16]。

在研发产品的同时，国内制造企业配套建立了完善的非金属复合材料连续管升下井作业施工流程，并开发出升下井装置及管线收放装置等连续作业设备[13]，集成建立了起下井车载作业系统(图 7)。该系统移动方便且配备吊车等辅助设备，无需组装即可直接作业，最高起下管速度为 15m/min，4 名施工人员 8h 内即可完成起下井作业并投产，降低作业成本的同时，作业效率也大幅提高。

图 7 非金属连续管车载作业系统

3 规范化研究建议

非金属复合材料连续管具有良好的防腐性能，可显著提高井下管柱的使用寿命，还可有效降低下井作业强度和生产成本。随着采出水井的不断增加以及无杆采油技术的不断推广，非金属复合材料连续管在井下注水、采油及采集气等领域将具有广阔的应用前景。但作业过程中的反复弯曲变形，井下高温、高压和腐蚀介质与固体介质的冲蚀，以及拉、压、扭、弯等复合载荷作用，对非金属复合材料连续管的性能和质量提出了极高要求。为了保障非金属复合材料连续管在井下应用的长期性、安全性和可靠性，非金属复合材料连续管规范化研究方向及建议如下：

3.1 建立规范化选材及结构设计技术

井下用非金属复合材料连续管除了具备足够的耐内压性能外，还应具备足够的拉伸强度、耐压溃强度、耐疲劳性能、耐摩擦磨损性能等，以保障全寿命周期的安全性和可靠性。由于应用于井下环境，与井液介质直接接触的内衬层要求耐温高、渗透率低、耐溶胀性能好；各结构层也应具备高温作用下的力学性能；外保护层要有较高的耐剥离、耐磨以及耐腐蚀性，以便满足下井和提升过程中牵引机履带的卡管牵引要求及外防腐要求。管材结构设计过程中还应考虑动力缆、信号缆等缆线敷设形式及结构，避免生产或应用过程中对各种缆线造成损伤。另外，建立非金属复合管与井口、配重管及其他井下附件合适的连接方式，也是保障管柱系统安全可靠的关键。因此，在研究确定不同下井深度的非金属复合材料连续管性能指标基础上，建立系统的管材结构设计（含连接）及选材技术是亟须规范的基础问题。

3.2 发展智能油管缆线复合及信号传输与控制技术

动力缆、信号缆等是非金属复合材料智能连续油管实现运行情况监测和智能控制采油的关键部件。为了避免与井筒的摩擦碰撞，保证线缆的绝缘等级且不被腐蚀，确保高温高压环境下服役至设计寿命，必须对各种线缆进行铠装保护，将其复合在内衬层或结构层中。在复合铠装过程中，要求复合的线缆与内衬或结构保护层之间有良好的配合，防止线缆产生局部变形。在合理敷设的基础上，如何通过内配缆线一方面实时了解并掌握井下电泵、井液参数、运行状况等基础信息，另一方面又可在监测到超预警范围信息时实现对井下电泵的智能控制，是实现智能化采油以及远程安全监控管理的关键环节。因此，建立规范化的缆线敷设及复合管制造技术，并实现对管材运行参数的全面感知、监测、数据收集分析以及反馈控制，是确保管材"智能"的关键。

3.3 建立基于全寿命周期的检验评价技术

与地面管线不同的是，井下用非金属复合连续管在服役过程中始终面临着高温及多相流介质综合作用以及拉、压、扭、弯等复杂载荷影响。另外,由于井下工具或潜油电泵需要定期检查,井下注水或采油管在全寿命周期内肯定会进行一定次数的起下井作业,应考虑重复起下井作业及重复盘卷对管道性能及使用寿命的影响。因此,应综合考虑运行工况、环境介质、受力状态、工程实际等,为井下用非金属复合连续管开发专用的检验试验设备、质量评价方法及寿命评估技术,是保障管材全寿命周期内服役安全性及可靠性的关键。

3.4 升级起下井施工工艺优化研究及装备

井下非金属复合材料连续管具有连接接头少、管柱连接方便的特点,可大幅降低人员劳动强度,提高起下作业效率。但与钢制油管相比,非金属复合材料连续管存在轴向抗拉强度和环刚度偏低问题,而复合管的多层结构在卡管牵引起下井过程中容易造成变形不一致问题。因此,针对不同的非金属复合材料连续管结构和性能特征,应持续研制出配套的收放线盘、牵引装置、导向装置、制动装置等起下井装备,确定起下井拉力、速度、夹持力等关键施工技术参数,进而建立更加完善高效的起下井施工工艺。另外,还需研制配套的打捞工具,实现对复合管快速有效的打捞及修复。

3.5 积极推进标准化研究工作

基于非金属复合材料连续管在井下注水领域成功的应用经验,2014年,国内首次颁布了SY/T 6662.6。该标准规定的产品主要适用于油气田的高压注水、污水处理等,井深不宜超过2000m。随着国内制造能力和应用技术的不断进步,非金属复合材料连续管在分层注水、无杆采油(智能)等领域的应用日益成熟,应用井深逐渐提升,使得现有标准存在了滞后问题。应积极推进井下非金属复合材料连续管标准化工作,为不同应用领域中产品的标准化设计、规范化制造和科学化应用保驾护航。

4 结束语

非金属复合材料连续管能有效解决钢制油管腐蚀、结蜡、结垢等问题,通过结构设计还可达到油井管柱要求的机械性能;连续缠绕成型工艺可实现智能化管材的设计制造,又同时具备单根长、接头少的特点,可大幅降低人员劳动强度,提高起下作业效率。随着我国油田开发力度的持续加大,非金属复合材料连续管将成为传统钢制油管的良好替代品,发展前景十分广阔。必须进一步认清该产品的材料特性和结构特征,攻关突破轴向抗拉强度和环刚度偏低、超深井非金属材料选材、线缆信号稳定传输、配套起下井工艺及装备等关键瓶颈问题,才能为该产品的推广应用奠定坚实基础。

参 考 文 献

[1] 李厚补,高珑,刘浪,等.螺纹制作工艺对高压玻璃钢油管连接性能的影响研究[J].玻璃钢/复合材料,2018,(12):53-56.

[2] 陈涛.浅议井下油管的腐蚀机理及防腐措施[J].油气田环境保护,2009,19(1):30-32.

[3] 李厚补,羊东明,戚东涛,等.增强热塑性塑料连续管标准现状及发展建议[J].塑料,2016,45(4):85-88.

[4] 李厚补,曾亚勤,戚东涛,等.石油天然气工业用非金属复合管 第6部分 井下用柔性复合连续管及接头:SY/T 6662.6—2014[S].北京:石油工业出版社,2014.

[5] MCCLATCHIE D W, REYNOLDS H A, WALSH T J, et al. Applications Enginnering For Composite Coiled Tubing[C]. SPE 54507, SPE/ICoTA Coiled Tubing Roundtable, Houston, Texas, 1999.

[6] SAS-JAWORSKY A, WILLIAMS J G. Development of Composite Coiled Tubing for Oilfield Services[C]. SPE

26536,68th Annual Technical Conference and Exhibition of the Society of Petroleum Engineers,Houston,Texas,1993.

[7] MICHAEL F.Field Experience with Composite Coiled Tubing[C].SPE 82045,SPE/ICoTA Coiled Tubing Conference,Houston,Texas,2003.

[8] 陈立人.国外连续管材料技术及其新进展[J].石油机械,2006,34(9):127-130.

[9] 王薇,王俊涛,魏向军,等.井下柔性复合管注水技术及应用[J].石油钻采工艺,2017,39(1):83-87.

[10] 胡美艳,申晓莉,于九政,等.注水井用非金属复合材料油管试验检测与评价[J].石油矿场机械,2014,43(1):49-52.

[11] 李召勇,崔晓轩,孟庆义.分层注水用柔性复合管系统设计及应用分析[J].机械研究与应用,2019,32(159):95-98.

[12] 毕红军,徐国强,付晶,等.井下用纤维增强连续复合管的性能评价[J].工程塑料应用,2014,42(9):81-84.

[13] 毕婷婷,连洪正,孙云鹏.纤维增强连续复合管在油田井下注水方面的应用[J].山东工业技术,2016,(13):84-85.

[14] 黄守志.智能复合油管技术在无杆采油中的应用[C].北京:IPPTC国际石油石化技术会议论文集,2017:522-523.

[15] Smartpipe Technologies:About smartpipe:new life for aging pipelines [EB/OL].[2019-10-21].http://smart_ pipe.com/.

[16] 李宁会.连续管缆采油试验及分析[J].石油管材与仪器,2018,4(5):67-69.

本论文原发表于《石油管材与仪器》2021年第7卷第2期。

四、其他

Analysis of Causes of Burst Failure of a Buffer Tank

Zhang Shuxin[1,2] Luo Jinheng[1] Wu Gang[1]
Li Lifeng[1] Zhang Penggang[3] Zheng Bin[3]

(1. Tubular Goods Research Institute, China National Petroleum Corporation & State Key Laboratory for Performance and Structure Safety of Petroleum Tubular Goods and Equipment Materials;
2. School of Civil Aviation, Northwestern Polytechnical University;
3. Tarim Oilfield Company, PetroChina Company Limited)

Abstract: An explosion failure occurred in a buffer tank at an oil transfer station. The explosion fragments flew out, causing 2 deaths and 1 injury. To analyze the root cause of the failure, visual inspection, thickness measurement, mechanical test, chemical composition test, corrosion product analysis, fracture analysis, pressure-bearing capacity check, explosion analysis, and operational condition investigation were conducted. The results showed that the chemical composition and mechanical properties of the buffer tank meet the standard requirements; The metallographic structure of the steel is ferrite, pearlite, and a small amount of Widmanstatite, and the structure has not deteriorated after a long time of service; The buffer tank has local corrosion, and the pressure-bearing capacity still meets the design requirements; The operational condition investigation revealed that when the buffer tank burst, the operating personnel were conducting pressure test on the newly-built pipeline outside the station. The pressure test medium used was air. The buffer tank was isolated from the pressure test equipment, and the valve connecting the pressure test pipeline and the buffer tank had internal leakage. Gas entered the buffer tank, causing an overpressure explosion of the tank, and the overpressure wave leads to the collapse of the wall and casualties. Explosion energy analysis showed that the location of casualties was within the range of overpressure wave casualties. In order to avoid such incidents, the use of gaseous media for pressure testing should be avoided.

Keywords: Buffer tank; Pressure test; Burst failure; Explosion energy analysis

1 Introduction

Vesselsand storage tanks are important process facilities in oil field stations, which have the

Corresponding author: Zhang Shuxin, wolfzsx@163.com.

functions of processing, storing, transferring, peak tuning, and metering. The main failure modes of tanks are internal and external corrosion perforation, stress corrosion cracking, etc. Yang et al. [1] research the fire accident caused by the chemical storage tank, they found the root cause of the accident was the spontaneous combustion of FeS which was formed by the internal corrosion of the roof with the medium. Ravi Kumar Sharma et al. [2] investigated a series of tank failures including Jaipur airport in India (2009), Puerto-Rico, USA (2009), Buncefield, UK (2005). Their studies focused on the consequences of the failure, and recommended that measures should be taken to avoid the huge cloud formation and ignition. The causes of the accidents were either overfill or leakage. Pablo. G. Cirimello et al. [3] analyze the collapse accident of a 2000m^3 crude oil tank. The root cause of the failure was thickness loss and leaks caused by internal and external corrosion. Sina Miladi et al. [4] investigated an un-anchored tank damaged in the earthquake of 2006 in Iran. The earthquake would cause the contained medium to slosh, the tank uplifted and buckled, further, the medium overfilled. Therefore, in this situation, the liquid level of the tank shall be lower to reduce the uplift value. Masahiro Kusano et al. [5] researched the degradation behavior of fiber reinforced plastics tanks with the medium of hydrochloric acid. When HCl diffused into FRP's reinforced layer, the strength decreased sharply, the lifetime of the tank decreased.

Few cases of tank explosion failure were reported. Because of its large volume, the storage medium is flammable and explosive. Once an explosion occurs, the failure consequences are very serious [6-10]. Therefore, the explosion failure of the storage tank should arouse people's attention.

The failure case in this study happened in 2014 was an explosion of a crude oil buffer tank in a transfer station yard of an oil field. The oil transfer station was built in October 1996 and played a role to undertake the metering, heating, and inhibitor dosing for the corrosive medium exploited from the oil/gas wells. The station produced 85m^3 of liquid per day, 24t, of oil per day, and water content in the medium is about 90%. The buffer tank was produced in June 1991, the dimension is ϕ2490mm×9200mm×12mm, the design pressure is 0.6MPa, the test pressure is 0.75MPa, the weight is 10210kg, and the material built is Q235-A [11]. The explosion of the storage tank caused 2 deaths and 1 injury. The schematic diagram of the explosion site and the scene after the explosion are shown in Fig. 1 and Fig. 2(a). It can be seen that the buffer tank burst into multiple fragments and flew away from the original position. The fragments near the medium inlet accounted for about 2/3 of the entire tank body, and they fell on the top of the pump room, marked as 1# fragment [Fig. 2(b), Fig. 2(c)]; Fragments contained the level gauge accounted for about 1/3 of the entire tank, and fell on the roof of storeroom, marked as 2# fragment [Fig. 2(d)]; the sewage outlet fragment fell on the nearby mountains, marked as 3# fragment; saddle support flew down 50m outside the gate, marked as 4# fragment.

In order to analyze the cause of the explosion, visual inspection, wall thickness measurement, mechanical test and chemical position test, corrosion product analysis, fracture analysis, pressure-bearing capacity check, explosion analysis, field condition investigation were conducted. Finally, a comprehensive analysis was carried out based on the results.

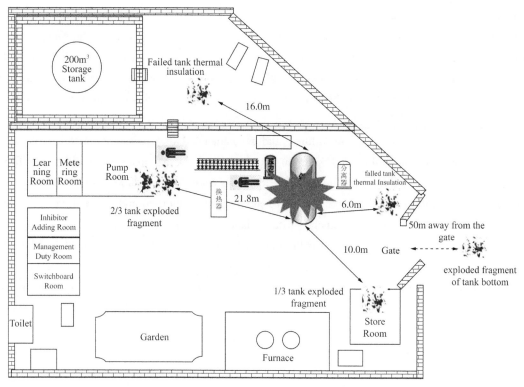

Fig. 1 Accident schematic diagram of the exploded buffer tank

(a) Explosion scene after the accident

(b) Fragment of the exploded buffer tank on the roof of the pump room

(c) 1# fragment of the exploded buffer tank on the roof of the pump room

(d) 2# fragment of the exploded buffer tank on the roof of the storeroom

Fig. 2 Accident scene of the exploded buffer tank

2 Experimental procedure

2.1 Visual inspection and thickness measurement

From Fig. 2(c) and Fig. 2(d), it can be seen that the buffer tank after the explosion has undergone severe plastic deformation, and there is obvious curling at the rupture of the tank. At the same time, it can be seen that there are longitudinal welds and circumferential welds on the tank body. It can be inferred that the tank was welded by steel plates. The crack propagation direction is shown in Fig. 3 as the red arrow, the crack converges to the bottom of the buffer tank, indicate the explosion source is from the bottom of the tank.

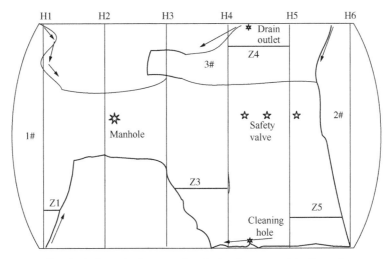

Fig. 3 Schematic diagram of explosion fragments of buffer tank
(H1~H6 represents circumferential welds, Z1~Z5 represents longitudinal welds)

Vernier caliperand ultrasonic thickness gauge were used to measure the residual thickness of 1#~3# fragments. The minimum wall thickness of the fracture is shown in Table 1. It can be seen that the buffer tank was corroded. Using the nominal wall thickness of 12mm to calculate the thinning rate, the maximum thinning rate of the tank reaches 69%. The thinning of the thickness seriously affects the pressure-bearing capacity of the buffer tank, laying hidden dangers for explosion failure.

Table 1 The minimum thickness of 1#~3# sample

Fragment	Description	Minimum thickness(mm)	Thickness thinning rate(%)
1#	Near the medium inlet -part1	5.29	56%
	Near the medium inlet -part2	3.72	69%
2#	Contained the level gauge	4.08	66%
3#	Sewage outlet	4.68	61%

Note: nominal thickness of 12mm is used to calculate the thinning rate.

2.2 Mechanical and chemical composition test

The chemical composition, tensile test, Charpy impact test, and metallographic analysis of the buffer tank were conducted to check whether the properties of the tank after years of servicing have deteriorated. The tensile test, chemical composition test specimens, the longitudinal weld specimen and the circumferential weld specimen were cut from the 1# fragment.

The ARL 4460 direct reading spectrometer was used to test the chemical composition of the buffer tank base material in accordance with the standard GB/T 4336—2016[12]. The test results are shown in Table 2. Since the allowable positive deviation of silicon element in GB/T 222—2006 [13] *Permissible tolerances of chemical composition for steel product* is +0.03%, the chemical composition of the buffer tank base material meets the requirements of GB/T 700—2006 *Carbon Structural Steels* [11].

Table 2 Chemical composition of buffer tank base metal (Wt. %)

Element	Content	Standard requirement
C	0.16	≤0.22
Si	0.38	≤0.35
Mn	1.40	≤1.40
P	0.012	≤0.045
S	0.011	≤0.050
Cr	0.013	≤0.30
Ni	0.0057	≤0.30
Cu	0.010	≤0.30

The UH-F500kNI tensile testing machine was used to test mechanical properties in accordance with the standard GB/T 228.1—2021[14]. The base material sample used is transverse, and the weld sample is taken perpendicular to the weld, and the weld is located in the middle of the sample. Before the tensile test, the sample was cold pressed and flattened. The results of the tensile test are shown in Table 3. Generally, pipe flattening changes mechanical properties, providing a conservative value of yield strength due to Bauschinger effect originating from the compressive pre-strain[15]. While the radius of curvature of the tank is large, the transverse sample was slightly flattened, and influence on the tensile test is relatively small.

It can be seen that the yield strength and elongation of the base material of the tank body meet the requirements of the standard GB/T 700—2006[11], and the tensile strength of the base material of the tank body, the longitudinal weld and the circumferential weld are slightly higher than the standard's upper limit requirements.

Table 3 The mechanical test results of the buffer tank

Sample	Width×Gauge (mm)	Tensile strength R_m(MPa)	Yield strength $R_{t0.5}$(MPa)	Elongation A(%)	Breaking area
Base material of the tank	38.1×50	534	431	36.0	—
Longitudinal weld	38.1	538	—	—	Base material
	38.1	541	—	—	Base material
Circumferential weld	38.1	534	—	—	Base material
	38.1	550	—	—	Base material
Requirement of GB/T 700 for Q235A		375~500	$R_{t0.5} \geqslant 235$	≥26	

The PIT752D-2 impact testing machine was used to perform the Charpy impact test in accordance with the standard GB/T 229—2020[16], the test temperature is 20°C. The sample size is 7.5mm×10mm×55mm, and the V-notch is perpendicular to the surface of the tank. The results of the impact test are shown in Table 4. The impact property of the base material and the longitudinal weld is better than the circumference weld which is in a brittle-ductile transition zone under 20℃。

Table 4　The charpy impact test results of the buffer tank

Sample		Dimension(mm)	Temperature(℃)	Absorbed energy $KV_2(J)$			Ductile area rate FA(%)		
No.									
Parent material		7.5×10×55	20	86	95	88	80	90	85
Longitudinal weld	Weld	7.5×10×55	20	72	109	100	70	80	80
	Heat affected zone	7.5×10×55	20	96	79	89	100	80	100
Circumferential weld	Weld	7.5×10×55	20	60	54	59	55	50	50
	Heat affected zone	7.5×10×55	20	91	83	95	95	90	95

The MeF3A metallurgical microscope, MeF4M metallurgical microscope and image analysis system were used to analyze the metallography of the base metal and welds. The results are shown in Table 5. The base metal structure of the tank body is ferrite, pearlite, and a small amount of Widmanstatite, and no abnormal structure appears.

Table 5　The buffer tank base material metallographic analysis results

Sample	Nonmetallic inclusion	Metallographic structure	Grain size
Base metal	A1.0, A0.5e, B0.5, D0.5	F+P+WF(Fig.4)	7.5

Fig.4　Metallography of the base metal of buffer tank

2.3　Corrosion product analysis

The corrosion products were taken from different regions of the 1# fragment, and X-ray diffraction (XRD) phase analysis was carried out to determine their composition. The analysis results are shown in Fig.5. The results show that the corrosion product of the inner wall of the tank is composed of Fe_2O_3 and Fe_3O_4.

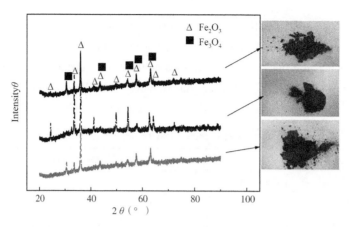

Fig. 5 XRD analysis of corrosion products in the different regions of 1# fragment

2.4 Fracture analysis

The macro fracture analysis of 1# fragments of buffer tank was carried out, as shown in Fig. 6. The results show that there are plastic and brittle fracture in different areas. Generally, the plastic fracture appeared when the residual thickness is far less than the nominal thickness of the buffer tank [Fig. 6(b) and Fig. 6(c)]. The brittle fracture has chevron cracks characteristic, and converged to thickness thinning region which is near the sewage outlet [Fig. 6(b)], weld between tank head and body [Fig. 6(a)], and the weld between tank body and support [Fig. 6(a)]. Considering the axial force is larger than the circumferential force, it can be inferred that the crack source originates from the thickness thinning region near the cleaning outlet.

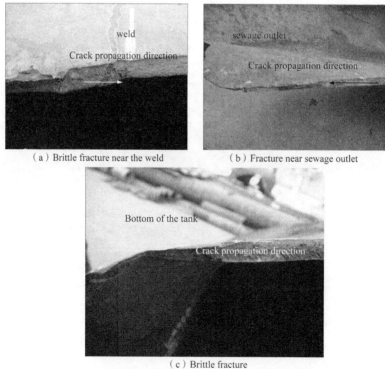

Fig. 6 Different fracture of 1# fragment

The scanning electron microscope (Tescan Vega Ⅱ) and Oxford INCA350 energy spectrum analyzer were adopted to analyze micromorphology of the crack source and to determine the composition of the corrosion product. The SEM and EDS results of the thinnest section of 1# fragment are shown in Fig. 7. The resuls show that there is a layer of corrosion product covering the fracture, the content of the corrosion product is mainly O, Fe, C elements, and a certain amount of S, Cl, and Ca, the fracture in the propagation zone has dimple characteristic.

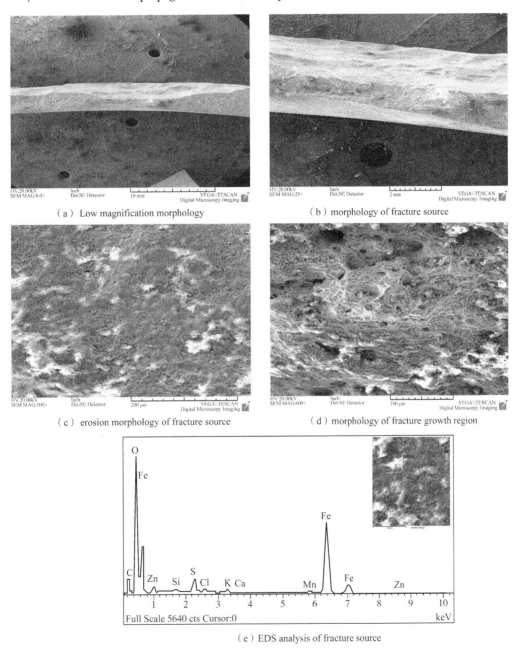

(a) Low magnification morphology (b) morphology of fracture source

(c) erosion morphology of fracture source (d) morphology of fracture growth region

(e) EDS analysis of fracture source

Fig. 7 SEM and EDS results of the thinnest section of 1# fragment

2.5 Pressure-bearing capacity check

Pressure bearing capacity was checked according to GB 150.1—2011 [17], the calculation formula of cylinder stress is shown in equation 1.

$$\sigma = \frac{p(D+\delta)}{2\delta} \tag{1}$$

In equation 1, p represents the internal pressure of the tank, MPa; σ represents the hoop stress generated by the tank under internal pressure, MPa; δ represents the effective wall thickness of the tank, mm; D represents the inner diameter of the tank, mm.

According to equation 1, the internal pressure p on the cylinder can be deduced, as shown in equation 2.

$$p = \frac{2\sigma\delta}{D+\delta} \tag{2}$$

The inner diameter of the buffer tank is 2400mm. The test results of the tensile properties of the base metal of the buffer tank show that the yield strength $R_{t0.5}$ is 431MPa, and the tensile strength σ_m is 534 MPa, and the flow stress is taken as σ_f was min $[1.1R_{t0.5}, 0.5(R_{t0.5} + \sigma_m)]$ = 474MPa. For axial cracking analysis, the value of circumferential stress σ is taken as σ_f. Table 6 shows the calculation results of the internal pressure of the buffer tank under different wall thickness measurements. It can be seen from Table 6 that the pressure bearing capacity of the buffer tank is 1.46MPa according to the minimum residual wall thickness of the fracture.

Table 6 Calculations of internal pressure capacity of buffer tank under different thickness

Fragment	Residual thickness(mm)	Pressure(MPa)	Note
1#-part1	5.29	2.08	
1#-part2	3.72	1.46	axial cracking
2#	4.08	1.60	
3#	4.68	1.84	

3 Discussion

The chemical composition and mechanical properties of the buffer tank meet the requirements of Q235-A in GB/T 700—2006. The water content of the liquid medium disposed by the oil transfer station is 90%, and it can be inferred from the energy spectrum composition and corrosion products at the fracture of the 1# fragment that there may be Cl^-, S^{2-}, Ca^{2+} and dissolved oxygen in the transferring liquid, causing erosion and corrosion to the inner wall of the buffer tank, forming a local wall thickness reduction area. Based on the calculation result of the 1.46MPa minimum pressure bearing capacity of the buffer tank, there is a risk of axial cracking in the local wall thickness reduction area of the 1# fragment near the sewage outlet. Since the minimum pressure bearing capacity is larger than the design pressure, the buffer tank would not explode, thus the operational condition was investigated.

3.1 Operational condition investigation

After further investigation, it was found that the pressure test for the pipeline outside the station was carried out by the operator, the valve between the buffer tank and pipeline leaked and this was confirmed by a seal test. The pressure of the pressure test was 2.5MPa, and the medium used was air. Thus, the compressed air flees into the buffer tank, and the internal pressure exceeded withstand pressure. The buffer tank explosion accident occurred.

Usually, the medium used in pressure test is water, and the main failure mode in the process of pressure test is leakage, rarely explosion. While in this case, due to the scarcity of water in field, and compressed air was easy to obtained, the medium used in pressure test was air, the failed consequence is serious. In order to quantify the consequence of the explosion failure, the explosion analysis was conducted.

3.2 Explosion analysis

The internal volume of the buffer tank is 40m³, the design pressure is 0.6MPa, and the pressure test pressure of the flexible composite pipe is 2.5MPa when the explosion occurred. The impact of the shock wave of the tank explosion on the surrounding environment is calculated as follows:

Assuming that medium is compressed gas, the blasting energy released is calculated as follows:

$$E_{gas} = \frac{pV}{k-1}\left[1-\left(\frac{0.1013}{p}\right)^{\frac{k-1}{k}}\right] \times 10^3 \qquad (3)$$

In the equation: E_{gas} represents the blasting energy, kJ; p represents the gas pressure in the storage tank, taken as 2.5MPa; V represents the storage tank volume, taken as 40m³; k represents the gas adiabatic index, taken as gas adiabatic index 1.4[18].

Assuming that water was used for hydraulic pressure test, if the valve leakage happened, the blasting energy released when a physical explosion occurs is calculated as follows:

$$E_{liquid} = \frac{1}{2}pV\beta \times 10^5 \qquad (4)$$

In the equation: E_{liquid} respresents blasting energy, kJ; p respresents liquid pressure in the storage tank, taken as 2.5MPa; V respresents storage tank volume, taken as 40m³; β respresents water compressibility index, taken as $4.5 \times 10^{-10} \mathrm{Pa}^{-1}$.

The blasting energy of gas and liquid was 1.71×10^5 kJ, 5.63×10^4 kJ, respectively. TNT blasting energy was taken as 4500kJ/kg, the physical explosion TNT equivalent of gas q_{gas} and liquid q_{liquid} was 39.1kg, 12.5kg, respectively.

According to Hopkinson-Cranz law[19], the shock wave overpressure was calculated.

$$\text{If } \alpha = \frac{R}{R_0} = \sqrt[3]{\frac{q}{q_0}} = \sqrt[3]{\frac{q}{1000}} = 0.1\sqrt[3]{q}, \qquad \Delta p = \Delta p_0 \qquad (5)$$

In the equation: R respresents the distance from the explosion center, m; R_0 respresents the distance from the 1000kg TNT explosion center, m; q respresents TNT equivalent, kg; q_0 respresents 1000kg TNT; Δp respresents overpressure of TNT equivalent explosion, MPa; Δp_0 respresents overpressure of 1000kg TNT explosion.

Substituting the TNT equivalents of gas and liquid explosion, α_{gas}, α_{liquid} is 0.339, 0.232.

The relationship between shock wave overpressure and distance from the explosion center when 1000kg TNT explodes is shown in Table 7. The shock wave overpressure generated during physical explosions was calculated.

Table 7 Shock wave overpressure of 1000kg TNT explosion

Distance R_0(m)	5	6	7	8	9	10	12	14	16
Overpressure Δp_0(MPa)	2.94	2.06	1.67	1.27	0.95	0.76	0.50	0.33	0.235
Distance R_0(m)	18	20	25	30	35	40	45	50	55
Overpressure $\Delta(p)_0$(MPa)	0.17	0.126	0.079	0.057	0.043	0.033	0.027	0.0235	0.0205

According to the shock wave overpressure criterion, the damage to human body is shown in Table 8. When buffer tank containing gas medium explodes, the overpressure of shock wave will cause severe injury or death within the range of 10.17m from the explosion center. For the liquid medium, the injury or death radius is 6.96m. The gas pressure test is riskier than liquid pressure test.

Table 8 The effect of shock wave overpressure on human

Overpressure Δp_0(MPa)	Damage
0.02~0.03	Minor injury
0.03~0.05	Damage to the hearing organ or viscera
0.05~0.10	Severe injury or death
>0.10	Most people died

The pipeline pressure test is to verify the pipeline construction quality and overall strength, determine its pressure-bearing capacity, and provide a basis for increasing the pipeline's capacity in the future, which is essential to ensure the safe operation of the pipeline.

Domestic and foreign standards tend to use water as the pressure test medium, and strictly regulate the gas pressure test application conditions and stress levels for theclass 3 and 4 locations which have a large population density. For example, GB 50251—2015 [20] and ASME B31.8—2020[21] stipulate that the maximum hoop stress of the pressure test pipe should be less than 0.5 times specified minimum yield strength (SMYS) for third level area and 0.4 times SMYS for fourth level area. CSA Z662—2020 [22] *Oil and Gas Pipeline System* stipulates that the hoop stress of the gas pressure test pipe should not exceed 0.8 times SMYS. In addition, the Chinese standard GB/T 16805—2017 [23] and API 1110—2007 [24] stipulate that liquid petroleum cannot be used for pressure test in the environmentally sensitive areas and densely populated areas, unless strict monitoring, unblocked communication, and measures were taken to respond the leaks.

The Australian Standard AS 2885.5—2020[25] stipulates strict limits for gas pressure testing: (1) The airtightness test should use a gas tracer for visual inspection; (2) The test plan should evaluate the consequences of leakage at any part in the pipeline; (3) Safety management study should be carried out, including gas directional energy diffusion, pipe fragmentation mechanism and secondary ejection protection, etc.

Therefore, the gas medium was not recommended tobe used for pressure test. The burst failure of buffer tank can be attributed to two factors. Firstly, the buffer tank itself has local corrosion and the pressure bearing capacity is reduced. When the valve leaks, the pressure bearing capacity of the buffer tank is less than the gas pressure test pressure, resulting in explosion. Secondly, the gas medium is improperly used for pressure test, the explosion energy far exceeds the energy contained in the liquid medium, resulting in casualties.

4 Conclusions and recommendations

An explosion failure occurred in a buffer tank at an oil transfer station. The explosion fragments flew out, causing 2 deaths and 1 injury. In order to analyze the root cause of the failure, visual inspection, thickness measurement, mechanical test, chemical composition test, corrosion product analysis, fracture analysis, pressure-bearing capacity check, explosion analysis, and operational condition investigation were conducted. The conclusions can be drawn as follows:

(1) The chemical composition and mechanical properties of the buffer tank meet the standard requirements, not deteriorated after a long time service.

(2) The buffer tank has local corrosion, and the pressure-bearing capacity still meets the design requirements;

(3) Operational condition investigation revealed that when the buffer tank burst, the operating personnel were conducting pressure test on the newly-built pipeline outside the station. The valve connecting the pressure test pipeline and the buffer tank had internal leakage which led to the overpressure explosion of the tank.

(4) Another reason for the casualties is that the pressure test medium used was air which has higher explosion energy than liquid.

In order to avoid such incidents, the use of gaseous media for pressure testing should be avoided.

Acknowledgements

The authors are grateful to the fund support of National Key R&D Program of China (2016YFC0801200, 2017YFC0805804), and the members in Tubular Goods Research Institute who assisted in carrying out this failure analysis study. Sincere thanks to Ms. Yan Xi and Zhang Jiahe for their tremendous support for me.

References

[1] Yang R, Wang Z, Jiang J, et al. Cause analysis and prevention measures of fire and explosion caused by sulfur corrosion, [J]. Engineering Failure Analysis. 2020, 108: 104342.

[2] Kusano M, Kanai T, Arao Y, et al, Degradation behavior and lifetime estimation of fiber reinforced plastics tanks for hydrochloric acid storage[J]. Engineering Failure Analysis, 2017, 79: 971-979.

[3] Pablo G, Cirimello J L, Otegui D, et al, A major leak in a crude oil tank: Predictable and unexpected root causes[J]. Engineering Failure Analysis, 2019, 100: 456-469.

[4] Miladi S, Razzaghi M, S. Failure analysis of an un-anchored steel oil tank damaged during the Silakhor earthquake of 2006 in Iran[J]. Engineering Failure Analysis, 2019, 96: 31-43.

[5] Kusano M, Kanai T, Arao Y, et al. Degradation behavior and lifetime estimation of fiber reinforced plastics tanks for hydrochloric acid storage[J]. Engineering Failure Analysis, 2017, 79: 971-979.

[6] Sharma R K, Gopalaswami N, Gurjar B R, et al. Assessment of Failure and Consequences Analysis of an Accident: A case study[J]. Engineering Failure Analysis, 2020, 109: 104192.

[7] Pouyakian M, Jafari M J, Laal, F, et al. A comprehensive approach to analyze the risk of floating roof storage tanks[J]. Process Safety and Environmental Protection, 2021, 146: 811-836.

[8] Qin R, Zhu J, Khakzad, N. Multi-hazard failure assessment of atmospheric storage tanks during hurricanes [J]. Journal of Loss Prevention in the Process Industries, 2020, 68: 104325.

[9] Wu D, Chen Z. Quantitative risk assessment of fire accidents of large-scale oil tanks triggered by lightning[J]. Engineering Failure Analysis, 2016, 63: 172-181.

[10] Kang J, Liang W, Zhang L, et al. A new risk evaluation method for oil storage tank zones based on the theory of two types of hazards[J]. Journal of Loss Prevention in the Process Industries, 2014, 29: 267-276.

[11] GB/T 700—2006. Carbon Structural Steels[S]. National Standard of the People Republic of China, 2006.

[12] GB/T 4336—2016. Standard test method for spark discharge atomic emission spectrometric analysis of carbon and low-Alloy steel (routine method)[S]. National Standard of the People Republic of China, 2016.

[13] GB/T 222—2006. Permissible tolerances for chemical composition of steel produts[S]. National Standard of the People Republic of China, 2006.

[14] GB/T 228.1—2021. Metallic materials - Tensile testing - Part 1 : Method of test at room temperature [S]. National Standard of the People Republic of China, 2021.

[15] Kang S, Speer J G, Van Tyne C J, Weeks TS, Effect of Pipe Flattening in API X65 Linepipe Steels Having Bainite vs. Ferrite/Pearlite Microstructures[J]. Metals, 2018, 8(5): 354.

[16] GB/T 229—2020. Metallic materials—Charpy pendulum impact test method[S]. National Standard of the People Republic of China, 2020.

[17] GB 150.1—2011. Pressure vessels—Part 1: General requirements[S]. National Standard of the People Republic of China, 2011.

[18] Crowl D A. Understanding Explosions - Appendix B: Equations for Determining the Energy of Explosion [M]. John Wiley & Sons, Inc, 2010.

[19] Karlos V, Solomos G. Calculation of blast loads for application to structural components[J]. Luxembourg: Publications Office of the European Union, 2013, 3.

[20] GB 50251—2015. Code for design of gas transmission pipeline engineering[S]. National Standard of the People Republic of China, 2015.

[21] ASME B31.8—2020. Gas Transmission and Distribution Piping Systems[S]. American Society of Mechanical Engineers, 2020.

[22] CSA Z662—2020 Oil and Gas Pipeline Systems[S]. Canadian Standards Association, 2020.

[23] GB/T 16805—2017. Pressure testing of liquid petroleum pipelines[S]. National Standard of the People Republic of China, 2017.

[24] API 1110—2007 Pressure Testing of Steel Pipelines for the Transportation of Gas, Petroleum Gas, Hazardous Liquids, Highly Volatile Liquids or Carbon Dioxide[S]. American Petroleum Institute, 2007.

[25] AS 2885.5—2020. Gas and liquid petroleum - Field pressure testing[S]. Standards Australia, 2020.

本论文原发表于《Engineering Failure Analysis》2022 年第 131 卷。

Study on the Effect of Cement Sheath on the Stress of Gas Storage Well

Song Chengli[1,2] Liu Xinbao[2] Li Guangshan[1] Wang Shuai[1]

(1. State key Laboratory for Performance and Structure Safety of Petroleum Tubular Goods and Equipment Materials, CNPC Tubular Goods Research Institute;
2. School of Chemical Engineering, Northwest University)

Abstract: Compressed natural gas (CNG) has been widely used as an automotive fuel in china, which the service security of CNG storage well (CSW), the main storage equipment in Chinese filling station, is becoming more and more prominent. In order to analyze the stress of CSW and its influencing factors, the mechanical model of CSW + cement sheath + stratum (CCS) processed by both elastic mechanics and finite element method was studied in this work. Using the most common well of ϕ177.8mm×10.36mm as a calculation case to obtain the analytic solutions and numerical solutions of CCS. The results indicated that the analytic solutions and numerical solutions are very close with relative deviation less than 3%, which verified their reliability each other. The calculation case can prove that the stress of CSW reduced evidently due to the effect of well cementation, which equivalent to the value of circular and axial stress of CSW strengthens to 18% and 20%. And when increasing the elastic modulus of cement sheath, the stress of CSW decreased that the support and potentiation to CSW by cement sheath becomes stronger.

Keywords: CNG; Storage well; Cement Sheath; Stratum; Stress

1 Introduction

As pollution issues threaten the widespread use of fossil fuels, compressed natural gas (CNG) has been widely used as an automobile fuel instead of gasoline (petrol) and diesel[1-2]. CNG storage well (CSW), a new and economical way to storage CNG, has a lot of advantages, namely low cost, high reliability, easy operation, long lifetime, less area occupied, fast filling, large storage capacity, etc[3-4]. At present, the number of Chinese CNG filling stations are increasing rapidly. As far as the CSW is concerned, CSW accounts for 95% of all the CNG filling station gas storage capacity in China[5].

Fig. 1 shows that CSW is buried under the ground and surrounded by a cement sheath. High

---

Corresponding author: Song Chengli, songcl@cnpc.com.cn.

internal pressure is extruded and supported by both cement sheath and stratum, which CSW simultaneously bears internal and external pressure. Furthermore, CSW is commonly built in large cities with a dense population, which shall lead to major potential hidden danger, such as leakage, ignition, well-channeling, well-sinking even explosion[6-7]. Accordingly, this paper concerns to the stress analysis of CSW since the failure of the high pressure vessel can result in a fatal disaster.

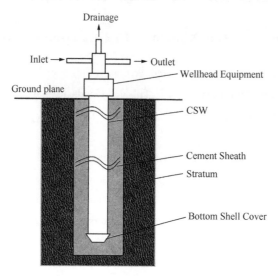

Fig. 1   Structure of CSW

2   Experimental details

2.1   Elastic mechanics analysis

Considering the cement sheath and stratum, the mechanical model of CCS was counted as a multi-layer thick-walled cylinder of 3 kinds of materials on the basis of the elastic mechanics theory[8]. Before calculation, some necessary simplifications and assumptions are listed as follows:

(1) The CSW is at the center of the well and owns uniform wall thickness;

(2) The CCS belongs to line elastomer;

(3) The consolidation between different materials is good without defect;

(4) Stratum stress is well-proportioned.

The cross-sectional area of CSW is much smaller thanthe axial length, so that the axial deformation has constraints each other under the internal pressure and the strain components (ε_x, ε_y, ε_{xy}) paralleled to the cross section are not zero. Therefore, the mechanical model of CCS was analyzed under plane strain principle, as shown in Fig. 2[9-10].

2.2   Finite element analysis

To verify the accuracy of elastic mechanics analysis, the finite element method was applied for establishing the mechanical model of CCS by ANSYS software. Meanwhile, the finite element method requires some necessary data, such as dimensions, material performance, force and constraint of CCS. There are four steps for numerical calculation, including 3D model establishing, mesh generation, boundary conditions loading (including displacement and force) and finite element calculation[11-12]. For numerical efficiency, the 1/4 model was used for finite element analysis due to its symmetry.

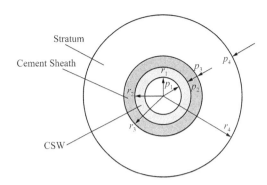

Fig. 2 Mechanical model of CCS

where r_1 is internal radius of CSW, r_2 is external radius of CSW, r_3 is external radius of cement sheath, r_4 is external radius of stratum, p_1 is internal pressure of CSW, p_4 is external force from stratum, p_2 and p_3 represents the first and the second interfacial pressure

3 Results

3.1 Parameters of an example

The most common CSW of $\phi 177.8\text{mm} \times 10.36$ mm with N80Q steel was taken as an instance to calculate both its analytic and numerical solutions. All the parameters required were shown in Tables 1—3. The internal pressure p_1 is the biggest operation pressure of CSW in Table 1. The external pressure p_4 is static pressure of underground liquid column of 150m length of CSW in Table 3.

Table 1 Parameters for analytic and numerical calculation of CSW

Parameters	Yield Strength σ_s(MPa)	Tensile Strength σ_b(MPa)	Elasticity Modulus E_1(MPa)	Poisson's Ratio μ_1	Internal Radius r_1(mm)	External Radius r_2(mm)	Internal Pressure p_1(MPa)
CSW	552	689	206	0.3	78.54	88.90	25

Table 2 Parameters for analytic and numerical calculation of cement sheath

Parameters	Elasticity Modulus E_2(MPa)	Poisson's Ratio μ_2	Internal Radius r_2(mm)	External Radius r_3(mm)
Cement heath	7	0.23	88.90	120.5

Table 3 Parameters for analytic and numerical calculation of stratum

Parameters	Elasticity Modulus E_3(MPa)	Poisson's Ratio μ_3	Internal Radius r_3(mm)	External Radius r_4(mm)	External Pressure p_4(MPa)
Stratum	3	0.2	120.5	602.5	2

3.2 Analytic solution

Displacement equations of single-layer thick-walled cylinder were listed as follows[13-14], including external wall radial displacement of CSW u_{11}, internal wall radial displacement of cement sheath u_{12}, external wall radial displacement of cement sheath u_{21}, and internal wall radial displacement of stratum u_{22}:

$$u_{11} = \frac{r_1^2 r_2 (2-\mu_1)}{E_1 (r_2^2 - r_1^2)} p_1 - \frac{r_2^3 (1-2\mu_1) + (1+\mu_1) r_1^2 r_2}{E_1 (r_2^2 - r_1^2)} p_2 = f_1 p_1 - f_2 p_2 \quad (1)$$

$$u_{12} = \frac{r_2^3(1-2\mu_2)+(1+\mu_2)r_3^2 r_2}{E_2(r_3^2-r_2^2)}p_2 - \frac{r_3^2 r_2(2-\mu_2)}{E_2(r_3^2-r_2^2)}p_3 = f_3 p_2 - f_4 p_3 \tag{2}$$

$$u_{21} = \frac{r_2^2 r_3(2-\mu_2)}{E_2(r_3^2-r_2^2)}p_2 - \frac{r_3^3(1-2\mu_2)+(1+\mu_2)r_2^2 r_3}{E_2(r_3^2-r_2^2)}p_3 = f_5 p_2 - f_6 p_3 \tag{3}$$

$$u_{22} = \frac{r_3^3(1-2\mu_3)(1+\mu_3)r_4^2 r_3}{E_3(r_4^2-r_3^2)}p_3 - \frac{r_4^2 r_3(2-\mu_3)}{E_3(r_4^2-r_3^2)}p_4 = f_7 p_3 - f_8 p_4 \tag{4}$$

The radial displacement of the first and the second interfacial were equal in terms of the model of CCS was continuous as shown in Eq. 5.

$$\begin{cases} f_1 p_1 - f_2 p_2 = f_3 p_2 - f_4 p_3 \\ f_5 p_2 - f_6 p_3 = f_7 p_3 - f_8 p_4 \end{cases} \tag{5}$$

where f_1—f_8 are assumptive calculated coefficient. The first and the second interfacial pressure p_2 and p_3 were calculated by the Eq. 5, as shown in Eq. 6.

$$\begin{cases} p_2 = \dfrac{f_1 p_1(f_6+f_7)+f_8 f_4 p_4}{(f_2+f_3)(f_6+f_7)-f_4 f_5} \\ p_3 = \dfrac{f_8 p_4(f_2+f_3)+f_1 f_5 p_1}{(f_2+f_3)(f_6+f_7)-f_4 f_5} \end{cases} \tag{6}$$

The radial stress σ_r, circular stress σ_θ and axial stress σ_z equations of CCS were obtained due to single-layer thick-walled cylinder stress equation, as shown in Eqs. 7—9[12].

For CSW($r_1 < r < r_2$),

$$\begin{cases} \sigma_{1r} = \dfrac{r_1^2 r_2^2}{r_2^2-r_1^2}\dfrac{p_2-p_1}{r^2}+\dfrac{r_1^2 p_1-r_2^2 p_2}{r_2^2-r_1^2} \\ \sigma_{1\theta} = -\dfrac{r_1^2 r_2^2}{r_2^2-r_1^2}\dfrac{p_2-p_1}{r^2}+\dfrac{r_1^2 p_1-r_2^2 p_2}{r_2^2-r_1^2} \\ \sigma_{1z} = 2\mu_1\dfrac{r_1^2 p_1-r_2^2 p_2}{r_2^2-r_1^2} \end{cases} \tag{7}$$

For cement sheath($r_2 < r < r_3$),

$$\begin{cases} \sigma_{2r} = \dfrac{r_2^2 r_3^2}{r_3^2-r_2^2}\dfrac{p_3-p_2}{r^2}+\dfrac{r_2^2 p_2-r_3^2 p_3}{r_3^2-r_2^2} \\ \sigma_{2\theta} = -\dfrac{r_2^2 r_3^2}{r_3^2-r_2^2}\dfrac{p_3-p_2}{r^2}+\dfrac{r_2^2 p_2-r_3^2 p_3}{r_3^2-r_2^2} \\ \sigma_{2z} = 2\mu_2\dfrac{r_2^2 p_2-r_3^2 p_3}{r_3^2-r_2^2} \end{cases} \tag{8}$$

For stratum ($r_3 < r < r_4$),

$$\begin{cases} \sigma_{3r} = \dfrac{r_3^2 r_4^2}{r_4^2-r_3^2}\dfrac{p_4-p_3}{r^2}+\dfrac{r_3^2 p_3-r_4^2 p_4}{r_4^2-r_3^2} \\ \sigma_{3\theta} = -\dfrac{r_3^2 r_4^2}{r_4^2-r_3^2}\dfrac{p_4-p_3}{r^2}+\dfrac{r_3^2 p_3-r_4^2 p_4}{r_4^2-r_3^2} \\ \sigma_{3z} = 2\mu_3 \dfrac{r_3^2 p_3-r_4^2 p_4}{r_4^2-r_3^2} \end{cases} \quad (9)$$

The stress analytic solutions of any point of CCS could be obtained by Eqs. 7—9. The stress analytic solutions of the above example were calculated, as shown in Table 4.

Table 4　Stress analytic solutions of CSW

Stress	Circular stress of internal wall (MPa)	Circular stress of external wall (MPa)	Axial stress (MPa)
Analytic solutions	151.32	131.97	37.90

3.3 Numerical solution

The 1/4 finite element model of the above example was studied. The model was divided into hexahedron eight - node element with the characteristics of plasticity, stress strengthening, expansion, creep, large strain and large deformation, and possessed three directions of degrees-of-freedom as shown in Fig. 3. Then the internal wall of CSW was loaded with the internal pressure of 25MPa (p_1) from CNG and the external wall of cement sheath was loaded with external pressure of 2MPa (p_4) from stratum. Ultimately, based on the fourth strength theory, the Mises equivalent stress distribution of CCS was obtained by ANSYS calculation, as shown in Fig. 4, where the maximal Mises equivalent stress was 170.22MPa. In addition, using ANSYS postprocessor, the circular and axial stress were obtained, which indicated the maximal stress of CCS occurred in the internal wall of CSW, as shown in Table 5, while the stress of cement sheath and stratum were much lower than CSW.

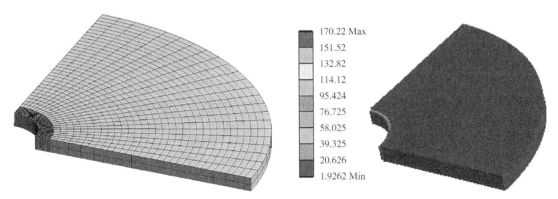

Fig. 3　Finite element model and mesh generation of CCS　　　Fig. 4　Stress distribution of CCS

Table 5　Stress numerical solutions of CSW

Stress	Circular stress of internal wall (MPa)	Circular stress of external wall (MPa)	Axial stress (MPa)
Numerical solutions	148.14	128.28	37.03

4 Discussion

The results of analytic calculation and numerical calculation were very close and their relative deviation was less than 3%, which verified their reliability. In addition, based on thenational standard of GB/T 150—2011, the safety factor of tensile strength (n_b) is 2.6 and the safety factor of yield strength (n_s) is 1.5, therefore the admissible stress ($[\sigma]$) of N80Q steel was:

$$[\sigma] = \min\left\{\frac{\sigma_s}{n_s}, \frac{\sigma_b}{n_b}\right\} = \min\left\{\frac{552}{1.5}, \frac{689}{2.6}\right\} = 265 \text{MPa} \quad (10)$$

The admissible stress (265MPa) was higher than the practical value of CSW (170.22MPa), indicating that CSW was still in elastic and secured state.

The results of the stress analysis and calculationfor CCS indicated that the well cementation could transmit the compression of crustal stress to CSW, and prevented the deformation of CSW at the same time. To analyze the potentiation of well cementation, the circular and axial stress of CSW internal wall before well cementation was calculated, which was 184.58MPa and 47.87MPa respectively by Eq. 7. Based on the Eq. 11, the potentiation factor by well cementation to CSW circular and axial stress was 18% and 20%, respectively.

$$\delta = \frac{\sigma' - \sigma}{\sigma'} \quad (11)$$

where σ' was the stress of CSW before well cementation and σ was the stress of CSW after well cementation.

The pressure fluctuation was very high at CNG storage time and gas fillingtime, even beyond the operation pressure sometimes. To ensure the safety operation of CSW, its load capacity should be enhanced. Compared to improving the material of CSW and the performance of stratum, changing the parameters of cement sheath was easier. This work chose the elasticity modulus of cement sheath between 3GPa and 28GPa to obtain the stress of CSW by Eq. 7, as shown in Table 6. It indicated that the stress of CSW decreased with the increase of elasticity modulus of cement sheath, while the support and potentiation to CSW by cement sheath became stronger for its rigidity helps CSW bear more load. For this reason, more and more scholars dedicated on the research to adjust the additive, water-cement ratio and grade of cement to improve well cementation technology for increasing its elasticity modulus.

Table 6 Circular stress of CSW with different elasticity modulus of cement sheath

Elasticity modulus of cement sheath(GPa)	The circular stress of internal wall(MPa)	The circular stress of external wall(MPa)
3	163.28	143.04
9	145.04	126.62
15	128.15	111.60
21	113.20	98.11
28	99.76	85.94

5 Summary

As for CSW, the stress not only comes from internal pressure but also extrusion force by stratum. The research on the stress distribution of CCS in light of a multi-layer thick-walled cylinder has been carried out in the present study. On the basis of the elastic mechanics theory and the finite element method of ANSYS, the analytic and numerical solutions of CSW have been obtained, which could calculate the stress at any point. The results indicate that well cementation could enhance the load capacity of CSW, and the potentiation factor by which to CSW circular and axial stress is 18% and 20%, respectively. The stress of CSW decreased with the increase of elasticity modulus of cement sheath. These findings can provide a way to improve the load capacity of CSW.

References

[1] SHI K, DUAN Z X, CHEN Z Z, et al. Current situation and prospect of underground gas storage well[J]. China Special Equipment Safety, 2014, 30(6): 5-10.

[2] TRIVEDI S, PRASAD R, MISHRA A, et al. Current scenario of CNG vehicular pollution and their possible abatement technologies: An overview[J]. Environmental Science and Pollution Research, 2020, 27(32): 39977-40000.

[3] ZHOU T Y, LI Y T, QIE Y H, et al. Study on fatigue life of CNG storage wall with thickness defect of wellbore [J]. Bulletin of Science and Technology, 2019, 35(6): 177-182.

[4] CHEN D H, CAO W F, ZHANG T. The gas injection control of CNG automobile based on bp neural network [J]. Applied Mechanics and Materials, 2013, 07: 626-630.

[5] HAN J L, LI Y S. Beijing necessity of development of natural gas buses[C]. Advanced Materials Research, 2011, 339: 509-516.

[6] ZHAO L, YAN Y F, WANG P, et al. A risk analysis model for underground gas storage well integrity failure [J]. Journal of Loss Prevention in the Process Industries, 2019, 62: 1-4.

[7] XIAO W, ZHANG Z C. Study on the production mode and leakage risk of gas storage well completion [C]. Proceedings of IOP Conference Series: Earth and Environmental Science, 2019, 233(4): 1-4.

[8] KIM E S, CHOI S K. Risk analysis of CNG composite pressure vessel via computer-aided method and fractography[J]. Engineering Failure Analysis, 2013, 27: 84-98.

[9] SONG C L, DAN Y. Stress analysis of the underground gas storage well stored CNG based on the finite element method[C]. Advanced Materials Research, 2013, 704: 338-342.

[10] CHU R G, SUN H F. Stress of Gas storage wells on the wall thickness under uniform terrestrial stress[J]. Oil Field Equipment, 2010, 39(2): 10-12.

[11] LI Z Y, SUN J F, LUO P Y. Research on the law of mechanical damage-induced deformation of cement sheaths of a gas storage well[J]. Journal of Natural Gas Science and Engineering, 2017, 43: 48-57.

[12] ZHANG H, SHEN R C, YUAN G J, et al. Analysis on the elastoplastic of cement sheath in underground gas storage wellbore[J]. Oil & Gas Storage and Transportation, 2018, 37(2): 150-156.

[13] YU F, YIN F, ZHANG D J. The effect of materials of formation and cement sheath on mechanical state of wellbore[J]. Journal of Yangtze University (Natrual Science Edition), 2019, 16(1): 34-37.

[14] WANG Z W, CAI R L. Design of Chemical Vessel[M]. Beijing: Chemical Industry Press, 2005: 178-182.

本论文原收录于 Journal of Physics：Conference Series 2021 年第 2002 卷第 1 期。

Q235B 原油储罐底板腐蚀穿孔原因

武 刚[1,2]　徐 帅[3]　张 楠[3]　张庶鑫[1,2]　孙冰冰[3]　周会萍[3]　马建朝[3]

(1. 中国石油集团石油管工程技术研究院；2. 石油管材及装备材料服役行为与结构安全国家重点实验室；3. 中石油管道有限责任公司西部分公司)

摘 要：某 Q235B 钢储罐底板发生腐蚀穿孔泄漏事故。通过宏观分析、微观分析、化学成分分析、力学性能测试、金相检验等方法，对储罐底板腐蚀等穿孔原因进行了分析。结果表明，底板涂层鼓包破损，在破损点发生氧腐蚀，进而导致储罐底板腐蚀穿孔。

关键词：原油储罐；底板；腐蚀穿孔；氧腐蚀

储罐是石油化工行业中非常重要的设备，它对石油化工装置的"安、稳、长、满、优"运行起到重要作用，还被广泛用于港口、石油化工企业和油库的液体原料、中间产品储存以及原油运输业[1-2]。原油储罐底板的腐蚀是储罐腐蚀比较严重的部位，而原油储罐底板内侧腐蚀问题主要是由罐底沉积水和沉积物引起的，沉积水主要是原油在开采、运输、储存等过程中所携带的水分在储存时通过沉降沉积出的[3]（由于储罐罐底排水管的结构限制，即使储罐经常进行罐底水排放也不能全部排出），导致罐底长期滞留有一定量的沉积水；沉积物主要是油泥。沉积水与沉积物的成分非常复杂，导致储罐内底板的严重腐蚀。国内某输油站对某泄压罐进行检测时发现罐底板腐蚀严重，其中中幅板发生腐蚀穿孔，随后生产单位进行了底板更换作业。该储罐为原油储罐，拱顶型，容积为 700m³，直径 10.2m，高 10.2m。储罐底板厚 8mm，材质为 Q235B。为找出储罐底板腐蚀穿孔的原因，防止此类事故的再次发生，笔者对其进行了相关的检验和分析。

1 理化检验

1.1 宏观分析

被腐蚀的储罐底板试样宏观形貌如图 1 所示。储罐底板上均覆盖有灰色涂层，涂层表面有大量鼓包，穿孔部位位于中幅板，附近底板的涂层鼓包已经剥落，露出锈蚀底板。罐底板壁厚测量检测最小值 7.8mm，如图 2 所示，符合 API 650—2013《焊接石油储罐》的技术要求；防腐层检测结果表明罐底板防腐层附着力 3 级，有起泡，涂层测厚值 300~560μm。

基金项目：国家"十三五"国家重点研发计划(2017YFC0805804)。
作者简介：武刚(1985—)，男，硕士，高级工程师，主要研究方向为油气管道及储存设施完整性。
E-mail：wugang010@cnpc.com.cn。

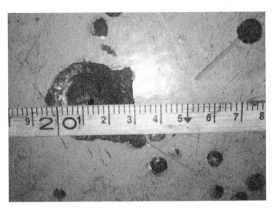

图1 腐蚀穿孔处宏观形貌

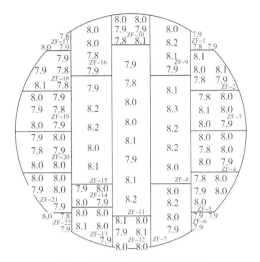

图2 储罐底板壁厚测试结果

1.2 微观分析

采用蔡司 Smartzoom 5 超景深三维显微镜对储罐底板上泄漏穿孔的腐蚀坑进行表征。图3中可以看到腐蚀坑已经贯穿底板，呈现"火山坑"样，表面覆盖黄褐色腐蚀产物，穿孔处有黑色"油泥状"物质，推测为油泥堆积。腐蚀坑深约6.5mm，穿孔直径约为1.5mm。取腐蚀穿孔部位试样进行组织分析，结果表明储罐底板腐蚀坑处组织为铁素体+珠光体，铁素体晶粒度8.5级，未见异常组织显示。

（a）宏观形貌

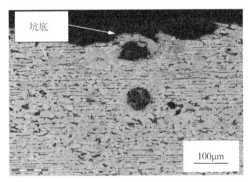

（b）显微组织形貌

图3 腐蚀坑宏观形貌及显微组织形貌

在罐底板上取腐蚀坑试样，采用 OXFORD INCA350 扫描电子显微镜（SEM）及能谱分析仪（EDS）对断口截面腐蚀产物进行形貌和能谱分析，结果如图4所示。可见，腐蚀坑底部有直径1mm的穿孔，腐蚀产物疏松，呈龟裂状，腐蚀产物主要元素为 O、S、Cl、Fe。

对现场采回的沉积水样进行成分分析，pH 值8.67，呈弱碱性，介质中 Cl^- 的质量浓度为44500mg/L、SO_4^{2-} 的质量浓度为2540mg/L，沉积水中存在较多的 Cl^-，极易诱发缝隙腐蚀和孔蚀等局部腐蚀，SO_4^{2-} 的存在易于引起硫酸盐还原菌（SRB）腐蚀。

1.3 化学成分分析

采用 ARL 4460 直读光谱仪，依据标准 GB/T 4336—2016《碳素钢和中低合金钢 多元素含量的测定 火花放电原子发射光谱法（常规法）》对远离腐蚀坑区域的底板取样进行了化学

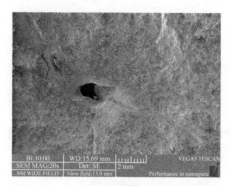

(a) SEM形貌

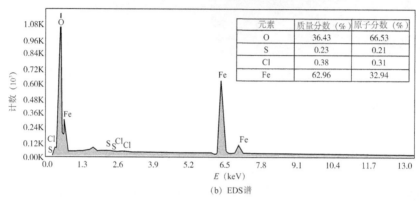

(b) EDS谱

图 4 腐蚀坑 SEM 形貌及 EDS 谱

成分测试,结果如表 1 所示。储罐底板化学成分分析结果符合 GB/T 700—2006《碳素结构钢》对 Q235B 钢的成分要求。

表 1 储罐底板的化学成分分析结果　　　　　　　　　　单位:%(质量分数)

元素	C	Si	Mn	P	S	Cr	Ni	Nb	V	Ti	Cu	Al
实测值	0.15	0.19	0.57	0.028	0.02	0.03	0.027	<0.0008	0.0035	0.001	0.07	0.0017
标准值	≤0.20	≤0.35	≤1.40	≤0.045	≤0.045	≤0.30	≤0.30	—	—	—	—	≤0.30

1.4 力学性能试验

采用 UTM5305 材料试验机依据 GB/T 228.1—2021《金属材料 拉伸试验 第 1 部分:室温拉伸试验方法》标准对远离腐蚀坑区域的底板取样进行室温拉伸性能测试,拉伸试样取板状试样,规格为 12.5mm×50mm(宽度×标距),结果如表 2 所示,可见拉伸性能符合 GB/T 700—2006 的技术要求。

表 2 远离腐蚀坑区域的储罐底板的拉伸试验结果

试样编号	宽度×标距(mm)	抗拉强度 R_m(MPa)	屈服强度 R_{eH}(MPa)	断面收缩率 A(%)
1#	12.5×50	462	324	40.5
2#		459	305	38.0
3#		456	295	43.0
GB/T 700—2006 中对 Q 235B 的拉伸性能要求		370~500	≥235	≥26

WZW-1000材料弯曲试验机依据GB/T 228.1—2021《金属材料 拉伸试验 第1部分：室温拉伸试验方法》标准对远离腐蚀坑区域的底板取样进行弯曲试验，由表3的试验结果可见，储罐底板的弯曲性能满足标准GB/T 700—2006的技术要求。

表3 远离腐蚀坑区域的储罐底板的弯曲试验结果

方向	板宽 b(mm)	壁厚 T(mm)	弯轴直径 d(mm)	弯曲角度 α(°)	弯曲方向	结果
横向	38	8	5	180	弯曲	未出现裂纹
横向	38	8	5	180	弯曲	未出现裂纹
纵向	38	8	10	180	弯曲	未出现裂纹
纵向	38	8	10	180	弯曲	未出现裂纹
GB/T 700—2006中对Q 235B的弯曲性能要求			横向试样冷弯弯轴直径为8mm，纵向试样冷弯弯轴直径为12mm			

1.5 金相检验

采用MEF3A金相显微镜、MEF4M金相显微及图像分析系统，依据GB/T 13298—2015《金属显微组织检验方法》、GB/T 6394—2017《金属平均晶粒度测定方法》、GB/T 10561—2005《钢中非金属夹杂物含量的测定标准 评级图显微检验方法》和GB/T 34474.1—2017《钢中带状组织的评定第1部分：标准评级图法》，对远离腐蚀坑区域的组织和非金属夹杂物进行分析。由表4中数据及图4可见，储罐底板组织均为铁素体+珠光体，铁素体晶粒度为8.0~8.5级，带状组织0.5级，未见异常组织显示。

表4 金相分析结果

非金属夹杂物	组织		
	储罐底板组织	铁素体晶粒度	带状组织
A0.5, B1.0, D0.5	F+P	8.0~8.5级	0.5

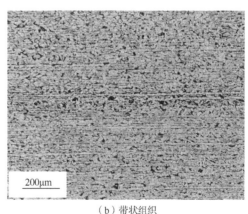

（a）组织　　　　　　　　　　　　　（b）带状组织

图5 远离腐蚀坑区域的储罐底板的金相组织

2 分析与讨论

储罐底板试样表面有大量的涂层鼓包，穿孔部位附近有大量鼓包破损，且露出基体，基体均存在腐蚀。结合现场提供的储罐底板壁厚测试结果，储罐整体壁厚并未发生减薄。因此可以推断，该腐蚀穿孔主要原因为涂层鼓包破损，在破损点发生点蚀，进而发生穿孔。

根据腐蚀产物分析能谱结果，腐蚀产物主要为Fe、O、S、Cl等元素，可以进一步明确，该储罐穿孔失效为氧腐蚀造成[4-6]。该储罐储存介质为原油，一般而言，原油本身并不具有腐蚀性，相反，原油还会在罐壁形成一层油膜，减弱储罐的腐蚀。造成原油储罐罐底板腐蚀的主要原因是罐底沉积水。原油在开采和运输过程中，会携带一定量的水分，当原油进出原油储罐静止储存时，原油携带的水分以及空气中的水蒸气就会凝结沉降在储罐底部。虽然一些原油储罐定期进行罐底沉积水的排放，但不一定能全部排出，罐底通常会存留200~300mm深度的沉积水。原油储罐罐底除了存积的沉积水外，还有固态的沉积物，由于沉积水的存在，一般沉积物下都容易形成适合发生局部腐蚀的条件。

沉积水是非常强的电解质溶液，含有大量的阴阳离子，矿化度非常高。其中以氯离子含量最高，容易造成点蚀；有的沉积水中钙、镁离子含量相对较高，易于形成沉积结垢，造成垢下腐蚀和氧浓差电池腐蚀，SO_4^{2-}、Cl^-的存在对腐蚀有促进作用[7-8]。覆盖储罐底板的油泥质地并不均匀，有的区域比较致密，黏稠性高，通透性差，可以充分地附着在底板表面，阻止腐蚀介质和氧与金属的接触，而其他区域的油泥质地相对疏松，并不能完全阻止腐蚀介质和氧与金属的接触，从而使部分区域发生腐蚀，而且质地疏松的油泥也在一定程度上影响了离子的扩散，造成闭塞原电池，再加上酸化自催化作用，使得局部腐蚀加剧。

3 结论及建议

储罐底板的化学成分、拉伸性能、弯曲性能均满足标准GB/T 700—2006对Q235B材质的要求。储罐底板腐蚀穿孔的主要原因为涂层鼓包破损，在破损点发生氧腐蚀，进而发生穿孔。

建议优选储罐防腐内涂层，增加牺牲阳极块阴极保护，定期排除储罐底部积水，加强储罐底板腐蚀检测及监测，实施储罐完整性管理。

参 考 文 献

[1] 刘晓斌.原油储罐防腐失效分析与研究[J].科技资讯.2014,12(14)：81-82.
[2] 曹华珍.原油储罐内底板腐蚀机理研究和防护措施[D].杭州：浙江工业大学，2002.
[3] 阎永贵，吴建华，陈光章.原油储罐内底板的腐蚀防护现状与展望[J].腐蚀与防护，2002(5)：199-201.
[4] 刘宇程，张寅龙，廖斯丞，等.$CaCl_2$型采出水对原油储罐底板的腐蚀影响分析[J].西南石油大学学报（自然科学版），2014，36(5)：155-159.
[5] 黄世海.Cl^-对原油罐底板腐蚀分析与防护[D].兰州：兰州理工大学，2006.
[6] 肖成磊.Q235B碳钢在原油储罐罐底沉积水中腐蚀行为研究[D].青岛：中国石油大学（华东），2016.
[7] 孙永泰.原油储罐底板发生腐蚀穿孔的原因分析[J].石油化工腐蚀与防护，2008，25(5)：63-64.
[8] 李循迹，张强，殷泽新，等.原油储罐罐底板腐蚀穿孔原因分析及对策[J].石油化工设备，2016，45(4)：89-94.

本论文原发表于《理化检验：物理分册》2021年第57卷第2期。

谐波沉降作用下大型原油储罐变形响应分析

武 刚[1]　徐 帅[2]　张 哲[3]　张庶鑫[1]　周会萍[2]　朱永斌[2]

(1. 中国石油集团石油管工程技术研究院·石油管材及装备材料服役行为与结构安全国家重点实验室；2. 国家管网集团西部管道公司；3. 中国石油新疆油田公司)

摘　要：地基沉降往往导致储罐罐壁产生较大变形，严重影响储罐的安全运行。为探究原油储罐在地基谐波沉降作用下的变形响应状态，采用有限元分析软件ABAQUS建立了谐波沉降作用下大型原油储罐的数值仿真模型。模型综合考虑了地基的影响，能够较准确模拟储罐的真实服役状态。基于所建数值仿真模型，定量研究了储罐罐壁径厚比、高径比、谐波数、谐波幅值及液位对储罐罐壁径向变形量的影响规律。结果表明：罐壁径厚比及液位对储罐罐壁顶端的径向变形量的影响较小，从工程应用角度可忽略其影响；储罐罐壁顶端径向变形量随着罐壁高径比及谐波幅值的增加近似呈线性增大；罐壁径向变形量随着谐波数的增加先增大后减小。

关键词：原油储罐；地基耦合；谐波沉降；变形响应；数值仿真

目前，我国大型原油储罐的建造数量日益增多。储罐罐容的增加使得罐体的薄壳结构特征更为明显。在各种服役荷的综合作用下，储罐基础通常会发生不同类型的沉降，严重影响了储罐的安全可靠运行。为有效评价地基沉降对储罐安全服役的影响，需深入开展地基非均匀沉降作用下大型储罐的力学响应分析，在此基础上制订科学有效的安全评价标准。

针对地基沉降作用下储罐的安全评价指标主要包括对径点沉降差、相邻点沉降差及不均匀沉降量。SY/T 5921—2017《立式圆筒形钢制焊接原油罐修理规程》规定储罐罐底10 m弧长内相邻点的沉降差不大于12 mm，对径点的沉降差不大于储罐外径的0.0035倍。API 653—2009《地上储罐检验、修理、改建》给出了储罐罐壁板所允许的最大地基不均匀沉降量的计算公式。以上标准在规定储罐最大允许沉降量时，忽略了地基的影响。而考虑地基的储罐应力状态分析能够较好地满足现场需求。储罐现场失效数据的统计分析表明，储罐的沉降值在满足现有标准要求的情况下依然存在失效问题。因此，现有标准由于忽略地基影响所得到的地基沉降量控制指标存在不合理性，有必要开展相关实验和数值仿真研究来获取相对准确的地基沉降作用下储罐力学响应状态。

国内外学者围绕地基沉降作用下大型原油储罐的力学响应开展了大量研究。Gong等利用对称的储罐罐壁有限元模型分析了大型储罐在不均匀沉降作用下的变形规律及屈曲行为特

基金项目：国家"十三五"重点研发计划资助项目"原油天然气储罐及附属管道、辅助设施安全评定与风险评价预警研究"，2017YFC0805804。

作者简介：武刚，男，1985生，高级工程师，2011年硕士毕业于北京科技大学机械设计及理论专业，现主要从事油气管道及储存设施完整性相关技术研究。地址：陕西省西安市雁塔区锦业二路89号石油管工程技术研究院，710077。电话：15191899113。E-mail: wugang010@cnpc.com.cn。

征，定量研究了储罐材料非线性、罐壁高径比、径厚比对储罐屈曲临界沉降量的影响，同时分析了罐顶抗风圈对储罐屈曲临界地基沉降量的影响[1-6]。Nassernia等[7-14]基于实验和理论分析方法研究了地基沉降作用下储罐的失效模式，同时考虑了储罐壁厚变化对罐体失效行为的影响。石磊等[15-19]基于大型原油储罐的全尺寸模型定量研究了谐波数、谐波幅值及液位对储罐罐壁径向变形的影响，并分析了地基单次谐波沉降的谐波次数、谐波幅值对大型原油储罐象足屈曲临界载荷的影响规律。范海贵等[20-21]采用有限元模拟方法模拟储罐在不均匀沉降作用下的变形响应。黎伟等[22]采用有限元方法分析了地基整体倾斜1.5°和地基局部不均匀沉降达到20mm工况下，大型石油储罐的变形形态以及应力分布特征。现有研究初步探明了地基沉降作用下储罐的力学响应状态，在一定程度上明确了储罐材料性能、尺寸参数及地基沉降形式对储罐变形及屈曲行为特征的影响规律，为地基沉降作用下储罐的安全评价提供了理论基础。但是现有的数值仿真研究多数基于对称模型开展，较少考虑底板与地基的耦合作用，将地基的沉降测量数据直接加载在储罐壁板或底板上，忽略了地基沉降作用下储罐底板与地基之间的脱离现象。在此基础上，部分学者对储罐数值仿真模型提出改进，建立了考虑地基耦合的储罐全尺寸有限元模型，初步开展了地基不均匀沉降作用下储罐的变形响应研究，提高了数值计算的准确性。但对沉降作用下储罐力学响应的影响因素缺乏全面系统的分析，定量研究较少，所得结论不能较好地指导现场储罐的安全评价。

在此建立了考虑地基与罐底板耦合作用的大型原油储罐有限元模型并开展了模型准确性验证分析。基于相对准确的储罐全尺寸数值仿真模型，定量研究了储罐罐壁径厚比、高径比、谐波数、谐波幅值及液位对谐波沉降作用下储罐罐壁变形响应的影响规律，综合考虑了各个影响因素之间的相互作用关系，并提出了保证谐波沉降作用下储罐安全服役的改进措施。

1 有限元模型建立

1.1 几何模型

考虑地基耦合作用的大型原油储罐数值仿真模型主要由壁板、底板、大角焊缝及地基组成。储罐外径为40m，高度为21.8m。储罐壁板及底板属于典型的薄壳结构，采用4节点壳单元（S4R）进行模拟。储罐壁板及底板连接部位的大角焊缝及环墙式地基采用三维8节点实体单元（C3D8R）进行模拟（图1）。整个数值仿真模型所包含的单元数量为52210个。

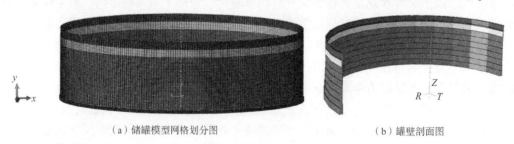

(a) 储罐模型网格划分图　　　　　　(b) 罐壁剖面图

图1　大型原油储罐有限元模型示意图

1.2 材料模型

为了满足储罐的强度要求，大型原油储罐第9层壁板及中幅板采用的材料类型为Q235-B；罐底边缘板及第1~7层壁板采用的材料类型为08MnNiVR；第8层壁板采用的材料类型为16MnR[16]。模型综合考虑了材料非线性，储罐材料采用Ramberg-Osgood模型。3种材料的应力应变关系曲线如下(图2)。

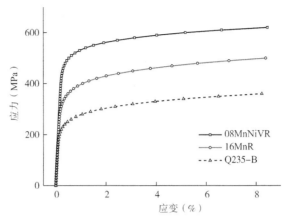

图 2　储罐材料的应力应变关系曲线

1.3　载荷及边界条件设定

大角焊缝与储罐底板的接触面之间设置 Tie 绑定约束。储罐底板与地基之间设置摩擦接触,摩擦因数为 0.2。储罐的初始运行载荷包括罐体及其附件的自重以及液体静压力(图3)。整个数值模拟过程包括重力分析步与沉降分析步。在重力分析步中,模型的边界条件设定如下:储罐地基下表面固定;模型施加重力荷载;罐壁及底板施加液体压力。罐壁的液压呈三角形分布,从罐底到罐顶逐渐减小。储罐罐底的液体压力以均布载荷的形式施加在储罐底板上。储罐底板径向的垂直位移基本可认为向中心线性递减[16]。在谐波沉降分析中,将砂土地基的环向外边缘节点的轴向位移设置为单次谐波沉降量,砂土地基径向节点的轴向位移沿半径方向线性增大。

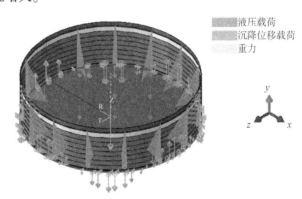

图 3　原油储罐载荷分布状况示意图

1.4　模型验证

基于现场 $10×10^4 m^3$ 大型原油储罐的几何尺寸建立数值仿真模型,通过对比分析现场应力实测结果与数值仿真结果验证有限元模型的准确性[16]。绘制液压与重力作用下储罐外壁轴向应力及环向应力随储罐高度的变化曲线(图4),其中蓝色三角形为 19.76 m 水位下储罐罐壁应力的实测值,红色曲线为文献[16]的仿真计算结果,绿色曲线为该模型有限元计算结果。结果表明:数值仿真结果与现场实测结果的变化规律基本一致,验证了地基沉降作用下储罐有限元模型的准确性。

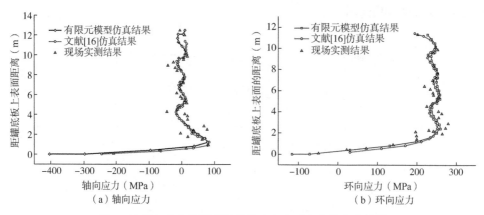

图 4 液压与重力作用下储罐外壁应力随高度的变化曲线

2 储罐变形响应分析

模拟得到相同液位下(液位高度设为 19.76m)储罐不同高度处的罐壁径向变形量随罐壁周向展开角的变化关系曲线(图 5)。由于罐壁径向变形量受到地基沉降载荷及液压载荷的影响，罐壁径向变形量沿储罐周向近似呈周期对称分布，周期数与谐波数相对应。罐壁较低位置处的径向变形量受液压载荷的影响较大，因此波峰位置的罐壁径向变形量为正值。随着罐壁高度的增加，液压对罐壁径向变形量的影响逐渐减小。当液压载荷的影响小于沉降载荷的影响时，波峰位置的罐壁径向变形量开始由正值转为负值。罐壁径向变形量沿储罐高度方向的变化关系曲线如下(图 6)，在相同地基谐波沉降作用下，罐壁径向变形量随储罐高度的增加而增大。

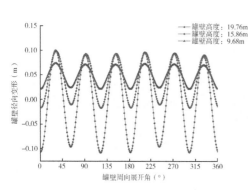

图 5 罐壁不同高度处罐壁径向变形随罐壁周向展开角的变化曲线

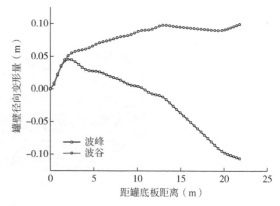

图 6 罐壁径向变形沿储罐高度方向的变化曲线

3 储罐变形的影响因素

3.1 罐壁径厚比

为了满足日益增长的原油储备需求，大型原油储罐的数量不断增加。随着单罐罐容的增加，储罐的薄壳结构特性随之退化。在相同地基沉降作用下，储罐罐体几何尺寸的变化将会显著影响储罐的变形响应状态。大型原油储罐的壁厚设计遵循等强度设计准则，沿储罐高度方向壁厚逐渐减薄。在保证储罐罐壁外径及各层壁板厚度的比值不变的前提下，通过改变储罐壁板的厚度来实现储罐模型的径厚比变化。储罐径厚比定义为罐体外半径与平均壁厚的比值(表1)。参考文献[6]中罐体径厚比的取值范围。通过改变储罐外径，定量计算不同径厚比储罐在特定单次谐波沉降作用下的变形响应，同时分析储罐液位对罐体变形的影响。具体计算工况参数如表1所示。

表1 不同径厚比储罐模型计算参数

谐波幅值(mm)	谐波数	高径比	径厚比	液位(m)
40	3	0.5	500, 800, 1 200, 1 500, 1 700, 2 000	0, 9.68, 19.76

不同径厚比下，由储罐罐壁顶端径向变形量随罐壁周向展开角的变化关系(图7)可知：当储罐液位为19.76m时，罐壁顶端径向变形量随罐壁径厚比的增加而增大。不同液位下，波谷位置的罐壁顶端径向变形量随罐壁径厚比的变化关系如下(图8)，波谷位置的罐壁顶端径向变形量随罐壁径厚比增加几乎维持不变，只有毫米级的增加；液位高度对罐壁径向变形量的影响较小，表明对于正常服役的大型储罐，壁厚的改变对谐波沉降下储罐径向变形的影响较小。

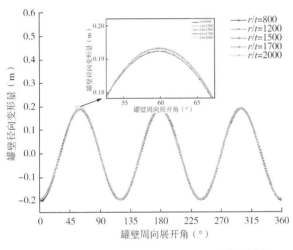
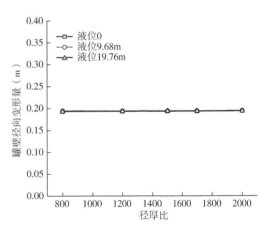

图7 不同径厚比下罐壁径向变形量随罐壁周向展开角的变化曲线(液位19.76 m)

图8 不同液位下罐壁径向变形量随径厚比的变化关系图

3.2 储罐高径比

不同类型储罐的高度往往具有较大差异，增加储罐高度能够在一定程度上增加储罐的储备能力。随着储罐高度的增加，地基沉降作用下储罐的刚度和稳定性将会发生显著变化。设定储罐外直径为80 m不变，储罐的高径比定义为储罐高度与罐体外径的比值，参考文献[6]中储罐罐体的高径比取值范围，定量计算不同高径比下储罐的变形响应结果(表2)。

表2 不同高径比储罐模型计算参数

谐波幅值(mm)	谐波数	高径比	径厚比	液位(m)
40	3	0.5, 0.8, 1.0, 1.2	2 200	0, 9.68, 19.76

不同高径比下，储罐罐壁顶端径向变形量随罐壁周向展开角的变化关系曲线如下

(图9),罐壁顶端径向变形量随罐壁高径比的增加而增大。由不同液位下波谷位置的罐壁顶端径向变形量随罐壁高径比的变化关系图(图10)可知,罐壁顶端径向变形量随罐壁高径比的增加近似呈线性增大。这是由于当储罐直径一定时,随着储罐高度的增加,储罐的稳定性逐渐降低,罐壁顶端更容易发生变形。液位对罐壁顶端径向变形量的影响相对较小,从工程应用角度看,其影响可忽略不计。在相同地基沉降作用下,适当减小储罐罐壁高径比能够降低储罐罐壁顶端的变形量。

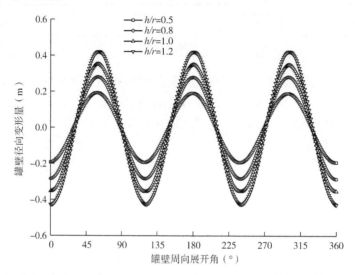

图9 不同高径比下罐壁径向变形量随罐壁周向展开角的变化曲线(液位19.76m)

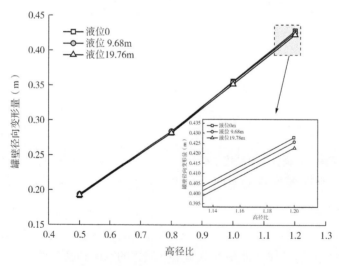

图10 不同液位下罐壁径向变形量随高径比的变化关系图

3.3 谐波数

现场地基不均匀沉降曲线可借助傅里叶级数分解为若干单次谐波沉降。研究单次谐波沉降作用下储罐的变形响应结果能够为不均匀沉降作用下储罐的变形响应分析提供一定的理论参考。研究表明,6次以内的谐波沉降能够较准确反映地基不均匀沉降的真实状态。谐波数 $n=0$、1 分别代表储罐的整体均匀沉降和平面倾斜沉降,其对储罐的安全运行影响相对较小。定量计算不同谐波数及液位下储罐的变形响应(表3)。

表 3　不同谐波数下储罐模型计算参数

谐波幅值(mm)	谐波数	高径比	径厚比	液位(m)
40	2,3,4,5,6	0.5	2 200	0, 9.68, 19.76

不同谐波数下，由储罐罐壁顶端径向变形量随罐壁周向展开角的变化关系图(图11)可知，储罐罐壁的径向变形量在罐体周向呈周期对称分布，周期数与谐波数一一对应。当液位较低时，罐壁最大径向位移随谐波数增加而增加(图12)。此时，罐底板与地基完全重合，地基谐波数的增加使得罐体需要在更小的环向宽度下产生同样的垂向位移，从而导致罐体变形随谐波的增加而一直增大。与液位较低工况不同，满罐时，罐体变形先增加后减小。这主要是由于罐内液体对罐壁板产生了较大的环向应力，使得罐体整体刚度增大，在谐波数大于4后，罐体底板与地基产生了脱离，脱离后谐波数增加使得罐体环向支撑点增加，其变形程度反而减小。

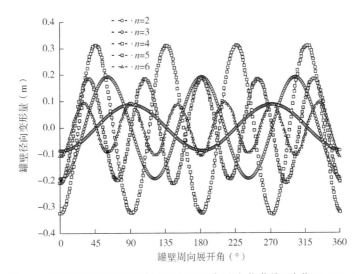

图 11　罐壁径向变形量随罐壁周向展开角的变化曲线(液位 19.76m)

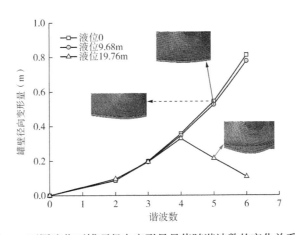

图 12　不同液位下罐顶径向变形量最值随谐波数的变化关系图

3.4　谐波幅值

依据储罐地基不均匀沉降的现场测试数据可知，低阶谐波组分的幅值相对较大，各阶单次谐波沉降的幅值一定程度上反映了储罐的实际沉降形式。谐波幅值对单次谐波沉降作用下储罐的变形响应状态具有显著影响。定量计算不同谐波幅值及液位下储罐的变形响应结果（表4）。

表4 不同谐波幅值下储罐模型计算参数

谐波幅值（mm）	谐波数	高径比	径厚比	液位（m）
20，40，60，80，100	2，3，4，5，6	0.5	2200	0，9.68，19.76

不同谐波幅值下，储罐罐壁顶端径向变形量随罐壁周向展开角的变化曲线（图13）可知，罐壁顶端径向变形量随谐波幅值的增加而增大，罐壁径向变形量沿储罐罐壁周向呈现周期对称分布。计算得到不同液位下，罐壁顶端径向变形量随谐波幅值的变化关系（图14），当谐波数为3时，波谷位置的罐壁径向变形量随谐波幅值的增加呈线性增大。这是由于谐波数为3时，储罐底板与地基之间未发生脱离，随着谐波幅值的增加，罐壁受地基沉降作用的影响越显著。在不同谐波大小下，液位对罐壁径向变形的影响仍然很小，可忽略。

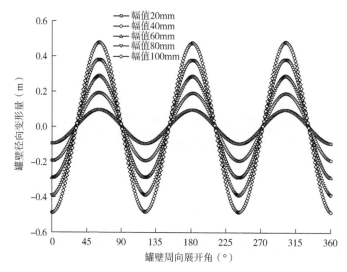

图13 不同谐波幅值下罐壁径向变形量随罐壁周向展开角的变化曲线（液位19.76 m）

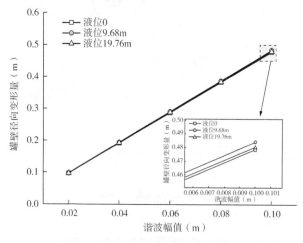

图14 罐壁径向变形量最值随谐波幅值的变化关系图

4 结论

采用非线性有限元软件建立了准确考虑罐体与地基耦合作用的谐波沉降作用下大型原油储罐的数值仿真模型,通过试验数据验证了模型的准确性。基于建立的数值计算模型,开展了影响因素分析,定量研究了储罐罐壁径厚比、高径比、谐波数、谐波幅值及液位对储罐罐壁径向变形量的影响规律。

(1)罐壁径厚比及液位对储罐罐壁顶端的径向变形量的影响相对较小;储罐罐壁顶部径向变形量随着罐壁高径比的增加近似呈线性增大。

(2)相同地基沉降作用下,适当减小储罐罐壁高径比能够降低罐壁顶端的变形量;罐壁顶端径向变形量随谐波幅值的增加而线性增大;在低液位条件下,储罐罐壁顶端的径向变形量随谐波数的增加而增加;在高液位条件下,储罐罐壁顶端径向变形随谐波数增大先增加,当谐波数增大导致底板与地基脱离时,储罐罐壁顶端径向变形随谐波数增大而减小。

参 考 文 献

[1] ZHAO Y, LEI X, WANG Z, et al. Buckling behavior of floating-roof steel tanks under measured differential settlement[J]. Thin-Walled Structures, 2013, 70: 70-80.

[2] CAO Q S, ZHAO Y. Buckling strength of cylindrical steel tanks under harmonic settlement[J]. Thin-Walled Structures, 2010, 48(6): 391-400.

[3] GONG J G, ZHOU Z Q, XUAN F Z, et al. Buckling strength of cylindrical steel tanks under measured differential settlement: harmonic components needed for consideration and its effect[J]. Thin-Walled Structures, 2017, 119: 345-355.

[4] ZHAO Y T, LIN Y M. Computer simulation model on tank's foundation settlement[C]. Dublin: 2019 International Conference on Artificial Intelligence and Advanced Manufacturing, 2019: 31-34.

[5] GONG J G, TAO J, ZHAO J, et al. Buckling analysis of open top tanks subjected to harmonic settlement[J]. Thin-Walled Structures, 2013, 63: 37-43.

[6] GONG J G, CUI W S, ZENG S, et al. Buckling analysis of large scale oil tanks with a conical roof subjected to harmonic settlement[J]. Thin-Walled Structures, 2012, 52: 143-148.

[7] NASSERNIA S, SHOWKATI H. Experimental Investigation to local settlement of steel cylindrical tanks with constant and variable thickness[J]. Engineering Failure Analysis, 2020, 118: 104916.

[8] IGNATOWICZ R, HOTALA E. Failure of cylindrical steel storage tank due to foundation settlements[J]. Engineering Failure Analysis, 2020, 115: 104628.

[9] GRGET G, RAVNJAK K, SZAVITS-NOSSAN A. Analysis of results of molasses tanks settlement testing[J]. Soils and Foundations, 2018, 58(5): 1260-1271.

[10] BOHRA H, GUZEY S. Fitness-for-service of open-top storage tanks subjected to differential settlement[J]. Engineering Structures, 2020, 225: 111277.

[11] 司刚强. 大型原油储罐的沉降观测及结果评价[J]. 石油化工设备, 2020, 49(4): 1-5.

[12] 杨勇. 薄壁圆柱壳在不均匀沉降下的试验研究[D]. 杭州: 浙江大学, 2011.

[13] 赵永涛, 赵星有, 林聿明. 储罐地基不均匀沉降造成壁板应变的数值模拟及试验验证研究[J]. 石油化工安全环保技术, 2019, 35(3): 17-21.

[14] GUNERATHNE S, SEO H, LAWSON W D, et al. Analysis of edge-to-center settlement ratio for circular storage tank foundation on elastic soil[J]. Computers and Geotechnics, 2018, 102: 136-147.

[15] 石磊, 帅健, 许葵. 基于 FEA 模型和 API653 的大型油罐基础沉降评价[J]. 中国安全科学学报,

2014，24(3)：114-119.

[16] 石磊. 大型原油储罐的强度与稳定性研究[D]. 北京：中国石油大学(北京)，2016.

[17] SHI L, SHUAI J, WANG X L, et al. Experimental and numerical investigation of stress in a large-scale steel tank with a floating roof[J]. Thin-Walled Structures, 2017, 117: 25-34.

[18] 石磊, 帅健, 许葵, 等. 大型非锚固变壁厚外浮顶原油储罐的应力测试[J]. 油气储运, 2017, 36(10)：1128-1132.

[19] 石磊, 帅健, 王晓霖, 等. 地基沉降对石油储罐象足屈曲的影响[J]. 中国安全科学学报, 2018, 28(9)：56-61.

[20] 范海贵, 陈志平, 徐烽, 等. 基于实测沉降的浮顶储罐变形分析[J]. 浙江大学学报(工学版)，2017，51(9)：1824-1833.

[21] CHEN Z P, FAN H G, CHENG J, et al. Buckling of cylindrical shells with measured settlement under axial compression[J]. Thin-Walled Structures, 2018, 123: 351-359.

[22] 黎伟, 宋伟, 宋金丽, 等. 不同沉降方式下大型储罐的变形与应力分析[J]. 空间结构，2016，22(4)：78-84.

本论文原发表于《油气储运》2021年第40卷第8期。

第二篇　成果篇

一、省部级(含社会力量)科技奖励

2021年获得省部级(含社会力量)科技奖励见表1。

表1 2021年获得省部级(含社会力量)科技奖励概览

序号	成果名称	颁奖机构	获奖等级	获奖类型
1	复杂油气井管柱工况模拟试验评价与应用技术	中国石油集团	一等奖	技术发明奖
2	高性能钻杆研发及检测评价技术研究	中国石油集团	一等奖	科技进步奖
3	油气管道缺陷修复质量检验评价和改进技术研究及应用	中国质量协会	一等奖	质量技术奖
4	苛刻油气井管柱服役安全可靠性评估技术及应用	陕西省人民政府	二等奖	科技进步奖
5	复杂油气井管柱优化设计与安全评价系列标准及应用	陕西省人民政府	二等奖	科技进步奖
6	超深井套管服役关键技术与特殊螺纹开发及应用	陕西省人民政府	二等奖	科技进步奖
7	集输管材应力导向氢致开裂腐蚀行为与选材图谱研究	中国石油集团	二等奖	科技进步奖
8	双金属复合管失效控制关键技术及应用	中国石油集团	二等奖	科技进步奖
9	特殊螺纹油套管检测评价及应用技术	中国石油集团	二等奖	科技进步奖
10	典型炼化装置高风险区成套防腐技术研究及工业应用	中国腐蚀与防护学会	二等奖	科技进步奖

1. 复杂油气井管柱工况模拟试验评价与应用技术

该项目围绕集团公司复杂油气田开发,形成系统的管柱工况模拟试验评价与应用技术。发明了非常规油气井筒工况模拟试验评价技术,实现了井筒非均匀外挤、剪切及复合加载模拟装备与试验能力,试验温度可达400℃、1000t拉压、1200t非均匀外挤、600t剪切、200MPa内压、40000N·m扭矩,形成技术标准和工业化软件,为复杂压裂套管选用评价提供了技术手段,填补了国内外技术空白。开发了基于应变设计的新型热采套管,实现工业产品推广,获得油田工程9年24轮次注汽零套损效果验证,获得中国石油自主创新产品认定,形成国家标准及评价软件,为稠油开发提供了产品及试验评价技术依据,在国内外首次实现了热采套管选用及评价技术标准化。发明了高速($30m^3/s$)气流摩阻系数测试装置与计算新方法、考虑振动(频率3~5kHz,强度<15G)的气密封螺纹分析及模拟试验评价方法,形成国家技术标准,为高压气井管柱动载环境下完整性评价提供了技术依据,填补了国内外技术空白。开发了钻柱旋转(140r/min)弯曲(20°/30m)疲劳试验装备,形成工况模拟试验能力,首次实现标准化应用,建立了抗硫钻杆选材及评价方法,实现了105ksi抗硫钻杆国产化,全面修订了ISO 11961国际标准,增加了抗硫钻杆新品种,新增D95及F105两个新钢级,成果为钻井工程,特别是含硫化氢环境钻杆选用及评价提供了技术依据,填补国内外产品空白。成果获得国家授权发明专利17件、实用新型专利3件,登记软件著作权2项,自主创新产品1项,制订标准6项,发表论文12篇,技术服务和相关制造企业借助技术研发成果增加产品销售额共计32.7亿元。

本成果获得2021年中国石油集团技术发明一等奖。

2. 高性能钻杆研发及检测评价技术研究

高性能钻杆是油气井通道打开的最重要依托载体,其承载能力及质量直接决定了钻井的能力、效率、安全和效益。项目组通过不断攻关核心技术,引领高性能钻杆向"高承载、长寿命、轻量化和安全可靠方向发展"。具体体现为:(1)研制了150ksi和160ksi钢级超高强韧钻杆材料,提高常规钻柱承载能力30%以上,现已成为我国超深井钻井管柱的主体材料。

(2)研制的 105~120ksi 系列钢级抗硫钻杆材料填补了国内技术空白,延长含硫环境下钻杆使用寿命80%以上。(3)发明了580MPa、620MPa级铝合金,720MPa、930MPa 和1200MPa级钛合金等轻量化钻杆材料,打破了国外技术垄断,降低常规钻柱悬重和钻机负载30%以上。(4)发明了高抗扭、高密封、耐疲劳等5种钢/铝/钛钻杆特殊螺纹接头结构,显著提升了钻杆的整体承载能力。(5)建立了钻杆接头螺纹检测新方法,发明的便携式螺纹锥度测量仪等4种螺纹测量装置,检测效率和精度提高逾50%。(6)形成了高性能钻杆选材评价方法和关键标准,解决了高性能钻杆规范化、使用极限及适用工况无据可依的问题。该成果创新性强,获授权国家专利14件,发表SCI/EI等高水平论文24篇,制定国家及行业标准11项;2011—2019年在塔里木油田应用80口井,在渤海能克钻杆等公司应用超8万吨,创效13.315亿元,经济和社会效益显著,应用前景广阔,技术总体达到国际先进水平。

本成果获得2021年中国石油集团科技进步一等奖。

3. 油气管道缺陷修复质量检验评价和改进技术研究及应用

本项目针对油气管道非动火和动火修复技术存在的工程缺陷、质量不稳定、无监督依据和检验指标等影响修复质量的关键技术问题,突破了油气管道缺陷修复质量检验评价和改进技术多项关键技术,取得了以下五个方面的创新成果:(1)在国内首次建立了系统的管道复合材料修复补强技术检验评价指标体系,包含补强修复设计参数、材料本身及施工质量三个部分检测评价内容,共4个一级指标和16个二级指标,并形成了3项企业标准。(2)在国内首次建立了系统的环氧钢套筒修复的评价指标体系,其中包括4项一级指标(套筒设计、材质、填充料和施工质量)和19项二级指标,根据现场施工方法建立了适合修复中或已服役环氧钢套筒的检测评价流程,形成了企业标准。(3)建立了扣帽、补疤、B型套筒等动火焊接修复技术评价指标和方法,包括现场检验评价方法、水压爆破验证、理化性能分析、动火焊接对主体管道影响程度,可有效检验焊接修复质量和提高修复可靠性。(4)建立了复合材料修复及环氧钢套筒修复质量飞检流程及检验指标体系,包括现场检验评价及室内检验评价,有效规范和提升了承包商的现场修复施工质量;(5)针对复合材料修复中存在的脱黏、分层等技术难题,研发了绝缘底漆、带锈转化、均匀加压固化等工艺改进技术,复合材料强度、弹性模量增加14%~20%,黏结力增加30%。采用内部预埋纤维布,自主设计研发了真空灌注浸渍工艺,避免了环氧钢套筒修复中树脂脆化开裂、空腔、固化气泡等问题,形成了新型复合环氧钢套筒技术和产品。

项目授权国家发明专利15件,制定标准规范7项,取得软件著作权1项,发表论文20余篇(SCI/EI 15篇)。研究成果在西气东输一线、二线和三线、陕京管线、西南油气田和长庆油田等地面管线共400余处缺陷隐患点进行了现场应用,避免了失效事故34起,取得直接经济效益3.94亿元,经济效益显著。研究成果将进一步丰富和完善油气管道失效控制和完整性管理技术,推动中国油气管道工程技术进步。

本成果获2021年中国质量协会质量技术一等奖。

二、授权发明专利目录

专利号	名　　　称	授权日期
ZL 2018116427132	一种油气管道凹陷变形应变场测量中应变片粘贴定位方法	2021-01-01
ZL 2020106140449	一种基于阵列探头的超声法应力测量方法	2021-08-31
ZL2017113978092	一种热采工况特殊螺纹接头密封接触压应力的预测方法	2021-04-30
ZL2018102205875	一种海底天然气水合物开采智能机器人	2021-01-01
ZL2018106532382	一种固化剂及其制备方法、防腐蚀涂料及其制备方法	2021-02-19
ZL2018107839896	一种埋弧焊钢管焊缝焊偏量参数的检测仪	2021-01-29
ZL2018110226638	低温输气钢管承压能力与韧脆转变行为全尺寸试验方法	2021-01-29
ZL2018110569407	一种注气驱注入井用高温抗氧型缓蚀剂	2021-01-01
ZL2018111414475	一种热塑性塑料最高使用温度的测试方法	2021-09-28
ZL2018112894605	一种双金属复合管端部处理结构的制造方法	2021-04-30
ZL2018113263200	一种石墨烯增强型渗铝油管的制备方法	2021-01-01
ZL2018114966523	一种石墨烯增强1200MPa级钛合金钻杆用钛合金及其管材制造方法	2021-01-01
ZL2018115735028	一种火驱稠油井用套管评价方法	2021-07-02
ZL2018116012952	用于油气开发的高强度高韧性钛合金管材及其制备方法	2021-11-23
ZL2018116076348	避免大壁厚高韧性X80钢管DWTT异常断口的DWTT试样及其方法	2021-06-01
ZL2018116076352	一种基于轮廓的石油管外螺纹参数测量定位方法	2021-04-30
ZL2018800039258	油井管接头的密封结构制造参量的确定方法、密封方法	2021-11-02
ZL2019100236801	一种管道环焊缝缺陷安全评价方法	2021-03-30
ZL2019101205124	一种钢骨架聚乙烯复合管相控阵成像检测装置及方法	2021-07-30
ZL2019101746110	一种用于小口径内防腐管道对焊的加工方法	2021-09-28
ZL201910239348.9	钢骨架聚乙烯复合管电熔焊焊接接头相控阵成像检测方法	2021-09-28
ZL2019102571715	一种铝合金钻杆用耐磨石墨烯改性微弧氧化涂层及其制备方法	2021-05-28
ZL2019102897917	一种油井管外螺纹接头端面密封圈结构	2021-01-29
ZL2019102898873	一种浇注双金属管坯的装置及其浇注方法	2021-03-30
ZL2019103684180	油气集输用双金属复合管环焊接头缺陷预测与控制方法	2021-07-02
ZL2019105824104	一种变齿宽螺纹接头	2021-09-28
ZL2019106904515	一种埋地非金属管道保护壳及基于其的管道铺设施工方法	2021-01-29
ZL2019107502247	一种用于X100输气管道玻璃纤维复合材料止裂器的设计方法	2021-03-30
ZL2019107649896	一种钛合金钻杆用低冲蚀钻井滤清器	2021-05-28
ZL2019109130412	一种电熔增材制造X100钢级三通管件材料及使用方法	2021-07-02
ZL2019113160232	一种石油管外螺纹顶径测量仪及方法	2021-07-02
ZL2020103646459	一种720MPa级高强度钛合金钻杆用管材及制造方法	2021-08-03
ZL2020103666024	一种930MPa级超高强度钛合金钻杆用管材及其制造方法	2021-08-31
ZL2020106130413	一种油田地面集输管线防垢下腐蚀用缓蚀剂及其制备方法	2021-11-02
ZL2020107609590	一种改性酚酯环氧涂料及其制备方法和应用	2021-11-29
ZL2020108648785	一种氧化石墨烯增强钛合金及其制备方法	2021-09-28
ZL2020205989459	一种增强塑料复合连续管盘绕性能检测试验机	2021-01-29